Sampling: Design and Analysis

Sampling:
Design and Analysis

Sharon L. Lohr
Arizona State University

Duxbury Press

An Imprint of Brooks/Cole Publishing Company

I⊤P® *An International Thomson Publishing Company*

Pacific Grove • Albany • Belmont • Bonn • Boston • Cincinnati • Detroit • Johannesburg • London
Madrid • Melbourne • Mexico City • New York • Paris • Singapore • Tokyo • Toronto • Washington

Sponsoring Editor: *Carolyn Crockett*
Marketing Representative: *Marcy Perman*
Marketing Manager: *Michele Mootz*
Editorial Assistant: *Kimberly Raburn*
Advertising Communications: *Kyrrha Sevco*
Production Editor: *Kirk Bomont*
Manuscript Editor: *Betty Duncan*

Permissions Editor: *May Clark*
Interior Design: *William Baxter*
Cover Design: *Laurie Albrecht*
Interior Illustrations: *Gloria Langer*
Art Editor: *Lisa Torri*
Typesetting: *Interactive Composition Corporation*
Printing and Binding: *RR Donnelley Crawfordsville*

For more information, contact:

BROOKS/COLE PUBLISHING COMPANY
511 Forest Lodge Road
Pacific Grove, CA 93950
USA

International Thomson Editores
Seneca 53
Col. Polanco
11560 México, D. F., México

International Thomson Publishing Europe
Berkshire House 168-173
High Holborn
London WC1V 7AA
England

International Thomson Publishing GmbH
Königswinterer Strasse 418
53227 Bonn
Germany

Thomas Nelson Australia
102 Dodds Street
South Melbourne, 3205
Victoria, Australia

International Thomson Publishing Asia
60 Albert Street #15-01
Albert Complex
Singapore 189969

Nelson Canada
1120 Birchmount Road
Scarborough, Ontario
Canada M1K 5G4

International Thomson Publishing Japan
Hirakawacho Kyowa Building, 3F
2-2-1 Hirakawacho
Chiyoda-ku, Tokyo 102
Japan

Printed in the United States of America

10 9 8 7 6 5 4 3 2 1

Library of Congress Cataloging-in-Publication Data

Lohr, Sharon L., (date)
    Sampling : design and analysis / Sharon L. Lohr.
        p.   cm.
    Includes bibliographical references and index.
    ISBN 0-534-35361-4 (alk. paper)
    1. Sampling (Statistics)   I. Title.
HA31.2.L64   1999
    519.5′2—dc21   98-44566
                                CIP

To Doug

Contents

CHAPTER 1 **Introduction** 1

1.1   A Sample Controversy   **1**

1.2   Requirements of a Good Sample   **2**

1.3   Selection Bias   **4**

1.4   Measurement Bias   **8**

1.5   Questionnaire Design   **10**

1.6   Sampling and Nonsampling Errors   **15**

1.7   Exercises   **17**

CHAPTER 2 **Simple Probability Samples** 23

2.1   Types of Probability Samples   **23**

2.2   Framework for Probability Sampling   **25**

2.3   Simple Random Sampling   **30**

2.4   Confidence Intervals   **35**

2.5   Sample Size Estimation   **39**

2.6   Systematic Sampling   **42**

2.7   Randomization Theory Results for Simple Random Sampling*   **43**

2.8   A Model for Simple Random Sampling*   **46**

2.9   When Should a Simple Random Sample Be Used?   **49**

2.10   Exercises   **50**

CHAPTER 3   Ratio and Regression Estimation   59

    3.1   Ratio Estimation   **60**

    3.2   Regression Estimation   **74**

    3.3   Estimation in Domains   **77**

    3.4   Models for Ratio and Regression Estimation*   **81**

    3.5   Comparison   **88**

    3.6   Exercises   **89**

CHAPTER 4   Stratified Sampling   95

    4.1   What Is Stratified Sampling?   **95**

    4.2   Theory of Stratified Sampling   **99**

    4.3   Sampling Weights   **103**

    4.4   Allocating Observations to Strata   **104**

    4.5   Defining Strata   **109**

    4.6   A Model for Stratified Sampling*   **113**

    4.7   Poststratification   **114**

    4.8   Quota Sampling   **115**

    4.9   Exercises   **118**

CHAPTER 5   Cluster Sampling with Equal Probabilities   131

    5.1   Notation for Cluster Sampling   **134**

    5.2   One-Stage Cluster Sampling   **136**

    5.3   Two-Stage Cluster Sampling   **145**

    5.4   Using Weights in Cluster Samples   **153**

    5.5   Designing a Cluster Sample   **154**

    5.6   Systematic Sampling   **159**

    5.7   Models for Cluster Sampling*   **163**

    5.8   Summary   **168**

    5.9   Exercises   **169**

CHAPTER 6 Sampling with Unequal Probabilities 179

6.1 Sampling One Primary Sampling Unit **181**

6.2 One-Stage Sampling with Replacement **184**

6.3 Two-Stage Sampling with Replacement **192**

6.4 Unequal-Probability Sampling Without Replacement **194**

6.5 Examples of Unequal-Probability Samples **199**

6.6 Randomization Theory Results and Proofs* **204**

6.7 Models and Unequal-Probability Sampling* **211**

6.8 Exercises **213**

CHAPTER 7 Complex Surveys 221

7.1 Assembling Design Components **221**

7.2 Sampling Weights **225**

7.3 Estimating a Distribution Function **229**

7.4 Plotting Data from a Complex Survey **235**

7.5 Design Effects **239**

7.6 The National Crime Victimization Survey **242**

7.7 Sampling and Experiment Design* **247**

7.8 Exercises **249**

CHAPTER 8 Nonresponse 255

8.1 Effects of Ignoring Nonresponse **256**

8.2 Designing Surveys to Reduce Nonsampling Errors **258**

8.3 Callbacks and Two-Phase Sampling **262**

8.4 Mechanisms for Nonresponse **264**

8.5 Weighting Methods for Nonresponse **265**

8.6 Imputation **272**

8.7 Parametric Models for Nonresponse* **278**

8.8 What Is an Acceptable Response Rate? **281**

8.9 Exercises **282**

CHAPTER 9   Variance Estimation in Complex Surveys*   289

9.1   Linearization (Taylor Series) Methods   290

9.2   Random Group Methods   293

9.3   Resampling and Replication Methods   298

9.4   Generalized Variance Functions   308

9.5   Confidence Intervals   310

9.6   Summary and Software   313

9.7   Exercises   315

CHAPTER 10   Categorical Data Analysis in Complex Surveys*   319

10.1   Chi-Square Tests with Multinomial Sampling   319

10.2   Effects of Survey Design on Chi-Square Tests   324

10.3   Corrections to Chi-Square Tests   329

10.4   Loglinear Models   336

10.5   Exercises   341

CHAPTER 11   Regression with Complex Survey Data*   347

11.1   Model-Based Regression in Simple Random Samples   348

11.2   Regression in Complex Surveys   352

11.3   Should Weights Be Used in Regression?   362

11.4   Mixed Models for Cluster Samples   368

11.5   Logistic Regression   370

11.6   Generalized Regression Estimation for Population Totals   372

11.7   Exercises   374

CHAPTER 12   Other Topics in Sampling*   379

12.1   Two-Phase Sampling   379

12.2   Capture-Recapture Estimation   387

12.3   Estimation in Domains, Revisited   396

12.4   Sampling for Rare Events   **400**

12.5   Randomized Response   **404**

12.6   Exercises   **407**

APPENDIX A   The SURVEY Program   413

APPENDIX B   Probability Concepts Used in Sampling   423

B.1   Probability   **423**

B.2   Random Variables and Expected Value   **426**

B.3   Conditional Probability   **430**

B.4   Conditional Expectation   **432**

APPENDIX C   Data Sets   437

APPENDIX D   Computer Codes Used for Examples   449

APPENDIX E   Statistical Table   457

References   459

Author Index   485

Subject Index   489

Preface

Surveys and samples sometimes seem to surround you. Many give valuable information; some, unfortunately, are so poorly conceived and implemented that it would be better for science and society if they were simply not done. This book gives you guidance on how to tell when a sample is valid or not, and how to design and analyze many different forms of sample surveys.

The book concentrates on the statistical aspects of taking and analyzing a sample. How to design and pretest a questionnaire, construct a sampling frame, and train field investigators are all important issues, but are not treated comprehensively in this book.

I have written the book to be accessible to a wide audience, and to allow flexibility in choosing topics to be read. To read most of Chapters 1 through 6, you need to be familiar with basic ideas of expectation, sampling distributions, confidence intervals, and linear regression—material covered in most introductory statistics classes. These chapters cover the basic sampling designs of simple random sampling, stratification, and cluster sampling with equal and unequal probabilities of selection. The optional sections on the statistical theory for these designs are marked with asterisks—these sections require you to be familiar with calculus or mathematical statistics. Appendix B gives a review of probability concepts used in the theory of probability sampling.

Chapters 7 through 12 discuss issues not found in many other sampling textbooks: how to analyze complex surveys such as those administered by the United States Bureau of the Census or by Statistics Canada, different approaches to analyzing sample surveys, what to do if there is nonresponse, and how to perform chi-squared tests and regression analyses using data from complex surveys. The National Crime Victimization Survey is discussed in detail as an example of a complex survey. Since many of the formulas used to find standard errors in simpler sampling designs are difficult to implement in complex samples, computer-intensive methods are discussed for estimating the variances.

The book is suitable for a first course in survey sampling. It can be used for a class of statistics majors, or for a class of students from business, sociology, psychology, or biology who want to learn about designing and analyzing data from sample surveys. Chapters 1 through 6 treat the building blocks of sampling, and the sections without asterisks in Chapters 1 through 6 would provide material for a one-quarter course on sampling. In my one-semester course, I cover sections without asterisks in Chapters 1 through 8, and selected topics from the other chapters. The material in Chapters 9 through 12 can be covered in almost any order, and topics chosen from those chapters to fit the needs of the students.

Exercises in the book are of three types: exercises involving critiquing and analyzing data from real surveys, or designing your own surveys, expose you to a variety of applications of sampling; mathematical exercises (indicated by asterisks) develop your theoretical knowledge of the subject; and exercises using SURVEY allow you to experiment with different sample designs without having to collect all the data in the field. The computer program SURVEY, developed by Professor Ted Chang of the University of Virginia (Chang, Lohr, and MacLaren, 1992), allows you to generate samples on the computer from a hypothetical population. The SURVEY exercises allow you to go through all the steps involved in sampling, rather than just plug numbers into a formula found earlier in the chapter. A disk that includes the data sets and the SURVEY program is provided with the book .

You must know how to use a statistical computer package or spreadsheet to be able to do the problems in this book. I encourage you to use a statistical package such as Splus, SAS, or Minitab, or to use a spreadsheet such as Excel, Quattro Pro, or Lotus 1-2-3 for the exercises. The package or spreadsheet you choose will depend on the length and level of the class. In a one-quarter class introducing the basic concepts of sampling, a spreadsheet will suffice for the computing. Some exercises in the later chapters require some computer programming; I have found that Splus is ideal for these exercises as it combines programming capability with existing functions for statistical analysis. Sampling packages such as SUDAAN (Shah et al., 1995) and WesVarPC (Brick et al., 1996), while valuable for the sampling practitioner, hide the structure behind the calculations from someone trying to learn the material. I have therefore not relied on any of the computer packages that exist for analyzing survey data in this book, although various packages are discussed in Section 9.6. Once you understand why the different designs and estimators used in survey sampling work the way they do, it is a small step to read the user's manual for the survey package and to use the software; however, if you have only relied on computer packages as a black box, it is difficult to know when you are performing an appropriate analysis.

Six main features distinguish this book from other texts intended for students from statistics and other disciplines who need to know about sampling methods.

■ The book is flexible for content and level. Many sampling courses have students with a wide range of statistical knowledge. By appropriate choice of sections, this book can be used for an audience of undergraduates who have had one introductory statistics course or for a first-year graduate course for statistics students. The book is also useful for a person doing survey research wanting to learn more about the statistical aspects of surveys and to learn about recent developments. The exercises are flexible as well. Some of the exercises emphasize mastering the mechanics. Many, however, encourage the student to think about the sampling issues involved, and to understand the structure of the sample design at a deeper level. Other exercises are open-ended, and encourage the student to explore the ideas further.

■ I have tried to use real data as much as possible—the Acme Widget Company never appears in this book. The examples and exercises come from social sciences, engineering, agriculture, ecology, medicine, and a variety of other disciplines, and are selected to illustrate the wide applicability of sampling methods. A number of the data sets have extra variables not specifically referred to in text; an instructor can use these for additional exercises or variations.

■ I have incorporated model-based as well as randomization-based theory into the text, with the goal of placing sampling methods within the framework used in other areas of statistics. Many of the important results in the last twenty years of sampling research have involved models, and an understanding of both approaches is essential for the survey practitioner. The model-based approach is introduced in Section 2.8 and further developed in successive chapters; however, those sections could be discussed at any time later in the course.

■ Many topics in this book, such as variance estimation and regression analysis of noindent complex surveys, are not found in other textbooks at this level. The comprehensive sampling reference *Model Assisted Survey Sampling*, by Särndal, Swensson, and Wretman is at a much higher mathematical level.

■ This book emphasizes the importance of graphing the data. Graphical analysis of survey data is often neglected because of the large sizes of data sets and the emphasis on randomization theory, and this neglect can lead to flawed data analyses.

■ Design of surveys is emphasized throughout, and is related to methods for analyzing the data from a survey. The philosophy presented in this book is that the design is by far the most important aspect of any survey: no amount of statistical analysis can compensate for a badly-designed survey. Models are used to motivate designs, and graphs presented to check the sensitivity of the design to model assumptions. For example, in Chapter 2, the usual formula for calculating sample size is presented. But a graph is also given so that the investigator can see the sensitivity of the sample size to the assumed population variance.

Many people have been generous with their encouragement and suggestions for this book. I am deeply in their debt, although I reserve any credit for the book's shortcomings for myself. The following persons reviewed or used various versions of the manuscript, and provided invaluable suggestions for improvement: Jon Rao, Elizabeth Stasny, Fritz Scheuren, Nancy Heckman, Ted Chang, Steve MacEachern, Mark Conaway, Ron Christensen, Michael Hamada, Partha Lahiri, and several anonymous reviewers: Dale Everson, University of Idaho; James Gentle, George Mason University; Ruth Mickey, University of Vermont; Sarah Nusser, Iowa State University; N. G. Narasimha Prasad, University of Alberta, Edmonton; and Deborah Rumsey, Kansas State University. I had many helpful discussions with, and encouragement from, Jon Rao, Fritz Scheuren, and Elizabeth Stasny. David Hubble and Marshall DeBerry provided much helpful advice on the National Crime Victimization Survey. Ted Chang first encouraged me to turn my class notes into a book, and generously allowed use of the SURVEY program in this book. Many thanks go to Alexander Kugushev, Carolyn Crockett, and the production staff at Brooks/Cole for their help, advice, and encouragement. Finally, I would like to thank Alastair Scott, whose inspiring class on sampling at the University of Wisconsin introduced me to the joys of the subject.

Sharon L. Lohr

Introduction

When statistics are not based on strictly accurate calculations, they mislead instead of guide. The mind easily lets itself be taken in by the false appearance of exactitude which statistics retain in their mistakes, and confidently adopts errors clothed in the form of mathematical truth.

—Alexis de Tocqueville, *Democracy in America*

1.1
A Sample Controversy

Shere Hite's book *Women and Love: A Cultural Revolution in Progress* (1987) had a number of widely quoted results:

- 84% of women are "not satisfied emotionally with their relationships" (p. 804).

- 70% of all women "married five or more years are having sex outside of their marriages" (p. 856).

- 95% of women "report forms of emotional and psychological harassment from men with whom they are in love relationships" (p. 810).

- 84% of women report forms of condescension from the men in their love relationships (p. 809).

The book was widely criticized in newspaper and magazine articles throughout the United States. The *Time* magazine cover story "Back Off, Buddy" (October 12, 1987), for example, called the conclusions of Hite's study "dubious" and "of limited value."

Why was Hite's study so roundly criticized? Was it wrong for Hite to report the quotes from women who feel that the men in their lives refuse to treat them as equals, who perhaps have never been given the chance to speak out before? Was it wrong to report the percentages of these women who are unhappy in their relationships with men?

Of course not. Hite's research allowed women to discuss how they viewed their experiences, and reflected the richness of these women's experiences in a way that a multiple-choice questionnaire could not. Hite's error was in generalizing these results to *all* women, whether they participated in the survey or not, and in claiming that the percentages applied to all women. The following characteristics of the survey make it unsuitable for generalizing the results to all women.

- The sample was self-selected—that is, recipients of questionnaires decided whether they would be in the sample or not. Hite mailed 100,000 questionnaires; of these, 4.5% were returned.

- The questionnaires were mailed to such organizations as professional women's groups, counseling centers, church societies, and senior citizens' centers. The members may differ in political views, but many have joined an "all-women" group, and their viewpoints may differ from other women in the United States.

- The survey has 127 essay questions, and most of the questions have several parts. Who will tend to return such a survey?

- Many of the questions are vague, using words such as *love*. The concept of love probably has as many interpretations as there are people, making it impossible to attach a single interpretation to any statistic purporting to state how many women are "in love." Such question wording works well for eliciting the rich individual vignettes that comprise most of the book but makes interpreting percentages difficult.

- Many of the questions are leading—they suggest to the respondent which response she should make. For instance: "Does your husband/lover see you as an equal? Or are there times when he seems to treat you as an inferior? Leave you out of the decisions? Act superior?" (p. 795).

Hite writes, "Does research that is not based on a probability or random sample give one the right to generalize from the results of the study to the population at large? If a study is large enough and the sample broad enough, and if one generalizes carefully, yes" (p. 778). Most survey statisticians would answer Hite's question with a resounding no. In Hite's survey, because the women sent questionnaires were purposefully chosen and an extremely small percentage of the women returned the questionnaires, statistics calculated from these data cannot be used to indicate attitudes of all women in the United States. The final sample is not *representative* of women in the United States, and the statistics can only be used to describe women who would have responded to the survey.

Hite claims that results from the sample could be generalized because characteristics such as the age, educational, and occupational profiles of women in the sample matched those for the population of women in the United States. But the women in the sample differed on one important aspect—they were willing to take the time to fill out a long questionnaire dealing with harassment by men and to provide intensely personal information to a researcher. We would expect that in every age group and socioeconomic class, women who choose to report such information would in general have had different experiences than women who choose not to participate in the survey.

1.2
Requirements of a Good Sample

In the movie *Magic Town*, the public opinion researcher played by James Stewart discovered a town that had exactly the same characteristics as the whole United States: Grandview had exactly the same proportion of people who voted Republican, the same proportion of people under the poverty line, the same proportion of auto mechanics, and so on, as the United States taken as a whole. All that Stewart's character had to do was to interview the people of Grandview, and he would know what public opinion was in the United States.

A perfect sample would be like Grandview: a scaled-down version of the population, mirroring every characteristic of the whole population. Of course, no such perfect sample can exist for complicated populations (even if it did exist, we would not know it was a perfect sample without measuring the whole population). But a good sample will reproduce the characteristics of interest in the population, as closely as possible. It will be **representative** in the sense that each sampled unit will represent the characteristics of a known number of units in the population.

Some definitions are needed to make the notion of a good sample more precise.

Observation unit An object on which a measurement is taken. This is the basic unit of observation, sometimes called an **element.** In studying human populations, observation units are often individuals.

Target population The complete collection of observations we want to study. Defining the target population is an important and often difficult part of the study. For example, in a political poll, should the target population be all adults eligible to vote? All registered voters? All persons who voted in the last election? The choice of target population will profoundly affect the statistics that result.

Sample A subset of a population.

Sampled population The collection of all possible observation units that might have been chosen in a sample; the population from which the sample was taken.

Sampling unit The unit we actually sample. We may want to study individuals but do not have a list of all individuals in the target population. Instead, households serve as the sampling units, and the observation units are the individuals living in the households.

Sampling frame The list of sampling units. For telephone surveys, the sampling frame might be a list of all residential telephone numbers in the city; for personal interviews, a list of all street addresses; for an agricultural survey, a list of all farms or a map of areas containing farms.

In an ideal survey, the sampled population will be identical to the target population, but this ideal is rarely met exactly. In surveys of people, the sampled population is usually smaller than the target population. As Figure 1.1 illustrates, not all persons in the target population are included in the sampling frame, and a number of persons will not respond to the survey.

In the Hite study, one characteristic of interest was the percentage of women who are harassed in their relationship. An individual woman was an element. The target population was all adult women in the United States. Hite's sampled population was women belonging to women's organizations who would return the questionnaire. Consequently, inferences can only be made to the sampled population, not to the population of all adult women in the United States.

The National Crime Victimization Survey is an ongoing survey to study victimization rates, administered by the U.S. Bureau of the Census and the Bureau of Justice

FIGURE   1.1

The target population and sampled population in a telephone survey of likely voters. Not all households will have telephones, so a number of persons in the target population of likely voters will not be associated with a telephone number in the sampling frame. In some households with telephones, the residents are not registered to vote and hence are ineligible for the survey. Some eligible persons in the sampling frame population do not respond because they cannot be contacted, some refuse to respond to the survey, and some may be ill and incapable of responding.

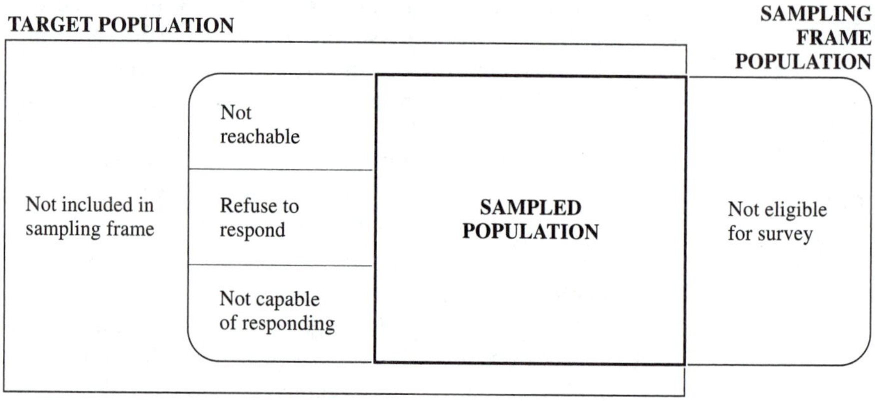

Statistics. If the characteristic of interest is the total number of households in the United States that were victimized by crime last year, the elements are households, the target population consists of all households in the United States, and the sampled population consists of households in the sampling frame, constructed from census information and building permits, that are "at home" and agree to answer questions.

The goal of the National Pesticide Survey, conducted by the Environmental Protection Agency, was to study pesticides and nitrate in drinking water wells nationwide. The target population was all community water systems and rural domestic wells in the United States. The sampled population was all community water systems (all are listed in the Federal Reporting Data System) and all identifiable domestic wells outside of government reservations that belonged to households willing to cooperate with the survey.

Public opinion polls are often taken to predict which candidate will win the next election. The target population is persons who will vote in the next election; the sampled population is often persons who can be reached by telephone and say they are likely to vote in the next election. Few national polls in the United States include Alaska or Hawaii or persons in hospitals, dormitories, or jails; they are not part of the sampling frame or of the sampled population.

1.3
Selection Bias

A good sample will be as free from selection bias as possible. **Selection bias** occurs when some part of the target population is not in the sampled population. If a survey designed to study household income omits transient persons, the estimates from the survey of the average or median household income are likely to be too large. A **sample of convenience** is often biased, since the units that are easiest to select or that are most likely to respond are usually not representative of the harder-to-select or nonresponding units. The following examples indicate some ways in which selection bias can occur.

■ Using a sample-selection procedure that, unknown to the investigators, depends on some characteristic associated with the properties of interest. For example, investigators took a convenience sample of adolescents to study how frequently adolescents talk to their parents and teachers about AIDS. But adolescents willing to talk to the investigators about AIDS are probably also more likely to talk to other authority figures about AIDS. The investigators, who simply averaged the amounts of time that adolescents in the sample said they spent talking with their parents and teachers, probably overestimated the amount of communication occurring between parents and adolescents in the population.

■ Deliberately or purposefully selecting a "representative" sample. If we want to estimate the average amount a shopper spends at the Mall of America and we sample shoppers who look like they have spent an "average" amount, we have deliberately selected a sample to confirm our prior opinion. This type of sample is sometimes called a **judgment sample**—the investigator uses his or her judgment to select the specific units to be included in the sample.

■ Misspecifying the target population. For instance, all the polls in the 1994 Democratic gubernatorial primary election in Arizona predicted that candidate Eddie Basha would trail the front-runner in the polls by at least 9 percentage points. In the election, Basha won 37% of the vote; the other two candidates won 35% and 28%, respectively. One problem is that many voters were undecided at the time the polls were taken. Another is that the target population for the polls was registered voters who had voted in previous primary elections and were interested in this one. In the primary election, however, Basha had heavy support in rural areas from demographic groups that had not voted before and hence were not targeted in the surveys.

■ Failing to include all the target population in the sampling frame, called **undercoverage.** Many large surveys use the U.S. decennial census to construct the sampling frame, but the census fails to enumerate a large number of housing units, producing undercounts of a number of population groups. Fay et al. (1988) estimate that the 1980 census missed 8% of all black males. So any survey that uses the 1980 census data as the only source for constructing a sampling frame will automatically miss that 8% of black males, and that error occurs before the survey has even started.

■ Substituting a convenient member of a population for a designated member who is not readily available. For example, if no one is at home in the designated household, a field representative might try next door. In a wildlife survey, the investigator might substitute an area next to a road for a less accessible area. In each case, the sampled units most likely differ on a number of characteristics from units not in the sample. The substituted household may be more likely to have a member who does not work outside the home than the originally selected household. The area by the road may have fewer frogs than the area that is harder to reach.

■ Failing to obtain responses from all the chosen sample. **Nonresponse** distorts the results of many surveys, even surveys that are carefully designed to minimize other sources of selection bias. Often, nonrespondents differ critically from the respondents, but the extent of that difference is unknown unless you can later obtain information about the nonrespondents. Many surveys reported in newspapers or research journals have dismal response rates—in some, the response rate is as low as 10%. It is difficult to see how results can be generalized to the population when 90% of the targeted sample cannot be reached or refuses to participate.

The Adolescent Health Database Survey was designed to obtain a representative sample of Minnesota junior and senior high school students in public schools (Remafedi et al. 1992). Overall, 49% of the school districts that were invited to participate in the survey agreed to participate. The response rate varied with the size of the school district:

Type of School District	Participation Rate (%)
Urban	100
Metropolitan suburban	25
Nonmetropolitan with more than 2000 students	62
Nonmetropolitan with 1000–1999 students	27
Nonmetropolitan with 500–999 students	61
Nonmetropolitan with fewer than 500 students	53

In each of the school districts that participated, surveys were distributed to students, and participation by the students was voluntary. Of the 52,553 surveys distributed to students, 36,741 were completed and returned, resulting in a student response rate of 70%. The survey asked questions about health habits, religious affiliation, psychosocial status, and sexual orientation. It seems likely that responding and nonresponding school districts have different levels of health and activity. It seems even more likely that students who respond to the survey will on average have a different health profile than students who do not respond to the survey.

Many studies comparing respondents and nonrespondents have found differences in the two groups. In the Iowa Women's Health Study, 41,836 women responded to a mailed questionnaire in 1986. Bisgard et al. (1994) compared those respondents to the 55,323 nonrespondents by checking records in the State Health Registry; they found that the age-adjusted mortality rate and the cancer attack rate were significantly higher for the nonrespondents than for the respondents.

■ Allowing the sample to consist entirely of volunteers. Such is the case in radio and television call-in polls, and the statistics from such surveys cannot be trusted. CBS News conducted a call-in poll immediately following President Bush's State of the Union Address on January 28, 1992. News anchors Dan Rather and Connie Chung were careful to say that this sample was "unscientific"; the broadcast, however, presented the percentages of viewers with various opinions as though they were from a statistically sound survey. Almost 315,000 callers responded to what the *New York Times* called "the largest biased sample in the history of instant polling," and many more tried to respond—AT&T computers recorded almost 25 million attempts to reach the toll-free telephone number. The Nielsen ratings estimated that about 9 million households had a television tuned to the CBS program, indicating that many individuals or organizations tried to call multiple times. The possibility always exists in a call-in survey that a determined organization will skew the results by monopolizing the toll-free number.

EXAMPLE 1.1   Many surveys have more than one of these problems. The *Literary Digest* (1932, 1936a, b, c) began taking polls to forecast the outcome of the U.S. presidential election in 1912, and their polls attained a reputation for accuracy because they forecast the correct winner in every election between 1912 and 1932. In 1932, for example, the poll predicted that Roosevelt would receive 56% of the popular vote and 474 votes in the electoral college; in the actual election, Roosevelt received 58% of the popular vote and 472 votes in the electoral college.

With such a strong record of accuracy, it is not surprising that the editors of the *Literary Digest* had a great deal of confidence in their polling methods by 1936. Launching the 1936 poll, they said:

> The Poll represents thirty years' constant evolution and perfection. Based on the "commercial sampling" methods used for more than a century by publishing houses to push book sales, the present mailing list is drawn from every telephone book in the United States, from the rosters of clubs and associations, from city directories, lists of registered voters, classified mail-order and occupational data. (1936a, 3)

On October 31, the poll predicted that Republican Alf Landon would receive 55% of the popular vote, compared with 41% for President Roosevelt. The article "Landon, 1,293,669; Roosevelt, 972,897: Final Returns in The Digest's Poll of Ten Million Voters" contained this statement: "We make no claim to infallibility. We did not coin the phrase 'uncanny accuracy' which has been so freely applied to our Polls" (1936b). It is a good thing they made no claim to infallibility; in the election, Roosevelt received 61% of the vote; Landon, 37%.

What went wrong? One problem may have been undercoverage in the sampling frame, which relied heavily on telephone directories and automobile registration lists—the frame was used for advertising purposes, as well as for the poll. Households with a telephone or automobile in 1936 were generally more affluent than other households, and opinion of Roosevelt's economic policies was generally related to the economic class of the respondent. But sampling frame bias does not explain all the discrepancy. Postmortem analyses of the poll by Squire (1988) and Calahan (1989) indicate that even persons with both a car and a telephone tended to favor Roosevelt, though not to the degree that persons with neither car nor telephone supported him.

The low response rate to the survey was likely the source of much of the error. *Ten million* questionnaires were mailed out, and 2.3 million were returned—an enormous sample but a response rate of less than 25%. In Allentown, Pennsylvania, for example, the survey was mailed to every registered voter, but the survey results for Allentown were still incorrect because only one-third of the ballots were returned. Squire (1988) reports that persons supporting Landon were much more likely to have returned the survey; in fact, many Roosevelt supporters did not even remember receiving a survey, even though they were on the mailing list.

One lesson to be learned from the *Literary Digest* poll is that the sheer size of a sample is no guarantee of its accuracy. The *Digest* editors became complacent because they sent out questionnaires to more than one quarter of all registered voters and obtained a huge sample of 2.3 million people. But large unrepresentative samples can perform as badly as small unrepresentative samples. A large unrepresentative sample may do more damage than a small one because many people think that large samples are always better than small ones. The design of the survey is far more important than the absolute size of the sample. ∎

What Good Are Samples with Selection Bias? We prefer to have samples with no selection bias, that serve as a microcosm of the population. When the primary interest is in estimating the total number of victims of violent crime in the United States or the percentage of likely voters in the United Kingdom who intend to vote for the Labour Party in the next election, serious selection bias can cause the sample estimates to be invalid.

Purposive or judgment samples can provide valuable information, though, particularly in the early stages of an investigation. Teichman et al. (1993) took soil samples along Interstate 880 in Alameda County, California, to determine the amount of lead in yards of homes and in parks close to the freeway. In taking the samples, they concentrated on areas where they thought children were likely to play and areas where soil might easily be tracked into homes. The purposive sampling scheme worked well for justifying the conclusion of the study, that "lead contamination of urban soil in the east bay area of the San Francisco metropolitan area is high and exceeds hazardous waste levels at many sites." A sampling scheme that avoided selection bias would only be needed for this study if the investigators wanted to generalize the estimated percentage of contaminated sites to the entire area.

1.4
Measurement Bias

A good sample has accurate responses to the items of interest. **Measurement bias** occurs when the measuring instrument has a tendency to differ from the true value in one direction. As with selection bias, measurement bias must be considered and minimized in the design stage of the survey; no amount of statistical analysis will disclose, for instance, that the scale erroneously added 5 kilograms to the weight of every person in a health survey.

Measurement bias is a concern in all surveys and can be insidious. In many surveys of vegetation, for example, areas to be sampled are divided into smaller plots. A sample of plots is selected, and the number of plants in each plot is recorded. When a plant is near the boundary of the region, the field researcher needs to decide whether to include the plant in the tally. A person who includes all plants near or on the boundary in the count is likely to produce an estimate of the total number of plants in the area that is too high because some plants may be counted twice. Duce et al. (1972) report concentrations of trace metals, lipids, and chlorinated hydrocarbons in the top 100 micrometers of Narragansett Bay that are 1.5 to 50 times as great as those in the water 20 centimeters below the surface. If studying the transport of pollutants from coastal waters to the deeper waters of the ocean, a sampling scheme that ignores this boundary effect may underestimate the amount transported.

Sometimes measurement bias is unavoidable. In the North American Breeding Bird Survey, observers stop every one-half mile on designated routes and count all birds heard singing or calling or sighted within a quarter-mile radius (Droege 1990). The count of birds for that point is almost always an underestimate of the number of birds in the area; statistical models may possibly be used to adjust for the measurement bias. If data are collected with the same procedure and with similarly skilled observers from year to year, the survey can be used to estimate trends in the population of different species—the biases from different years are expected to be similar and may cancel when year-to-year differences are calculated.

Obtaining accurate responses is challenging in all types of surveys, but particularly so in surveys of people:

- People sometimes do not tell the truth. In an agricultural survey, farmers in an area with food-aid programs may underreport crop yields, hoping for more food aid. Obtaining truthful responses is a particular challenge in surveys involving sensitive subject matter, such as surveys about drug use.

- People do not always understand the questions. Many persons in the United States were shocked by the results of a 1993 Roper poll reporting that 25% of Americans did not believe the Holocaust really happened. When the double-negative structure of the question was eliminated and the question reworded, only 1% thought it was "possible . . . the Nazi extermination of the Jews never happened."

- People forget. One problem faced in the design of the National Crime Victimization Survey is that of **telescoping:** Persons are asked about experiences as a crime victim that took place in the last six months, but some include victimizations that occurred more than six months ago.

- People give different answers to different interviewers. Schuman and Converse (1971) employed both white and black interviewers to interview black residents of Detroit. To the question "Do you personally feel that you can trust most white people, some white people, or none at all?" the response of 35% of those interviewed by a white person was that they could trust most white people. The percentage was 7% for those interviewed by a black person.

- People may say what they think an interviewer wants to hear or what they think will impress the interviewer. In experiments done with questions beginning with "Do you agree or disagree with the following statement?" it has been found that a subset of the population tends to agree with any statement regardless of its content. Lenski and Leggett (1960) found that about one-tenth of their sample agreed with both of the following statements:

It is hardly fair to bring children into the world, the way things look for the future.

Children born today have a wonderful future to look forward to.

Some commentators speculate that the "shame factor" may have played a part in the polls before the British general election of 1992, in which the Conservative Party government won the election but almost all polls predicted that Labour would win: "People may *say* they would prefer better public services, but in the end they will *vote* for tax cuts. At least some of them had the decency to feel too ashamed to admit it" (Harris 1992).

- A particular interviewer may affect the accuracy of the response by misreading questions, recording responses inaccurately, or antagonizing the respondent. In a survey about abortion, a poorly trained interviewer with strong feelings against abortion may encourage the respondent to provide one answer rather than another.

- Certain words mean different things to different people. A simple question such as "Do you own a car?" may be answered yes or no depending on the respondent's interpretation of *you* (does it refer to just the individual or to the household?), *own* (does it count as ownership if you are making payments to a finance company?), or *car* (are pickup trucks included?).

- Question wording and order have a large effect on the responses obtained. Two surveys were taken in late 1993 and early 1994 about Elvis Presley. One survey asked, "In the past few years, there have been a lot of rumors and stories about whether Elvis Presley is really dead. How do you feel about this? Do you think there is any possibility that these rumors are true and that Elvis Presley is still alive, or don't you think so?" The other survey asked, "A recent television show examined various theories about Elvis Presley's death. Do you think it is possible that Elvis is alive or not?" To the first survey, 8% of the respondents said it is possible that Elvis is still alive; to the second survey, 16% of the respondents said it is possible that Elvis is still alive.

Excellent discussions of these problems can be found in Groves (1989) and Asher (1992). In some cases, accuracy can be increased by careful questionnaire design.

1.5
Questionnaire Design

This section, a very brief introduction to writing and testing questions, provides some general guidelines and examples. If you are writing a questionnaire, however, consult one of the more comprehensive references on questionnaire design listed in the References. Much recent research has been done in the area of using results from cognitive psychology when writing questionnaires; Tanur (1993) and Blair and Presser (1993) are useful references on the topic.

- *Decide what you want to find out; this is the most important step in writing a questionnaire.* Write down the goals of your survey and be precise. "I want to learn something about the homeless" won't do. Instead, write down specific questions such as "What percentage of persons using homeless shelters in Chicago between January and March 1996 are under 16 years old?" Then, write or select questions that will elicit accurate answers to the research questions and that will encourage persons in the sample to respond to the questions.

- *Always test your questions before taking the survey.* Ideally, the questions would be tested on a small sample of members of the target population. Try different versions for the questions and ask respondents in your pretest how they interpret the questions.

The National Crime Victimization Survey (NCVS) was tested for several years before it was conducted on a national scale (Lehnen and Skogan 1981). The pretests were used to help decide on a recall period (it was decided to ask respondents about victimizations that had occurred in the previous six months), to test the effects of different interviewing procedures and questions, and to compare information from selected interviews with information found in the police report about the victimization. As a result of the pretests, some of the long and repetitious questions were shortened and more specific wording introduced.

The questionnaire was revised in 1985 and again in 1991 to make use of recent research in cognitive psychology and to include topics, such as victim and bystander behavior, that were not found in the earlier versions. All revisions are tested extensively in the field before being used (Taylor 1989). In the past, for example, the NCVS has been criticized for underreporting the crime of rape; when the questionnaire was designed in the early 1970s, there was worry that asking about rape directly would be perceived as insensitive and embarrassing and would provoke congressional outrage. The original NCVS questionnaire asked a series of specific questions intended to prompt the memory of respondents. These included questions such as "Did anyone take something directly from you by using force, such as by a stickup, mugging or threat?" The last question in the violent-crime screening section of the questionnaire was "Did anyone try to attack you in some other way?" If the respondent mentioned in response that he or she was raped, then a rape was reported. Not surprisingly, the victimization rate for rape reported for the 1990 and earlier NCVS is very low: It is reported that about 1 per 1000 females aged 12 and older were raped in 1990. The latest version of the NCVS questionnaire asks about rape directly; as a result, estimates of the prevalence of rape have doubled.

You will not necessarily catch misinterpretations of questions by trying them out on friends or colleagues; your friends and colleagues may have backgrounds similar to yours and may not have the same understanding of words as persons in your target population. Belson (1981) demonstrates that each of 29 questions about television viewing was misinterpreted by some respondents. The question "Do you think that the television news programmes are impartial about politics?" was tested on 56 people. Of these, 13 interpreted the question as intended, 18 respondents narrowed the term *news programmes* to mean "news bulletins," 21 narrowed it to "political programmes," and 1 interpreted it as "newspapers." Only 25 persons interpreted *impartial* as intended; 5 inferred the opposite meaning, "partial"; 11, as "giving too much or too little attention to"; and the others were simply unfamiliar with the word.

- *Keep it simple and clear.* Questions that seem clear to you may not be clear to someone listening to the whole question over the telephone or to a person with a different native language. Belson (1981, 240) tested the question "What proportion of your evening viewing time do you spend watching news programmes?" on 53 people. Only 14 people correctly interpreted the word *proportion* as "percentage," "part," or "fraction." Others interpreted it as "how long do you watch" or "which news programs do you watch."

- *Use specific questions instead of general ones, if possible.* Strunk and White advise writers to "prefer the specific to the general, the definite to the vague, the concrete to the abstract" (1959, 15). Good questions result from good writing.

    Instead of asking "Did anyone attack you in the last six months?" the NCVS asks a series of specific questions detailing how one might be attacked. The NCVS question is "Has anyone attacked or threatened you in any of these ways: (a) With any weapon, for instance, a gun or knife, (b) With anything like a baseball bat, frying pan, scissors, or stick. . . ."

- *Relate your questions to the concept of interest.* This seems obvious but is forgotten or ignored in many surveys. In some disciplines, a standard set of questions has been developed and tested, and these are then used by subsequent researchers. Often, use of a common survey instrument allows results from different studies to be compared. In some cases, however, the standard questions are inappropriate for addressing the research hypotheses.

    Pincus (1993) criticizes early research that concluded that persons with arthritis were more likely to have psychological problems than persons without arthritis. In those studies, persons with arthritis were given the Minnesota Multiphasic Personality Inventory, a test of 566 true/false questions commonly used in psychological research. Patients with rheumatoid arthritis tended to have high scores on the scales of hypochondriasis, depression, and hysteria. Part of the reason they scored high on those scales is clear when the actual questions are examined. A person with arthritis can truthfully answer false to questions such as "I am about as able to work as I ever was," "I am in just as good physical health as most of my friends," and "I have few or no pains" without being either hysterical or a hypochondriac.

- *Decide whether to use open or closed questions.* An **open question** (the respondent is not prompted with categories for responses) allows respondents to form their own response categories; in a **closed question** (multiple choice), the respondent chooses from a set of categories read or displayed on a card. Each has advantages. A closed question may prompt the respondent to remember responses that might otherwise be forgotten and is in accordance with the principle that specific questions are better than general ones. If the subject matter has been thoroughly pretested and responses of interest are known, a well-written closed question will usually elicit more accurate responses, as in the NCVS question "Has anyone attacked or threatened you with anything like a baseball bat, frying pan, scissors, or stick?" If the survey is exploratory or questions are sensitive, though, it is often better to use an open question. Bradburn and Sudman (1979) note that respondents reported higher frequency of drinking alcoholic beverages when asked an open question than a closed question with categories "never" through "daily."

    The survey by Skelly et al. (1968) on women's attitudes toward fabrics used in clothing gave about half the sample an open version of the questionnaire and the other half a closed version of the questionnaire, to study the difference in responses. The first question in the open questionnaire was "What difficulties and problems do you run into most often when buying clothes, any kind of clothes, for yourself?"

The corresponding question in the closed version of the questionnaire was "Which of these reasons best describes the difficulties and problems you run into most often when buying clothes, any kind of clothes, for yourself? Any others?" The respondent was asked to indicate the statements on Card A that apply to her.

Card A

1. I am short waisted.
2. I am long waisted.
3. I need a short length.
4. I need a long length.
5. I have a small waist.
6. I have a large waist.
7. Doesn't fit around the shoulders.

8. I have wide hips.
9. Limited styles, selections.
10. I have problems with necklines.
11. Can't find correct sizes.
12. Sizes don't run true.
13. Poor workmanship.

Of the women given the closed questionnaire, 25% mentioned that they were short waisted, whereas only 9% of the women given the open questionnaire mentioned that they were short waisted. A higher percentage of women mentioned each of the difficulties on the card in the closed group than in the open group. However, 10% of the women in the open group mentioned the difficulty that the price is too high; in the closed group, only 1% of the respondents mentioned high price, perhaps because the card emphasized fitting problems and focused on the woman's figure rather than other difficulties.

If using a closed question, always have an "other" category. In one study of sexual activity among adolescents, adolescents were asked from whom they felt the most pressure to have sex. Categories for the closed question were "friends of same sex," "boyfriend/girlfriend," "friends of opposite sex," "TV or radio," "don't feel pressure," and "other." The response "parents" or "father" was written in by a number of the adolescent respondents, a response that had not been anticipated by the researchers.

■ *Report the actual question asked.* Public opinion is complex, and you inevitably leave a distorted impression of it when you compress the results of your careful research into a summary statement "$x\%$ of Americans favor affirmative action."

The results of three surveys in spring 1995, all purportedly about affirmative action, emphasize the importance of reporting the question. A *Newsweek* poll asked, "Should there be special consideration for each of the following groups to increase their opportunities for getting into college and getting jobs or promotions?" and asked about these groups: blacks, women, Hispanics, Asians, and Native Americans. The poll found that 62% of blacks but only 25% of whites answered yes to the question about blacks. A USA Today–CNN–Gallup poll asked the question "What is your opinion on affirmative action programs for women and minorities: do you favor them or oppose them?" and reported that 55% of respondents favored such programs. A Harris poll asking "Would you favor or oppose a law limiting affirmative action programs in your state?" reported 51% of respondents favoring such a law. These questions are clearly addressing different concepts because the differences in percentages obtained are too great to be ascribed to the different samples of people taken by the three organizations. Yet all three polls' results were described in newspapers in terms of percentages of persons who support affirmative action.

- *Avoid questions that prompt or motivate the respondent to say what you would like to hear.* These are often called **leading, or loaded, questions.** The May 17, 1994, issue of the *Wall Street Journal* reported the following question asked by the Gallup Organization in a survey commissioned by the American Paper Institute: "It is estimated that disposable diapers account for less than 2 percent of the trash in today's landfills. In contrast, beverage containers, third-class mail and yard waste are estimated to account for about 21 percent of trash in landfills. Given this, in your opinion, would it be fair to tax or ban disposable diapers?"

- *Use forced-choice, rather than agree/disagree, questions.* As noted earlier, some persons will agree with almost any statement. Schuman and Presser (1981, 223) report the following differences from an experiment comparing agree/disagree with forced-choice versions:

Q1: Do you agree or disagree with this statement: Most men are better suited emotionally for politics than are most women.

Q2: Would you say that most men are better suited emotionally for politics than are most women, that men and women are equally suited, or that women are better suited than men in this area?

	Years of Schooling		
	0–11	12	13+
Q1: percent "agree"	57	44	39
Q2: percent "men better suited"	33	38	28

- *Ask only one concept in each question.* In particular, avoid what are sometimes called **double-barreled questions**—so named because if one barrel of the shotgun does not get you, the other one will.

    The question "Do you agree with Bill Clinton's $50 billion bailout of Mexico?" appeared on a survey distributed by a member of the U.S. House of Representatives to his constituents. The question is really confusing two opinions of the respondent: the opinion of Bill Clinton and the opinion of the Mexico policy. Disapproval of either one will lead to a "disagree" answer to the question. Note also the loaded content of the word *bailout,* which will almost certainly elicit more negative responses than the term *aid package* would.

- *Pay attention to question-order effects.* If you ask more than one question on a topic, it is usually (but not always) better to ask the more general question first and follow it by the specific questions. McFarland (1981) conducted an experiment in which half of the respondents were given general questions first (for example, "How interested would you say you are in religion: very interested, somewhat interested, or not very interested?"), followed by specific questions on the subject ("Did you, yourself, happen to attend church in the last seven days?"); the other half were asked the specific questions first and then asked the general questions. When the general question was asked first, 56% reported that they were "very interested in religion"; the percentage rose to 64% when the specific question was asked first.

Serdula et al. (1995) found that in the years in which a respondent of a health survey was asked to report his or her weight and then immediately asked "Are you trying to lose weight?" 28.8% of men and 48.0% of women reported that they were trying to lose weight. When "Are you trying to lose weight?" was asked in the middle of the survey and the self-report question on weight was asked at the end of the survey, 26.5% of the men and 40.9% of the women reported that they were trying to lose weight. The authors speculate that respondents who are reminded of their weight status may overreport trying to lose weight.

1.6
Sampling and Nonsampling Errors

Most opinion polls that you see report a *margin of error.* Many merely say that the margin of error is 3 percentage points. Others give more detail, as in this excerpt from a *New York Times* poll: "In theory, in 19 cases out of 20 the results based on such samples will differ by no more than three percentage points in either direction from what would have been obtained by interviewing all Americans." The margin of error given in polls is an expression of **sampling error,** the error that results from taking one sample instead of examining the whole population. If we took a different sample, we would most likely obtain a different sample percentage of persons who visited the public library last week. Sampling errors are usually reported in probabilistic terms, as done above by the *New York Times.* (We discuss the calculation of sampling errors for different survey designs in Chapters 2 through 7.)

Selection bias and inaccuracy of responses are examples of **nonsampling errors,** which are any errors that cannot be attributed to the sample-to-sample variability. In many surveys, the sampling error that is reported for the survey may be negligible compared with the nonsampling errors; you often see surveys with a 30% response rate proudly proclaiming their 3% margin of error, while ignoring the tremendous selection bias in their results.

The goal of this chapter is to sensitize you to various forms of selection bias and inaccurate responses. We can reduce some forms of selection bias by using probability sampling methods, as described in the next chapter. Accurate responses can often be achieved through careful design and testing of the survey instrument, training of interviewers, and pretesting the survey. We will return to nonsampling errors in Chapter 8.

Why Sample at All?   With the abundance of poorly done surveys, it is not surprising that some people are skeptical of *all* surveys. "After all," some say, "my opinion has never been asked, so how can the survey results claim to represent me?" Public questioning of the validity of surveys intensifies after a survey makes a large mistake in predicting the results of an election, such as in the *Literary Digest* survey of 1936 or in the 1948 U.S. presidential election in which most pollsters predicted that Dewey would defeat Truman. A public backlash against survey research occurred again after the British general election of 1992, when the Conservative government won reelection despite the predictions from all but one of the major polling organizations that it would be a dead heat or that Labour would win. One member of Parliament expressed his opinion that "extrapolating what tens of millions are thinking from a tiny sample of opinions affronts human intelligence and negates true freedom of thought."

Some people insist that only a complete census, in which every element of the population is measured, will be satisfactory; this objection to sampling has a long history. When Anders Kiaer (1897), director of Norwegian statistics, proposed using sampling for collecting official government statistics, his proposal was by no means universally well received. Opponents of sampling argued that it was dangerous and that samples could never replace a census. Within a few years, however, the international statistical community was largely persuaded that representative samples are a good thing, although probability samples were not widely used until the 1930s and 1940s.

For small populations, a census may of course be practical. For example, if you want to know about the employment history of 1990 Arizona State University graduates who majored in mathematics, it would be sensible to try to contact all of them. If all graduates respond, then estimates from the survey will have no sampling error. The estimates will have nonsampling errors, however, if the questions are poorly written or if respondents give inaccurate information. If some of the graduates do not return the questionnaire, then the estimates will likely be biased because of nonresponse.

In general, taking a complete census of a population uses a great deal of time and money and does not eliminate error. The biggest causes of error in a survey are often undercoverage, nonresponse, and sloppiness in data collection. Most of us have kept a checkbook register at some time, which is essentially a census of all check and deposit amounts. How many of us can say that we have never made an error in our checkbooks? It is usually much better to take a high-quality sample and allocate resources elsewhere, for instance, by being more careful in collecting or recording data, doing follow-up studies, or measuring more variables.

After all, the *Literary Digest* poll (see Example 1.1) predicted the vote wrong even in some counties in which it attempted to take a census. The decennial census, which attempts to enumerate every U.S. resident, misses segments of the population. For the year 2000 census, a panel from the National Academy of Sciences has recommended that enumeration be combined with sampling to improve the accuracy of the census. Congress is currently debating this proposal.

There are three main justifications for using sampling:

- Sampling can provide reliable information at far less cost than a census. With probability samples (described in the next chapter), you can quantify the sampling error from a survey. In some instances, an observation unit must be destroyed to be measured, as when a cookie must be pulverized to determine the fat content. In such a case, a sample provides reliable information about the population; a census destroys the population and, with it, the need for information about it.

- Data can be collected more quickly, so estimates can be published in a timely fashion. An estimate of the unemployment rate for 1994 is not very helpful if it takes until 2004 to interview every household.

- Finally, and less well known, estimates based on sample surveys are often more accurate than those based on a census because investigators can be more careful when collecting data. A complete census often requires a large administrative organization and involves many persons in the data collection. With the administrative complexity and the pressure to produce timely estimates, many types of errors can be easily injected into the census. In a sample, more attention can be devoted to data quality through training personnel and following up on nonrespondents. It is far better to have good measurements on a representative sample than unreliable or biased measurements on the whole population.

Deming says, "Sampling is not mere substitution of a partial coverage for a total coverage. Sampling is the science and art of controlling and measuring the reliability of useful statistical information through the theory of probability" (1950, 2). In the remaining chapters of this book, we will explore this science and art in detail.

1.7
Exercises

For each of the following surveys, describe the target population, sampling frame, sampling unit, and observation unit. Discuss any possible sources of selection bias or inaccuracy of responses.

1 The article "What Readers Say About Marijuana" reports that "more than 75% of the readers who took part in an informal PARADE telephone poll say marijuana should be as legal as alcoholic beverages" (*Parade*, 31 July 1994, 16). The telephone poll was announced on page 5 of the June 12 issue; readers were instructed to "call 1-900-773-1200, at 75 cents a call, if you would like to answer the following questions. Use touch-tone phones only. To participate, call between 8 a.m. EDT [Eastern Daylight Time] on Saturday, June 11, and midnight EDT on Wednesday, June 15."

2 A student wants to estimate the percentage of mutual funds whose shares went up in price last week. She selects every tenth fund listing in the mutual fund pages of the newspaper and calculates the percentage of those in which the share price increased.

3 Potential jurors in some jurisdictions are chosen from a list of county residents who are registered voters or licensed drivers over age 18. In the fourth quarter of 1994, there were 100,300 jury summons mailed to Maricopa County, Arizona, residents. Approximately 23,000 of those were returned from the post office as undeliverable. Approximately 7000 persons were unqualified for service because they were not citizens, were under age 18, were convicted felons, or had other reasons that disqualified them from serving on a jury. An additional 22,000 were excused from jury service because of illness, financial hardship, military service, or other acceptable reason. The final sample consists of persons who appear for jury duty; some unexcused jurors fail to appear.

4 A sample of 8 architects was chosen in a city with 14 architects and architectural firms. To select a survey sample, each architect was contacted by telephone in order of appearance in the telephone directory. The first 8 agreeing to be interviewed formed the sample.

5 To estimate how many books in the library need rebinding, a librarian uses a random number table to randomly select 100 locations on library shelves. He then walks to each location, looks at the book that resides at that spot, and records whether the book needs rebinding or not.

6   Many scholars and policymakers are interested in the proportion of homeless people who are mentally ill. Wright (1988) estimates that 33 percent of all homeless people are mentally ill, by sampling homeless persons who received medical attention from one of the clinics in the Health Care for the Homeless (HCH) project. He argues that selection bias is not a serious problem because the clinics were easily accessible to the homeless and because the demographic profiles of HCH clients were close to those of the general homeless population in each city in the sample. Do you agree?

7   Approximately 16,500 women returned the Healthy Women Survey that appeared in the September 1992 issue of *Prevention*. The May 1993 issue, reporting on the survey, stated that "ninety-two percent of our readers rated their health as excellent, very good or good."

8   A survey is conducted to find the average weight of cows in a region. A list of all farms is available for the region, and 50 farms are selected at random. Then the weight of each cow at the 50 selected farms is recorded.

9   The Arizona Intrastate Travel Committee commissioned a study to identify in-state travel patterns of Phoenix and Tucson residents and to evaluate different sources of vacation planning information. They conducted 400 interviews with Phoenix residents and 400 interviews with Tucson residents. Telephone numbers with Phoenix and Tucson exchanges were generated randomly so that listed and unlisted telephone numbers could be reached. "Respondents were limited to heads of household and quotas were established in order to have an equal representation of male and female respondents. Additionally, income and age brackets were monitored in order to maintain the same proportions as the general population bases of metropolitan Phoenix and Tucson" (Arizona Office of Tourism 1991).

10   The following letter to the editor appeared in the December 10, 1995, issue of the Appleton *Post-Crescent:* "Paul Harvey, God bless him, has started a nationwide survey being conducted by independent radio stations through their talk show hosts to determine the real sentiments of the American people relative to the sending of troops to Bosnia. So far, the results from one end of the nation to the other average out to over 90% against."

11   To study nutrient content of menus in boarding homes for the elderly in Washington State, Goren et al. (1993) mailed surveys to all 184 licensed homes in the state, directed to the administrator and food service manager. Of those, 43 were returned by the deadline and included menus.

12   The June 1994 issue of *PC World* (on newsstands, May 1994) included a report on reliability and service support for personal computers (PCs). One of the conclusions, "25% of new PC's have problems," formed the top headline of the May 23, 1994, issue of *USA Today*. Every issue of *PC World* since October 1993 had included a survey form asking questions about users' hardware troubles. Survey respondents for each month were entered in a drawing to win a new PC, and over 45,000 responses were received.

13   In lawsuits about trademarks, a plaintiff claiming that another company is infringing on its trademarks must often show that the marks have a *secondary meaning* in the marketplace—that is, potential users of the product associate the trademarks with the plaintiff even when the company's name is missing. In the court case *Harlequin Enterprises Ltd v. Gulf & Western Corporation* (503 F. Supp. 647, 1980), the publisher of Harlequin Romances persuaded the court that the cover design for "Harlequin Presents" novels had acquired secondary meaning. Part of the evidence presented was a survey of 500 women from three cities who identified themselves as readers of romance fiction. They were shown copies of unpublished "Harlequin Presents" novels with the Harlequin name hidden; over 50% identified the novel as a Harlequin product.

14 Ann Landers (1976) asked readers of her column to respond to this question: "If you had it to do over again, would you have children?" About 70% of the readers who responded said no. She received over 10,000 responses, 80% of those from women.

15 The August 1996 issue of *Consumer Reports* contained satisfaction ratings for various health maintenance organizations (HMOs) used by readers of the magazine. Describing the survey, the editors say that "the Ratings are based on more than 20,000 responses to our 1995 Annual Questionnaire about experiences in HMOs between May 1994 and April 1995. Those results reflect experiences of *Consumer Reports* subscribers, who are a more affluent and educated cross-section of the U.S. population" (p. 40). Answer the general questions about target population, sampling frame, and units for this survey. Also, do you think this survey provides valuable information for comparing health plans? If you were selecting an HMO for yourself, which information would you rather have: results from this survey or results from customer-satisfaction surveys conducted by the individual HMOs?

16 Mutations of the BRCA1 gene on chromosome 17 have been shown to be associated with higher risk of breast and ovarian cancer. Ford et al. (1994) studied cancer risks in BRCA1-mutation carriers, using a sample of 33 families in North America and Western Europe. The families were selected by researchers who study breast cancer. Each family in the sample had at least four persons who had been diagnosed with breast or ovarian cancer before age 60. The researchers estimated breast and ovarian cancer risk from the occurrence of second cancers in individuals with breast cancers and estimated a "cumulative risk of breast cancer in gene carriers of 87% by age 70." They concluded: "This study confirms that *BRCA1*-gene carriers have a lifetime risk of either breast or ovarian cancer of close to 100%, and that carriers previously with one cancer have a high risk of developing a second breast or ovarian cancer and need to be managed accordingly" (p. 694). Based on the high calculated risks from this analysis and samples with similar designs, many physicians have recommended that women with a family history of breast cancer have genetic testing; some women have undergone prophylactic mastectomies after discovering they are likely to have the gene.

a Answer the general questions about target population, sampling frame, and units for this sample.

b Does this study provide an estimate of the probability that a woman having the gene will develop breast or ovarian cancer? Explain.

17 The following questions, quoted in Kinsley (1981), are from a survey conducted by Cambridge Reports and financed by Union Carbide. Critique these questions.

> Some people say that granting companies tax credits for the taxes they actually pay to foreign nations could increase these companies' international competitiveness. If you knew for a fact that the tax credits for taxes paid to foreign countries would increase the

money available to U.S. companies to expand and modernize their plants and create more jobs, would you favor or oppose such a tax policy?

Do you favor or oppose changing environmental regulations so that while they still protect the public, they cost American businesses less and lower product costs?

18 The following article, "Abortion-Rights Groups Surveying Voters' Views," by Jack Coffman, appeared in the December 26, 1989, issue of the *St. Paul Pioneer Press Dispatch*[1]. Critique the survey described in this article.

What has been called the biggest survey of abortion-rights sentiment in the nation has become even bigger than its organizers expected, leaders of the survey effort say.

Since Nov. 20, more than 7,000 volunteers have operated six telephone centers in the Twin Cities metropolitan area and Duluth with an additional 1,000 volunteers waiting to begin work in January after a two-week holiday break that started Dec. 15. Another 2,000 volunteers next month will begin telephone operations in their own homes in rural Minnesota.

Since it started, the effort has contacted nearly 160,000 Minnesota families about their views on abortion and lined up county chairmen or chairwomen in 74 of the state's 87 counties. The announced goal of the survey is to contact the families of every registered voter to determine how the voters feel about abortion.

"It's bigger than any political campaign by far," said Marlene Kayser, president of the board of Planned Parenthood of Minnesota, the leading abortion-rights group in the state and key group connected with the survey.

The survey is expected to play an important role in the 1990 legislative session when lawmakers, who traditionally have had anti-abortion leanings, will have to grapple with the volatile abortion issue. The issue has been made even more sensitive because of the decision last summer by the U.S. Supreme Court upholding a Missouri law that increases abortion restrictions and appears to open the way for action by state legislatures.

Anti-abortion forces are gearing up for new abortion restrictions. Backers of the abortion-rights survey plan to use the results in part to try to head off any further anti-abortion legislation.

So far the results of the survey are "overwhelmingly pro-choice," said Kayser.

Results from the calls made since November are being tabulated and will be made available during the next legislative session, which begins Feb. 12. The survey of 1 million Minnesotans is expected to end March 10.

The survey, sponsored by several abortion-rights groups, is being conducted under a contract with Nancy Brataas Associates Inc., a consulting firm owned by state Sen. Nancy Brataas, IR-Rochester. It is expected to cost $250,000.

"It's the most wonderful outpouring of volunteers I have ever seen, and that includes campaigns for president and governor," said Brataas. (The Minnesota presidential campaign of Massachusetts Gov. Michael Dukakis involved about 8,000 volunteers, according to a campaign official.)

Brataas said she believes the strong volunteer response results from "people who are pro-choice and have depended on the Supreme Court and are suddenly very worried and concerned about a woman's right to choose."

---

[1] Reproduced with permission of the *St. Paul Pioneer Press*.

Recruiting volunteers "wasn't hard," said Mary Stringer, co-chairwoman of the St. Paul survey center in the Griggs–Midway building, where 865 volunteers have operated 15 telephones trying to meet a goal of 1,625 calls a day.

Stringer, who described the outpouring of volunteers as "incredible," pointed to two boxes of forms filled out by volunteers who haven't been called yet. Calls also are being made from phone banks in Bloomington, St. Louis Park, White Bear Lake, Minneapolis and Duluth.

Jackie Schwietz, co-director of Minnesota Citizens Concerned for Life, said the survey is "biased" and "dishonest" because the questions don't mention abortion.

She said the MCCL has a "definite plan" for "more restrictive" abortion legislation to be pushed in the 1990 Legislature. However, she declined to describe the group's proposals, which she said will be the subject of a public announcement before the session begins.

When volunteers telephone registered voters, they ask this question: "Do you agree or disagree with the following statement: The decision to terminate a pregnancy is a private matter between a woman, her family and doctor . . . and not a decision to be made by government and politicians."

If the person questioned answers yes, the person is then asked: "In light of current government threats to safe, legal abortion . . . will this issue influence your opinion of politicians in the future?"

If the answer to the original question was no, the person is then asked: "Are you opposed to abortion in cases of rape . . . incest . . . serious fetal deformity . . . or to save the life of a woman?"

19 On March 21, 1993, NBC televised "The First National Referendum—Government Reform Presented by Ross Perot." During the show, 1992 U.S. presidential candidate Perot asked viewers to express their opinions by mailing in the National Referendum on Government Reform, printed in the March 20 issue of *TV Guide*. Some of the questions on the survey were the following:

Do you believe that for every dollar of tax increase there should be $2.00 in spending cuts with the savings earmarked for deficit and debt reduction?

Should the President present an overall plan including spending cuts, spending increases, and tax increases and present the net result of the overall plan, so that the people can know the net result before paying more taxes?

Should the electoral college be replaced with a popular vote for the Presidential election?

Was this TV forum worthwhile? Do you wish to continue participating as a voting member of United We Stand America?

20 Read the following article that describes a proposal for using sampling in the year 2000 U.S. census: W. Roush, 1996. "A census in which all Americans count," *Science* 274: 713–714. What are the main arguments for using sampling in 2000? Against? What do you think?

21 (For students of U.S. history.) Eighty-five letters appeared in New York City newspapers in 1787 and 1788, with the purpose of drawing support in the state for the newly drafted Constitution. Collectively, these letters are known as *The Federalist*. Read number 54 of *The Federalist*, in which the author (widely thought to be James

Madison) discusses using a population census to apportion elected representatives and taxes among the states. This article explains part of Article I, Section II, of the U.S. Constitution.

Write a short paper discussing Madison's view of a population census. What is the target population and sampling frame? What sources of bias does Madison mention, and how does he propose to reduce bias? What is your reaction to Madison's plan, from a statistical point of view? Where do you think Madison would stand today on the issue of using sampling versus complete enumeration to obtain population estimates?

22   Find a recent survey reported in a newspaper or popular magazine. Describe the survey. What are the target population and sampled population? What conclusions are drawn about the survey in the article? Do you think those conclusions are justified? What are possible sources of bias for the survey?

23   Find a survey on the Internet. For example, SurveyNet (www.survey.net) allows you to participate in surveys on a variety of subjects; you can find other surveys by searching online for *survey* or *take survey*. Participate in one of the surveys yourself and write a paragraph or two describing the survey and its results (most online surveys allow you to see the statistics from all persons who have taken the survey). What are the target population and sampled population? What biases do you think might occur in the results?

2

Simple Probability Samples

[Kennedy] read every fiftieth letter of the thirty thousand coming weekly to the White House, as well as a statistical summary of the entire batch, but he knew that these were often as organized and unrepresentative as the pickets on Pennsylvania Avenue.

—Theodore Sorensen, *Kennedy*

The examples of bad surveys in Chapter 1—for example, the *Literary Digest* survey, Example 1.1—had major flaws that resulted in unrepresentative samples. In this chapter, we discuss how to use **probability sampling** to conduct surveys. In a probability sample, each unit in the population has a known probability of selection, and a chance method such as using numbers from a random number table is used to choose the specific units to be included in the sample. If a probability sampling design is implemented well, an investigator can use a relatively small sample to make inferences about an arbitrarily large population.

In Chapters 2 through 6, we explore survey design and properties of estimates for the three major design components used in a probability sample: simple random sampling, stratified sampling, and cluster sampling. We will integrate all these ideas in Chapter 7 and show how they are combined in complex surveys such as the U.S. National Crime Victimization Survey. To simplify presentation of the concepts, we assume for now that the sampled population is the target population, that the sampling frame is complete, that there is no nonresponse or missing data, and that all measurements are accurate. We return to nonsampling errors in Chapter 8.

As you might suppose, you need to know some probability to be able to understand probability sampling. You may want to review the material in Sections B.1 and B.2 of Appendix B while reading this chapter.

2.1
Types of Probability Samples

The terms *simple random sample*, *stratified sample,* and *cluster sample* are basic to any discussion of sample surveys, so let's define them now.

- A **simple random sample** (SRS) is the simplest form of probability sample. An SRS of size n is taken when every possible subset of n units in the population has the same chance of being the sample. SRSs are the focus of this chapter and the foundation for more complex sampling designs. In taking a random sample, the investigator is in effect mixing up the population before grabbing n units. The investigator does not need to examine every member of the population for the same reason that a medical technician does not need to drain you of blood to measure your red blood cell count: Your blood is sufficiently well mixed that any sample should be representative. SRSs are discussed in Section 2.3, after we present the basic framework for probability samples in Section 2.2.

- In a **stratified random sample,** the population is divided into subgroups called *strata.* Then an SRS is selected from each stratum, and the SRSs in the strata are selected independently. The strata are often subgroups of interest to the investigator—for example, the strata might be different ethnic or age groups in a survey of people, different types of terrain in an ecological survey, or sizes of firms in a business survey. Elements in the same stratum often tend to be more similar than randomly selected elements from the whole population, so stratification often increases precision, as we will see in Chapter 4.

- In a **cluster sample,** observation units in the population are aggregated into larger sampling units, called *clusters.* Suppose you want to survey Lutheran church members in Minneapolis but do not have a list of all church members in the city, so you cannot take an SRS of church members. However, you do have a list of all the Lutheran churches. You can then take an SRS of the churches and then subsample all or some church members in the selected churches. In this case, the churches form the clusters, and the church members are the observation units. It is more convenient to sample at the church level; however, members of the same church may have more similarities than Lutherans selected at random in Minneapolis, so a cluster sample of 500 Lutherans may not provide as much information as an SRS of 500 Lutherans. We will explore this idea further in Chapter 5.

Suppose you want to estimate the average amount of time that professors at your university spent grading homework in a specific week. To take an SRS, construct a list of all professors and randomly select n of them to be your sample. Now ask each professor in your sample how much time he or she spent grading homework that week—you would of course have to define the words *homework* and *grading* carefully in your questionnaire. In a stratified sample, you might classify faculty by college: engineering, liberal arts and sciences, business, nursing, and fine arts. You would then take an SRS of faculty in the engineering college, a separate SRS of faculty in liberal arts and sciences, and so on. For a cluster sample, you might randomly select 10 of the 60 academic departments in the university and ask each faculty member in those departments how much time he or she spent grading homework.

All these methods—SRS, stratified random sampling, and cluster sampling—involve random selection of units to be in the sample. In an SRS, the observation units themselves are selected at random from the population of observation units; in a stratified random sample, observation units within each stratum are randomly selected; in a cluster sample, the clusters are randomly selected from the population of all clusters. Each method is a form of probability sampling, which we will discuss in the next section.

2.2
Framework for Probability Sampling

To show how probability sampling works, we need to be able to list the N units in the finite population. The finite **population,** or **universe,** of N units is denoted by the index set

$$\mathcal{U} = \{1, 2, \ldots, N\}. \tag{2.1}$$

Out of this population we can choose various samples, which are subsets of $\mathcal{U}$. The particular sample chosen is denoted by $\mathcal{S}$, a subset consisting of n of the units in $\mathcal{U}$.

Suppose the population has four units: $\mathcal{U} = \{1, 2, 3, 4\}$. Six different samples of size 2 could be chosen from this population:

$$\mathcal{S}_1 = \{1, 2\} \qquad \mathcal{S}_4 = \{2, 3\}$$
$$\mathcal{S}_2 = \{1, 3\} \qquad \mathcal{S}_5 = \{2, 4\}$$
$$\mathcal{S}_3 = \{1, 4\} \qquad \mathcal{S}_6 = \{3, 4\}$$

In probability sampling, each possible sample $\mathcal{S}$ from the population has a known probability $P(\mathcal{S})$ of being chosen, and the probabilities of the possible samples sum to 1. One possible sample design for a probability sample of size 2 would have $P(\mathcal{S}_1) = 1/3$, $P(\mathcal{S}_2) = 1/6$, and $P(\mathcal{S}_6) = 1/2$, and $P(\mathcal{S}_3) = P(\mathcal{S}_4) = P(\mathcal{S}_5) = 0$. The probabilities $P(\mathcal{S}_1)$, $P(\mathcal{S}_2)$, and $P(\mathcal{S}_6)$ of the possible samples are known before the sample is drawn. One way to select the sample is to place six labeled balls in a box; two of the balls are labeled 1, one is labeled 2, and three are labeled 6. Now choose one at random; if a ball labeled 6 is chosen, then $\mathcal{S}_6$ is the sample.

In a probability sample, since each possible sample has a known probability of being the chosen sample, each unit in the population has a known probability of appearing in our selected sample. We calculate

$$P(\text{unit } i \text{ in sample}) = \pi_i$$

by summing the probabilities of all possible samples that contain unit i. In probability sampling, the π_i are known before the survey commences, and we assume that $\pi_i > 0$ for every unit in the population. For the sample design described above, $\pi_1 = P(\mathcal{S}_1) + P(\mathcal{S}_2) + P(\mathcal{S}_3) = 1/2$, $\pi_2 = P(\mathcal{S}_1) + P(\mathcal{S}_4) + P(\mathcal{S}_5) = 1/3$, $\pi_3 = P(\mathcal{S}_2) + P(\mathcal{S}_4) + P(\mathcal{S}_6) = 2/3$, and $\pi_4 = P(\mathcal{S}_3) + P(\mathcal{S}_5) + P(\mathcal{S}_6) = 1/2$.

Of course, we never write all possible samples down and calculate the probability with which we would choose every possible sample—this would take far too long. But such enumeration underlies all of probability sampling. Investigators using a probability sample have much less discretion about which units are included in the sample, so using probability samples helps us avoid some of the selection biases described in Chapter 1. In a probability sample, the interviewer cannot choose to substitute a friendly looking person for the grumpy person selected to be in the sample by the random selection method. A forester taking a probability sample of trees cannot simply measure the trees near the road but must measure the trees designated for inclusion in the sample. Taking a probability sample is much harder than taking a convenience sample, but a probability sampling procedure guarantees that each unit in the population could appear in the sample and provides information that can be used to assess the precision of statistics calculated from the sample.

Within the framework of probability sampling, we can quantify how likely it is that our sample is a "good" one. A single probability sample is not guaranteed to be representative of the population with regard to the characteristics of interest, but we can quantify how often samples will meet some criterion of representativeness. The notion is the same as that of confidence intervals: We do not know whether the particular 95% confidence interval we construct for the mean contains the true value of the mean. We do know, however, that if the assumptions for the confidence interval procedure are valid and if we repeat the procedure over and over again, we can expect 95% of the resulting confidence intervals to contain the true value of the mean.

Let y_i be a characteristic associated with the ith unit in the population. We consider y_i to be a fixed quantity; if farm 723 is included in the sample, then the amount of corn produced on farm 723, y_{723}, is known exactly.

EXAMPLE 2.1   To illustrate these concepts, let's look at an artificial situation in which we know the value of y_i for each of the $N = 8$ units in the whole population. The index set for the population is

$$\mathcal{U} = \{1, 2, 3, 4, 5, 6, 7, 8\}.$$

The values of y_i are

i	1 2 3 4 5 6 7 8
y_i	1 2 4 4 7 7 7 8

There are 70 possible samples of size 4 that can be drawn without replacement from this population; the samples are listed in file samples.dat on the data disk. If the sample consisting of units $\{1, 2, 3, 4\}$ were chosen, the corresponding values of y_i would be 1, 2, 4, and 4. The values of y_i for the sample $\{2, 3, 6, 7\}$ are 2, 4, 7, and 7. Define $P(\mathcal{S}) = 1/70$ for each distinct subset of size 4 from $\mathcal{U}$. As you will see after you read Section 2.3, this design is an SRS without replacement. Each unit is in exactly 35 of the possible samples, so $\pi_i = 1/2$ for $i = 1, 2, \ldots, 8$.

A random mechanism is used to select one of the 70 possible samples. One possible mechanism for this example, because we have listed all possible samples, is to generate a random number between 1 and 70 and select the corresponding sample. With large populations, the number of samples is so great that in practice the units themselves are randomly selected according to prespecified probabilities. ■

Most results in sampling rely on the **sampling distribution** of a statistic, the distribution of different values of the statistic obtained by the process of taking all possible samples from the population. A sampling distribution is an example of a discrete probability distribution.

Suppose we want to use a sample to estimate a population quantity—say, the population total $t = \sum_{i=1}^{N} y_i$. One estimate we might use for t is $\hat{t}_{\mathcal{S}} = N \bar{y}_{\mathcal{S}}$, where $\bar{y}_{\mathcal{S}}$ is the average of the y_i's in $\mathcal{S}$, the chosen sample. In our example, $t = 40$. If the sample $\mathcal{S}$ consists of units 1, 3, 5, and 6, then $\hat{t}_{\mathcal{S}} = 8 \times (1+4+7+7)/4 = 38$. Since we know the whole population here, we can find $\hat{t}_S$ for each of the 70 possible samples. The probabilities of selection for the samples give the sampling distribution of $\hat{t}$:

$$P\{\hat{t} = k\} = \sum_{S:\hat{t}_S=k} P(S).$$

The summation is over all samples S for which $\hat{t}_S = k$. We know the probability $P(S)$ with which we select a sample S because we take a probability sample.

EXAMPLE **2.2**　The sampling distribution of $\hat{t}$ for the population and sampling design in Example 2.1 derives entirely from the probabilities of selection for the various samples. Four samples ({3, 4, 5, 6}, {3, 4, 5, 7}, {3, 4, 6, 7}, and {1, 5, 6, 7}) result in the estimate $\hat{t} = 44$, so $P\{\hat{t} = 44\} = 4/70$. For this example, we can write out the sampling distribution of $\hat{t}$ because we know the values for the entire population.

k	22	28	30	32	34	36	38	40	42	44	46	48	50	52	58
$P\{\hat{t} = k\}$	$\frac{1}{70}$	$\frac{6}{70}$	$\frac{2}{70}$	$\frac{3}{70}$	$\frac{7}{70}$	$\frac{4}{70}$	$\frac{6}{70}$	$\frac{12}{70}$	$\frac{6}{70}$	$\frac{4}{70}$	$\frac{7}{70}$	$\frac{3}{70}$	$\frac{2}{70}$	$\frac{6}{70}$	$\frac{1}{70}$

Figure 2.1 displays the sampling distribution. ■

The **expected value** of $\hat{t}$, $E[\hat{t}]$, is the mean of the sampling distribution of $\hat{t}$:

$$E[\hat{t}] = \sum_{S} P(S)\hat{t}_S \tag{2.2}$$

$$= \sum_{k} kP(\hat{t} = k).$$

The expected value of the statistic is the weighted average of the possible sample values of the statistic, with weights the probability that that particular value of the statistic would occur.

The **estimation bias** of the estimator $\hat{t}$ is

$$\text{Bias}[\hat{t}] = E[\hat{t}] - t. \tag{2.3}$$

FIGURE **2.1**

Sampling distribution of the sample total in Example 2.2.

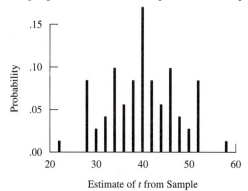

Estimate of *t* from Sample

If Bias[$\hat{t}$] = 0, we say that the estimator $\hat{t}$ is **unbiased** for t. For the data in Example 2.1, the expected value of $\hat{t}$ is

$$E[\hat{t}] = \frac{1}{70}(22) + \frac{6}{70}(28) + \cdots + \frac{1}{70}(58) = 40.$$

Thus, the estimator is unbiased.

Note that the mathematical definition of bias in Equation (2.3) is *not* the same thing as the selection or measurement bias described in Chapter 1. All indicate a systematic deviation from the population value, but from different sources. Selection bias is due to the method of selecting the sample—often, the investigator acts as though every possible sample S has the same probability of being selected, but some subsets of the population actually have a different probability of selection. With undercoverage, for example, the probability of including a unit not in the sampling frame is zero. Measurement bias means that the y_i's are not really the quantities of interest, so although $\hat{t}$ may be unbiased in the sense of (2.3) for $t = \sum_{i=1}^{N} y_i$, t itself would not be the true total of interest. Estimation bias means that the estimator chosen results in bias—for example, if we used $\hat{t}_S = \sum_{i \in S} y_i$ and did not take a census, $\hat{t}$ would be biased. To illustrate these distinctions, suppose you want to estimate the average height of male actors belonging to the Screen Actor's Guild. Selection bias would occur if you took a convenience sample of actors on the set—perhaps taller actors are more or less likely to be working. Measurement bias would occur if your tape measure inaccurately added 3 centimeters (cm) to each actor's height. Estimation bias would occur if you took an SRS from the list of all actors in the Guild but estimated mean height by the average height of the six shortest men in the sample—the sampling procedure is good, but the estimator is bad.

The **variance** of the sampling distribution of $\hat{t}$ is

$$V[\hat{t}] = E[(\hat{t} - E[\hat{t}])^2] \tag{2.4}$$
$$= \sum_{\substack{\text{all possible} \\ \text{samples } S}} P(S)(\hat{t}_S - E[\hat{t}])^2.$$

For the data in Example 2.1,

$$V[\hat{t}] = \frac{1}{70}(22 - 40)^2 + \cdots + \frac{1}{70}(58 - 40)^2 = \frac{3840}{70} = 54.86.$$

Because we sometimes use biased estimators, we often use the **mean squared error** (MSE) rather than variance to measure the accuracy of an estimator:

$$\begin{aligned}
\text{MSE}[\hat{t}] &= E[(\hat{t} - t)^2] \\
&= E[(\hat{t} - E[\hat{t}] + E[\hat{t}] - t)^2] \\
&= E[(\hat{t} - E[\hat{t}])^2] + (E[\hat{t}] - t)^2 + 2E[(\hat{t} - E[\hat{t}])(E[\hat{t}] - t)] \\
&= V[\hat{t}] + \left(\text{Bias}[\hat{t}]\right)^2.
\end{aligned}$$

Thus, an estimator $\hat{t}$ of t is *unbiased* if $E[\hat{t}] = t$, *precise* if $V[\hat{t}] = E[(\hat{t} - E[\hat{t}])^2]$ is small, and *accurate* if $\text{MSE}[\hat{t}] = E[(\hat{t} - t)^2]$ is small. A badly biased estimate may be precise, but it will not be accurate; accuracy (MSE) is how close the estimate is to the true value, whereas precision (variance) measures how close estimates from different samples are to each other. Figure 2.2 illustrates these concepts.

FIGURE 2.2

Unbiased, precise, and accurate archers. Archer A is unbiased—the average position of all arrows is at the bull's-eye. Archer B is precise but not unbiased—all arrows are close together but systematically away from the bull's-eye. Archer C is accurate—all arrows are close together and near the center of the target.

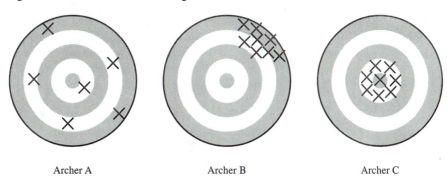

Archer A   Archer B   Archer C

In summary, the finite population $\mathcal{U}$ consists of units $\{1, 2, \ldots, N\}$ whose measured values are $\{y_1, y_2, \ldots, y_N\}$. We select a sample $\mathcal{S}$ of n units from $\mathcal{U}$ using the probabilities of selection that define the sampling design. The y's are fixed but unknown quantities—unknown unless that unit happens to appear in our sample $\mathcal{S}$. Unless we make additional assumptions, the only information we have about the set of y's in the population is in the set $\{y_i : i \in \mathcal{S}\}$.

You may be interested in many different population quantities from your population. Historically, however, the main impetus for developing theory for sample surveys has been estimating population means and totals. Suppose we want to estimate the total number of persons in Canada who have diabetes, or the average number of oranges produced per orange tree. The population total is

$$t = \sum_{i=1}^{N} y_i,$$

and the mean of the population is

$$\bar{y}_U = \frac{1}{N} \sum_{i=1}^{N} y_i.$$

Almost all populations exhibit some variability; for example, households have different incomes and trees have different diameters. Define the **variance** of the population values about the mean as

$$S^2 = \frac{1}{N-1} \sum_{i=1}^{N} (y_i - \bar{y}_U)^2. \tag{2.5}$$

The population **standard deviation** is $S = \sqrt{S^2}$.

The population standard deviation is often related to the mean. A population of trees might have a mean height of 10 meters (m) and a standard deviation of 1 m. A population of small cacti, however, with a mean height of 10 cm, might have a standard deviation of 1 cm. The **coefficient of variation** (CV) is a measure of relative variability, which can be defined when $\bar{y}_U \neq 0$ as

$$CV(y) = \frac{S}{\bar{y}_U}.$$

If tree height is measured in meters, then $\bar{y}_U$ and S are also in meters. The coefficient of variation does not depend on the unit of measurement. In this example, the trees and the cacti have the same coefficient of variation.

It is sometimes helpful to have a special notation for proportions. The proportion of units having a characteristic is simply a special case of the mean, obtained by letting $y_i = 1$ if the ith unit has the characteristic of interest, and $y_i = 0$ if the ith unit does not have the characteristic. Let

$$p = \frac{\text{number of units with the characteristic in the population}}{N}.$$

EXAMPLE 2.3 For the population in Example 2.1, let

$$y_i = \begin{cases} 1 & \text{if the } i\text{th unit has the value 7} \\ 0 & \text{if the } i\text{th unit does not have the value 7} \end{cases}$$

Let $\hat{p}_S = \sum_{i \in S} y_i / 4$, the proportion of 7s in the sample. The list of all possible samples in the data file samples.dat has 5 samples with no 7s, 30 samples with exactly one 7, 30 samples with exactly two 7s, and 5 samples with three 7s. Since one of the possible samples is selected with probability 1/70, the sampling distribution of $\hat{p}$ is[1]:

k	0	$\frac{1}{4}$	$\frac{1}{2}$	$\frac{3}{4}$
$P(\hat{p} = k)$	$\frac{5}{70}$	$\frac{30}{70}$	$\frac{30}{70}$	$\frac{5}{70}$

∎

2.3
Simple Random Sampling

Simple random sampling is the most basic form of probability sampling and provides the theoretical basis for the more complicated forms. There are two ways of taking a simple random sample: with replacement, in which the same unit may be included more than once in the sample, and without replacement, in which all units in the sample are distinct.

A **simple random sample with replacement** (SRSWR) of size n from a population of N units can be thought of as drawing n independent samples of size 1. One unit is randomly selected from the population to be the first sampled unit, with probability $1/N$. Then the sampled unit is replaced in the population, and a second unit is randomly selected with probability $1/N$. This procedure is repeated until the sample has n units, which may include duplicates from the population.

In finite population sampling, however, sampling the same person twice provides no additional information. We usually prefer to sample without replacement so that

---

[1] An alternative derivation of the sampling distribution is in Exercise B.2 (p. 427).

the sample contains no duplicates. A **simple random sample without replacement** (SRS) of size n is selected so that every possible subset of n distinct units in the population has the same probability of being selected as the sample. There are $\binom{N}{n}$ possible samples (see Appendix B), and each equally likely, so the probability of selecting any individual sample $\mathcal{S}$ of n units is

$$P(\mathcal{S}) = \frac{1}{\binom{N}{n}} = \frac{n!(N-n)!}{N!}.$$

As a consequence of this definition, the probability that any given unit appears in the sample is n/N, as shown later in Equation (2.18).

To take an SRS, you need a list of all observation units in the population; this list is the sampling frame. In an SRS, the sampling unit and observation unit coincide. Each unit is assigned a number, and a sample is selected so that (1) each unit has the same chance of occurring in the sample and (2) the selection of a unit is not influenced by which other units have already been selected. This can be thought of as drawing numbers out of a hat; in practice, computer-generated pseudorandom numbers are usually used to select a sample.

EXAMPLE 2.4 The U.S. government conducts a Census of Agriculture every five years, collecting data on all farms (defined as any place from which $1000 or more of agricultural products were produced and sold) in the 50 states.[2] The Census of Agriculture provides data on number of farms, the total acreage devoted to farms, farm size, yield of different crops, and a wide variety of other agricultural measures for each of the $N = 3078$ counties and county-equivalents in the United States. The file agpop.dat on the data disk contains the 1982, 1987, and 1992 information on the number of farms, acreage devoted to farms, number of farms with fewer than 9 acres, and number of farms with more than 1000 acres for the population.

To take an SRS of size 300 from this population, I generated 300 random numbers between 0 and 1 on the computer, multiplied each by 3078, and rounded the result up to the next highest integer. This procedure generates an SRSWR. If the population is large relative to the sample, it is likely that each unit in the sample occurs only once in the list. In this case, however, 13 of the 300 numbers were duplicates. The duplicates were discarded and replaced with new randomly generated numbers between 1 and 3078, until all 300 numbers were distinct; the set of random numbers generated is in the file selectrs.dat, and the data set for the SRS is in agsrs.dat. Other methods that might be used to select an SRS are described in the exercises and in Appendix E.

The counties selected to be in the sample may not "feel" very random at first glance. For example, counties 2840, 2841, and 2842 are all in the sample, whereas none of the counties between 2740 to 2787 appear. The sample contains 18% of Virginia counties, but no counties in Alaska, Arizona, Connecticut, Delaware, Hawaii, Rhode Island, Utah, or Wyoming. There is a quite natural temptation to want to "adjust" the random

---

[2]The Census of Agriculture was formerly conducted by the U.S. Bureau of the Census; currently, it is conducted by the U.S. National Agricultural Statistics Service (NASS). More information about the census and selected data is available on the Internet through the NASS material on www.fedstats.gov.

FIGURE   2.3

Histogram: number of acres devoted to farms in 1992, for an SRS of 300 counties. Note the skewness of the data. Most of the counties have fewer than 500,000 acres in farms; some counties, however, have more than 1.5 million acres in farms.

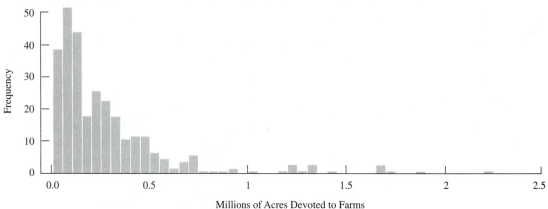

Millions of Acres Devoted to Farms

number list, to spread it out a bit more. If you want a random sample, you must resist this temptation. Research, beginning with Neyman (1934), repeatedly demonstrates that purposive samples often do not represent the population on key variables. If you deliberately substitute other counties for those in the randomly generated sample, you may be able match the population on one particular characteristic such as geographic distribution; however, you will likely fail to match the population on characteristics of interest such as number of farms or average farm size. If you want to ensure that all states are represented, do not adjust your randomly selected sample purposively but take a stratified sample (to be discussed in Chapter 4).

Let's look at the variable *acres92*, the number of acres devoted to farms in 1992. A small number of counties in the population are missing that value—in some cases, the data are withheld to prevent disclosing data on individual farms. Thus, we first check to see the extent of the missing data in our sample. Fortunately, our sample has no missing data (see Exercise 7 to see how likely such an occurrence is). Figure 2.3 displays a histogram of the acreage devoted to farms in each of the 300 counties.   ∎

For estimating the population mean $\bar{y}_U$ in an SRS, we use the sample mean

$$\bar{y}_S = \frac{1}{n} \sum_{i \in S} y_i. \tag{2.6}$$

In the following, we use $\bar{y}$ to refer to the sample mean and drop the subscript S unless it is needed for clarity. As will be shown in Section 2.7, $\bar{y}$ is an unbiased estimator of the population mean $\bar{y}_U$, and the variance of $\bar{y}$ is

$$V(\bar{y}) = \frac{S^2}{n} \left(1 - \frac{n}{N} \right) \tag{2.7}$$

for S^2 defined in Equation (2.5). The variance $V(\bar{y})$ measures the variability among estimates of $\bar{y}_U$ from different samples.

The factor $(1 - n/N)$ is called the **finite population correction** (fpc). Intuitively, we make this correction because with small populations the greater our **sampling fraction** n/N, the more information we have about the population and thus the smaller the variance. If $N = 10$ and we sample all 10 observations, we would expect the variance of $\bar{y}$ to be 0 (which it is). If $N = 10$, there is only one possible sample $\mathcal{S}$ of size 10 without replacement, with $\bar{y}_S = \bar{y}_U$, so there is no variability due to taking a sample. For a census, the fpc, and hence $V(\bar{y})$, is 0. When the sampling fraction n/N is large in an SRS without replacement, the sample is closer to a census, which has no sampling variability.

For most samples that are taken from extremely large populations, the fpc is approximately 1. For large populations it is the size of the sample taken, not the percentage of the population sampled, that determines the precision of the estimator: If your soup is well stirred, you need to taste only one or two spoonfuls to check the seasoning, whether you have made 1 liter or 20 liters of soup. A sample of size 100 from a population of 100,000 units has almost the same precision as a sample of size 100 from a population of 100 million units:

$$V[\bar{y}] = \frac{S^2}{100}\frac{99,900}{100,000} = \frac{S^2}{100}(0.999) \qquad \text{for } N = 100,000$$

$$V[\bar{y}] = \frac{S^2}{100}\frac{99,999,900}{100,000,000} = \frac{S^2}{100}(0.999999) \quad \text{for } N = 100,000,000$$

The population variance S^2, which depends on the values for the entire population, is still unknown. We estimate it by the sample variance:

$$s^2 = \frac{1}{n-1}\sum_{i \in \mathcal{S}}(y_i - \bar{y})^2. \tag{2.8}$$

An unbiased estimator of the variance of $\bar{y}$ is (see Section 2.7)

$$\hat{V}[\bar{y}] = \left(1 - \frac{n}{N}\right)\frac{s^2}{n}. \tag{2.9}$$

We usually report not the estimated variance of $\bar{y}$ but its square root, the **standard error** (SE):

$$\text{SE}[\bar{y}] = \sqrt{\left(1 - \frac{n}{N}\right)\frac{s^2}{n}}. \tag{2.10}$$

The estimated coefficient of variation of an estimate gives a measure of the relative variability of an estimate. It is the standard error divided by the mean (defined only when the mean is nonzero):

$$\widehat{\text{CV}}(\bar{y}) = \frac{\text{SE}[\bar{y}]}{\bar{y}}. \tag{2.11}$$

All these results apply to the estimation of a population total, t, since

$$t = \sum_{i=1}^{N} y_i = N\bar{y}_U.$$

To estimate t, we use the unbiased estimator

$$\hat{t} = N\bar{y}. \tag{2.12}$$

Then, from Equation (2.7),

$$V[\hat{t}] = N^2 V[\bar{y}] = N^2 \left(1 - \frac{n}{N}\right) \frac{S^2}{n} \tag{2.13}$$

and

$$\hat{V}[\hat{t}] = N^2 \left(1 - \frac{n}{N}\right) \frac{s^2}{n}. \tag{2.14}$$

EXAMPLE 2.5   For the data in Example 2.4, $N = 3078$ and $n = 300$, so the sampling fraction is $300/3078 = 0.097$. The sample statistics are $\bar{y} = 297{,}897$, $s = 344{,}551.9$, and $\hat{t} = N\bar{y} = 916{,}927{,}110$. Standard errors are

$$\text{SE}[\bar{y}] = \sqrt{\frac{s^2}{n}\left(1 - \frac{300}{3078}\right)} = 18{,}898.434428$$

and

$$\text{SE}[\hat{t}] = (3078)(18{,}898.434428) = 58{,}169{,}381$$

and the estimated coefficient of variation is

$$\begin{aligned}
\widehat{\text{CV}}[\hat{t}] &= \widehat{\text{CV}}[\bar{y}] \\
&= \frac{\text{SE}[\bar{y}]}{\bar{y}} \\
&= \frac{18{,}898.434428}{297{,}897} \\
&= 0.06344.
\end{aligned}$$

Since these data are so highly skewed, we should also report the median number of farm acres in a county, which is 196,717. ∎

We might also want to estimate the proportion of counties in Example 2.4 with fewer than 200,000 acres in farms. Since estimating a proportion is a special case of estimating a mean, the results in Equations (2.6)–(2.11) hold for proportions as well, and they take a simple form. Suppose we want to estimate the proportion of units in the population that have some characteristic—call this proportion p. Define y_i to be 1 if the unit has the characteristic and to be 0 if the unit does not have that characteristic. Then $p = \sum_{i=1}^{N} y_i/N = \bar{y}_U$, and p is estimated by $\hat{p} = \bar{y}$. Consequently, $\hat{p}$ is an unbiased estimator of p. For the response y_i, taking on values 0 or 1,

$$S^2 = \frac{\sum_{i=1}^{N}(y_i - p)^2}{N-1} = \frac{\sum_{i=1}^{N} y_i^2 - 2p\sum_{i=1}^{N} y_i + Np^2}{N-1} = \frac{N}{N-1}p(1-p).$$

Thus, (2.7) implies that

$$V[\hat{p}] = \left(\frac{N-n}{N-1}\right)\frac{p(1-p)}{n}. \tag{2.15}$$

Also,

$$s^2 = \frac{1}{n-1} \sum_{i \in S} (y_i - \hat{p})^2 = \frac{n}{n-1} \hat{p}(1 - \hat{p}).$$

So from (2.9),

$$\hat{V}[\hat{p}] = \left(1 - \frac{n}{N}\right) \frac{\hat{p}(1 - \hat{p})}{n-1}. \qquad \text{(2.16)}$$

EXAMPLE 2.6   For the sample described in Example 2.4, the estimated proportion of counties with fewer than 200,000 acres in farms is

$$\hat{p} = \frac{153}{300} = 0.51$$

with standard error

$$\text{SE}(\hat{p}) = \sqrt{\left(1 - \frac{300}{3078}\right) \frac{(0.51)(0.49)}{299}} = 0.0275. \quad \blacksquare$$

2.4
Confidence Intervals

When you take a sample survey, it is not sufficient to simply report the average height of trees or the sample proportion of voters who intend to vote for candidate B in the next election. You also need to give an indication of how accurate your estimates are. In statistics, **confidence intervals** (CIs) are used to indicate the accuracy of an estimate.

A 95% CI is often explained heuristically: If we take samples from our population over and over again and construct a confidence interval using our procedure for each possible sample, we expect 95% of the resulting intervals to include the true value of the population parameter.

In probability sampling from a finite population, only a finite number of possible samples exist, and we know the probability with which each will be chosen; if we could generate all possible samples from the population, we could calculate the exact confidence level for a confidence interval procedure.

EXAMPLE 2.7   Return to Example 2.1, in which the entire population is known. Let's choose an arbitrary procedure for calculating a confidence interval, constructing interval estimates for t as

$$\text{CI}(\mathcal{S}) = [\hat{t}_\mathcal{S} - 4s_\mathcal{S}, \hat{t}_\mathcal{S} + 4s_\mathcal{S}].$$

There is no theoretical reason to choose this procedure, but it will illustrate the concept of a confidence interval. Define $u(\mathcal{S})$ to be 1 if $\text{CI}(\mathcal{S})$ contains the true population value 40, and 0 if $\text{CI}(\mathcal{S})$ does not contain 40. Since we know the population, we can calculate the confidence interval $\text{CI}(\mathcal{S})$ and the value of $u(\mathcal{S})$ for each possible sample $\mathcal{S}$. Some of the 70 confidence intervals are shown in Table 2.1 (all entries are rounded to two decimals).

TABLE   2.1

Confidence intervals for possible samples from small population

Sample S	$y_i, i \in S$	$\hat{t}_S$	s_S	CI(S)	$u(S)$
{1, 2, 3, 4}	1, 2, 4, 4	22	1.50	[16.00, 28.00]	0
{1, 2, 3, 5}	1, 2, 4, 7	28	2.65	[17.42, 38.58]	0
{1, 2, 3, 6}	1, 2, 4, 7	28	2.65	[17.42, 38.58]	0
{1, 2, 3, 7}	1, 2, 4, 7	28	2.65	[17.42, 38.58]	0
{1, 2, 3, 8}	1, 2, 4, 8	30	3.10	[17.62, 42.38]	1
{1, 2, 4, 5}	1, 2, 4, 7	28	2.65	[17.42, 38.58]	0
{1, 2, 4, 6}	1, 2, 4, 7	28	2.65	[17.42, 38.58]	0
{1, 2, 4, 7}	1, 2, 4, 7	28	2.65	[17.42, 38.58]	0
{1, 2, 4, 8}	1, 2, 4, 8	30	3.10	[17.62, 42.38]	1
{1, 2, 5, 6}	1, 2, 7, 7	34	3.20	[21.19, 46.81]	1
$\vdots$	$\vdots$	$\vdots$	$\vdots$	$\vdots$	$\vdots$
{2, 3, 4, 8}	2, 4, 4, 8	36	2.52	[25.93, 46.07]	1
{2, 3, 5, 6}	2, 4, 7, 7	40	2.45	[30.20, 49.80]	1
{2, 3, 5, 7}	2, 4, 7, 7	40	2.45	[30.20, 49.80]	1
{2, 3, 5, 8}	2, 4, 7, 8	42	2.75	[30.98, 53.02]	1
{2, 3, 6, 7}	2, 4, 7, 7	40	2.45	[30.20, 49.80]	1
{2, 3, 6, 8}	2, 4, 7, 8	42	2.75	[30.98, 53.02]	1
$\vdots$	$\vdots$	$\vdots$	$\vdots$	$\vdots$	$\vdots$
{4, 5, 6, 7}	4, 7, 7, 7	50	1.50	[44.00, 56.00]	0
{4, 5, 6, 8}	4, 7, 7, 8	52	1.73	[45.07, 58.93]	0
{4, 5, 7, 8}	4, 7, 7, 8	52	1.73	[45.07, 58.93]	0
{4, 6, 7, 8}	4, 7, 7, 8	52	1.73	[45.07, 58.93]	0
{5, 6, 7, 8}	7, 7, 7, 8	58	0.50	[56.00, 60.00]	0

Each individual confidence interval either does or does not contain the population total 40. The probability statement in the confidence interval is made about the collection of all possible samples; for this confidence interval procedure and population, the confidence level is

$$\sum_S P(S)u(S) = 0.77.$$

This means that if we take an SRS of four elements without replacement from this population of eight elements, there is a 77% chance that our sample is one of the "good" ones whose confidence interval contains the true value 40. This procedure thus creates a 77% confidence interval.

Of course, in real life, we take only one sample and do not know the value of the population total t. Without further investigation, we have no way of knowing whether the sample we obtained is one of the "good" ones, such as $S = \{2, 3, 5, 6\}$, or one of the "bad" ones, such as $S = \{4, 6, 7, 8\}$. The confidence interval gives us only a probabilistic statement of how often we expect to be right.   ∎

In practice, we do not know the values of statistics from all possible samples, so we cannot calculate the exact confidence coefficient for a procedure as done in Example 2.7. In your introductory statistics class, you relied largely on **asymptotic** (as the sample size goes to infinity) results to construct confidence intervals for an unknown mean μ. The central limit theorem says that if we have a random sample with replacement, then the probability distribution of $\sqrt{n}(\bar{y} - \mu)$ converges to a normal distribution as the sample size n approaches infinity.

In most sample surveys, though, we only have a finite population. To use asymptotic results in finite population sampling, we pretend that our population is itself part of a larger **superpopulation;** the superpopulation is itself a subset of a larger superpopulation, and so on, until the superpopulations are as large as we could wish. Our population is embedded in a series of increasing finite populations. This embedding can give us properties such as consistency and asymptotic normality. One can imagine the superpopulations as "alternative universes" in a science fiction sense—what might have happened if circumstances were slightly different.

Hájek (1960) proves a central limit theorem for simple random sampling without replacement. In practical terms, Hájek's theorem says that if certain technical conditions hold and if n, N, and $N - n$ are all "sufficiently large," then the sampling distribution of

$$\frac{\bar{y} - \bar{y}_U}{\sqrt{\left(1 - \dfrac{n}{N}\right)} \dfrac{S}{\sqrt{n}}}$$

is approximately normal (Gaussian) with mean 0 and variance 1. A large-sample $100(1 - \alpha)\%$ CI for the population mean is

$$\left[\bar{y} - z_{\alpha/2}\sqrt{1 - \frac{n}{N}} \frac{S}{\sqrt{n}}, \ \bar{y} + z_{\alpha/2}\sqrt{1 - \frac{n}{N}} \frac{S}{\sqrt{n}}\right],$$

where $z_{\alpha/2}$ is the $(1 - \alpha/2)$th percentile of the standard normal distribution. Usually, S is unknown, so in large samples s is substituted for S with little change in the approximation; the large-sample confidence interval is

$$[\bar{y} - z_{\alpha/2}\text{SE}(\bar{y}), \ \bar{y} + z_{\alpha/2}\text{SE}(\bar{y})].$$

In simple random sampling without replacement, 95% of the possible samples that could be chosen will give a 95% CI for $\bar{y}_U$ that contains the true value of $\bar{y}_U$. When $n/N \approx 0$, this confidence interval is the same as the one taught in introductory statistics classes for sampling with replacement.

The imprecise term *sufficiently large* in the theorem occurs because the adequacy of the normal approximation depends on n and on how closely the population $\{y_i, i = 1, \ldots, N\}$ resembles a population generated from the normal distribution. The "magic number" of $n = 30$, often cited in introductory statistics books as a sample size that is "sufficiently large" for the central limit theorem to apply, often does not suffice in finite population sampling problems. Many populations we sample are highly skewed—we may be measuring income, number of acres on a farm that are devoted to corn, or the concentration of mercury in Minnesota lakes. For these examples, we expect most of the observations to be relatively small but a few to be very, very large, so that a smoothed histogram of the entire population would look like this:

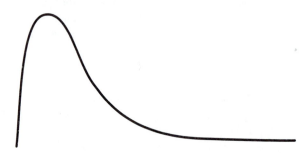

Thinking of observations as generated from some distribution is useful in deciding whether or not it is safe to use the central limit theorem. If you can think of the generating distribution as being somewhat close to normal, it is probably safe to use the central limit theorem with a sample size as small as 50. If the sample size is too small and the sampling distribution of $\bar{y}$ is not approximately normal, we would need to use another method, relying on distributional assumptions, to obtain a confidence interval for $\bar{y}_U$. Such methods fit in with a model-based perspective for sampling (Section 2.8) and are described in the References section on page 460, under "Mathematical Statistics and Probability."

EXAMPLE **2.8**   The histogram in Figure 2.3 exhibits an underlying distribution for farm acreage that is far from normal. Is the sample size large enough so that we can apply the Hájek central limit theorem? For this example, the sample probably is sufficiently large for the sampling distribution of $\bar{y}$ to be approximately normal. (See Exercise 14.)

For the data in Example 2.4, an approximate 95% CI for $\bar{y}_U$ is

$$[297,897 - (1.96)(18,898.434428), \quad 297,897 + (1.96)(18,898.434428)]$$
$$= [260,856, \ 334,938].$$

For the population total t, an approximate 95% CI is

$$[916,927,110 - 1.96(58,169,381), 916,927,110 + 1.96(58,169,381)]$$
$$= [802,915,123, \ 1,030,939,097].$$

For estimating proportions, the usual criterion that the sample size is large enough to use the normal distribution if both $np \geq 5$ and $n(1 - p) \geq 5$ is a useful guideline. A 95% CI for the proportion of counties with fewer than 200,000 acres in farms is

$$0.51 \pm 1.96(0.0275), \quad \text{or} \quad [0.456, 0.564].$$

To find a 95% CI for the total number of counties with fewer than 200,000 acres in farms, we only need to multiply all quantities by N, so the point estimate is $3078(0.51) = 1570$, with standard error $3078 \times \text{SE}(\hat{p}) = 84.65$ and 95% CI [1404, 1736]. ▪

2.5
Sample Size Estimation

An investigator often measures several variables and has a number of goals for a survey. Anyone designing an SRS must decide what amount of sampling error in the estimates is tolerable and must balance the precision of the estimates with the cost of the survey. Even though many variables may be measured, an investigator can often focus on one or two responses that are of primary interest in the survey and use these for estimating a sample size.

For a single response, follow these steps to estimate the sample size:

1 Ask "What is expected of the sample, and how much precision do I need?" What are the consequences of the sample results? How much error is tolerable? If your survey measures the unemployment rate every month, you would like your estimates to be very precise indeed so that you can detect changes in unemployment rates from month to month. A preliminary investigation, however, often needs less precision than an ongoing survey.

   Instead of asking about required precision, many people ask, "What percentage of the population should I include in my sample?" This is usually the wrong question to be asking. Except in very small populations, precision is obtained through the absolute size of the sample, not the proportion of the population covered. We saw in Section 2.3 that the fpc has little effect on the variance of the estimate in large populations.

2 Find an equation relating the sample size n and your expectations of the sample.

3 Estimate any unknown quantities and solve for n.

4 If you are relatively new at designing surveys, at this point you will find that the sample size you calculated in step 3 is much larger than you can afford. Go back and adjust some of your expectations for the survey and try again. In some cases, you will find that you cannot even come close to the precision you need with the resources that are available; in that case, perhaps you should consider whether you should even conduct your study.

Specify the Tolerable Error   Only the investigators in the study can say how much precision is needed. The desired precision is often expressed in absolute terms, as

$$P(|\bar{y} - \bar{y}_U| \le e) = 1 - \alpha.$$

The investigator must decide on reasonable values for α and e; e is called the **margin of error** in many surveys. For many surveys of people in which a proportion is measured, $e = 0.03$ and $\alpha = 0.05$.

Sometimes you would like to achieve a desired relative precision. In that case, the precision may be expressed as

$$P\left(\left|\frac{\bar{y} - \bar{y}_U}{\bar{y}_U}\right| \le e\right) = 1 - \alpha.$$

Find an Equation   The simplest equation relating the precision and sample size comes from the confidence intervals in the previous section. To obtain absolute precision, find a value of n that satisfies

$$e = z_{\alpha/2} \sqrt{\left(1 - \frac{n}{N}\right)} \frac{S}{\sqrt{n}}.$$

Solving for n, we have

$$n = \frac{z_{\alpha/2}^2 S^2}{e^2 + \dfrac{z_{\alpha/2}^2 S^2}{N}} = \frac{n_0}{1 + \dfrac{n_0}{N}}, \tag{2.17}$$

where $n_0 = z_{\alpha/2}^2 S^2 / e^2$. The value n_0 is the sample size for an SRSWR.

In surveys in which one of the main responses of interest is a proportion, using that response in setting the sample size is often easiest. For large populations, $S^2 \approx p(1-p)$, which attains its maximal value when $p = 1/2$. So using $n_0 = 1.96^2/(4e^2)$ will result in a 95% CI with width at most $2e$.

To calculate a sample size to obtain a specified relative precision, substitute $e\bar{y}_U$ for e in Equation (2.17). This results in sample size

$$n = \frac{z_{\alpha/2}^2 S^2}{(e\bar{y}_U)^2 + \dfrac{z_{\alpha/2}^2 S^2}{N}} = \frac{z_{\alpha/2}^2 \, \text{CV}^2(y)}{e^2 + \dfrac{z_{\alpha/2}^2 \, \text{CV}^2(y)}{N}}.$$

To achieve a specified relative precision, the sample size may be determined using only the coefficient of variation.

EXAMPLE 2.9   Suppose we want to estimate the proportion of recipes in the Better Homes & Gardens *New Cook Book* that do not involve animal products. We plan to take an SRS of the $N = 1251$ test kitchen–tested recipes, and we want to use a 95% CI with margin of error 0.03. Then,

$$n_0 = \frac{(1.96)^2 \left(\dfrac{1}{2}\right)\left(1 - \dfrac{1}{2}\right)}{(0.03)^2} \approx 1067.$$

The sample size—ignoring the fpc—is large compared with the population size, so in this case we would make the fpc adjustment and use

$$n = \frac{n_0}{1 + \dfrac{n_0}{N}} = \frac{1067}{1 + \dfrac{1067}{1251}} = 576. \quad \blacksquare$$

EXAMPLE 2.10   Many public opinion polls specify using a sample size of about 1100. That number comes from rounding the value of n_0 in Example 2.9 up to the next hundred and then noting that the population size is so large relative to the sample that the fpc should be ignored. For large populations, it is the size of the sample, not the proportion of the population that is sampled, that determines the precision. $\blacksquare$

Estimate Unknown Quantities   When interested in a proportion, we can use 1/4 as an upper bound for S^2. For other quantities, S^2 must be estimated or guessed at. Some methods for estimating S^2 include:

1 Use sample quantities obtained when pretesting your survey. This is probably the best method, because your pretest should be similar to the survey you take. A **pilot sample,** a small sample taken to provide information and guidance for the design of the main survey, can be used to estimate quantities needed for setting the sample size.

2 Use previous studies or data available in the literature. You are rarely the first person in the world to study anything related to your investigation. You may be able to find estimates of variances that have been published in related studies; use these as a starting point for estimating your sample size. You have no control over the quality or design of those studies, however, and their estimates may be unreliable or may not apply to your study. In addition, estimates may change over time and vary in different geographic locations.

Sometimes you can use the coefficient of variation (CV), the ratio of the standard deviation to the mean, in obtaining estimates of variability. The CV of a quantity is a measure of relative error and tends to be more stable over time and location than the variance. If we take a random sample of houses for sale in the United States today, we will find that the variability will be much greater than if we had taken a similar survey in 1930. But the average price of a house has also increased from 1930 to today. We would probably find that the CV today is close to the CV in 1930.

3 If nothing else is available, guess the variance. Sometimes a hypothesized distribution of the data will give us information about the variance. For example, if you believe the population to be normally distributed, you may not know what the variance is, but you may have an idea of the range of the data. You could then estimate S by range/4 or range/6, because approximately 95% of values from a normal population are within 2 standard deviations of the mean, and 99.7% of the values are within 3 standard deviations of the mean.

EXAMPLE 2.11   Before taking the sample of size 300 in Example 2.4, a pilot sample of size 30 was taken from the population. One county in the pilot sample of size 30 was missing the value of *acres92*; the sample standard deviation of the remaining 29 observations was 519,085. Using this value and a desired margin of error of 60,000,

$$n_0 = (1.96)^2 \frac{519,085^2}{60,000^2} = 288.$$

We took a sample of size 300 in case the estimated standard deviation from the pilot sample is too low. Also, we ignored the fpc in the sample size calculations; in most populations, the fpc will have little effect on the sample size.

You may also view possible consequences of different sample sizes graphically. Figure 2.4 shows the value of $(1.96)s/\sqrt{n}$, for a range of sample sizes between 50 and 700, and for two possible values of the standard deviation s. The plot shows that if we ignore the fpc and if the standard deviation is about 500,000, a sample of size 300 will give a margin of error of about 60,000.   ■

Determining the sample size is one of the early steps that must be taken in an investigation, and no magic formula will tell you the perfect sample size for your investigation (you only know that in hindsight, after you have completed the study!). Choosing a sample size is somewhat like deciding how much food to take on a picnic.

FIGURE   **2.4**

The plot of $(1.96)s/\sqrt{n}$ vs. n, for two possible values of the standard deviation s

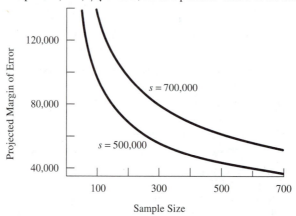

You have a rough idea of how many people will attend but do not know how much food you should have brought until after the picnic is over. You also need to bring extra food to allow for unexpected happenings, such as 2-year-old Freddie feeding a bowl of potato salad to the ducks or cousin Ted bringing along some extra guests. But you do not want to bring too much extra food, or it will spoil and you will have wasted money. Of course, the more picnics you have organized and the better acquainted you are with the picnic guests, the better you become at bringing the right amount of food. It is comforting to know that the same is true of determining sample sizes—experience and knowledge about the population make you much better at designing surveys.

The results in this section can give you some guidance in choosing the size of the sample, but the final decision is up to you. In general, the larger the sample, the smaller the sampling error. Remember, though, that in most surveys you also need to worry about nonsampling errors and need to budget resources to control selection and measurement bias. In many cases, nonsampling errors are greater when a larger sample is taken—with a large sample, it is easy to introduce additional sources of error (for example, it becomes more difficult to control the quality of the interviewers or to follow up on nonrespondents) or to become more relaxed about selection bias.

2.6
Systematic Sampling

Sometimes **systematic sampling** is used as a proxy for simple random sampling, when no list of the population exists or when the list is in roughly random order. To obtain a systematic sample, choose a sample size n and let k be the next integer after N/n. Then find a random integer R between 1 and k, which determines the sample to be the units numbered $R, R+k, R+2k, \ldots, R+(n-1)k$. For example, to select a sample of 45 students from the list of 45,000 students at Arizona State University, the sampling interval k is 1000. Suppose the random integer we choose is 597. Then the students numbered 597, 1597, 2597, ..., 44,597 would be in the sample.

If the names of the students are in alphabetical order, we will probably obtain a sample that will behave much like an SRS—it is unlikely that a person's position in the alphabet is associated with the characteristic of interest. However, systematic sampling is not the same as simple random sampling; it does not have the property that every possible group of n units has the same probability of being the sample. In the preceding example, it is impossible to have students 345 and 346 both appear in the sample. Systematic sampling is technically a form of cluster sampling, as will be discussed in Chapter 5.

Most of the time, a systematic sample gives results comparable to those of an SRS, and SRS methods can be used in the analysis. If the population is in random order, the systematic sample will be much like an SRS. The population itself can be thought of as being mixed. In the quote at the beginning of the chapter, Sorensen reports that President Kennedy used to read a systematic sample of letters written to him at the White House. This systematic sample most likely behaved much like a random sample. Note that Kennedy was well aware that the letters he read, although representative of letters written to the White House, were not at all representative of public opinion.

Systematic sampling does not necessarily give a representative sample, though, if the listing of population units is in some periodic or cyclical order. If male and female names alternate in the list, for example, and k is even, the systematic sample will contain either all men or all women—this cannot be considered a representative sample. In ecological surveys done on agricultural land, a ridge-and-furrow topography may be present that would lead to a periodic pattern of vegetation. If a systematic sampling scheme follows the same cycle, the sample will not behave like an SRS.

On the other hand, some populations are in increasing or decreasing order. A list of accounts receivable may be ordered from largest amount to smallest amount. In this case, estimates from the systematic sample may have smaller (but unestimable) variance than comparable estimates from the SRS. A systematic sample from an ordered list of accounts receivable is forced to contain some large amounts and some small amounts. It is possible for an SRS to contain all small amounts or all large amounts, so there may be more variability among the sample means of all possible SRSs than there is among the sample means of all possible systematic samples.

In systematic sampling, we must still have a sampling frame and be careful when defining the target population. Sampling every 20th student to enter the library will not give a representative sample of the student body. Sampling every 10th person exiting an airplane, though, will probably give a representative sample of the persons on that flight. The sampling frame for the airplane passengers is not written down, but it exists all the same.

2.7
Randomization Theory Results for Simple Random Sampling*[1]

In this section we show that $\bar{y}$ is an unbiased estimator of $\bar{y}_U$: $\bar{y}_U$ is the average of all possible values of $\bar{y}_S$ if we could examine all possible SRSs S that could be chosen. We also calculate the variance of $\bar{y}$ given in Equation (2.7) and show that the estimator in Equation (2.9) is unbiased over repeated sampling.

---

[1] An asterisk (*) indicates a section, chapter, or exercise that requires more mathematical background.

No distributional assumptions are made about the y_i's in order to ascertain that $\bar{y}$ is unbiased for estimating $\bar{y}_U$. We do not, for instance, assume that the y_i's are normally distributed with mean μ. In the **randomization theory** (also called **design-based**) approach to sampling, the y_i's are considered to be fixed but unknown numbers—any probabilities used arise from the probabilities of selecting units to be in the sample. The randomization theory approach provides a **nonparametric** approach to inference—we need not make any assumptions about the distribution of random variables.

Let's see how the randomization theory works for deriving properties of the sample mean in simple random sampling. As done in Cornfield (1944), define

$$Z_i = \begin{cases} 1 & \text{if unit } i \text{ is in the sample} \\ 0 & \text{otherwise} \end{cases}$$

Then

$$\bar{y} = \sum_{i \in \mathcal{S}} \frac{y_i}{n} = \sum_{i=1}^{N} Z_i \frac{y_i}{n}.$$

The Z_i's are the *only* random variables in the above equation because, according to randomization theory, the y_i's are fixed quantities. When we choose an SRS of n units out of the N units in the population, $\{Z_1, \ldots, Z_N\}$ are identically distributed Bernoulli random variables with

$$\pi_i = P(Z_i = 1) = P(\text{select unit } i \text{ in sample}) = \frac{n}{N}. \tag{2.18}$$

The probability in (2.18) follows from the definition of an SRS. To see this, note that if unit i is in the sample, then the other $n-1$ units in the sample must be chosen from the other $N-1$ units in the population. A total of $\binom{N-1}{n-1}$ possible samples of size $n-1$ may be drawn from a population of size $N-1$, so

$$P(Z_i = 1) = \frac{\text{number of samples including unit } i}{\text{number of possible samples}} = \frac{\binom{N-1}{n-1}}{\binom{N}{n}} = \frac{n}{N}.$$

As a consequence of Equation (2.18),

$$E[Z_i] = E[Z_i^2] = \frac{n}{N}$$

and

$$E[\bar{y}] = E\left[\sum_{i=1}^{N} Z_i \frac{y_i}{n}\right] = \sum_{i=1}^{N} \frac{n}{N} \frac{y_i}{n} = \sum_{i=1}^{N} \frac{y_i}{N} = \bar{y}_U.$$

The variance of $\bar{y}$ is also calculated using properties of the random variables $Z_1, \ldots, Z_N$. Note that

$$V(Z_i) = E[Z_i^2] - (E[Z_i])^2 = \frac{n}{N} - \left(\frac{n}{N}\right)^2 = \frac{n}{N}\left(1 - \frac{n}{N}\right).$$

For $i \neq j$,

$$
\begin{aligned}
E[Z_i Z_j] &= P(Z_i = 1 \text{ and } Z_j = 1) \\
&= P(Z_j = 1 \mid Z_i = 1) P(Z_i = 1) \\
&= \left(\frac{n-1}{N-1}\right)\left(\frac{n}{N}\right).
\end{aligned}
$$

Because the population is finite, the Z_i's are not quite independent—if we know that unit i is in the sample, we do have a small amount of information about whether unit j is in the sample, reflected in the conditional probability $P(Z_j = 1 \mid Z_i = 1)$. Consequently, for $i \neq j$,

$$
\begin{aligned}
\mathrm{Cov}(Z_i, Z_j) &= E[Z_i Z_j] - E[Z_i] E[Z_j] \\
&= \frac{n-1}{N-1}\frac{n}{N} - \left(\frac{n}{N}\right)^2 \\
&= -\frac{1}{N-1}\left(1 - \frac{n}{N}\right)\left(\frac{n}{N}\right).
\end{aligned}
$$

We use the covariance (Cov) of Z_i and Z_j to calculate the variance of $\bar{y}$; see Appendix B for properties of covariances. The negative covariance of Z_i and Z_j is the source of the fpc.

$$
\begin{aligned}
V(\bar{y}) &= \frac{1}{n^2} V\left(\sum_{i=1}^{N} Z_i y_i\right) \\
&= \frac{1}{n^2} \mathrm{Cov}\left(\sum_{i=1}^{N} Z_i y_i, \sum_{j=1}^{N} Z_j y_j\right) \\
&= \frac{1}{n^2}\left[\sum_{i=1}^{N} y_i^2 V(Z_i) + \sum_{i=1}^{N}\sum_{j \neq i}^{N} y_i y_j \,\mathrm{Cov}(Z_i, Z_j)\right] \\
&= \frac{1}{n^2}\left[\frac{n}{N}\left(1 - \frac{n}{N}\right)\sum_{i=1}^{N} y_i^2 - \sum_{i=1}^{N}\sum_{j \neq i}^{N} y_i y_j \frac{1}{N-1}\left(1 - \frac{n}{N}\right)\left(\frac{n}{N}\right)\right] \\
&= \frac{1}{n^2}\frac{n}{N}\left(1 - \frac{n}{N}\right)\left[\sum_{i=1}^{N} y_i^2 - \frac{1}{N-1}\sum_{i=1}^{N}\sum_{j \neq i}^{N} y_i y_j\right] \\
&= \frac{1}{n}\left(1 - \frac{n}{N}\right)\frac{1}{N(N-1)}\left[(N-1)\sum_{i=1}^{N} y_i^2 - \left(\sum_{i=1}^{N} y_i\right)^2 + \sum_{i=1}^{N} y_i^2\right] \\
&= \frac{1}{n}\left(1 - \frac{n}{N}\right)\frac{1}{N(N-1)}\left[N\sum_{i=1}^{N} y_i^2 - \left(\sum_{i=1}^{N} y_i\right)^2\right] \\
&= \left(1 - \frac{n}{N}\right)\frac{S^2}{n}
\end{aligned}
$$

To show that the estimator in (2.9) is an unbiased estimator of the variance, we need to show that $E[s^2] = S^2$. The argument proceeds much like the previous one. Since $S^2 = \sum_{i=1}^{N}(y_i - \bar{y}_U)^2/(N-1)$, it makes sense when trying to find an unbiased estimator to find the expected value of $\sum_{i \in \mathcal{S}}(y_i - \bar{y})^2$ and then find the multiplicative constant that will give the unbiasedness:

$$E\left[\sum_{i \in S}(y_i - \bar{y})^2\right] = E\left[\sum_{i \in S}\{(y_i - \bar{y}_U) - (\bar{y} - \bar{y}_U)\}^2\right]$$

$$= E\left[\sum_{i \in S}(y_i - \bar{y}_U)^2 - n(\bar{y} - \bar{y}_U)^2\right]$$

$$= E\left[\sum_{i=1}^{N} Z_i(y_i - \bar{y}_U)^2\right] - nV(\bar{y})$$

$$= \frac{n}{N}\sum_{i=1}^{N}(y_i - \bar{y}_U)^2 - \left(1 - \frac{n}{N}\right)S^2$$

$$= \frac{n(N-1)}{N}S^2 - \frac{N-n}{N}S^2$$

$$= (n-1)S^2.$$

Thus,

$$E\left[\frac{1}{n-1}\sum_{i \in S}(y_i - \bar{y})^2\right] = E[s^2] = S^2.$$

2.8
A Model for Simple Random Sampling*

Unless you have studied randomization theory in the design of experiments, the proofs in the preceding section probably seemed strange to you. The random variables in randomization theory are not concerned with the responses y_i: They are simply random variables that tell us whether the ith unit is in the sample or not. In a design-based, or randomization theory, approach to sampling inference, the only relationship between units sampled and units not sampled is that the nonsampled units could have been sampled had we used a different starting value for the random number generator.

In Section 2.7 we found properties of the sample mean $\bar{y}$ using randomization theory: $y_1, y_2, \ldots, y_N$ were considered to be fixed values, and $\bar{y}$ is unbiased because the average of $\bar{y}_S$ for all possible samples S equals $\bar{y}_U$. The only probabilities used in finding the expected value and variance of $\bar{y}$ are the probabilities that units are included in the sample.

In your basic statistics class, you learned a different approach to inference. There, you had random variables $\{Y_i\}$ that followed some probability distribution, and the actual sample values were realizations of those random variables. Thus you assumed, for example, that $Y_1, Y_2, \ldots, Y_n$ were independent and identically distributed from a normal distribution with mean μ and variance σ^2 and used properties of independent random variables and the normal distribution to find expected values of various statistics.

We can extend this approach to sampling by thinking of random variables $Y_1, Y_2, \ldots, Y_N$ generated from some model. The actual values for the finite population,

$y_1, y_2, \ldots, y_N$, are one realization of the random variables. The joint probability distribution of $Y_1, Y_2, \ldots, Y_N$ supplies the link between units in the sample and units not in the sample in this **model-based** approach—a link that is missing in the randomization approach. Here, we sample $\{y_i, i \in S\}$ and use these data to predict the unobserved values $\{y_i, i \notin S\}$. Thus, problems in finite population sampling may be thought of as prediction problems.

In an SRS, a simple model to adopt is

$$Y_1, Y_2, \ldots, Y_N \text{ independent with } E_M[Y_j] = \mu \text{ and } V_M[Y_j] = \sigma^2. \qquad \textbf{(2.19)}$$

The subscript M indicates that the expectation uses the model, not the randomization distribution used in Section 2.7. Here, μ and σ^2 represent unknown infinite population parameters, not the finite population quantities in Section 2.7. We take a sample S and observe the values y_i for $i \in S$; that is, we see realizations of the random variables Y_i for $i \in S$. The other observations in the population $\{y_i, i \notin S\}$ are also realizations of random variables, but we do not see those. The finite population total t can be written as

$$t = \sum_{i=1}^{N} y_i = \sum_{i \in S} y_i + \sum_{i \notin S} y_i$$

and is one possible value that can be taken on by the random variable

$$T = \sum_{i=1}^{N} Y_i = \sum_{i \in S} Y_i + \sum_{i \notin S} Y_i.$$

We know the values $\{y_i, i \in S\}$. To estimate t from the sample, we need to find estimates of the y values not in the sample. This is where our model of the common mean μ comes in. The least squares estimator of μ from the sample is $\bar{Y}_S = \sum_{i \in S} Y_i / n$, and this is the best linear unbiased predictor (under the model) of the unobserved values, so that

$$\hat{T} = \frac{N}{n} \sum_{i \in S} Y_i.$$

The estimator $\hat{T}$ is *model-unbiased*: If the model is true, then the average of $\hat{T} - T$ over repeated realizations of the population is

$$E_M[\hat{T} - T] = \frac{N}{n} \sum_{i \in S} E_M[Y_i] - \sum_{i=1}^{N} E_M[Y_i] = 0.$$

(Notice the difference between finding expectations under the model-based approach and under the design-based approach. In the model-based approach, the Y_i's are the random variables, and the sample has no information for calculating expected values. In the design-based approach, the random variables are contained in the sample S.)

The mean squared error is also calculated as the average squared deviation between the estimate and the finite population total. For any given realization of the random variables, the squared error is

$$\left[\frac{N}{n} \sum_{i \in S} y_i - \sum_{i=1}^{N} y_i \right]^2.$$

Averaging this quantity over all possible realizations of the random variables gives the mean squared error under the model assumptions:

$$E_M[(\hat{T} - T)^2] = E_M\left[\left(\frac{N}{n}\sum_{i \in S} Y_i - \sum_{i=1}^{N} Y_i\right)^2\right]$$

$$= E_M\left[\left\{\left(\frac{N}{n} - 1\right)\sum_{i \in S} Y_i - \sum_{i \notin S} Y_i\right\}^2\right]$$

$$= E_M\left[\left(\frac{N}{n} - 1\right)^2\left(\sum_{i \in S} Y_i - n\mu\right)^2 + \left(\sum_{i \notin S} Y_i - (N - n)\mu\right)^2\right]$$

$$= \left(\frac{N}{n} - 1\right)^2 n\sigma^2 + (N - n)\sigma^2$$

$$= N^2\frac{\sigma^2}{n}\left(1 - \frac{n}{N}\right).$$

In practice, if the model in Equation (2.19) were adopted, you would estimate σ^2 by the sample variance s^2. Thus, the design-based approach and the model-based approach—with the model in (2.19)—lead to the same estimate of the population total and the same variance estimate. If a different model were adopted, however, the estimates might differ. We will see in Chapters 3 and 11 how a design-based approach and a model-based approach can lead to different inferences.

The design-based approach and the model-based approach with the model in (2.19) also lead to the same confidence interval for the mean. These confidence intervals have different interpretations, however. The design-based confidence interval for $\bar{y}_U$ may be interpreted as follows: If we take all possible SRSs of size n from the finite population of size N and construct a 95% confidence interval for each sample, 95% of all confidence intervals constructed will include the true population value $\bar{y}_U$. Thus, the design-based confidence interval has a *repeated sampling* interpretation.

The model-based confidence interval for the parameter μ is interpreted in terms of the model in (2.19). The confidence interval procedure results in two random variables: $LL = \bar{Y}_S - 1.96S/\sqrt{n}$ and $UL = \bar{Y}_S + 1.96S/\sqrt{n}$. Then, using the model to infer that $\bar{Y}_S$ is approximately normally distributed with mean μ and variance S^2/n,

$$P(LL \leq \mu \leq UL) = 0.95.$$

This model-based confidence interval is also commonly interpreted using repeated samples in introductory statistics courses: If we generate values for the population over and over again using the model in (2.19) and construct a confidence interval for each resulting sample, we expect that 95% of the confidence intervals will contain the true value of μ. Although both the design-based and model-based confidence intervals may be interpreted using repeated samples, there is a difference between them. The design-based confidence level gives the expected proportion of confidence intervals that will include $\bar{y}_U$, from the set of all confidence intervals that could be constructed by taking an SRS of size n from the finite population of fixed values

$\{y_1, y_2, \dots, y_N\}$. The model-based confidence level gives the expected proportion of confidence intervals that will include μ, from the set of all samples that could be generated from the model in (2.19).

A Note on Notation Some books (for example, Cochran 1977) and journal articles use Y to represent the population total (t in this book) and $\bar{Y}$ to represent the population mean (our $\bar{y}_U$). In this book, we reserve Y and T to represent random variables in a model-based approach. Our usage is consistent with other areas of statistics, in which capital letters near the end of the alphabet usually represent random variables. However, you should be aware that notation in the survey sampling literature is not uniform.

2.9
When Should a Simple Random Sample Be Used?

Simple random sampling without replacement is the simplest of all probability sampling methods, and estimates are all computed very much as you learned in your introductory statistics class. The estimates are:

Population Quantity	Estimate	Standard Error of Estimate
Population mean, $\bar{y}_U$	$\bar{y} = \dfrac{1}{n} \displaystyle\sum_{i \in \mathcal{S}} y_i$	$\sqrt{\left(1 - \dfrac{n}{N}\right) \dfrac{s^2}{n}}$
Population proportion, p	$\hat{p}$	$\sqrt{\left(1 - \dfrac{n}{N}\right) \dfrac{\hat{p}(1 - \hat{p})}{n - 1}}$
Population total, t	$\hat{t} = N\bar{y}$	$N\, \text{SE}(\bar{y})$

The only feature found in the estimates for without-replacement random samples that does not occur in with-replacement random samples is the finite population correction, $(1 - n/N)$, which decreases the standard error when the sample size is large relative to the population size. In most surveys done in practice, the fpc is so close to 1 that it can be ignored.

For "sufficiently large" sample sizes, an approximate 95% CI is given by

$$\text{estimate} \pm 1.96\, \text{SE(estimate)}.$$

The margin of error of an estimate is the half-width of the confidence interval—that is, $1.96 \times \text{SE(estimate)}$.

SRSs are usually easy to design and analyze. But they are not the best design to use in the following situations:

- Before taking an SRS, consider whether a survey sample is the best method for studying your research question. If you want to study whether a certain brand of bath oil is an effective mosquito repellent, you should perform a controlled experiment, not take a survey. You should take a survey if you want to estimate how many people use the bath oil as a mosquito repellent or if you want to estimate how many mosquitoes are in an area.

- You may not have a list of the observation units, or it may be expensive in terms of travel time to take an SRS. If interested in the proportion of mosquitoes in southwestern Wisconsin that carry an encephalitis virus, you cannot construct a sampling frame of the individual mosquitoes. You would need to sample different areas and then examine some or all of the mosquitoes found in those areas, using a sampling technique known as cluster sampling. (Cluster sampling will be discussed in Chapters 5 and 6.)

- You may have additional information that can be used to design a more cost-effective sampling scheme. In a survey to estimate the total number of mosquitoes in an area, an entomologist would know what terrain would be likely to have high mosquito density and what areas would be likely to have low mosquito density, before any samples were taken. You would save effort in sampling by dividing the area into strata, groups of similar units, and then sampling plots within each stratum. (Stratified sampling will be discussed in Chapter 4.)

You should use an SRS in these situations:

- Persons analyzing the data insist on using SRS formulas, whether they are appropriate or not. Some persons will not be swayed from the belief that one should only estimate the mean by taking the average of the sample—in that case, design a sample in which averaging the sample values is the right thing to do. SRSs are often recommended when sample evidence is used in legal actions; sometimes, when a more complicated sampling scheme is used, an opposing counsel will try to persuade the jury that the sample results are not valid.

- Little extra information is available that can be used when designing the survey. If your sampling frame is merely a list of university students' names in alphabetic order and you have no additional information such as major or year, SRS or systematic sampling is probably the best probability sampling strategy.

- The primary interest is in multivariate relationships such as regression equations that hold for the whole population, and there are no compelling reasons to take a stratified or cluster sample. Multivariate analyses can be done in complex samples, but they are much easier to perform and interpret in an SRS.

2.10
Exercises

1 Let $N = 6$ and $n = 3$. For purposes of studying sampling distributions, we assume that all population values are known.

$$y_1 = 98 \quad y_3 = 154 \quad y_5 = 190$$
$$y_2 = 102 \quad y_4 = 133 \quad y_6 = 175$$

We are interested in $\bar{y}_U$, the population mean. Two sampling plans are proposed.

- Plan 1 Eight possible samples may be chosen.

Sample Number	Sample, $\mathcal{S}$	$P(\mathcal{S})$
1	$\{1, 3, 5\}$	$\dfrac{1}{8}$
2	$\{1, 3, 6\}$	$\dfrac{1}{8}$
3	$\{1, 4, 5\}$	$\dfrac{1}{8}$
4	$\{1, 4, 6\}$	$\dfrac{1}{8}$
5	$\{2, 3, 5\}$	$\dfrac{1}{8}$
6	$\{2, 3, 6\}$	$\dfrac{1}{8}$
7	$\{2, 4, 5\}$	$\dfrac{1}{8}$
8	$\{2, 4, 6\}$	$\dfrac{1}{8}$

- Plan 2 Three possible samples may be chosen.

Sample Number	Sample, $\mathcal{S}$	$P(\mathcal{S})$
1	$\{1, 4, 6\}$	$\dfrac{1}{4}$
2	$\{2, 3, 6\}$	$\dfrac{1}{2}$
3	$\{1, 3, 5\}$	$\dfrac{1}{4}$

a What is the value of $\bar{y}_U$?

b Let $\bar{y}$ be the mean of the sample values. For each sampling plan, find:

   i $E[\bar{y}]$

   ii $V[\bar{y}]$

   iii Bias($\bar{y}$)

   iv MSE($\bar{y}$)

c Which sampling plan do you think is better? Why?

2 For the population in Example 2.1, consider the following sampling scheme:

$\mathcal{S}$	$P(\mathcal{S})$
$\{1, 3, 5, 6\}$	$\dfrac{1}{8}$
$\{2, 3, 7, 8\}$	$\dfrac{1}{4}$
$\{1, 4, 6, 8\}$	$\dfrac{1}{8}$
$\{2, 4, 6, 8\}$	$\dfrac{3}{8}$
$\{4, 5, 7, 8\}$	$\dfrac{1}{8}$

a   Find the probability of selection π_i for each unit i.

b   What is the sampling distribution of $\hat{t} = 8\bar{y}$?

3   For the population in Example 2.1, find the sampling distribution of $\bar{y}$ for:

a   an SRS of size 3 without replacement.

b   an SRS of size 3 with replacement.

For each, draw the histogram of the sampling distribution of $\bar{y}$. Which sampling distribution has the smaller variance, and why?

4   One way of selecting an SRS is to assign a number to every unit in the population, then use a random number table to select units from the list. A page from a random number table is given in Appendix E. Explain why each of the following methods will or will not result in an SRS.

a   The population has 742 units and we want to take an SRS of size 30. Divide the random list into segments of size 3 and throw out any sequences of three digits *not* between 001 and 742. If a number occurs that has already been included in the sample, ignore it. If we used this method with the first line of random numbers in Appendix E, the sequence of three-digit numbers would be

749   700   699   611   136   ...

We would include units 700, 699, 611, and 136 in the sample.

b   For the situation in part (a), when a random three-digit number is larger than 742, eliminate only the first digit and start the sequence with the next digit. With this procedure, the first five numbers would be 497, 006, 611, 136, and 264.

c   The population has 170 items. Using the procedures described in part (a) or (b), we would throw away many of the numbers from the list. To avoid this waste, divide every random three-digit number by 170 and use the rounded remainder as the unit in the sample. If the remainder is 0, use unit 170. As in parts (a) and (b), eliminate the duplicates. For the sequence in the first row of the random number table, the numbers generated would be

69   20   19   101   136   ...

d   The population has 200 items. Take two-digit sequences of random numbers and put a decimal point in front of each to obtain the sequence

.74   .97   .00   .69   .96   ...

Then multiply each decimal by 200 to get the units for the sample (convert .00 to 200):

148   194   200   138   192   ...

e   A school has 20 homeroom classes; each homeroom class contains between 20 and 40 students. To select a student for the sample, draw a random number between 1 and 20, then select a student at random from the chosen class. Do not include duplicates in your sample.

f   For the situation described in part (e), select a random number between 1 and 20 to choose a class. Then select a second random number between 1 and 40. If the number corresponds to a student in the class, then select that student; if the second random number is larger than the class size, then ignore this pair of random numbers and start again. As usual, eliminate duplicates from your list.

5 Mayr et al. (1994) took an SRS of 240 children aged 2 to 6 years who visited their pediatric outpatient clinic. They found the following frequency distribution for free (unassisted) walking among the children:

Age (months)	9	10	11	12	13	14	15	16	17	18	19	20
Number of children	13	35	44	69	36	24	7	3	2	5	1	1

a Construct a histogram of the distribution of age at walking. Is the shape normally distributed? Do you think the sampling distribution of the sample average will be normally distributed? Why, or why not?

b Find the mean, standard error, and a 95% CI for the average age for onset of free walking.

c Suppose the researchers want to do another study in a different region and want a 95% CI for the mean age of onset of walking to have margin of error 0.5. Using the estimated standard deviation for these data, what sample size would they need to take?

6 One quantity that is often of interest for a medical clinic is the percentage of patients that are overdue for a vaccination. Some clinics examine every record to determine that percentage; in a large practice, though, taking a census of the records can be time-consuming. Cullen (1994) took a sample of the 580 children served by an Auckland family practice to estimate the proportion of interest.

a What sample size in an SRS (without replacement) would be necessary to estimate the proportion with 95% confidence and margin of error 0.10?

b Cullen actually took an SRSWR of size 120, of whom 27 were *not* overdue for vaccination. Give a 95% CI for the proportion of children not overdue for vaccination.

*7 (Requires probability.) In the population used in Example 2.4, 19 of the 3078 counties in the population are missing the value of *acres92*. What is the probability that an SRS of size 300 would have no missing data for that variable?

8 At one university there were 807 faculty members and research specialists in the College of Liberal Arts and Science in 1993; the list of faculty and their reported publications for 1992–1993 were available on the computer system. For each faculty member, the number of refereed publications was recorded. This number is not directly available on the database, so the investigator is required to examine each record separately. A frequency table for number of refereed publications is given for an SRS of 50 faculty members.

Refereed publications	0	1	2	3	4	5	6	7	8	9	10
Faculty members	28	4	3	4	4	2	1	0	2	1	1

a Plot the data using a histogram. Describe the shape of the data.

b Estimate the mean number of publications per faculty member and give a standard error for your estimate.

c Do you think that $\bar{y}$ from part (b) will be approximately normally distributed? Why, or why not?

d Estimate the proportion of faculty members with no publications and give a 95% CI for your estimate.

9 Define a confidence interval procedure by

$$\text{CI}(\mathcal{S}) = [\hat{t}_{\mathcal{S}} - 1.96 \text{ SE}(\hat{t}_{\mathcal{S}}), \hat{t}_{\mathcal{S}} + 1.96 \text{ SE}(\hat{t}_{\mathcal{S}})].$$

Using the method illustrated in Example 2.7, find the exact confidence level for a confidence interval based on an SRS without replacement of size 4 from the population in Example 2.1. Does your confidence level equal 95%?

10 A letter in the December 1995 issue of *Dell Champion Variety Puzzles* stated: "I've noticed over the last several issues there have been no winners from the South in your contests. You always say that winners are picked at random, so does this mean you're getting fewer entries from the South?" In response, the editors took a random sample of 1000 entries from the last few contests and found that 175 of those came from the South.

a Find a 95% CI for the percentage of entries that come from the South.

b According to *Statistical Abstract of the United States*, 30.9% of the U.S. population live in states that the editors considered to be in the South. Is there evidence from your confidence interval that the percentage of entries from the South differs from the percentage of persons living in the South?

11 The data set agsrs.dat also contains information on other variables. For each of the following quantities, plot the data and estimate the population mean for that variable, along with its standard error. Give a 95% CI for your estimate.

a Number of acres devoted to farms in 1987

b Number of farms, 1992

c Number of farms with 1000 acres or more, 1992

d Number of farms with 9 acres or fewer, 1992

12 The Special Census of Maricopa County, Arizona, gave 1995 populations for the following cities:

City	Population
Buckeye	4,857
Gilbert	59,338
Gila Bend	1,724
Phoenix	1,149,417
Tempe	153,821

Suppose you want to estimate the percentage of persons who have been immunized against polio in each city and can take an SRS of persons. What should be your sample size in each of the five cities if you want the estimate from each city to have margin of error of 4 percentage points? For which cities does the finite population correction make a difference?

FIGURE   2.5

Histogram of the means of 1000 samples of size 300, taken with replacement from the data in Example 2.4

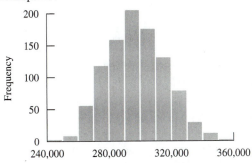

Estimated Sampling Distribution of $\bar{y}$

*13 *Decision theoretic approach for sample-size estimation.* (Requires calculus.) In a decision theory approach, two functions are specified:

$$L(n) = \text{ loss or "cost" of a bad estimate}$$
$$C(n) = \text{ cost of taking the sample}$$

Suppose for some constants c_0, c_1, and k,

$$L(n) = kV(\bar{y}_S) = k\left(1 - \frac{n}{N}\right)\frac{S^2}{n}$$
$$C(n) = c_0 + c_1 n.$$

What sample size n minimizes the total cost $L(n) + C(n)$?

14 (Requires computing.) If you have a large SRS, you can estimate the sampling distribution of $\bar{y}_S$ by repeatedly taking samples of size n with replacement from the list of sample values. A histogram of the means from 1000 samples of size 300 with replacement from the data in Example 2.4 is displayed in Figure 2.5; the shape may be slightly skewed but still appears approximately normal. Would a sample of size 100 from this population be sufficiently large to use the central limit theorem? Take 500 samples with replacement of size 100 from the variable *acres92* in agsrs.dat and draw a histogram of the 500 means. The approach described in this exercise is known as the *bootstrap* (see Efron and Tibshirani 1993); we discuss the bootstrap further in Section 9.3.

15 The Internet site www.golfcourse.com lists 14,938 golf courses by state. It gives a variety of information about each course, including greens fees, course rating, par for the course, and facilities. Data from an SRS of 120 of the golf courses are in the file golfsrs.dat on the data disk.

    a Display the data in a histogram for the weekday greens fees for nine holes of golf. How would you describe the shape of the data?

    b Find the average weekday greens fee to play nine holes of golf and give the standard error for your estimate.

16   Repeat Exercise 15 for the back-tee yardage.

17   For the data in golfsrs.dat, estimate the proportion of golf courses that have 18 holes and give a 95% CI for your estimate.

18   In an SRS, each possible subset of n units has probability $1/\binom{N}{n}$ of being chosen as the sample; in this chapter, we showed that this definition implies that each unit has probability n/N of appearing in the sample. The converse is not true, however. Exhibit a sampling design for which the selection probability for each unit is n/N, but the design is not an SRS.

*19   (Requires probability.) A typical opinion poll surveys about 1000 adults. Suppose the sampling frame contains 100 million adults, including yourself, and that an SRS of 1000 adults is chosen from the frame without replacement.

    a   What is the probability that you are selected to be in the sample?

    b   Now suppose that 2000 such samples are selected, each sample selected independently of the others. What is the probability that you will *not* be in any of the samples?

    c   How many samples must be selected for you to have a .5 probability of being in at least one sample?

*20   (Requires probability.) In an SRSWR, a population unit can appear in the sample anywhere between 0 and n times. Let

$$Q_i = \text{number of times unit } i \text{ appears in the sample}$$

and

$$\hat{t} = \frac{N}{n} \sum_{i=1}^{N} Q_i y_i.$$

    a   Argue that the joint distribution of $Q_1, Q_2, \ldots, Q_N$ is multinomial with n trials and $p_1 = p_2 = \cdots = p_N = 1/N$.

    b   Using part (a), show that $E[\hat{t}] = t$.

    c   Using part (a), find $V[\hat{t}]$.

*21   (Requires probability.) Suppose you would like to take an SRS of size n from a list of N units but do not know the population size N in advance. Consider the following procedure:

    a   Set $S_0 = \{1, 2, \ldots, n\}$ so that the initial sample for consideration consists of the first n units on the list.

    b   For $k = 1, 2, \ldots$, generate a random number u_k between 0 and 1. If $u_k > n/(n+k)$, then set S_k equal to S_{k-1}. If $u_k \leq n/(n+k)$, then select one of the units in S_{k-1} at random and replace it by unit $(n+k)$ to form S_k.

    Show that S_{N-n} from this procedure is an SRS of size n from the population.

22   Take a small SRS of something you're interested in. Explain what it is you decide to study and carefully describe how you chose your random sample (give the random numbers generated and explain how you translated them into observations), report your data, and give a a point estimate and the standard error for the quantity or quantities of interest.

The data collection for this exercise should not take a great deal of effort, as you are surrounded by things waiting to be sampled. Some examples: mutual fund data in the financial section of today's newspaper, actual weights of 1-pound bags of carrots at the supermarket, cost of an item at various stores, and time it takes to wait until your modem connects you to the computer system.

23 How trustworthy is information found on the Internet? Choose a topic you are knowl- edgeable about for which there is some controversy. Use a search engine to generate a sampling frame of contributions on the subject. If you are familiar with medical treat- ments for asthma, for example, you might do a search on "asthma treatment." Now select an SRS of those contributions, using the numbers assigned by the search engine to the contributions to select your sample. Estimate the proportion of contributions in the list that give incorrect information and give a 95% CI for your proportion.

SURVEY Exercises

The following exercises use the SURVEY program described in Appendix A.

24 Why is the following procedure *not* suitable for drawing an SRS of addresses in Lockhart City?

   a Randomly select a district between 51 and 75.

   b Randomly select a house from those in the chosen district.

   c Reject both district and house selection if the house is already in the sample.

   d Repeat parts (a)–(c) until the desired sample size is achieved.

25 No district in Lockhart City has more than 1313 houses. Prove that the following procedure produces an SRS of houses in Lockhart City:

   a Randomly select a district between 51 and 75.

   b Randomly select a random number (the potential house selection) between 1 and 1313.

   c Reject the two random numbers from parts (a) and (b) if the number in part (b) exceeds the number of houses in the district or if the house is already in the sample. Otherwise, add that house to your sample.

   d Repeat parts (a)–(c) until the desired sample size is achieved.

26 Use the random number table in Appendix E to select an SRS of size 10 from Lockhart City. Report the list of the random numbers you selected and the addresses to which they correspond. Describe exactly how you converted a random number to an address.

27 Use the SURVEY program to obtain the answers to the questionnaire for your ten ran- domly selected addresses. Hand in a printout of the output file. Estimate the following from your sample of ten households. Give standard errors for your estimates.

   a The average number of TVs per household in Lockhart City

   b The average price a household in Lockhart City is willing to pay for cable TV service

Actually, we only know for each sampled household the price it is willing to pay for service, rounded down to the nearest $5. Recognizing this limitation to question 4 of the survey questionnaire, use the answers to that question as the prices that the sampled houses are willing to pay.

28 Use the program ADDGEN to generate 200 random addresses in Lockhart City and then the program SURVEY to obtain the responses of these houses. Estimate the following:

a   The average price a household is willing to pay for cable TV

b   The average number of TVs in a household in Lockhart City

c   The proportion of houses willing to pay at least $10 for cable service.

Be sure to give standard errors for all estimates. (Use the fpc, even though it may not be strictly necessary.) Make sure you save the sample you obtained for this exercise—you will use it again in the next chapter.

29 Using your sample of size 200, estimate the average assessed valuation in Lockhart City. Does a 95% CI include the known value of $71,117? Estimating a known quantity is often used to check the representativeness of a sample.

30 Draw a histogram or stem-and-leaf diagram of the responses to question 8 of the survey (number of hours watching children's TV) using the sample you drew in Exercise 28. Does the distribution of number of hours spent watching children's TV for households in Lockhart City appear normal? Find an approximate 95% CI for the mean number of hours spent watching children's TV. Based on your histogram, is constructing a confidence interval an appropriate thing to do? Why, or why not? HINT: Do you think that the sampling distribution of the mean viewing time for children's TV could be normal?

3

Ratio and Regression Estimation

The registers of births, which are kept with care in order to assure the condition of the citizens, can serve to determine the population of a great empire without resorting to a census of its inhabitants, an operation which is laborious and difficult to do with exactness. But for this it is necessary to know the ratio of the population to the annual births. The most precise means for this consists of, first, choosing subdivisions in the empire that are distributed in a nearly equal manner on its whole surface so as to render the general result independent of local circumstances; second, carefully enumerating the inhabitants of several communes in each of the subdivisions, for a specified time period; third, determining the corresponding mean number of annual births, by using the accounts of births during several years preceding and following this time period. This number, divided by that of the inhabitants, will give the ratio of the annual births to the population, in a manner that will be more reliable as the enumeration becomes larger.

—Pierre-Simon Laplace, *Essai Philosophique sur les Probabilités* (trans. S. Lohr)

France had no population census in 1802, and Laplace wanted to estimate the number of persons living there (Cochran 1978; Laplace 1814). He obtained a sample of 30 communes spread throughout the country. These communes had a total of 2,037,615 inhabitants on September 23, 1802. In the 3 years preceding September 23, 1802, a total of 215,599 births were registered in the 30 communes. Laplace determined the annual number of registered births in the 30 communes to be $215,599/3 = 71,866.33$. Dividing 2,037,615 by 71,866.33, Laplace estimated that each year there was one registered birth for every 28.352845 persons. Reasoning that communes with large populations are also likely to have large numbers of registered births and judging that the ratio of population to annual births in his sample would likely be similar to that throughout France, he concluded that one could estimate the total population of France by multiplying the total number of annual births in all of France by 28.352845. (For some reason, Laplace decided not to use the actual number of registered births in France in the year prior to September 22, 1802, in his calculation but instead multiplied the ratio by 1 million.)

Laplace was not interested in the total number of registered births for its own sake but used it as auxiliary information for estimating the total population of France. We often have auxiliary information in surveys; few investigators go to the expense of taking a good sample and then measure only one quantity. Often the sampling frame gives us extra information about each unit that can be used to improve the precision of our estimates. Ratio and regression estimation use variables that are correlated with the variable of interest to improve the precision of estimates of the mean and total of a population.

3.1
Ratio Estimation

For ratio estimation to apply, two quantities y_i and x_i must be measured on each sample unit; x_i is often called an **auxiliary variable** or **subsidiary variable.** In the population of size N

$$t_y = \sum_{i=1}^{N} y_i, \ t_x = \sum_{i=1}^{N} x_i$$

and their ratio[1] is

$$B = \frac{t_y}{t_x} = \frac{\bar{y}_U}{\bar{x}_U}.$$

In the simplest use of ratio estimation, a simple random sample (SRS) of size n is taken, and the information in both x and y is used to estimate B, t_y, or $\bar{y}_U$.

Ratio and regression estimation both take advantage of the correlation of x and y in the population; the higher the correlation, the better they work. Define the **population correlation coefficient** of x and y to be

$$R = \frac{\sum_{i=1}^{N} (x_i - \bar{x}_U)(y_i - \bar{y}_U)}{(N - 1)S_x S_y}. \tag{3.1}$$

Here, S_x is the population standard deviation of the x_i's, S_y is the population standard deviation of the y_i's, and R is simply the Pearson correlation coefficient of x and y for the N units in the population.

EXAMPLE **3.1**   Suppose the population consists of agricultural fields of different sizes. Let

$$y_i = \text{bushels of grain harvested in field } i$$
$$x_i = \text{acreage of field } i.$$

Then

$$B = \text{average yield in bushels per acre}$$
$$\bar{y}_U = \text{average yield in bushels per field}$$
$$t_y = \text{total yield in bushels.} \quad \blacksquare$$

---

[1] Why use the letter B to represent the ratio? As we will see in Section 3.4, ratio estimation is motivated by a regression model: $Y_i = \beta x_i + \varepsilon_i$, with $E[\varepsilon_i] = 0$ and $V[\varepsilon_i] = \sigma^2 x_i$. Thus, the ratio of t_y and t_x is actually a regression coefficient.

If an SRS is taken, natural estimators for B, t_y, and $\bar{y}_U$ are

$$\hat{B} = \frac{\bar{y}}{\bar{x}} = \frac{\hat{t}_y}{\hat{t}_x}$$

$$\hat{t}_{yr} = \hat{B} t_x \tag{3.2}$$

$$\hat{\bar{y}}_r = \hat{B} \bar{x}_U,$$

where t_x and $\bar{x}_U$ are assumed known.

3.1.1 Why Use Ratio Estimation?

1 Sometimes we simply want to estimate a ratio. In Example 3.1, B—the average yield per acre—is of interest and is estimated by the ratio of the sample means $\hat{B} = \bar{y}/\bar{x}$. If the fields differ in size, both numerator and denominator are random quantities; if a different sample is selected, both $\bar{y}$ and $\bar{x}$ are likely to change. In other survey situations, ratios of interest might be the ratio of liabilities to assets, the ratio of the number of fish caught to the number of hours spent fishing, or the per capita income of household members in Australia.

Some ratio estimates appear disguised because the denominator looks like it is just a regular sample size. To determine whether you need to use ratio estimation for a quantity, ask yourself, "If I took a different sample, would the denominator be a different number?" If yes, then you are using ratio estimation. Suppose you are interested in the percentage of pages in *Good Housekeeping* magazine that contain at least one advertisement. You might take an SRS of ten issues of the magazine and for each issue measure the following:

$$x_i = \text{total number of pages in issue } i$$

$$y_i = \text{total number of pages in issue } i$$
$$\text{that contain at least one advertisement.}$$

The proportion of interest can be estimated as

$$\hat{B} = \frac{\sum\limits_{i \in S} y_i}{\sum\limits_{i \in S} x_i}.$$

The denominator is the total number of pages in the ten issues and will likely be different if a different sample of issues is taken.

Technically, we are using ratio estimation every time we take an SRS and estimate a mean or proportion for a subpopulation, as will be discussed in Section 3.3.

2 Sometimes we want to estimate a population total, but the population size N is unknown. Then we cannot use the estimator $\hat{t}_y = N\bar{y}$ from Chapter 2. But we know that $N = t_x / \bar{x}_U$ and can estimate N by $t_x / \bar{x}$. We thus use another measure of size, t_x, instead of the population count N.

To estimate the total number of fish in a haul that are longer than 12 cm, you could take a random sample of fish, estimate the proportion that are longer than 12 cm, and multiply that proportion by the total number of fish, N. Such a procedure cannot be used if N is unknown. You can, however, weigh the total haul of fish and use the fact that having a length of more than 12 cm (y) is related to weight (x), so

$$\hat{t}_{yr} = \bar{y}\,\frac{t_x}{\bar{x}}.$$

The total weight of the haul, t_x, is easily measured, and $t_x/\bar{x}$ estimates the total number of fish in the haul.

3 Ratio estimation is often used to increase the precision of estimated means and totals. Laplace used ratio estimation for this purpose in the example at the beginning of the chapter, and increasing precision will be the main use discussed in the chapter.

In Laplace's use of ratio estimation,

$$y_i = \text{number of persons in commune } i$$
$$x_i = \text{number of registered births in commune } i.$$

Laplace could have estimated the total population of France by multiplying the average number of persons in the 30 communes ($\bar{y}$) by the total number of communes in France (N). He reasoned that the ratio estimate would attain more precision: on average, the larger the population of a commune, the higher the number of registered births. Thus, the population correlation coefficient R, defined in Equation (3.1), is likely to be positive. Since $\bar{y}$ and $\bar{x}$ are then also positively correlated (see Equation (B.11) in Appendix B), the sampling distribution of $\bar{y}/\bar{x}$ will have less variability than the sampling distribution of $\bar{y}/\bar{x}_U$. So if

$$t_x = \text{total number of registered births}$$

is known, the mean squared error (MSE) of $\hat{t}_{yr} = \hat{B}t_x$ is likely to be smaller than the MSE of $N\bar{y}$, an estimator that does not use the auxiliary information of registered births.

4 Ratio estimation is used to adjust estimates from the sample so that they reflect demographic totals. An SRS of 400 students taken at a university with 4000 students may contain 240 women and 160 men, with 84 of the sampled women and 40 of the sampled men planning to follow careers in teaching. Using only the information from the SRS, you would estimate that

$$\frac{4000}{400} \times 124 = 1240$$

students plan to be teachers. Knowing that the college has 2700 women and 1300 men, a better estimate of the number of students planning teaching careers might be

$$\frac{84}{240} \times 2700 + \frac{40}{160} \times 1300 = 1270.$$

Ratio estimation is used within each gender: In the sample, 60% are women, but 67.5% of the population are women, so we adjust the estimate of the total number of students planning a career in teaching accordingly. To estimate the total number of women who plan to follow a career in teaching, let

$$y_i = \begin{cases} 1 & \text{if woman and plans career in teaching} \\ 0 & \text{otherwise} \end{cases}$$

$$x_i = \begin{cases} 1 & \text{if woman} \\ 0 & \text{otherwise.} \end{cases}$$

Then $(84/240) \times 2700 = (\sum_{i \in \mathcal{S}} y_i / \sum_{i \in \mathcal{S}} x_i)t_x$ is a ratio estimate of the total number of women planning a career in teaching. Similarly, $(40/160) \times 1300$ is a ratio estimate of the total number of men planning a teaching career.

This use of ratio estimation, called *poststratification*, will be discussed in Section 4.7 and Chapters 7 and 8.

5 Ratio estimation is used to adjust for nonresponse, as will be discussed in Chapter 8. Suppose a sample of businesses is taken; let y_i be the amount spent on health insurance by business i and x_i be the number of employees in business i. Assume that x_i is known for every business in the population. We expect that the amount a business spends on health insurance will be related to the number of employees. Some businesses may not respond to the survey, however. One method of adjusting for nonresponse when estimating total insurance expenditures is to multiply the ratio $\bar{y}/\bar{x}$ (using data only from the respondents) by the population total t_x. If companies with few employees are less likely to respond to the survey and if y_i is proportional to x_i, then we would expect the estimate $N\bar{y}$ to overestimate the population total t_y. In the ratio estimate $t_x \bar{y}/\bar{x}$, $t_x/\bar{x}$ is likely to be smaller than N because companies with many employees are more likely to respond to the survey. Thus, the ratio estimate of total health-care insurance expenditures adjusts for the nonresponse of companies with few employees.

EXAMPLE 3.2   Let's return to the data from the U.S. Census of Agriculture, described in Example 2.4. The file agsrs.dat contains data from an SRS of 300 of the 3078 counties.

For this example, suppose we know the population totals for 1987 but only have 1992 information on the SRS of 300 counties. When the same quantity is measured at different times, the response of interest at an earlier time often makes an excellent auxiliary variable. Let

$$y_i = \text{total acreage of farms in county } i \text{ in 1992}$$
$$x_i = \text{total acreage of farms in county } i \text{ in 1987.}$$

In 1987 a total of $t_x = 964{,}470{,}625$ acres were devoted to farms in the United States. The average acreage per county for the population is then $\bar{x}_U = 964{,}470{,}625/3078 = 313{,}343.3$ acres of farms per county. The data, and the line through the origin with slope $\hat{B}$, are plotted in Figure 3.1.

A portion of a spreadsheet with the 300 values of x_i and y_i is given in Table 3.1. Cells C304 and D304 contain the sum of y and x, respectively, for the sample, so

$$\hat{B} = \frac{\bar{y}}{\bar{x}} = \frac{C304}{D304} = 0.986565,$$

$$\hat{\bar{y}}_r = \hat{B}\bar{x}_U = (\hat{B})(313{,}343.283) = 309{,}133.6,$$

$$\hat{t}_{yr} = \hat{B}t_x = (\hat{B})(964{,}470{,}625) = 951{,}513{,}191.$$

Note that $\bar{y}$ for these data is 297,897.0, so $\hat{t}_{y\text{SRS}} = (3078)(\bar{y}) = 916{,}927{,}110$. In this example, $\bar{x}_\mathcal{S} = 301{,}953.7$ is smaller than $\bar{x}_U = 313{,}343.3$. This means that our SRS of size 300 slightly underestimates the true population mean of the x's; if the

FIGURE   3.1

The plot of acreage, 1992 vs. 1987, for an SRS of 300 counties. The line in the plot goes through the origin and has slope $\hat{B} = 0.9866$. Note that the variability about the line increases with x.

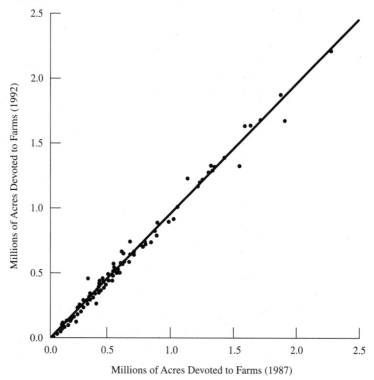

Millions of Acres Devoted to Farms (1987)

sampling distribution of $\bar{x}$ is normally distributed, our particular sample value of $\bar{x}_S$ may be approximately in the position given below:

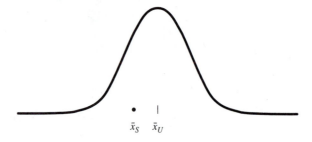

Since the x's and y's are positively correlated, we have reason to believe that $\bar{y}_S$ may also underestimate the population value $\bar{y}_U$. Ratio estimation gives a more precise estimate of $\bar{y}_U$ by expanding $\bar{y}_S$ by the factor $\bar{x}_U/\bar{x}_S$. Figure 3.2 shows the ratio and SRS estimates of $\bar{y}_U$ on a graph of the center part of the data. ∎

TABLE 3.1

Part of the Spreadsheet for the Census of Agriculture Data

	A	B	C	D	E
	County	State	acres92 (y)	acres87 (x)	Residual
1					
2					
3	COFFEE COUNTY	AL	175209	179311	−1693.00
4	COLBERT COUNTY	AL	138135	145104	−5019.56
5	LAMAR COUNTY	AL	56102	59861	−2954.78
6	MARENGO COUNTY	AL	199117	220526	−18446.29
7	MARION COUNTY	AL	89228	105586	−14939.48
8	TUSCALOOSA COUNTY	AL	96194	120542	−22728.55
⋮	⋮	⋮	⋮	⋮	⋮
298	OZAUKEE COUNTY	WI	78772	85201	−5284.34
299	ROCK COUNTY	WI	343115	357751	−9829.70
300	KANAWHA COUNTY	WV	19956	21369	−1125.91
301	PLEASANTS COUNTY	WV	15650	15716	145.14
302	PUTNAM COUNTY	WV	55827	55635	939.44
303					
304	Column sum		89369114	90586117	3.96176E-09
305	Column average		297897.0467	301953.7233	
306	Column standard deviation		344551.8948	344829.5964	31657.21817
307	$\hat{B}$ = C304/D304 =		0.986565237		

FIGURE 3.2

Detail of the center portion of Figure 3.1. Here, $\bar{x}_U$ is larger than $\bar{x}_S$, so $\hat{\bar{y}}_r$ is larger than $\bar{y}_S$.

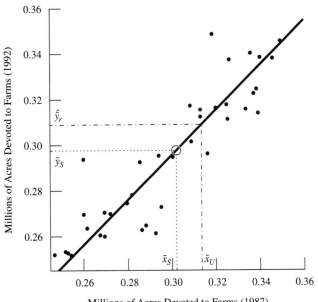

Millions of Acres Devoted to Farms (1987)

3.1.2   Bias and Mean Squared Error of Ratio Estimators

Unlike the estimators $\bar{y}$ and $N\bar{y}$ in an SRS, ratio estimators are usually *biased* for estimating $\bar{y}_U$ and t_y. We start with the unbiased estimate $\bar{y}$—if we calculate $\bar{y}_S$ for each possible SRS S, then the average of all sample means from the possible samples is the population mean $\bar{y}_U$. The estimation bias in ratio estimation arises because $\bar{y}$ is multiplied by $\bar{x}_U/\bar{x}$; if we calculate $\hat{\bar{y}}_r$ for all possible SRSs S, then the average of all the values of $\hat{\bar{y}}_r$ from the different samples will be close to $\bar{y}_U$ but will usually not equal $\bar{y}_U$ exactly.

The reduced variance of the ratio estimator usually compensates for the presence of bias—although $E[\hat{\bar{y}}_r] \neq \bar{y}_U$, the value of $\hat{\bar{y}}_r$ for any individual sample is likely to be closer to $\bar{y}_U$ than is the sample mean $\bar{y}_S$. After all, we take only one sample in practice; most people would prefer to say that their particular estimate from the sample is likely to be close to the true value—rather than that their particular value of $\bar{y}_S$ may be quite far from $\bar{y}_U$, but that the average deviation $\bar{y}_S - \bar{y}_U$, averaged over all possible samples S that could be obtained, is zero. For large samples, the sampling distributions of both $\bar{y}$ and $\hat{\bar{y}}_r$ will be approximately normal; if x and y are highly positively correlated, the following illustrates the relative bias and variance of the two estimators:

Sampling Distribution of $\bar{y}$ Sampling Distribution of $\hat{\bar{y}}_r$

$\bar{y}_U$ $\bar{y}_U$

The calculation of both bias and variance for ratio estimation uses the identity

$$\hat{t}_{yr} - t_y = \frac{\hat{t}_y}{\hat{t}_x} t_x - t_y = \hat{t}_y\left(1 - \frac{\hat{t}_x - t_x}{\hat{t}_x}\right) - t_y.$$

Since $E[\hat{t}_y] = t_y$,

$$E[\hat{t}_{yr} - t_y] = E[\hat{t}_y] - t_y - E\left[\frac{\hat{t}_y}{\hat{t}_x}(\hat{t}_x - t_x)\right]$$

$$= -E\left[\hat{B}(\hat{t}_x - t_x)\right] \qquad \text{(3.3)}$$

$$= -\text{Cov}(\hat{B}, \hat{t}_x),$$

and $E[\hat{B} - B] = E[\hat{t}_{yr} - t_y]/t_x = -\text{Cov}(\hat{B}, \bar{x})/\bar{x}_U$. Consequently, as shown by Hartley and Ross (1954),

$$\frac{|\text{Bias}(\hat{B})|}{[V(\hat{B})]^{1/2}} = \left|\frac{\text{Corr}(\hat{B}, \bar{x})}{\bar{x}_U}\right|\left(\frac{V(\hat{B})V(\bar{x})}{V(\hat{B})}\right)^{1/2} \leq \frac{V[\bar{x}]^{1/2}}{\bar{x}_U} = \text{CV}(\bar{x}).$$

In an SRS, then, the absolute value of the bias of the ratio estimator is small relative to the standard deviation of the estimator if $\text{CV}(\bar{x})$ is small.

We can use an argument similar to that used in Section 2.7 (see Exercise 16 on page 91) to show that

$$E[\hat{B} - B] \approx \left(1 - \frac{n}{N}\right) \frac{1}{n\bar{x}_U^2} (BS_x^2 - RS_x S_y)$$

$$= \frac{1}{\bar{x}_U^2} [B \, V(\bar{x}) - \text{Cov}(\bar{x}, \bar{y})],$$

(3.4)

with R the correlation between x and y. The last equality uses the derivation of the covariance of $\bar{x}$ and $\bar{y}$ in Equation (B.10) in Appendix B. The bias of $\hat{B}$ is thus small if

- The sample size n is large.
- The sampling fraction n/N is large.
- $\bar{x}_U$ is large.
- S_x is small.
- The correlation R is close to 1.

For estimating the MSE of $\hat{B}$, the same identity used in the calculation of the bias gives

$$E[(\hat{B} - B)^2] = E\left[\left(\frac{\bar{y} - B\bar{x}}{\bar{x}}\right)^2\right]$$

$$= E\left[\left\{\frac{\bar{y} - B\bar{x}}{\bar{x}_U}\left(1 - \frac{\bar{x} - \bar{x}_U}{\bar{x}}\right)\right\}^2\right]$$

$$= E\left[\left(\frac{\bar{y} - B\bar{x}}{\bar{x}_U}\right)^2 + \left(\frac{\bar{y} - B\bar{x}}{\bar{x}_U}\right)^2\left\{\left(\frac{\bar{x} - \bar{x}_U}{\bar{x}}\right)^2 - 2\frac{\bar{x} - \bar{x}_U}{\bar{x}}\right\}\right].$$

The denominator of the first term is a constant, not a random variable. It can be shown that the second term is generally small compared with the first term, so the variance and MSE are approximated by

$$E[(\hat{B} - B)^2] \approx E\left[\left(\frac{\bar{y} - B\bar{x}}{\bar{x}_U}\right)^2\right] = \frac{1}{\bar{x}_U^2} E\left[(\bar{y} - B\bar{x})^2\right].$$

Let

$$d_i = y_i - Bx_i.$$

Then, $\bar{y} - B\bar{x} = \bar{d}$, so

$$E\left[(\bar{y} - B\bar{x})^2\right] = V(\bar{d}) = \left(1 - \frac{n}{N}\right)\frac{1}{n}\sum_{i=1}^{N}\frac{d_i^2}{N-1}$$

(3.5)

and

$$E[(\hat{B} - B)^2] \approx \frac{1}{\bar{x}_U^2} V(\bar{d}).$$

Note the method used here: We approximate $\hat{B} - B$ by $(\bar{y} - B\bar{x})/\bar{x}_U$, which contains no sampled quantity in the denominator. Then we rewrite the numerator as the sample mean of a new variable. An alternative expression, algebraically equivalent to (3.5), is

$$\frac{1}{\bar{x}_U^2} E\left[(\bar{y} - B\bar{x})^2\right] = \left(1 - \frac{n}{N}\right)\frac{1}{n\bar{x}_U^2}(S_y^2 - 2BRS_x S_y + B^2 S_x^2). \tag{3.6}$$

(See Exercise 12.)

From (3.5) and (3.6), the approximated MSE will be small when

- The sample size n is large.
- The sampling fraction n/N is large.
- The deviations about the line $y = Bx$ are small.
- The correlation between x and y is close to $+1$.
- $\bar{x}_U$ is large.

In practice, B is unknown, so we cannot calculate d_i for the sampled values. Instead, use

$$e_i = y_i - \hat{B}x_i,$$

which is the ith residual from fitting the line $y = \hat{B}x$. Estimate the variance of $\hat{B}$ by

$$\hat{V}[\hat{B}] = \left(1 - \frac{n}{N}\right)\frac{s_e^2}{n\bar{x}_U^2} = \left(1 - \frac{n}{N}\right)\frac{1}{n\bar{x}_U^2}\frac{\sum\limits_{i \in \mathcal{S}}(y_i - \hat{B}x_i)^2}{n - 1}. \tag{3.7}$$

If $\bar{x}_U$ is unknown, we can substitute $\bar{x}_{\mathcal{S}}$ for it in (3.7).

It follows from (3.2) and (3.7) that

$$\hat{V}[\hat{t}_{yr}] = \hat{V}[t_x \hat{B}] = N^2\left(1 - \frac{n}{N}\right)\frac{s_e^2}{n} \tag{3.8}$$

and

$$\hat{V}[\hat{\bar{y}}_r] = \hat{V}[\bar{x}_U \hat{B}] = \left(1 - \frac{n}{N}\right)\frac{s_e^2}{n}. \tag{3.9}$$

If the sample sizes are sufficiently large, 95% confidence intervals (CIs) can be constructed as

$$\hat{B} \pm 1.96\, \text{SE}[\hat{B}], \quad \hat{\bar{y}}_r \pm 1.96\, \text{SE}[\hat{\bar{y}}_r], \quad \text{or} \quad \hat{t}_{yr} \pm 1.96\, \text{SE}[\hat{t}_{yr}].$$

In large samples, the bias of the estimator is typically small relative to the standard error (SE), so we can ignore the effect of bias in the confidence intervals (see Exercise 14).

Note that if all x's are the same value ($S_x = 0$), then the simple random sampling estimator is the same as the ratio estimator: $\hat{\bar{y}}_r = \bar{y}$ and $\text{SE}[\hat{\bar{y}}_r] = \text{SE}[\bar{y}]$.

EXAMPLE 3.3 Let's return to the sample taken from the Census of Agriculture. In the spreadsheet in Table 3.1, we created column E, containing the residuals $e_i = y_i - \hat{B}x_i$. The sample standard deviation of column E, calculated in cell E306, is s_e. Thus, using (3.8),

$$\text{SE}(\hat{t}_{yr}) = 3078\sqrt{1 - \frac{300}{3078}}\,\frac{s_e}{\sqrt{300}} = 5{,}344{,}568.$$

An approximate 95% CI for the total farm acreage, using the ratio estimator, is

$$951,513,191 \pm 1.96(5,344,568) = [941,037,838, \ 961,988,544].$$

In contrast, the standard error of $N\bar{y}_S$ is more than ten times as large:

$$SE(N\bar{y}_S) = 3078\sqrt{\left(1 - \frac{300}{3078}\right)}\frac{s_y}{\sqrt{300}} = 58,169,381.$$

The estimated coefficient of variation (CV) for the ratio estimator is $5,344,568/951,513,191 = 0.0056$, as compared with the CV of 0.0634 for the SRS estimator $N\bar{y}$ that does not use the auxiliary information. Including the 1987 information through the ratio estimator has greatly increased the precision. If all quantities to be estimated were highly correlated with the 1987 acreage, we could dramatically reduce the sample size and still obtain high precision by using ratio estimators rather than $N\bar{y}$. ∎

EXAMPLE 3.4 Let's take another look at the hypothetical population used in Example 2.1 to exhibit the sampling distribution of $\hat{t}_{yr}$. Now suppose we also have an auxiliary measurement x for each unit in the population; the population values are the following:

Unit Number	x	y
1	4	1
2	5	2
3	5	4
4	6	4
5	8	7
6	7	7
7	7	7
8	5	8

Note that x and y are positively correlated. We can calculate population quantities since we know the entire population and sampling distribution:

$$t_x = 47 \qquad\qquad t_y = 40$$
$$S_x = 1.3562027 \qquad S_y = 2.618615$$
$$R = 0.6838403 \qquad B = 0.8510638$$

Part of the sampling distribution for $\hat{t}_{yr}$ is given in Table 3.2. Figure 3.3 gives histograms for the sampling distributions of two estimates of t_y: $\hat{t}_{SRS} = N\bar{y}$, the estimate used in Chapter 2; and $\hat{t}_{yr}$. The sampling distribution for the ratio estimate is not spread out as much as the sampling distribution for $N\bar{y}$; it is also skewed rather than symmetric. The skewness leads to the slight estimation bias of the ratio estimate. The population total is $t_y = 40$; the mean value of the sampling distribution of $\hat{t}_{yr}$ is 39.85063.

The mean value of the sampling distribution of $\hat{B}$ is 0.8478857, resulting in a bias of -0.003178. Using the population quantities above, the approximate bias from (3.4) is

$$\left(1 - \frac{n}{N}\right)\frac{1}{n\bar{x}_U^2}(BS_x^2 - RS_xS_y) = -0.003126.$$

TABLE 3.2
Sampling Distribution for $\hat{t}_{yr}$.

Sample Number	Sample, S	$\bar{x}_S$	$\bar{y}_S$	$\hat{B}$	$\hat{t}_{SRS}$	$\hat{t}_{yr}$
1	{1, 2, 3, 4}	5.00	2.75	0.55	22.00	25.85
2	{1, 2, 3, 5}	5.50	3.50	0.64	28.00	29.91
3	{1, 2, 3, 6}	5.25	3.50	0.67	28.00	31.33
4	{1, 2, 3, 7}	5.25	3.50	0.67	28.00	31.33
5	{1, 2, 3, 8}	4.75	3.75	0.79	30.00	37.11
6	{1, 2, 4, 5}	5.75	3.50	0.61	28.00	28.61
⋮	⋮	⋮	⋮	⋮	⋮	⋮
67	{4, 5, 6, 8}	6.50	6.50	1.00	52.00	47.00
68	{4, 5, 7, 8}	6.50	6.50	1.00	52.00	47.00
69	{4, 6, 7, 8}	6.25	6.50	1.04	52.00	48.88
70	{5, 6, 7, 8}	6.75	7.25	1.07	58.00	50.48

FIGURE 3.3
Sampling distributions for (a) $\hat{t}_{SRS}$ and (b) $\hat{t}_{yr}$.

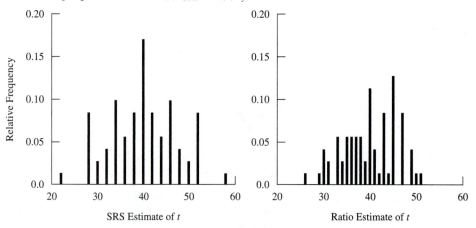

The variance of the sampling distribution of $\hat{B}$, calculated using the definition of variance in (2.4), is 0.015186446; the approximation in (3.6) is

$$\left(1 - \frac{n}{N}\right) \frac{1}{n\bar{x}_U^2}(S_y^2 - 2BRS_xS_y + B^2S_x^2) = 0.01468762. ∎$$

3.1.2.1 Accuracy of the MSE Approximation

Example 3.4 demonstrates that the approximation to the MSE in (3.6) is in fact only an approximation; it happens to be a good approximation in that example even though the population and sample are both small.

For (3.6) to be a good approximation to the MSE, the bias should be small, and the terms discarded in the approximation of the variance should be small. If the coefficient of variation of $\bar{x}$ is small—that is, if $\bar{x}_U$ is estimated with high relative precision—the bias is small relative to the square root of the variance. If we form a confidence interval using $\hat{t}_{yr} \pm 1.96 \, \text{SE}[\hat{t}_{yr}]$, using (3.8) to find the estimated variance and standard error, then the bias will not have any great effect on the coverage probability of the confidence interval. A small $\text{CV}(\bar{x})$ also means that $\bar{x}$ is stable from sample to sample and that $\bar{x}$ is likely to be nonzero—a desirable result since we divide by $\bar{x}$ when forming the ratio estimate. In some of the complex sampling designs to be discussed in subsequent chapters, though, the bias may be a matter of concern—we will return to this issue in Chapters 9 and 12.

For (3.6) to be a good approximation of MSE, we want a large sample size (n larger than 30 or so) and $\text{CV}(\bar{x}) \le .1$, $\text{CV}(\bar{y}) \le .1$. If these conditions are not met, then (3.6) may severely underestimate the true MSE.

3.1.2.2 Advantages of Ratio Estimation

What do we gain from using ratio estimation? If the deviations of y_i from $\hat{B}x_i$ are smaller than the deviations of y_i from $\bar{y}$, then $\hat{V}[\hat{\bar{y}}_r] \le \hat{V}[\bar{y}]$. Recall from Chapter 2 that

$$\text{MSE}[\bar{y}] = V[\bar{y}] = \left(1 - \frac{n}{N}\right)\frac{S_y^2}{n}.$$

Using the approximation in (3.6),

$$\text{MSE}[\hat{\bar{y}}_r] \approx \left(1 - \frac{n}{N}\right)\frac{1}{n}(S_y^2 - 2BRS_xS_y + B^2S_x^2).$$

Thus,

$$\text{MSE}[\hat{\bar{y}}_r] - \text{MSE}[\bar{y}] \approx \left(1 - \frac{n}{N}\right)\frac{1}{n}(S_y^2 - 2BRS_xS_y + B^2S_x^2 - S_y^2)$$

$$= \left(1 - \frac{n}{N}\right)\frac{1}{n}S_xB(-2RS_y + BS_x).$$

So to the accuracy of the approximation,

$$\text{MSE}[\hat{\bar{y}}_r] \le \text{MSE}[\bar{y}] \quad \text{if and only if} \quad R \ge \frac{BS_x}{2S_y} = \frac{\text{CV}(x)}{2\text{CV}(y)}.$$

If the coefficients of variation are approximately equal, then it pays to use ratio estimation when the correlation between x and y is larger than 1/2.

Ratio estimation is most appropriate if a straight line through the origin summarizes the relationship between x_i and y_i and if the variance of y_i about the line is proportional to x_i. Under these conditions, $\hat{B}$ is the weighted least squares regression slope for the line through the origin with weights proportional to $1/x_i$—the slope $\hat{B}$ minimizes the sum of squares

$$\sum_{i \in \mathcal{S}} \frac{1}{x_i}(y_i - \hat{B}x_i)^2.$$

3.1.3   Ratio Estimation with Proportions

Ratio estimation works the same way when the quantity of interest is a proportion.

EXAMPLE **3.5**   Peart (1994) collected the data shown in Table 3.3 as part of a study evaluating the effects of feral pig activity and drought on the native vegetation on Santa Cruz Island, California. She counted the number of woody seedlings in pig-protected areas under each of ten sampled oak trees in March 1992, following the drought-ending rains of 1991. She put a flag by each seedling, then determined how many were still alive in February 1994. The data (courtesy of Diann Peart) are plotted in Figure 3.4.

When most people who have had one introductory statistics course see data like these, they want to find the sample proportion of the 1992 seedlings that are still alive in 1994 and then use the formula for the variance of a binomial random variable to calculate the standard error of their estimate. Using the binomial standard error is *incorrect* for these data since the binomial distribution requires that trials be independent; in this example, that assumption is inappropriate. Seedling survival depends on many factors, such as local rainfall, amount of light, and predation. Such factors are likely to affect seedlings in the same plot to a similar degree, leading different plots to have, in general, different survival rates. The sample size in this example is 10, not 206.

The design is actually a **cluster sample;** the clusters are the plots associated with each tree, and the observation units are individual seedlings in those plots. To look at this example from the framework of ratio estimation, let

$$y_i = \text{number of seedlings near tree } i \text{ that are alive in 1994}$$
$$x_i = \text{number of seedlings near tree } i \text{ that are alive in 1992.}$$

TABLE   **3.3**
Santa Cruz Island Seedling Data

Tree	$x =$ Number of Seedlings, 3/92	$y =$ Seedlings Alive, 2/94
1	1	0
2	0	0
3	8	1
4	2	2
5	76	10
6	60	15
7	25	3
8	2	2
9	1	1
10	31	27
Total	206	61
Average	20.6	6.1
Standard deviation	27.4720	8.8248

FIGURE   3.4

The plot of seedlings that survived (February 1994) vs. seedlings alive (March 1992), for ten oak trees.

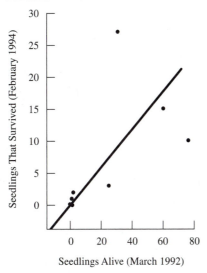

Seedlings Alive (March 1992)

Then, the ratio estimate of the proportion of seedlings still alive in 1994 is

$$\hat{B} = \hat{p} = \frac{\bar{y}}{\bar{x}} = \frac{6.1}{20.6} = 0.2961.$$

Using (3.7) and ignoring the finite population correction (fpc),

$$SE[\hat{B}] = \sqrt{\frac{1}{(10)(20.6)^2} \cdot \frac{\sum_{i \in \mathcal{S}}(y_i - 0.2961165x_i)^2}{9}}$$

$$= \sqrt{\frac{56.3778}{(10)(20.6)^2}}$$

$$= 0.115.$$

Had we used the binomial formula, we would have calculated a standard error of

$$\sqrt{\frac{(0.2961)(0.7039)}{206}} = .0318,$$

which is much too small and gives a misleading impression of precision.

The approximation to the variance of $\hat{B}$ in this example may not be particularly good because the sample size is small; although the estimated variance of $\hat{B}$ is likely an underestimate, it will still be better than the variance calculation using the binomial distribution, because the seedlings are not independent. ∎

3.2
Regression Estimation

3.2.1 Using a Straight-Line Model

Ratio estimation works best if the data are well fit by a straight line through the origin. Sometimes, data appear to be evenly scattered about a straight line that does not go through the origin—that is, the data look as though the usual straight-line regression model

$$y = B_0 + B_1 x$$

would provide a good fit.

Suppose we know $\bar{x}_U$, the population mean for the x's. Then the regression estimator of $\bar{y}_U$ is the predicted value of y from the fitted regression model when $x = \bar{x}_U$:

$$\hat{\bar{y}}_{\text{reg}} = \hat{B}_0 + \hat{B}_1 \bar{x}_U = \bar{y} + \hat{B}_1 (\bar{x}_U - \bar{x}), \tag{3.10}$$

where $\hat{B}_0$ and $\hat{B}_1$ are the ordinary least squares regression coefficients of the intercept and slope, respectively. For this model,

$$\hat{B}_1 = \frac{\displaystyle\sum_{i \in S} (x_i - \bar{x})(y_i - \bar{y})}{\displaystyle\sum_{i \in S} (x_i - \bar{x})^2} = \frac{r s_y}{s_x},$$

$$\hat{B}_0 = \bar{y} - \hat{B}_1 \bar{x},$$

and r is the sample correlation coefficient of x and y.

Like the ratio estimator, the regression estimator is biased. Let B_1 be the least squares regression slope calculated from all the data in the population:

$$B_1 = \frac{\displaystyle\sum_{i=1}^{N} (x_i - \bar{x}_U)(y_i - \bar{y}_U)}{\displaystyle\sum_{i=1}^{N} (x_i - \bar{x}_U)^2} = \frac{R S_y}{S_x}.$$

Then, using (3.10), the bias of $\hat{\bar{y}}_{\text{reg}}$ is given by

$$E[\hat{\bar{y}}_{\text{reg}} - \bar{y}_U] = E[\bar{y} - \bar{y}_U] + E[\hat{B}_1(\bar{x}_U - \bar{x})] = -\text{Cov}(\hat{B}_1, \bar{x}). \tag{3.11}$$

If the regression line goes through all points (x_i, y_i) in the population, then the bias is zero: In that situation, $\hat{B}_1 = B_1$ for every sample, so $\text{Cov}(\hat{B}_1, \bar{x}) = 0$.

As with ratio estimation, for large SRSs the MSE for regression estimation is approximately equal to the variance (see Exercise 18); the bias can often be disregarded in large samples.

The method used in approximating the MSE in ratio estimation can also be applied to regression estimation. Let $d_i = y_i - [\bar{y}_U + B_1(x_i - \bar{x}_U)]$. Then,

$$\text{MSE}(\hat{\bar{y}}_{\text{reg}}) = E[\{\bar{y} + \hat{B}_1(\bar{x}_U - \bar{x}) - \bar{y}_U\}^2]$$

$$\approx V[\bar{d}] \tag{3.12}$$

$$= \left(1 - \frac{n}{N}\right) \frac{S_d^2}{n}.$$

Using the relation $B_1 = RS_y/S_x$, it may be shown that

$$\left(1 - \frac{n}{N}\right)\frac{S_d^2}{n} = \left(1 - \frac{n}{N}\right)\frac{1}{n}\sum_{i=1}^{N}\frac{(y_i - \bar{y}_U - B_1[x_i - \bar{x}_U])^2}{N-1}$$

$$= \left(1 - \frac{n}{N}\right)\frac{1}{n}S_y^2(1 - R^2). \qquad (3.13)$$

(See Exercise 17.) Thus, the approximate MSE is small when

- n is large.
- n/N is large.
- S_y is small.
- The correlation R is close to -1 or $+1$.

The standard error can be calculated by finding the sample variance of the residuals. Let $e_i = y_i - (\hat{B}_0 + \hat{B}_1 x_i)$; then,

$$\mathrm{SE}(\hat{\bar{y}}_{\mathrm{reg}}) = \sqrt{\left(1 - \frac{n}{N}\right)\frac{s_e^2}{n}}. \qquad (3.14)$$

EXAMPLE 3.6   To estimate the number of dead trees in an area, we divide the area into 100 square plots and count the number of dead trees on a photograph of each plot. Photo counts can be made quickly, but sometimes a tree is misclassified or not detected. So we select an SRS of 25 of the plots for field counts of dead trees. We know that the population mean number of dead trees per plot from the photo count is 11.3. The data—plotted in Figure 3.5—and selected SAS output are as follows:

Photo	10 12 7 13 13 6 17 16 15 10 14 12 10
Field	15 14 9 14 8 5 18 15 13 15 11 15 12

Photo	5 12 10 10 9 6 11 7 9 11 10 10
Field	8 13 9 11 12 9 12 13 11 10 9 8

Simple Statistics

Variable	N	Mean	Std Dev	Sum	Minimum	Maximum
PHOTO	25	10.6000	3.0687	265.0000	5.0000	17.0000
FIELD	25	11.5600	3.0150	289.0000	5.0000	18.0000

Dependent Variable: FIELD

Analysis of Variance

Source	DF	Sum of Squares	Mean Square	F Value	Prob>F
Model	1	84.99982	84.99982	14.682	0.0009
Error	23	133.16018	5.78957		
C Total	24	218.16000			

(Output continued on page 76)

FIGURE   3.5

The plot of photo and field tree-count data, along with the regression line. Note that $\hat{\bar{y}}_{\text{reg}}$ is the predicted value from the regression equation when $x = \bar{x}_U$.

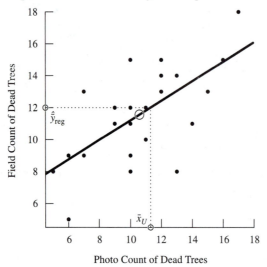

Photo Count of Dead Trees

Root MSE	2.40615	R-square	0.3896
Dep Mean	11.56000	Adj R-sq	0.3631
C.V.	20.81447		

Parameter Estimates

Variable	DF	Parameter Estimate	Standard Error	T for H0: Parameter=0	Prob > |T|
INTERCEP	1	5.059292	1.76351187	2.869	0.0087
PHOTO	1	0.613274	0.16005493	3.832	0.0009

Using (3.10), the regression estimate of the mean is

$$\hat{\bar{y}}_{\text{reg}} = 5.06 + 0.613(11.3) = 11.99.$$

From the SAS output, s_e^2 can be calculated from the residual sum of squares; $s_e^2 = 133.16018/24 = 5.54834$ (alternatively, you could use the MSE of the residuals, which divides by $n - 2$ rather than $n - 1$). Thus, the standard error is, from (3.14),

$$SE[\hat{\bar{y}}_{\text{reg}}] = \sqrt{\left(1 - \frac{25}{100}\right)\frac{5.54834}{25}} = 0.408.$$

Again, the standard error is less than that for $\bar{y}$:

$$SE[\bar{y}] = \sqrt{\left(1 - \frac{25}{100}\right)\frac{s_y^2}{25}} = 0.522.$$

We expect regression estimation to increase the precision in this example because the variables *photo* and *field* are positively correlated ($r = 0.62$). To estimate the total number of dead trees, use

$$\hat{t}_{yreg} = (100)(11.99) = 1199;$$
$$SE[\hat{t}_{yreg}] = (100)(0.408) = 40.8.$$

An approximate 95% confidence interval for the total number of dead trees is given by

$$1199 \pm (2.07)(40.8) = [1114, 1283].$$

Because of the relatively small sample size, we used the *t*-distribution percentile (with $n - 2 = 23$ degrees of freedom) of 2.07 rather than the normal distribution percentile of 1.96. ∎

3.2.2 Difference Estimation

Difference estimation is a special case of regression estimation, used when the investigator "knows" that the slope B_1 is 1. Difference estimation is often recommended in accounting when an SRS is taken. A list of accounts receivable consists of the book value for each account—the company's listing of how much is owed on each account. In the simplest sampling scheme, the auditor scrutinizes a random sample of the accounts to determine the audited value—the actual amount owed—in order to estimate the error in the total accounts receivable. The quantities considered are

$$y_i = \text{audited value for company } i$$
$$x_i = \text{book value for company } i.$$

Then, $\bar{y} - \bar{x}$ is the mean difference for the audited accounts.

The estimated total difference is $\hat{t}_y - \hat{t}_x = N(\bar{y} - \bar{x})$; the estimated audited value for accounts receivable is

$$\hat{t}_{ydiff} = t_x + (\hat{t}_y - \hat{t}_x).$$

Again, define the residuals from this model: Here, $e_i = y_i - x_i$. The variance of $\hat{t}_{ydiff}$ is

$$V(\hat{t}_{ydiff}) = V[t_x + (\hat{t}_y - \hat{t}_x)] = V(\hat{t}_e),$$

where $\hat{t}_e = (N/n) \sum_{i \in S} e_i$. If the variability in the residuals e_i is smaller than the variability among the y_i's, then difference estimation will increase precision.

Difference estimation works best if the population and sample have a large fraction of nonzero differences that are roughly equally divided between overstatements and understatements, and if the sample is large enough so that the sampling distribution of $(\bar{y} - \bar{x})$ is approximately normal.

In auditing, it is possible that all audited values in the sample are the same as the corresponding book values. Then, $\bar{y} = \bar{x}$, and the standard error of $\hat{t}_y$ would be calculated as zero. In such a situation, where most of the differences are zero, more sophisticated modeling is needed.

3.3
Estimation in Domains

Often we want separate estimates for subpopulations; the subpopulations are called **domains** or **subdomains.** We may want to take an SRS of visitors who fly to New York City on September 18 and to estimate the proportion of out-of-state visitors who intend to stay longer than 1 week. For that survey, there are two domains of study: visitors from in-state and visitors from out-of-state. We do not know which persons in the population belong to which domain until they are sampled, though. Thus, the number of persons in an SRS who fall into each domain is a random variable, with value unknown at the time the survey is designed.

Suppose there are D domains. Let $\mathcal{U}_d$ be the index set of the units in the population that are in domain d and let $\mathcal{S}_d$ be the index set of the units in the sample that are in domain d, for $d = 1, 2, \ldots, D$. Let N_d be the number of population units in $\mathcal{U}_d$, and n_d be the number of sample units in $\mathcal{S}_d$. Suppose we want to estimate

$$\bar{y}_{\mathcal{U}_d} = \sum_{i \in \mathcal{U}_d} \frac{y_i}{N_d}.$$

A natural estimator of $\bar{y}_{\mathcal{U}_d}$ is

$$\bar{y}_d = \sum_{i \in \mathcal{S}_d} \frac{y_i}{n_d}, \tag{3.15}$$

which looks at first just like the sample means studied in Chapter 2.

The quantity n_d is a random variable, however: If a different SRS is taken, we will very likely have a different value for n_d. Different samples from New York City would have different numbers of out-of-state visitors. Technically, (3.15) is a ratio estimate. To see this, let

$$u_i = \begin{cases} y_i & \text{if } i \in \mathcal{U}_d \\ 0 & \text{if } i \notin \mathcal{U}_d \end{cases}$$

$$x_i = \begin{cases} 1 & \text{if } i \in \mathcal{U}_d \\ 0 & \text{if } i \notin \mathcal{U}_d. \end{cases}$$

Then, $\bar{x}_U = N_d/N$, $\bar{y}_{\mathcal{U}_d} = \sum_{i=1}^{N} u_i / \sum_{i=1}^{N} x_i$, and

$$\bar{y}_d = \hat{B} = \frac{\bar{u}}{\bar{x}} = \frac{\displaystyle\sum_{i \in \mathcal{S}} u_i}{\displaystyle\sum_{i \in \mathcal{S}} x_i}.$$

Because we are estimating a ratio, we use (3.7) to calculate the standard error:

$$\mathrm{SE}(\bar{y}_d) = \sqrt{\left(1 - \frac{n}{N}\right) \frac{1}{n\bar{x}_U^2} \frac{\displaystyle\sum_{i \in \mathcal{S}} (u_i - \hat{B}x_i)^2}{n-1}}$$

$$= \sqrt{\left(1 - \frac{n}{N}\right) \frac{1}{n\bar{x}_U^2} \frac{\displaystyle\sum_{i \in \mathcal{S}_d} (y_i - \hat{B})^2}{n-1}}$$

$$= \sqrt{\left(1 - \frac{n}{N}\right) \frac{1}{n} \left(\frac{N}{N_d}\right)^2 \frac{(n_d - 1)s_{yd}^2}{n-1}}$$

$$\approx \sqrt{\left(1 - \frac{n}{N}\right) \frac{s_{yd}^2}{n_d}}.$$

The approximation in the last line depends on a large sample size in domain d; if the sample is large enough, then we will expect that $n_d/n \approx N_d/N$ and $(n_d - 1)/(n - 1) \approx$

n_d/n. In a large sample, the standard error of $\bar{y}_d$ is approximately the same as if we used formula (2.10). Thus, in a sufficiently large sample, the technicality that we are using a ratio estimator makes little difference in practice for estimating a domain mean.

The situation is a little more complicated when estimating a domain total. If N_d is known, estimation is simple: Use $N_d \bar{y}_d$. If N_d is unknown, though, we need to estimate it by $N n_d/n$. Then,

$$\hat{t}_{yd} = N \frac{n_d}{n} \frac{\displaystyle\sum_{i \in \mathcal{S}} u_i}{n_d} = N\bar{u}.$$

The standard error is

$$\text{SE}(\hat{t}_{yd}) = N\,\text{SE}(\bar{u}) = N\sqrt{\left(1 - \frac{n}{N}\right)\frac{s_u^2}{n}}.$$

EXAMPLE 3.7 In the SRS of size 300 from the Census of Agriculture (see Example 2.4), 39 counties are in western states.[2] What is the estimated total number of acres devoted to farming in the West?

The sample mean of the 39 counties is $\bar{y}_d = 598{,}680.6$, with sample standard deviation $s_{yd} = 516{,}157.7$. Thus,

$$\text{SE}(\bar{y}_d) = \sqrt{\left(1 - \frac{300}{3078}\right)\frac{516{,}157.7}{\sqrt{39}}} = 78{,}520.$$

Thus, $\widehat{\text{CV}}[\bar{y}_d] = 0.1312$, and an approximate 95% confidence interval for the mean farm acreage for counties in the western United States is [444,781, 752,580].

For estimating the total number of acres devoted to farming in the West, suppose we do not know how many counties in the population are in the western United States. Define

$$u_i = \begin{cases} y_i & \text{if county } i \text{ is in the western United States} \\ 0 & \text{otherwise} \end{cases}.$$

Then,

$$\hat{t}_{yd} = N\bar{u} = 3078(77{,}828.48) = 239{,}556{,}051.$$

The standard error is

$$\text{SE}(\hat{t}_{yd}) = 3078\sqrt{\left(1 - \frac{300}{3078}\right)\frac{273{,}005.4}{\sqrt{300}}} = 46{,}090{,}460.$$

The estimated coefficient of variation for $\hat{t}_{yd}$ is $\widehat{\text{CV}}[\hat{t}_{yd}] = 46{,}090{,}460/239{,}556{,}051 = 0.1924$; had we known the number of counties in the western United States and been able to use that value in the estimate, the coefficient of variation for the estimated total would have been 0.1312, the coefficient of variation for the estimated mean. ∎

EXAMPLE 3.8 An SRS of 1500 licensed boat owners in a state was sampled from a list of 400,000 names with currently licensed boats; 472 of the respondents said they owned an open motorboat longer than 16 feet. The 472 respondents with large motorboats reported

[2]Alaska (AK), Arizona (AZ), California (CA), Colorado (CO), Hawaii (HI), Idaho (ID), Montana (MT), Nevada (NV), New Mexico (NM), Oregon (OR), Utah (UT), Washington (WA), and Wyoming (WY).

having the following numbers of children:

Number of Children	Number of Respondents
0	76
1	139
2	166
3	63
4	19
5	5
6	3
8	1
Total	472

To estimate the percentage of large-motorboat owners who have children, we can use $\hat{p} = 396/472 = 0.839$. This is a ratio estimator, but in this case, as explained above, the standard error is approximately what you would think it would be. Ignoring the fpc,

$$\text{SE}(\hat{p}) = \sqrt{\frac{.839(1 - .839)}{471}} = 0.017.$$

To look at the average number of children per household among registered boat owners who register a motorboat more than 16 feet long, note that the average number of children for the 472 respondents in the domain is 1.667373, with variance 1.398678. Thus, an approximate 95% confidence interval for the average number of children in large-motorboat households is

$$1.667 \pm 1.96 \sqrt{\frac{1.398678}{472}} = [1.56, 1.77].$$

To estimate the total number of children in the state whose parents register a large motorboat, we create a new variable u for the respondents that takes on the value number of children if respondent has a motorboat, and zero otherwise. The frequency distribution for the variable u is then

Number of Children	Number of Respondents
0	1104
1	139
2	166
3	63
4	19
5	5
6	3
8	1
Total	1500

Now, $\bar{u} = 0.52466$ and $s_u^2 = 1.0394178$, so $\hat{t}_{yd} = 400{,}000(.524666) = 209{,}867$ and

$$\text{SE}(\hat{t}_{yd}) = \sqrt{(400{,}000)^2 \frac{1.0394178}{1500}} = 10{,}529.5.$$

In this example, the variable u_i simply counts the number of children in household i who belong to a household with a large open motorboat. ∎

In this section, we have shown that estimating domain means is a special case of ratio estimation because the sample size in the domain varies from sample to sample. If the sample size for the domain in an SRS is sufficiently large, we can use SRS formulas for inference about the domain mean.

Inference about totals depends on whether the population size of the domain, N_d, is known. If N_d is known, then the estimated total is $N_d \bar{y}_d$. If N_d is unknown, then define a new variable u_i that equals y_i for observations in the domain and zero for observations not in the domain; then use $\hat{t}_u$ to estimate the domain total.

The results of this section are only for SRSs. In Section 12.3, we will discuss estimating domain means if the data are collected using other sampling designs.

3.4
Models for Ratio and Regression Estimation*

Many statisticians have proposed that (1) if a regression model provides a good fit to survey data, the model should be used to estimate the total for y and its standard error and that (2) how one obtains the data is not as important as the model that is fit. In this section we discuss models that give the point estimates in Equations (3.2) and (3.10) for ratio and regression estimation. The variances under a model-based approach, however, are slightly different, as we will see.

3.4.1   A Model for Ratio Estimation

We stated earlier that ratio estimation is most appropriate in an SRS when a straight line through the origin fits well and when the variance of the observations about the line is proportional to x. We can state these conditions as a linear regression model: Assume that $x_1, x_2, \ldots, x_N$ are known (and all are greater than zero) and that $Y_1, Y_2, \ldots, Y_N$ are independent and follow the model

$$Y_i = \beta x_i + \varepsilon_i, \tag{3.16}$$

where $E_M[\varepsilon_i] = 0$ and $V_M[\varepsilon_i] = \sigma^2 x_i$. The independence of observations in the model is an explicit statement that the sampling design gives no information that can be used in estimating quantities of interest; the sampling procedure has no effect on the validity of the model. Under the model, $T_y = \sum_{i=1}^{N} Y_i$ is a random variable, and the population total of interest, t_y, is one realization of the random variable T_y (this is in contrast to the randomization approach, in which t_y is considered to be a fixed but unknown quantity and the only random variables are the sample indicators Z_i). If S represents the set of units in our sample, then

$$t_y = \sum_{i \in S} y_i + \sum_{i \notin S} y_i.$$

We observe the values of y_i for units in the sample and predict those for units not in the sample as $\hat{\beta} x_i$, where $\hat{\beta} = \bar{y}/\bar{x}$ is the weighted least squares estimate of β under the model in (3.16). Then, a natural estimate of t_y is

$$\hat{t}_y = \sum_{i \in S} y_i + \hat{\beta} \sum_{i \notin S} x_i = n\bar{y} + \frac{\bar{y}}{\bar{x}} \sum_{i \notin S} x_i = \frac{\bar{y}}{\bar{x}} \sum_{i=1}^{N} x_i = \frac{\bar{y}}{\bar{x}} t_x.$$

This is simply the ratio estimate of t_y.

In many common sampling schemes, we find that if we adopt a model consistent with the reasons we would adopt a certain sampling scheme or method of estimation, the point estimators obtained using the model are very close to the design-based estimators. The model-based variance, though, may differ from the variance from the randomization theory. In **randomization theory,** or **design-based sampling,** the *sampling design* determines how sampling variability is estimated. In **model-based sampling,** the *model* determines how variability is estimated, and the sampling design is irrelevant—as long as the model holds, you could choose any n units you want to from the population.

The model-based estimator

$$\hat{T}_y = \sum_{i \in S} Y_i + \hat{\beta} \sum_{i \notin S} x_i$$

is model-unbiased since

$$E_M[\hat{T}_y - T] = E_M \left[\hat{\beta} \sum_{i \notin S} x_i - \sum_{i \notin S} Y_i \right] = 0.$$

The model-based variance is

$$V_M[\hat{T}_y - T] = V_M \left[\hat{\beta} \sum_{i \notin S} x_i - \sum_{i \notin S} Y_i \right]$$

$$= V_M \left[\hat{\beta} \sum_{i \notin S} x_i \right] + V_M \left[\sum_{i \notin S} Y_i \right]$$

because $\hat{\beta}$ and $\sum_{i \notin S} Y_i$ are independent under the model assumptions. The model (3.16) does not depend on which population units are selected to be the sample S, so S can be treated as though it is fixed. Consequently, using (3.16),

$$V_M \left[\sum_{i \notin S} Y_i \right] = V_M \left[\sum_{i \notin S} (\beta x_i + \varepsilon_i) \right] = V_M \left[\sum_{i \notin S} \varepsilon_i \right] = \sigma^2 \left(\sum_{i \notin S} x_i \right),$$

and, similarly,

$$V_M \left[\hat{\beta} \sum_{i \notin S} x_i \right] = \left(\sum_{i \notin S} x_i \right)^2 V_M \left[\frac{\sum_{i \in S} Y_i}{\sum_{i \in S} x_i} \right] = \left(\sum_{i \notin S} x_i \right)^2 \frac{\sigma^2}{\sum_{i \in S} x_i}.$$

Combining the two terms gives

$$V_M[\hat{T}_y - T] = \frac{\sigma^2 \sum_{i \notin S} x_i}{\sum_{i \in S} x_i} \left(\sum_{i \notin S} x_i + \sum_{i \in S} x_i \right).$$

$$= \frac{\sigma^2 \sum_{i \notin S} x_i}{\sum_{i \in S} x_i} t_x \qquad (3.17)$$

$$= \left(1 - \frac{\sum_{i \in S} x_i}{t_x}\right) \frac{\sigma^2 t_x^2}{\sum_{i \in S} x_i}.$$

Note that if the sample size is small relative to the population size, then

$$V_M[\hat{T}_y - T] \approx \frac{\sigma^2 t_x^2}{\sum_{i \in S} x_i}.$$

The quantity $(1 - \sum_{i \in S} x_i/t_x)$ serves as an fpc in the model-based approach to ratio estimation.

EXAMPLE 3.9 Let's perform a model-based analysis of the data from the Census of Agriculture, used in Examples 3.2 and 3.3. We already plotted the data in Figure 3.1, and it looked as though a straight line through the origin would fit well and that the variability about the line was greater for observations with larger values of x. For the data points with x positive, we can run a regression analysis in SAS or S-PLUS with no intercept and with weight variable $1/x$. In SAS, we add two lines to the bottom of the data file to obtain predicted values, as shown in Appendix E.

```
Model: MODEL1
NOTE: No intercept in model. R-square is redefined.
Dependent Variable: ACRES92
```

                      Analysis of Variance

Source	DF	Sum of Squares	Mean Square	F Value	Prob>F
Model	1	88168461.147	88168461.147	41487.306	0.0001
Error	298	633306.99655	2125.19126		
U Total	299	88801768.143			

Root MSE	46.09980	R-square	0.9929
Dep Mean	38097.06433	Adj R-sq	0.9928
C.V.	0.12101		

                      Parameter Estimates

Variable	DF	Parameter Estimate	Standard Error	T for H0: Parameter=0	Prob > |T|
ACRES87	1	0.986565	0.00484360	203.684	0.0001

(Output continued on page 84)

Obs	Weight	Dep Var ACRES92	Predict Value	Std Err Predict	Lower95% Mean	Upper95% Mean	Residual
1	5.577E-6	175209	176902	868.511	175193	178611	-1693.0
2	6.892E-6	138135	143155	702.826	141771	144538	-5019.6
3	0.000017	56102.0	59056.8	289.943	58486.2	59627.4	-2954.8
4	4.535E-6	199117	217563	1068.140	215461	219665	-18446.3
5	9.471E-6	89228.0	104167	511.416	103161	105174	-14939.5
6	8.296E-6	96194.0	118923	583.857	117774	120072	-22728.5
7	0.000015	57253.0	65414.2	321.155	64782.2	66046.2	-8161.2
8	4.472E-6	210692	220590	1083.000	218459	222721	-9898.1
9	0.000012	78498.0	79188.6	388.781	78423.5	79953.7	-690.6
10	4.262E-6	219444	231453	1136.333	229217	233689	-12009.1
⋮	⋮	⋮	⋮	⋮	⋮	⋮	⋮
299	0.000064	15650.0	15504.9	76.122	15355.1	15654.7	145.1
300	0.000018	55827.0	54887.6	269.474	54357.2	55417.9	939.4
301	0		309134	1517.709	306147	312120	.
302	0		9.5151E8	4671509	9.4232E8	9.6071E8	.

The slope, 0.986565, and the model-based estimate of the total, 9.5151×10^8, are the same as the design-based estimates obtained in Example 3.2. The model-based standard error of the estimated total, using (3.17), is

$$\sqrt{\hat{\sigma}^2 \frac{t_x - \sum\limits_{i \in S} x_i}{\sum\limits_{i \in S} x_i} t_x}.$$

We can use the weighted residuals (for nonzero x_i)

$$r_i = \frac{y_i - \hat{\beta} x_i}{\sqrt{x_i}}$$

to estimate σ^2: If the model assumptions hold, $\hat{\sigma}^2 = \sum r_i^2/(n-1)$ (given as the MSE in the SAS ANOVA table) estimates σ^2. Thus,

$$SE_M[\hat{T}_y] = \sqrt{(2125.19126)\left(\frac{964,470,625 - 90,586,117}{90,586,117}\right)(964,470,625)}$$

$$= 4,446,719.$$

A model-based analysis is easier if we ignore the fpc. Then the standard error for the estimated total is the standard error for the mean response when x is set equal to t_x. If we ignore the fpc, the model-based standard error is exactly that given as the "Std Err Predict" in the SAS output (in SAS, this is the standard error of the mean predicted value), which is

$$\sqrt{MSE \frac{t_x^2}{\sum\limits_{i \in S} x_i}} = 4,671,509.$$

Note that, for this example, the model-based standard error is smaller than the standard error we calculated using randomization inference, which was 5,344,568. ■

When adopting a model for a set of data, we need to check the assumptions of the model. The assumptions for any linear regression model are as follows:

1 The model is correct.

2 The variance structure is as given.

3 The observations are independent.

Typically, assumptions 1 and 2 are checked by plotting the data and examining residuals from the model. Assumption 3, however, is difficult to check in practice and requires knowledge of how the data were collected. Generally, if you take a random sample, then you may assume independence of the observations.

We can perform some checks on the appropriateness of a model with a straight line through the origin for these data: If the variance of y_i about the line is proportional to x_i, then a plot of the weighted residuals

$$\frac{y_i - \hat{\beta}x_i}{\sqrt{x_i}}$$

against x_i or $\log x_i$ should not exhibit any patterns. This plot is given for the agriculture census data in Figure 3.6; nothing appears in the plot to make us doubt the adequacy of this model for the observations in our sample.

FIGURE **3.6**

The plot of weighted residuals vs. x, for the random sample from the agricultural census. A few counties may be outliers; overall, though, scatter appears to be fairly random.

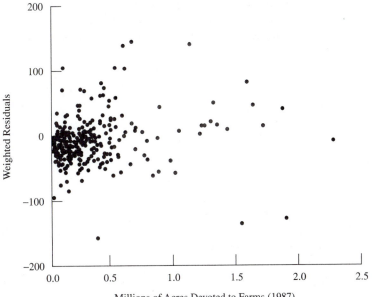

Millions of Acres Devoted to Farms (1987)

3.4.2   A Model for Regression Estimation

A similar result occurs for regression estimation; for that, the model is

$$Y_i = \beta_0 + \beta_1 x_i + \varepsilon_i,$$

where the ε_i's are independent and identically distributed with mean 0 and constant variance σ^2. The least squares estimators of β_0 and β_1 in this model are

$$\hat{\beta}_1 = \frac{\displaystyle\sum_{i \in S}(x_i - \bar{x}_S)(Y_i - \bar{Y}_S)}{\displaystyle\sum_{i \in S}(x_i - \bar{x}_S)^2}$$

$$\hat{\beta}_0 = \bar{Y}_S - \hat{\beta}_1 \bar{x}_S.$$

Then, using the predicted values in place of the units not sampled,

$$\hat{T}_y = \sum_{i \in S} Y_i + \sum_{i \notin S}(\hat{\beta}_0 + \hat{\beta}_1 x_i)$$

$$= n\bar{Y}_S + \sum_{i \notin S}(\hat{\beta}_0 + \hat{\beta}_1 x_i)$$

$$= n(\hat{\beta}_0 + \hat{\beta}_1 \bar{x}_S) + \sum_{i \notin S}(\hat{\beta}_0 + \hat{\beta}_1 x_i)$$

$$= \sum_{i=1}^{N}(\hat{\beta}_0 + \hat{\beta}_1 x_i)$$

$$= N(\hat{\beta}_0 + \hat{\beta}_1 \bar{x}_U).$$

The regression estimator of T_y is thus N times the predicted value under the model at $\bar{x}_U$.

In practice, if the sample size is small relative to the population size and we have an SRS, we can simply ignore the fpc and use the standard error for estimating the mean value of a response. From regression theory (see one of the regression books listed in the references for Chapter 11), the variance of $(\hat{\beta}_0 + \hat{\beta}_1 \bar{x}_U)$ is

$$\sigma^2 \left[\frac{1}{n} + \frac{(\bar{x}_U - \bar{x})^2}{\displaystyle\sum_{i \in S}(x_i - \bar{x})^2} \right].$$

Thus, if n/N is small,

$$V_M[\hat{T}_y - T] \approx N^2 \sigma^2 \left[\frac{1}{n} + \frac{(\bar{x}_U - \bar{x}_S)^2}{\displaystyle\sum_{i \in S}(x_i - \bar{x}_S)^2} \right]. \tag{3.18}$$

EXAMPLE   3.10   In Example 3.6, the predicted value when $x = 11.3$ is the regression estimator for $\bar{y}_U$. The predicted value is easily obtained from SAS as 11.9893:

Obs	Dep Var FIELD	Predict Value	Std Err Predict	Lower95% Mean	Upper95% Mean	Residual
1	15.0000	11.1920	0.491	10.1769	12.2072	3.8080

2	14.0000	12.4186	0.531	11.3205	13.5167	1.5814
3	9.0000	9.3522	0.751	7.7992	10.9052	-0.3522
⋮	⋮	⋮	⋮	⋮	⋮	⋮
24	9.0000	11.1920	0.491	10.1769	12.2072	-2.1920
25	8.0000	11.1920	0.491	10.1769	12.2072	-3.1920
26	.	11.9893	0.494	10.9672	13.0114	.

Substituting estimates into (3.18),

$$
\text{SE}_M[\hat{\bar{Y}}_{\text{reg}}] = \sqrt{\hat{\sigma}^2 \left[\frac{1}{n} + \frac{(\bar{x}_U - \bar{x}_S)^2}{\sum_{i \in S} (x_i - \bar{x}_S)^2} \right]}
$$

$$
= \sqrt{5.79 \left[\frac{1}{25} + \frac{(11.3 - 10.6)^2}{226.006} \right]} = 0.494.
$$

The value 0.494 is easy to compute using standard software but does not incorporate the fpc. Exercise 21 examines the fpc in model-based regression. ∎

3.4.3 Differences Between Model-Based and Design-Based Estimates

Why aren't standard errors the same as in randomization theory? That is, how can we have two different variances for the same estimator? The discrepancy is due to the different definitions of *variance:* In design-based sampling, the variance is the average squared deviation of the estimate from its expected value, averaged over all samples that could be obtained using a given design. If we are using a model, the variance is again the average squared deviation of the estimate from its expected value, but here the average is over all possible samples that could be generated from the population model. Thompson (1997) discusses inference using regression estimators and provides references for further reading.

If you were absolutely certain that your model was correct, you could minimize the model-based variance of the regression estimator by including only the members of the population with the largest and smallest values of x to be in the sample and excluding units with values of x between those extremes. No one would recommend such a design in practice, of course, because one never has that much assurance in a model. However, nothing in the model says that you should take an SRS (or any other type of probability sample) or that the sample needs to be representative of the population—*as long as the model is correct.*

What if the model is wrong? The model-based estimates are only model-unbiased— that is, they are unbiased only within the structure of that particular model. If the model is wrong, the model-based estimators will be biased, but, from within the model, we will not necessarily be able to tell how big the bias is. Thus, if the model is wrong, the model-based estimate of the variance will underestimate the MSE. When using model-based inference in sampling, you need to be very careful to check the assumptions of the model by examining residuals and using other diagnostic tools. Be very careful with the assumption of independence, for that typically is the most difficult to check. You can (and should!) perform diagnostics to check some assumptions of the model for the sampled data; however, you are making a strong, untestable assumption that the model applies to population units you did not observe.

The randomization-based estimate of the MSE may be used whether or not any given model fits the data because randomization inference depends only on *how* the sample was selected. But even the most die-hard randomization theorist relies on models for nonresponse and for designing the survey. Hansen et al. (1983) point out that generally randomization theory samplers have a model in mind when designing the survey and take that model into account to improve efficiency.

We will return to this issue in Chapter 11.

3.5
Comparison

Both ratio and regression estimation provide a way of using an auxiliary variable that is highly correlated with the variable of interest. We "know" that y is correlated with x, and we know how far $\bar{x}$ is from $\bar{x}_U$, so we use this information to adjust $\bar{y}$ and (we hope) increase the precision of our estimate. The estimators in ratio and regression estimation come from models that we hope describe the data, but the randomization theory properties of the estimators do not depend on these models.

As will be seen in Chapter 11, the ratio and regression estimators discussed in this chapter are special cases of a generalized regression estimator. All three estimators of the population total discussed so far—$\hat{t}_y$, $\hat{t}_{yr}$, and $\hat{t}_{yreg}$—can be expressed in terms of regression coefficients. For an SRS of size n, the estimators are given in the following table:

	Estimator	e_i
SRS	$\hat{t}_y$	$y_i - \bar{y}$
Ratio	$\hat{t}_y \left(\dfrac{t_x}{\hat{t}_x} \right)$	$y_i - \hat{B} x_i$
Regression	$N[\bar{y} + \hat{B}_1(\bar{x}_U - \bar{x})]$	$y_i - \hat{B}_0 - \hat{B}_1 x_i$

For each, the estimated variance is

$$N^2 \left(1 - \frac{n}{N} \right) \frac{s_e^2}{n}$$

for the particular e_i in the table; s_e^2 is the sample variance of the e_i's.

Ratio or regression estimators give greater precision than $\hat{t}_y$ when $\sum e_i^2$ for the method is smaller than $\sum (y_i - \bar{y})^2$. Ratio estimation is especially useful in cluster sampling, as we will see in Chapters 5 and 6.

In this chapter, we discussed ratio and regression estimation using just one auxiliary variable x. In practice, you may have several auxiliary variables you want to use to improve the precision of your estimates. The principles for using multiple regression models will be the same; we will present the theory for general surveys in Section 11.6.

3.6
Exercises

1   For each of the following situations, indicate how you might use ratio or regression estimation.

   a   Estimate the proportion of time devoted to sports in television news broadcasts in your city.

   b   Estimate the average number of fish caught per hour by anglers visiting a lake in August.

   c   Estimate the average amount that undergraduate students spent on textbooks at your university in the fall semester.

   d   Estimate the total weight of usable meat (discarding bones, fat, and skin) in a shipment of chickens.

2   The data set agsrs.dat also contains information on the number of farms in 1987 for the sample of 300 counties. In 1987 the United States had a total of 2,087,759 farms.

   a   Plot the data.

   b   Use ratio estimation to estimate the total number of acres devoted to farming in 1992, using the number of farms in 1987 as the auxiliary variable.

   c   Repeat part (b), using regression estimation.

   d   Which method gives the most precision: ratio estimation with auxiliary variable *acres87*, ratio estimation with auxiliary variable *farms87*, or regression estimation with auxiliary variable *farms87*? Why?

3   Using the data set agsrs.dat, estimate the total number of acres devoted to farming in 1992 for each of two domains: (a) counties with fewer than 600 farms and (b) counties with 600 or more farms. Give standard errors for your estimates.

4   Foresters want to estimate the average age of trees in a stand. Determining age is cumbersome because one needs to count the tree rings on a core taken from the tree. In general, though, the older the tree, the larger the diameter, and diameter is easy to measure. The foresters measure the diameter of all 1132 trees and find that the population mean equals 10.3. They then randomly select 20 trees for age measurement.

Tree No.	Diameter, x	Age, y	Tree No.	Diameter, x	Age, y
1	12.0	125	11	5.7	61
2	11.4	119	12	8.0	80
3	7.9	83	13	10.3	114
4	9.0	85	14	12.0	147
5	10.5	99	15	9.2	122
6	7.9	117	16	8.5	106
7	7.3	69	17	7.0	82
8	10.2	133	18	10.7	88
9	11.7	154	19	9.3	97
10	11.3	168	20	8.2	99

a   Plot the data.

b   Estimate the population mean age of trees in the stand and give an approximate standard error for your estimate. Label your estimate on your graph. Why did you use the method of estimation that you chose?

5   The data set counties.dat contains information on land area, population, number of physicians, unemployment, and a number of other quantities for an SRS of 100 of the 3141 counties in the United States (U.S. Bureau of the Census 1994). The total land area for the United States is 3,536,278 square miles; the 1993 population was estimated to be 255,077,536.

a   Draw a histogram of the number of physicians for the 100 counties.

b   Estimate the total number of physicians in the United States, along with its standard error, using $N\bar{y}$.

c   Plot the number of physicians versus population for each county. Which method do you think is more appropriate for these data: ratio estimation or regression estimation? Why?

d   Using the method you chose in part (c), use the auxiliary variable population to estimate the total number of physicians in the United States, along with the standard error.

e   The "true" value for total number of physicians in the population is 532,638. Which method of estimation came closer?

6   Repeat Exercise 5, with y = farm population and x = land area.

7   Repeat Exercise 5, with y = number of veterans and x = population.

8   Use the data in golfsrs.dat for this problem. Using the 18-hole courses only, estimate the average greens fee to play 18 holes on a weekend. Give a standard error for your estimate.

9   For the 18-hole courses in golfsrs.dat, plot the weekend 18-hole greens fee versus the back-tee yardage. Estimate the regression parameters for predicting weekend greens fees from back-tee yardage. Is there a strong relationship between the two variables?

10   Use the data in golfsrs.dat for this problem.

a   Estimate the mean weekday greens fee to play 9 holes, for courses with a golf professional available.

b   Estimate the mean weekday greens fee to play 9 holes, for courses without a golf professional.

c   Perform a hypothesis test to compare the mean weekday greens fee for golf courses with a professional to golf courses without a professional.

*11   Refer to the situation in Exercise 5. Use a model-based analysis to estimate the total number of physicians in the United States. Which model did you choose, and why? What are the assumptions for the model? Do you think they are met? Be sure to examine the residual plots for evidence of the inadequacy of the model. How do your results differ from those you obtained in Exercise 5?

*12 (Requires probability.) Use covariances derived in Appendix B to show formula (3.6).

13 Some books use the formula

$$\hat{V}[\hat{B}] = \left(1 - \frac{n}{N}\right)\frac{1}{n\bar{x}_U^2}(s_y^2 - 2\hat{B}rs_xs_y + \hat{B}^2s_x^2),$$

where r is the sample correlation coefficient of x and y for the values in the sample, to estimate the variance of a ratio.

a Show that this formula is algebraically equivalent to (3.7).

b It often does not work as well as (3.7) in practice, however: If s_x and s_y are large, many computer packages will truncate some of the significant digits so that the subtraction will be inaccurate. For the data in Example 3.2, calculate the values of s_y^2, s_x^2, r, and $\hat{B}$. Use the preceding formula to calculate the estimated variance of $\hat{t}_{yr}$. Is it exactly the same as the value from (3.7)?

*14 Recall from Section 2.2 that MSE = variance + (Bias)^2. Using (3.4) and other approximations in Section 3.1, show that $(E[\hat{B} - B])^2$ is small compared to MSE$[\hat{B}]$, when n is large.

*15 Show that if we consider approximations to the MSE in (3.6) and (3.12) to be accurate, then the variance of $\hat{\bar{y}}_r$ from ratio estimation is at least as large as the variance of $\hat{\bar{y}}_{\text{reg}}$ from regression estimation. HINT: Look at $V(\hat{\bar{y}}_r) - V(\hat{\bar{y}}_{\text{reg}})$ using the formulas in (3.6) and (3.12) and show that the difference is nonnegative.

*16 Prove Equations (3.4) and (3.11).

*17 Prove (3.13).

*18 Let $d_i = y_i - [\bar{y}_U + B_1(x_i - \bar{x}_U)]$. Show that for regression estimation,

$$E[\hat{\bar{y}}_{\text{reg}} - \bar{y}_U] \approx -\frac{1 - \dfrac{n}{N}}{nS_x^2} \sum_{i=1}^{N} \frac{d_i(x_i - \bar{x}_U)^2}{N - 1}.$$

As in Exercise 14, show that $(E[\hat{\bar{y}}_{\text{reg}} - \bar{y}_U])^2$ is small compared to MSE$[\hat{\bar{y}}_{\text{reg}}]$, when n is large.

*19 (Requires knowledge of linear models.) Suppose we have a stochastic model

$$Y_i = \beta x_i + \varepsilon_i,$$

where the ε_i's are independent with mean 0 and variance $\sigma^2 x_i$, and all $x_i > 0$. Show that the weighted least squares estimator of β is $\bar{Y}/\bar{x}$ and thus that $\hat{\beta}$ can be calculated by using weighted least squares. Is the standard error for $\hat{\beta}$ that comes from weighted least squares the same as that in (3.7)?

*20 (Requires knowledge of linear models.) Suppose the model in (3.16) misspecifies the variance structure and that a better model has $V[\varepsilon_i] = \sigma^2$.

a What is the weighted least squares estimator of β if $V[\varepsilon_i] = \sigma^2$? What is the corresponding estimator of the population total for y?

b Derive $V[\hat{T}_y - T_y]$.

c Apply your estimators to the data in agsrs.dat. How do these estimates compare with those in Examples 3.2 and 3.9?

*21 Equation (3.18) gives the model-based variance for a population total when it is assumed that the sample size is small relative to the population size. Derive the variance incorporating the finite population correction.

22 The quantity B used in ratio estimation is sometimes called the *ratio-of-means estimator.* In some situations, one might prefer to use a *mean-of-ratios estimator:* Let $b_i = y_i/x_i$ for unit i; then the mean-of-ratios estimator is

$$\bar{b} = \frac{1}{n} \sum_{i \in S} b_i$$

with standard error

$$SE[\bar{b}] = \sqrt{\left(1 - \frac{n}{N}\right) \frac{s_b^2}{n}}$$

from SRS theory.

a Do you think the mean-of-ratios estimator is appropriate for the data in Example 3.5? Why, or why not?

*b (Requires knowledge of linear models.) Show that $\bar{b}$ is the weighted least squares estimate of β under the model

$$Y_i = \beta x_i + \varepsilon_i$$

when ε_i has mean 0 and variance $\sigma^2 x_i^2$.

*23 (Requires computing.)

a Generate 500 data sets, each with 30 pairs of observations (x_i, y_i). Use a bivariate normal distribution with mean 0, standard deviation 1, and correlation 0.5 to generate each pair (x_i, y_i). For each data set, calculate $\bar{y}$ and $\hat{\bar{y}}_{reg}$, using $\bar{x}_U = 0$. Graph a histogram of the 500 values of $\bar{y}$ and another histogram of the 500 values of $\hat{\bar{y}}_{reg}$. What do you see?

b Repeat part (a) for 500 data sets, each with 60 pairs of observations.

24 Find a dictionary of a language you have studied. Choose 30 pages at random from the dictionary. For each, record

$$x = \text{number of words on the page}$$
$$y = \text{number of words that you know on the page (be honest!).}$$

How many words do you estimate are in the dictionary? How many do you estimate that you know? What percentage of the words do you know? Give standard errors for all your estimates.

SURVEY Exercises

25 Using the same sample of size 200, repeat Exercise 28 in Chapter 2, using a ratio estimate with assessed value of the house as the auxiliary variable. Which estimate of the mean gives greater precision? How are your results related to the SURVEY program assumptions? Be sure to include an appropriate plot of the data.

26 Using your sample of size 200, estimate the average number of adults per household in Lockhart City households willing to pay at least $10 for cable service. Give the standard error and the estimated coefficient of variation of your estimate.

27 Using your sample of size 200, estimate the total number of adults in Lockhart City who live in households willing to pay at least $10 for cable service. Give the standard error and the estimated coefficient of variation of your estimate.

4

Stratified Sampling

One of the things she [Mama] taught me should be obvious to everyone, but I still find a lot of cooks who haven't figured it out yet. Put the food on first that takes the longest to cook.

—Pearl Bailey, *Pearl's Kitchen*

4.1
What Is Stratified Sampling?

Often, we have supplementary information that can help us design our sample. For example, we would know before undertaking an income survey that men generally earn more than women, that New York City residents pay more for housing than residents of Des Moines, or that rural residents shop for groceries less frequently than urban residents.

If the variable we are interested in takes on different mean values in different subpopulations, we may be able to obtain more precise estimates of population quantities by taking a **stratified random sample.** The word *stratify* comes from Latin words meaning "to make layers"; we divide the population into H subpopulations, called **strata.** The strata do not overlap, and they constitute the whole population so that each sampling unit belongs to exactly one stratum. We draw an independent probability sample from each stratum, then pool the information to obtain overall population estimates.

We use stratified sampling for one or more of the following reasons:

1 We want to be protected from the possibility of obtaining a really bad sample. When taking a simple random sample (SRS) of size 100 from a population of 1000 male and 1000 female students, obtaining a sample with no or very few males is theoretically possible, although such a sample is not likely to occur. Most people would not consider such a sample to be representative of the population and would worry that men and women might respond differently on the item of interest. In a stratified sample, one could take an SRS of 50 males and an independent SRS of 50 females, guaranteeing that the proportion of males in the sample is the same as that in the population. With this design, a sample with no or few males cannot be selected.

2   We may want data of known precision for subgroups. These subgroups should be the strata, which then coincide with the domains of study. McIlwee and Robinson (1992) sampled graduates from electrical and mechanical engineering programs at public universities in southern California. They were interested in comparing the educational and workforce experiences of male and female graduates, so they stratified their sampling frame by gender and took separate random samples of male graduates and female graduates. Because there were many more male than female graduates, they sampled a higher fraction of female graduates than male graduates in order to obtain comparable precisions for the two groups.

3   A stratified sample may be more convenient to administer and may result in a lower cost for the survey. For example, different sampling approaches may be used for different strata. In a survey of businesses, a mail survey might be used for large firms, whereas a personal or telephone interview is used for small firms. In other surveys, different sampling methods may be needed in urban and rural strata.

4   Stratified sampling, if done correctly, will give more precise (having lower variance) estimates for the whole population. Persons of different ages tend to have different blood pressures, so in a blood pressure study it would be helpful to stratify by age groups. If studying the concentration of plants in an area, one would stratify by type of terrain; marshes would have different plants than woodlands. Stratification works for lowering the variance because the variance within each stratum is often lower than the variance in the whole population. Prior knowledge can be used to save money in the sampling procedure.

EXAMPLE **4.1**   Refer to Example 2.4, in which we took an SRS to estimate the average number of farm acres per county. In Example 2.4, we noted that, even though we scrupulously generated a random sample, some areas were overrepresented and others not represented at all. Taking a stratified sample can provide some balance in the sample on the stratifying variable.

The SRS in Example 2.4 exhibited a wide range of values for y_i, the number of acres devoted to farms in county i in 1992. You might conjecture that part of the large variability arises because counties in the western United States are larger, and thus tend to have larger values of y, than counties in the eastern United States.

For this example, we use the four census regions of the United States—Northeast, North Central, South, and West—as strata. The SRS in Example 2.4 sampled about 10% of the population; to compare the results of the stratified sample with the SRS, we also sample about 10% of the counties in each stratum. (We discuss other stratified sampling designs later in the chapter.)

Stratum	Number of Counties in Stratum	Number of Counties in Sample
Northeast	220	21
North Central	1054	103
South	1382	135
West	422	41
Total	3078	300

FIGURE 4.1

The boxplot of data from Example 4.1. The thick line for each region is the median of the sample data from that region; the other horizontal lines in the boxes are the 25th and 75th percentiles. The Northeast region has a relatively low median and small variance; the West region, however, has a much higher median and variance. The distribution of farm acreage appears to be positively skewed in each of the regions.

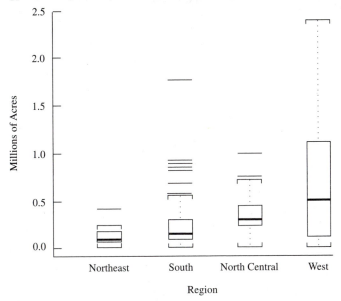

We select four separate SRSs, one from each of the four strata. To select the SRS from the Northeast stratum, we number the counties in that stratum from 1 to 220 and select 21 numbers randomly from {1, ..., 220}. We follow a similar procedure for the other three strata, selecting 103 counties at random from the 1054 in the North Central region, 135 counties from the 1382 in the South, and 41 counties from the 422 in the West. The four SRSs are independent: Knowing which counties are in the sample from the Northeast tells us nothing about which counties are in the sample from the South.

The data sampled from all four strata are in data file agstrat.dat. A boxplot, showing the data for each stratum, is in Figure 4.1. Summary statistics for each stratum are given below:

Region	Sample Size	Average	Variance
Northeast	21	97,629.8	7,647,472,708
North Central	103	300,504.2	29,618,183,543
South	135	211,315.0	53,587,487,856
West	41	662,295.5	396,185,950,266

Since we took an SRS in each stratum, we can use Equations (2.12) and (2.14) to estimate the population quantities for each stratum. We use

$$(220)(97,629.81) = 21,478,558.2$$

to estimate the total number of acres devoted to farms in the Northeast, with estimated variance

$$(220)^2 \left(1 - \frac{21}{220}\right) \frac{7,647,472,708}{21} = 1.594316 \times 10^{13}.$$

The following table gives estimates of the total number of farm acres and estimated variance of the total for each of the four strata:

Stratum	Estimated Total of Farm Acres	Estimated Variance of Total
Northeast	21,478,558	1.59432×10^{13}
North Central	316,731,379	2.88232×10^{14}
South	292,037,391	6.84076×10^{14}
West	279,488,706	1.55365×10^{15}
Total	909,736,034	2.5419×10^{15}

We can estimate the total number of acres devoted to farming in the United States by adding the totals for each stratum; as sampling was done independently in each stratum, the variance of the U.S. total is the sum of the variances of the population stratum totals. Thus, we estimate the total number of acres devoted to farming as 909,736,034, with standard error $\sqrt{2.5419 \times 10^{15}} = 50,417,248$. We would estimate the average number of acres devoted to farming per county as $909,736,034/3078 = 295,560.7649$, with standard error $50,417,248/3078 = 16,379.87$.

For comparison, the estimate of the total in Example 2.4, using an SRS of size 300, was 916,927,110, with standard error 58,169,381. For this example, stratified sampling ensures that each region of the United States is represented in the sample and produces an estimate with a slightly smaller standard error than an SRS with the same number of observations. The sample variance in Example 2.4 was $s^2 = 1.1872 \times 10^{11}$. Only the West had sample variance larger than s^2; the sample variance in the Northeast was only 7.647×10^9.

Observations within many strata tend to be more homogeneous than observations in the population as a whole, and the reduction in variance in the individual strata often leads to a reduced variance for the population estimate. In this example, the relative gain from stratification can be estimated by the ratio

$$\frac{\text{estimated variance from stratification, with } n = 300}{\text{estimated variance from SRS, with } n = 300} = \frac{2.5419 \times 10^{15}}{3.3837 \times 10^{15}} = 0.75.$$

If these figures were the population variances, we would expect that we would need only $(300)(0.75) = 225$ observations with a stratified sample to obtain the same precision as from an SRS of 300 observations.

Of course, no law says that you must sample the same fraction of observations in every stratum. In this example, there is far more variability from county to county in the western region; if acres devoted to farming were the primary variable of interest, you would reduce the variance of the estimated total even further by taking a higher sampling fraction in the western region than in the other regions. You will explore an alternative sampling design in Exercise 12.   ∎

4.2
Theory of Stratified Sampling

We divide the population of N sampling units into H "layers," or strata, with N_h sampling units in the hth stratum. For stratified sampling to work, we must know the values of $N_1, N_2, \ldots, N_H$ and must have

$$N_1 + N_2 + \cdots + N_H = N,$$

where N is the total number of units in the entire population.

In **stratified random sampling,** the simplest form of stratified sampling, we independently take an SRS from each stratum so that n_h observations are randomly selected from the population units in stratum h. Define $\mathcal{S}_h$ to be the set of n_h units in the SRS for stratum h.

Notation for Stratification   The population quantities are:

$y_{hj} = $ value of jth unit in stratum h

$$t_h = \sum_{j=1}^{N_h} y_{hj} = \text{population total in stratum } h$$

$$t = \sum_{h=1}^{H} t_h = \text{population total}$$

$$\bar{y}_{hU} = \frac{\displaystyle\sum_{j=1}^{N_h} y_{hj}}{N_h} = \text{population mean in stratum } h$$

$$\bar{y}_{U} = \frac{t}{N} = \frac{\displaystyle\sum_{h=1}^{H}\sum_{j=1}^{N_h} y_{hj}}{N} = \text{overall population mean}$$

$$S_h^2 = \sum_{j=1}^{N_h} \frac{(y_{hj} - \bar{y}_{hU})^2}{N_h - 1} = \text{population variance in stratum } h$$

Corresponding quantities for the sample, using SRS estimates within each stratum, are:

$$\bar{y}_h = \frac{\displaystyle\sum_{j\in\mathcal{S}_h} y_{hj}}{n_h}$$

$$\hat{t}_h = \frac{N_h}{n_h} \sum_{j\in\mathcal{S}_h} y_{hj} = N_h \bar{y}_h$$

$$s_h^2 = \sum_{j\in\mathcal{S}_h} \frac{(y_{hj} - \bar{y}_h)^2}{n_h - 1}$$

Suppose we only sampled the hth stratum. In effect, we have a population of N_h units and take an SRS of n_h units. Then we would estimate $\bar{y}_{hU}$ by $\bar{y}_h$, and t_h by $\hat{t}_h = N_h \bar{y}_h$. The population total is $t = \sum_{h=1}^{H} t_h$, so we estimate t by

$$\hat{t}_{\text{str}} = \sum_{h=1}^{H} \hat{t}_h = \sum_{h=1}^{H} N_h \bar{y}_h. \qquad (4.1)$$

To estimate $\bar{y}_U$, then, use

$$\bar{y}_{\text{str}} = \frac{\hat{t}_{\text{str}}}{N} = \sum_{h=1}^{H} \frac{N_h}{N} \bar{y}_h. \qquad (4.2)$$

This is a weighted average of the sample stratum averages; the weights are the relative sizes of the strata. To use stratified sampling, the sizes or relative sizes of the strata must be known.

The properties of these estimators follow directly from the properties of SRS estimators:

- **Unbiasedness.** $\bar{y}_{\text{str}}$ and $\hat{t}_{\text{str}}$ are unbiased estimators of $\bar{y}_U$ and t. This is true because

$$E\left[\sum_{h=1}^{H} \frac{N_h}{N} \bar{y}_h \right] = \sum_{h=1}^{H} \frac{N_h}{N} E[\bar{y}_h] = \sum_{h=1}^{H} \frac{N_h}{N} \bar{y}_{hU} = \bar{y}_U.$$

- **Variance of the estimators.** Since we are sampling independently from the strata and we know $V(\hat{t}_h)$ from SRS theory, the properties of expected value (p. 427) and Equation (2.13) imply that

$$V(\hat{t}_{\text{str}}) = \sum_{h=1}^{H} V(\hat{t}_h) = \sum_{h=1}^{H} \left(1 - \frac{n_h}{N_h} \right) N_h^2 \frac{S_h^2}{n_h}. \qquad (4.3)$$

- **Variance estimates for stratified samples.** We can obtain an unbiased estimator of $V(\hat{t}_{\text{str}})$ by substituting the sample estimates s_h^2 for the population quantities S_h^2. Note that, to estimate the variances, we need to sample at least two units from each stratum:

$$\hat{V}(\hat{t}_{\text{str}}) = \sum_{h=1}^{H} \left(1 - \frac{n_h}{N_h} \right) N_h^2 \frac{s_h^2}{n_h}, \qquad (4.4)$$

$$\hat{V}(\bar{y}_{\text{str}}) = \frac{1}{N^2} \hat{V}(\hat{t}_{\text{str}}) = \sum_{h=1}^{H} \left(1 - \frac{n_h}{N_h} \right) \left(\frac{N_h}{N} \right)^2 \frac{s_h^2}{n_h}. \qquad (4.5)$$

As always, the standard error of an estimator is the square root of the estimated variance: $\text{SE}(\bar{y}_{\text{str}}) = \sqrt{\hat{V}(\bar{y}_{\text{str}})}$.

- **Confidence intervals for stratified samples.** If either (1) the sample sizes within each stratum are large or (2) the sampling design has a large number of strata, an approximate $100(1 - \alpha)\%$ confidence interval (CI) for the mean is

$$\bar{y}_{\text{str}} \pm z_{\alpha/2} \, \text{SE}(\bar{y}_{\text{str}}).$$

The central limit theorem used for constructing this confidence interval is stated in Krewski and Rao (1981). Some survey researchers use the percentile of a

t distribution with $n - H$ degrees of freedom (df) rather than the percentile of the normal distribution.

EXAMPLE **4.2**   Siniff and Skoog (1964) used stratified random sampling to estimate the size of the Nelchina herd of Alaskan caribou in February 1962. In January and early February, several sampling techniques were field-tested. The field tests told the investigators that several of the proposed sampling units, such as equal-flying-time sampling units, were difficult to implement in practice and that an equal-area sampling unit of 4 square miles (mi^2) would work well for the survey. The biologists used preliminary estimates of caribou densities to divide the area of interest into six strata; each stratum was then divided into a grid of 4-mi^2 sampling units. Stratum A, for example, contained $N_1 = 400$ sampling units; $n_1 = 98$ of these were randomly selected to be in the survey. The following data were reported:

Stratum	N_h	n_h	$\bar{y}_h$	s_h^2
A	400	98	24.1	5,575
B	30	10	25.6	4,064
C	61	37	267.6	347,556
D	18	6	179.0	22,798
E	70	39	293.7	123,578
F	120	21	33.2	9,795

With the data in this form, using a spreadsheet to do the calculations necessary for stratified sampling is easy. The spreadsheet shown in Table 4.1 simplifies the calculations that the estimated total number of caribou is 54,497 with standard error 5840. An approximate 95% CI for the total number of caribou is

$$54{,}497 \pm 1.96(5840) = [43{,}051, 65{,}943].$$

TABLE   **4.1**

Spreadsheet for Calculations in Example 4.2

	A	B	C	D	E	F	G
1	Stratum	N_h	n_h	$\bar{y}_h$	s_h^2	$\hat{t}_h = N_h \bar{y}_h$	$\left(1 - \dfrac{n_h}{N_h}\right) N_h^2 \dfrac{s_h^2}{n_h}$
2	A	400	98	24.1	5,575	9,640	6,872,040.82
3	B	30	10	25.6	4,064	768	243,840.00
4	C	61	37	267.6	347,556	16,324	13,751,945.51
5	D	18	6	179.0	22,798	3,222	820,728.00
6	E	70	39	293.7	123,578	20,559	6,876,006.67
7	F	120	21	33.2	9,795	3,984	5,541,171.43
8	total		211			54,497	34,105,732.43
9	sqrt(total)						5,840.01

Of course, this confidence interval only reflects the uncertainty due to sampling error; if the field procedure for counting caribou tends to miss animals, then the entire confidence interval will be too low. ■

Stratified Sampling for Proportions   As we observed in Section 2.3, a proportion is a mean of a variable that takes on values 0 and 1. To make inferences about proportions, we just use Equations (4.1)–(4.5), with $\bar{y}_h = \hat{p}_h$ and $s_h^2 = [n_h/(n_h - 1)]\hat{p}_h(1 - \hat{p}_h)$. Then,

$$\hat{p}_{\text{str}} = \sum_{h=1}^{H} \frac{N_h}{N} \hat{p}_h \tag{4.6}$$

and

$$\hat{V}(\hat{p}_{\text{str}}) = \sum_{h=1}^{H} \left(1 - \frac{n_h}{N_h}\right)\left(\frac{N_h}{N}\right)^2 \frac{\hat{p}_h(1 - \hat{p}_h)}{n_h - 1}. \tag{4.7}$$

Estimating the total number of population units having a specified characteristic is similar:

$$\hat{t}_{\text{str}} = \sum_{h=1}^{H} N_h \hat{p}_h.$$

Thus, the estimated total number of population units with the characteristic is the sum of the estimated totals in each stratum. Similarly, $\hat{V}(\hat{t}_{\text{str}}) = N^2 \hat{V}(\hat{p}_{\text{str}})$.

EXAMPLE **4.3**   The American Council of Learned Societies (ACLS) used a stratified random sample of selected ACLS societies in seven disciplines to study publication patterns and computer and library use among scholars who belong to one of the member organizations of the ACLS (Morton and Price 1989). The data are shown in Table 4.2.

Ignoring the nonresponse for now (we'll return to the nonresponse in Exercise 9 in Chapter 8) and supposing there are no duplicate memberships, let's use the stratified sample to estimate the percentage and number of respondents of the major societies in those seven disciplines who are women. Here, let N_h be the membership figures

TABLE **4.2**
Data from ACLS Survey

Discipline	Membership	Number Mailed	Valid Returns	Female Members (%)
Literature	9,100	915	636	38
Classics	1,950	633	451	27
Philosophy	5,500	658	481	18
History	10,850	855	611	19
Linguistics	2,100	667	493	36
Political science	5,500	833	575	13
Sociology	9,000	824	588	26
Totals	44,000	5,385	3,835	

and let n_h be the number of valid surveys. Thus,

$$\hat{p}_{\text{str}} = \sum_{h=1}^{7} \frac{N_h}{N} \hat{p}_h = \frac{9100}{44,000} 0.38 + \cdots + \frac{9000}{44,000} 0.26 = 0.2465$$

and

$$\text{SE}(\hat{p}_{\text{str}}) = \sqrt{\sum_{h=1}^{7} \left(1 - \frac{n_h}{N_h}\right) \left(\frac{N_h}{N}\right)^2 \frac{\hat{p}_h(1 - \hat{p}_h)}{n_h - 1}} = 0.0071.$$

The estimated total number of female members in the societies is $\hat{t}_{\text{str}} = 44,000 \times (0.2465) = 10,847$, with $\text{SE}(\hat{t}_{\text{str}}) = 44,000(.0071) = 312$. ∎

4.3
Sampling Weights

The stratified sampling estimator $\hat{t}_{\text{str}}$ can be expressed as a weighted sum of the individual sampling units. Using (4.1),

$$\hat{t}_{\text{str}} = \sum_{h=1}^{H} \sum_{j \in \mathcal{S}_h} \frac{N_h}{n_h} y_{hj}.$$

The **sampling weight** $w_{hj} = (N_h/n_h)$ can be thought of as the number of units in the population represented by the sample member (h, j). If the population has 1600 men and 400 women and the stratified sample design specifies sampling 200 men and 200 women, then each man in the sample has weight 8 and each woman has weight 2. Each woman in the sample represents herself and 1 other woman not selected to be in the sample, and each man represents himself and 7 other men not in the sample. Note that the probability of selecting the jth unit in the hth stratum to be in the sample is $\pi_{hj} = n_h/N_h$, the sampling fraction in the hth stratum. Thus, the sampling weight is simply the reciprocal of the probability of selection:

$$w_{hj} = \frac{1}{\pi_{hj}}. \tag{4.8}$$

The sum of the sampling weights equals the population size N; each sampled unit "represents" a certain number of units in the population, so the whole sample "represents" the whole population. This identity provides a check on whether you have constructed your weight variable correctly: If the sum of the weights for your sample is something other than N, then you have made a mistake somewhere.

The stratified estimate of the population total may thus be written as

$$\hat{t}_{\text{str}} = \sum_{h=1}^{H} \sum_{j \in \mathcal{S}_h} w_{hj} y_{hj}, \tag{4.9}$$

and the estimate of the population mean as

$$\bar{y}_{\text{str}} = \frac{\displaystyle\sum_{h=1}^{H} \sum_{j \in \mathcal{S}_h} w_{hj} y_{hj}}{\displaystyle\sum_{h=1}^{H} \sum_{j \in \mathcal{S}_h} w_{hj}}. \tag{4.10}$$

EXAMPLE 4.4   For the caribou survey in Example 4.2, the weights are

Stratum	N_h	n_h	w_{hj}
A	400	98	4.08
B	30	10	3.00
C	61	37	1.65
D	18	6	3.00
E	70	39	1.79
F	120	21	5.71

In stratum A, each sampling unit of 4 mi^2 represents 4.08 sampling units in the stratum (including itself); in stratum B, a sampling unit in the sample represents itself and 2 other sampling units that are not in the sample. To estimate the population total, then, a new variable of weights could be constructed. This variable would contain the value 4.08 for every observation in stratum A, 3.00 for every observation in stratum B, and so on. ■

EXAMPLE 4.5   The sample in Example 4.1 was designed so that each county in the United States would have approximately the same probability of appearing in the sample. To estimate the total number of acres devoted to agriculture in the United States, we can create a column in the data set (column 17 in the file agstrat.dat) consisting of the sampling weights. The weight column contains the value 220/21 for counties in the Northeast stratum, 1054/103 for the North Central counties, 1382/135 for the South counties, and 422/41 for the West counties. We can use (4.9) to estimate the population total by forming a new column containing the product of the variables *weight* and *acres92,* then calculating the sum of the new column. In doing so, we calculate $\hat{t}_{str} = 909,736,035$, the same estimate (except for roundoff error) as obtained in Example 4.1.

The variable *weight* in column 17 can be used to estimate the population total for every variable measured in the sample. Note, however, that you cannot calculate the standard error of $\hat{t}_{str}$ unless you know the stratification. Equation (4.4) requires that you calculate the variance separately within each stratum; the weights do not tell you the stratum membership of the observations. ■

4.4
Allocating Observations to Strata

So far we have simply analyzed data from a survey that someone else has designed. Designing the survey is the most important part of using a survey in research: If the survey is badly designed, then no amount of analysis will yield the needed information. In this section, different methods of allocating observations to strata are discussed.

4.4.1 Proportional Allocation

If you are taking a stratified sample to ensure that the sample reflects the population with respect to the stratification variable and you would like your sample to be a miniature version of the population, you should use proportional allocation when designing the sample.

In **proportional allocation,** so called because the number of sampled units in each stratum is proportional to the size of the stratum, the probability of selection $\pi_{hj} = n_h/N_h$ is the same ($= n/N$) for all strata; in a population of 2400 men and 1600 women, proportional allocation with a 10% sample would mean sampling 240 men and 160 women. Thus, the probability that an individual will be selected to be in the sample, n/N, is the same as in an SRS, but many of the "bad" samples that could occur in an SRS (for example, a sample in which all 400 persons are men) cannot be selected in a stratified sample with proportional allocation.

If proportional allocation is used, each unit in the sample represents the same number of units in the population: In our example, each man in the sample represents 10 men in the population, and each woman represents 10 women in the population. The sampling weight for every unit in the sample thus equals 10, and the stratified sampling estimate of the population mean is simply the average of all observations. When every unit in the sample has the same weight and represents the same number of units in the population, the sample is called **self-weighting.** The sample in Example 4.1 was designed to be self-weighting. In a self-weighting sample, $\bar{y}_{str}$ is the average of all observations in the sample.

When the strata are large enough, the population variance of $\bar{y}_{str}$ under proportional allocation is usually at most as large as the population variance of $\bar{y}$, using the same number of observations but collected in a random sample. This is true no matter how silly the stratification scheme may be. To see why this might be so, let's display the between-strata and within-strata variances, for proportional allocation, in an ANOVA table for the population (Table 4.3).

In a stratified sample of size n with proportional allocation, since $n_h/N_h = n/N$, Equation (4.3) implies that

$$V_{prop}(\hat{t}_{str}) = \sum_{h=1}^{H}\left(1 - \frac{n_h}{N_h}\right)N_h^2\frac{S_h^2}{n_h}$$

$$= \left(1 - \frac{n}{N}\right)\frac{N}{n}\sum_{h=1}^{H}N_h S_h^2$$

$$= \left(1 - \frac{n}{N}\right)\frac{N}{n}\left(\text{SSW} + \sum_{h=1}^{H}S_h^2\right).$$

TABLE 4.3

Population ANOVA Table

Source	df	Sum of Squares
Between strata	$H - 1$	$\text{SSB} = \sum\limits_{h=1}^{H}\sum\limits_{j=1}^{N_h}(\bar{y}_{hU} - \bar{y}_U)^2 = \sum\limits_{h=1}^{H}N_h(\bar{y}_{hU} - \bar{y}_U)^2$
Within strata	$N - H$	$\text{SSW} = \sum\limits_{h=1}^{H}\sum\limits_{j=1}^{N_h}(y_{hj} - \bar{y}_{hU})^2 = \sum\limits_{h=1}^{H}(N_h - 1)S_h^2$
Total, about $\bar{y}_U$	$N - 1$	$\text{SSTO} = \sum\limits_{h=1}^{H}\sum\limits_{j=1}^{N_h}(y_{hj} - \bar{y}_U)^2 = (N - 1)S^2$

The sums of squares add up, with $\text{SSTO} = \text{SSB} + \text{SSW}$, so

$$
\begin{aligned}
V_{\text{SRS}}(\hat{t}) &= \left(1 - \frac{n}{N}\right) N^2 \frac{S^2}{n} \\
&= \left(1 - \frac{n}{N}\right) \frac{N^2}{n} \frac{\text{SSTO}}{N-1} \\
&= \left(1 - \frac{n}{N}\right) \frac{N^2}{n(N-1)} (\text{SSW} + \text{SSB}) \\
&= V_{\text{prop}}(\hat{t}_{\text{str}}) + \left(1 - \frac{n}{N}\right) \frac{N}{n(N-1)} \left[N(\text{SSB}) - \sum_{h=1}^{H} (N - N_h) S_h^2 \right].
\end{aligned}
$$

This result shows us that proportional allocation with stratification always gives an equal or smaller variance than SRS *unless*

$$
\text{SSB} < \sum_{h=1}^{H} \left(1 - \frac{N_h}{N}\right) S_h^2. \tag{4.11}
$$

This rarely happens when the N_h's are large; generally, the large population sizes of the strata will force $N_h(\bar{y}_{hU} - \bar{y}_U)^2 > S_h^2$. In general, the variance of the estimator of t from proportional allocation will be smaller than the variance of the estimator of t from simple random sampling. The more unequal the stratum means $\bar{y}_{hU}$, the more precision you will gain by using proportional allocation. Of course, this result only holds for population variances; it is possible for a variance estimate from proportional allocation to be larger than that from an SRS merely because the sample selected resulted in a large sample variance.

4.4.2 Optimal Allocation

If the variances S_h^2 are more or less equal across all the strata, proportional allocation is probably the best allocation for increasing precision. In cases where the S_h^2's vary greatly, **optimal allocation** can result in smaller costs. In practice, when we are sampling units of different sizes, the larger units are likely to be more variable than the smaller units, and we would sample them at a higher rate. For example, if we were to take a sample of American corporations and our goal was to estimate the amount of trade with Europe, the variation among the larger corporations would be greater than the variation among smaller ones. As a result, we would sample a higher percentage of the larger corporations. Optimal allocation works well for sampling units such as corporations, cities, and hospitals, which vary greatly in size. It is also effective when some strata are much more expensive to sample than others.

Neter (1978) tells of a study done by the Chesapeake and Ohio (C&O) Railroad Company to determine how much revenue they should get from interline freight shipments, since the total freight from a shipment that traveled along several railroads was divided among the different railroads. The C&O took a stratified sample of waybills, the documents that detailed the goods, route, and charges for the shipments. The waybills were stratified by the total freight charges, and all waybills with charges of over $40 were sampled, whereas only 1% of the waybills with charges less than $5 were sampled. The justification was that there was little variability among the amounts due the C&O in the stratum of the smallest total freight charges, whereas the variability in the stratum with charges of over $40 was much higher.

EXAMPLE 4.6   How are musicians paid when their compositions are performed? In the United States, many composers are affiliated with the American Society of Composers, Authors, and Publishers (ASCAP). Television networks, local television and radio stations, services such as Muzak, symphony orchestras, restaurants, nightclubs, and other operations pay ASCAP an annual license fee, based largely on the size of the audience, that allows them to play compositions in the ASCAP catalog. ASCAP then distributes royalties to composers whose works are played.

Theoretically, an ASCAP member should get royalties every time one of his or her compositions is played. Taking a census of every piece of music played in the United States, however, would be impractical; to estimate the amount of royalties due to members, ASCAP uses sampling. According to Dobishinski (1991) and "The ASCAP Advantage" (1992), ASCAP relies on television producers' cue sheets, which provide details on the music used in a program, to identify and tabulate musical pieces played on network television and major cable channels. About 60,000 hours of tapes are made from radio broadcasts each year, and experts identify the musical compositions aired in these broadcasts.

Stratified sampling is used to sample radio stations for the survey. Radio stations are grouped into strata based on the license fee paid to ASCAP, the type of community the station is in, and the geographic region. As stations paying higher license fees contribute more money for royalties, they are more likely to be sampled; once in the sample, high-fee stations are taped more often than low-fee stations. ASCAP thus uses a form of optimal allocation in taping: Strata with the highest radio fees, and thus with the highest variability in royalty amounts, have larger sampling fractions than strata containing radio stations that pay small fees.   ■

The objective in sampling is to gain the most information for the least cost. A simple cost function is given below: Let C represent total cost, c_0 represent overhead costs such as maintaining an office, and c_h represent the cost of taking an observation in stratum h so that

$$C = c_0 + \sum_{h=1}^{H} c_h n_h. \tag{4.12}$$

We want to allocate observations to strata in order to minimize $V(\bar{y}_{\mathrm{str}})$ for a given total cost C or, equivalently, to minimize C for a fixed $V(\bar{y}_{\mathrm{str}})$. Suppose the costs $c_1, c_2, \ldots, c_H$ are fixed. To minimize the variance for a fixed cost, we can prove, using calculus, that the optimal allocation has n_h proportional to

$$\frac{N_h S_h}{\sqrt{c_h}} \tag{4.13}$$

for each h (see Exercise 22). Thus, the optimal sample size in stratum h is

$$n_h = \left(\frac{\dfrac{N_h S_h}{\sqrt{c_h}}}{\displaystyle\sum_{l=1}^{H} \dfrac{N_l S_l}{\sqrt{c_l}}} \right) n.$$

We then sample heavily within a stratum if

- The stratum accounts for a large part of the population.
- The variance within the stratum is large; we sample more heavily to compensate for the heterogeneity.
- Sampling in the stratum is inexpensive.

EXAMPLE **4.7**   **Dollar stratification** is often used in accounting. The recorded book amounts are used to stratify the population. If you are auditing the loan amounts for a financial institution, stratum 1 might consist of all loans of more than $1 million, stratum 2 might consist of loans between $500,000 and $999,999, and so on, down to the smallest stratum of loans less than $10,000. Optimal allocation is often an efficient strategy for such a stratification: S_h^2 will be much larger in the strata with the large loan amounts, so optimal allocation will prescribe a higher sampling fraction for those strata. If the goal of the audit is to estimate the dollar discrepancy between the audited amounts and the amounts in the institution's books, an error in the recorded amount of one of the $3 million loans is likely to contribute more to the audited difference than an error in the recorded amount of one of the $3000 loans. In a survey such as this, you may even want to use sample size N_1 in stratum 1 so that each population unit in stratum 1 has probability 1 of appearing in the sample.   ∎

If all variances and costs are equal, proportional allocation is the same as optimal allocation. If we know the variances within each stratum and they differ, optimal allocation gives a smaller variance for the estimate of $\bar{y}_U$ than proportional allocation. But optimal allocation is a more complicated scheme; often the simplicity and self-weighting property of proportional allocation are worth the extra variance. In addition, the optimal allocation will differ for each variable being measured, whereas the proportional allocation depends only on the number of population units in each stratum.

Neyman allocation is a special case of optimal allocation, used when the costs in the strata (but not the variances) are approximately equal. Under Neyman allocation, n_h is proportional to $N_h S_h$. If the variances S_h^2 are specified correctly, Neyman allocation will give an estimator with smaller variance than proportional allocation.

EXAMPLE **4.8**   The caribou survey in Example 4.2 used a form of optimal allocation to determine the n_h. Before taking the survey, the investigators obtained approximations of the caribou densities and distribution and then constructed strata to be relatively homogeneous in terms of population density. They set the total sample size as $n = 225$. They then used the estimated count in each stratum as a rough estimate of the standard deviation, with the result shown in Table 4.4. The first row contains the names of the spreadsheet columns, and the second row contains the formulas used to calculate the table. The investigators wanted the sampling fraction to be at least $1/3$ in smaller strata, so they used the optimal allocation sample sizes in column E as a guideline for determining the sample sizes they actually used, in column F.   ∎

4.4.3   Allocation for Specified Precision Within Strata

Sometimes you are less interested in the precision of the estimate of the population total or mean for the whole population than in comparing means or totals among different strata. In that case, you would determine the sample size needed for the individual strata, using the guidelines in Section 2.5.

TABLE   4.4

Quantities Used for Designing the Caribou Survey in Example 4.8

	A	B	C	D	E	F
1	Stratum	N_h	s_h	$N_h s_h$	n_h	Sample size
2				B*C	D*225/D9	
3	A	400	3,000	1,200,000	96.26	98
4	B	30	2,000	60,000	4.81	10
5	C	61	9,000	549,000	44.04	37
6	D	18	2,000	36,000	2.89	6
7	E	70	12,000	840,000	67.38	39
8	F	120	1,000	120,000	9.63	21
9	total	699		2,805,000	225	211

EXAMPLE   4.9   The U.S. Postal Service often conducts surveys asking postal customers about their perceptions of the quality of mail service. The population of residential postal service customers is stratified by geographic area, and it is desired that the precision be ±3 percentage points, at a 95% confidence level, within each area. If there were no nonresponse, such a requirement would lead to sampling at least 1067 households in each stratum, as calculated in Example 2.9. Such an allocation is neither proportional, because the number of residential households varies a great deal from stratum to stratum, nor optimal in the sense of providing the greatest efficiency for estimating percentages for the whole population. It does, however, provide the desired precision within each stratum.   ∎

4.4.4 Determining Sample Sizes

The different methods of allocating observations to strata give the relative sample sizes n_h/n. After strata are constructed (see Section 4.5) and observations allocated to strata, Equation (4.3) can be used to determine the sample size necessary to achieve a prespecified variance. Because

$$V(\hat{t}_{str}) \le \frac{1}{n} \sum_{h=1}^{H} \frac{n}{n_h} N_h^2 S_h^2 = \frac{v}{n},$$

an approximate 95% CI if the fpc's can be ignored and if the normal approximation is valid will be $\hat{t}_{str} \pm z_{\alpha/2}\sqrt{v/n}$. Set $n = z_{\alpha/2}^2 v/e^2$ to achieve a desired confidence interval half-width e.

This approach requires knowledge of the values of S_h^2. An alternative approach, which works for any survey design, will be discussed in Section 7.5.

4.5
Defining Strata

One might wonder, since stratified sampling almost always gives higher precision than simple random sampling, why anyone would ever take SRSs. The answer is that stratification adds complexity to the survey, and the added complexity may not be worth a small gain in precision. In addition, to carry out a stratified sample, we need more information: For each stratum, we need to know how many and which members of the population belong to that stratum. In general, we want stratification to be very efficient, or the strata to be subgroups we are interested in, before we will be willing to incur the additional administrative costs and complexity associated with stratifying.

Remember, stratification is most efficient when the stratum means differ widely; then the between sum of squares is large, and the variability within strata will be smaller. Consequently, when constructing strata we want the strata means to be as different as possible. Ideally, we would stratify by the values of y; if our survey is to estimate total business expenditures on advertising, we would like to put businesses that spent the most on advertising in stratum 1, businesses with the next highest level of advertising expenditures in stratum 2, and so on, until the last stratum contained businesses that spent nothing on advertising. The problem with this scheme is that we do not know the advertising expenditures for all the businesses while designing the survey—if we did, we would not need to do a survey at all! Instead, we try to find some variable closely related to y. For estimating total business expenditures on advertising, we might stratify by number of employees or size of the business and by the type of product or service. For farm income, we might use the size of the farm as a stratifying variable, since we expect that larger farms would have higher incomes.

Most surveys measure more than one variable, so any stratification variable should be related to many characteristics of interest. The U.S. Census Bureau's Current Population Survey, which measures characteristics relating to employment, stratifies the primary sampling units by geographic region, population density, racial composition, principal industry, and similar variables. In the Canadian Survey of Employment, Payrolls, and Hours, business establishments are stratified by industry, province, and estimated number of employees. The Nielsen television ratings stratify by geographic region, county size, and cable penetration, among other variables. If several stratification variables are available, use the variables associated with the most important responses.

The number of strata you choose depends on many factors—for example, the difficulty in constructing a sampling frame with stratifying information and the cost of stratifying. A general rule to keep in mind is: The less information, the fewer strata you should use. Thus, you should use an SRS when little prior information about the target population is available.

You can often collect preliminary data that can be used to stratify your design. If you are taking a survey to estimate the number of fish in a region, you can use physical features of the area that are related to fish density, such as depth, salinity, and water temperature. Or you can use survey information from previous years or data from a preliminary cruise to aid in constructing strata. In this situation, according to Saville, "Usually there will be no point in designing a sampling scheme with more than 2 or 3 strata, because our knowledge of the distribution of fish will be rather imprecise. Strata may be of different size, and each stratum may be composed of several distinct areas in different parts of the total survey area" (1977, 10). In a survey with more precise prior information, we will want to use more strata—many surveys are stratified to the point that only two sampling units are observed in each stratum.

For many surveys, stratification can increase precision dramatically and often well repays the effort used in constructing the strata. Example 4.10 describes how strata were constructed in one large-scale survey, the National Pesticide Survey.

EXAMPLE **4.10**   Between 1988 and 1990, the U.S. Environmental Protection Agency (1990a, b) sampled drinking water wells to estimate the prevalence of pesticides and nitrate. When designing the National Pesticide Survey (NPS), the EPA scientists wanted a sample that was representative of drinking water wells in the United States. In particular, they wanted to guarantee that wells in the sample would have a wide range of levels of pesticide use and susceptibility to groundwater pollution. They also wanted to study two categories of wells: *community water systems* (CWSs), defined as "systems of piped drinking water with at least 15 connections and/or 25 or more permanent residents of the service area that have at least one working well used to obtain drinking water"; and *rural domestic wells,* "drinking water wells supplying occupied housing units located in rural areas of the United States, except for wells located on government reservations."

The following selections from the EPA describe how it chose the strata for the survey:

> In order to determine how many wells to visit for data collection, EPA first needed to identify approximately how many drinking water wells exist in the United States. This process was easier for community water systems than for rural domestic wells because a list of all public water systems, with their addresses, is contained in the Federal Reporting Data System (FRDS), which is maintained by EPA. From FRDS, EPA estimated that there were approximately 51,000 CWSs with wells in the United States. EPA did not have a comprehensive list of rural domestic wells to serve as the foundation for well selection, as it did for CWSs. Using data from the Census Bureau for 1980, EPA estimated that there were approximately 13 million rural domestic wells in the country, but the specific owners and addresses of these rural domestic wells were not known.
>
> EPA chose a survey design technique called "stratification" to ensure that survey data would meet its objectives. This technique was used to improve the precision of the estimates by selecting extra wells from areas with substantial agricultural activity and high susceptibility to ground-water pollution (vulnerability). EPA developed criteria for separating the population of CWS wells and rural domestic wells into four categories of pesticide use and three relative ground-water vulnerability measures. This design ensures that the range of variability that exists nationally with respect to the agricultural use of pesticides and ground-water vulnerability is reflected in the sample of wells.
>
> EPA identified five subgroups of wells for which it was interested in obtaining information. These subgroups were community water system wells in counties with relatively high average ground-water vulnerability; rural domestic wells in counties with relatively high average ground-water vulnerability; rural domestic wells in counties with high pesticide use; rural domestic wells in counties with both high pesticide use and relatively high average ground-water vulnerability; and rural domestic wells in "cropped and vulnerable" parts of counties (high pesticide use and relatively high ground-water vulnerability).
>
> Two of the most difficult design questions were determining how many wells to include in the Survey and determining the level of precision that would be sought for the NPS national estimates. These two questions were connected, because greater precision is usually obtained by collecting more data. Resolving these questions would have been simpler if the Survey designers had known in advance what proportion of

wells in the nation contained pesticides, but answering that question was one of the purposes of the Survey. Although many State studies have been conducted for specific pesticides, no reliable national estimates of well water contamination existed. EPA evaluated alternative precision requirements and costs for collecting data from different numbers of wells to determine the Survey size that would meet EPA's requirements and budget.

The Survey designers ultimately selected wells for data collection so that the Survey provided a 90 percent probability of detecting the presence of pesticides in the CWS wells sampled, assuming 0.5 percent of all community water system wells in the country contained pesticides. The rural domestic well Survey design was structured with different probabilities of detection for the several subgroups of interest, with the greatest emphasis placed on the cropped and vulnerable subcounty areas, where EPA was interested in obtaining very precise estimates of pesticide occurrence. EPA assumed that 1 percent of rural domestic wells in these areas would contain pesticides and designed the Survey to have about a 97 percent probability of detection in "cropped and vulnerable" areas if the assumption proved accurate. EPA concluded that sampling approximately 1,300 wells (564 public wells and 734 private wells) would meet the Survey's accuracy specifications and provide a representative national assessment of the number of wells containing pesticides.

Selecting Wells for the Survey. Because the exact number and location of rural domestic wells was unknown, EPA chose a survey design composed of several steps (stages) for those wells. The design began with a sampling of counties, and then characterized pesticide use and ground-water vulnerability for subcounty areas. This eventually allowed small enough geographic areas to be delineated to enable the sampling of individual rural domestic wells. This procedure was not needed for community water system wells, because their number and location were known.

The first step in well selection was common to both CWS wells and rural domestic wells. Each of the 3,137 counties or county equivalents in the U.S. was characterized according to pesticide use and ground-water vulnerability to ensure that the variability in agricultural pesticide use and ground-water vulnerability was reflected in the Survey. EPA used data on agricultural pesticide use obtained from a marketing research source and information on the proportion of the county area that was in agricultural production to rank agricultural pesticide use for each county as high, medium, low, or uncommon. Ground-water vulnerability of each county was estimated using a numerical classification system called Agricultural DRASTIC, which assesses seven factors: (depth of water, recharge, aquifer media, soil media, topography, impact of unsaturated zone, conductivity of the aquifer). The model was modified for the Survey to evaluate the vulnerability of aquifers to pesticide and nitrate contamination, and one of the subsidiary purposes of the Survey was to assess the effectiveness of the DRASTIC classification. Each area was evaluated and received a score of high, moderate, or low, based on information obtained from U.S. Geological Survey maps, U.S. Department of Agriculture soil survey maps and other resources from State agencies, associations, and universities. (1990a)

The procedure resulted in 12 strata for counties, as given in Table 4.5.

Stratification provides several advantages in this survey. It allows for more precise estimates of pesticide and nitrate concentrations in the United States as a whole, as

TABLE 4.5
Strata for National Pesticide Survey

Stratum	Pesticide Use	Groundwater Vulnerability (as Estimated by DRASTIC)	Number of Counties
1	High	High	106
2	High	Moderate	234
3	High	Low	129
4	Moderate	High	110
5	Moderate	Moderate	204
6	Moderate	Low	267
7	Low	High	193
8	Low	Moderate	375
9	Low	Low	404
10	Uncommon	High	186
11	Uncommon	Moderate	513
12	Uncommon	Low	416

SOURCE: Adapted from U.S. EPA 1990a, 3.

it is expected that the wells within a stratum are more homogeneous than the entire population of wells. Stratification ensures that wells for each level of pesticide use and groundwater vulnerability are included in the sample and allows estimation of pesticide concentration with a prespecified sample size in each stratum. The factorial design, with four levels of the factor *pesticide use* and three levels of the factor *groundwater vulnerability,* allows investigation of possible effects of each factor separately, and the interaction of the factors, on pesticide concentrations. ∎

4.6
A Model for Stratified Sampling*

The one-way ANOVA model with fixed effects provides an underlying structure for stratified sampling. Here,

$$Y_{hj} = \mu_h + \varepsilon_{hj},$$

where the ε_{hj}'s are independent with mean 0 and variance σ_h^2. Then, as in Section 2.8, the least squares estimator of μ_h from units in the sample is the average of the sampled observations in stratum h.

Let the random variable

$$T_h = \sum_{j=1}^{N_h} Y_{hj}$$

represent the total in stratum h_j and the random variable

$$T = \sum_{h=1}^{H} T_h$$

represent the overall total.

From Section 2.8, the best linear unbiased estimator for T_h is

$$\hat{T}_h = \frac{N_h}{n_h} \sum_{j \in S_h} Y_{hj}.$$

Then, from the results shown for simple random sampling in Section 2.8,

$$E_M[\hat{T}_h - T_h] = 0$$

and

$$E_M[(\hat{T}_h - T_h)^2] = N_h^2 \frac{\sigma_h^2}{n_h}\left(1 - \frac{n_h}{N_h}\right).$$

Since we sample independently in the strata,

$$
\begin{aligned}
E_M[(\hat{T} - T)^2] &= E_M\left[\left\{\sum_{h=1}^{H}(\hat{T}_h - T_h)\right\}^2\right] \\
&= E_M\left[\sum_{h=1}^{H}(\hat{T}_h - T_h)^2 + \sum_{h=1}^{H}\sum_{k \neq h}(\hat{T}_h - T_h)(\hat{T}_k - T_k)\right] \\
&= E_M\left[\sum_{h=1}^{H}(\hat{T}_h - T_h)^2\right] \\
&= \sum_{h=1}^{H}\left(1 - \frac{n_h}{N_h}\right)N_h^2 \frac{\sigma_h^2}{n_h}.
\end{aligned}
$$

The theoretical variance σ_h^2 can be estimated by s_h^2. Adopting this model results in the same estimates for t and its standard error as found under randomization theory in Equations (4.1) and (4.4). If a different model is used, however, then different estimates are obtained.

4.7
Poststratification

Suppose a sampling frame lists all households in an area, and you would like to estimate the average amount spent on food in a month. One desirable stratification variable might be household size because large households might be expected to have higher food bills than smaller households. From U.S. census data, the distribution of household size in the region is known:

Number of Persons in Household	Percentage of Households
1	25.75
2	31.17
3	17.50
4	15.58
5+	10.00

The sampling frame, however, does not include information on household size—it only lists the households.

Without additional information, you cannot design an intelligent stratified sampling plan. You can, however, take an SRS and record the amount spent on food as well as the household size for each household in your sample. If n, the size of the SRS, is large enough, then the sample is likely to resemble a stratified sample with proportional allocation: We would expect about 26% of the sample to be one-person households, about 31% to be two-person households, and so on.

Considering the different household-size groups to be different domains, we can use the methods from Section 3.3 to estimate the average amount spent on groceries for each domain: Take an SRS of size n. Let $n_1, n_2, \ldots, n_H$ be the numbers of units sampled in the various household-size groups (domains) and $\bar{y}_1, \ldots, \bar{y}_H$ be the sample means for the groups.

After the observations are taken, form a "stratified" estimate of $\bar{y}_U$:

$$\bar{y}_{\text{post}} = \sum_{h=1}^{H} \frac{N_h}{N} \bar{y}_h. \tag{4.14}$$

If (1) N_h/N is known, (2) n_h is reasonably large (≥ 30 or so), and (3) n is large, then we can use the variance for proportional allocation as a good approximation to the variance:

$$\hat{V}(\bar{y}_{\text{post}}) \approx \left(1 - \frac{n}{N}\right) \sum_{h=1}^{H} \frac{N_h}{N} \frac{s_h^2}{n}. \tag{4.15}$$

WARNING: Poststratification can be *dangerous* if you indulge in data snooping: You can obtain arbitrarily small variances if you choose the strata after seeing the data, just as you can always obtain statistical significance if you decide on your null and alternative hypotheses after looking at the data. Poststratification is most often used to correct for the effects of differential nonresponse in the poststrata (see Chapter 8).

4.8
Quota Sampling

Many samples that masquerade as stratified random samples are actually quota samples. In **quota sampling,** the population is divided into different subpopulations just as in stratified random sampling, but with one important difference: Probability sampling is not used to choose individuals in the subpopulation for the sample. In extreme versions of quota sampling, choice of units in the sample is entirely at the discretion of the interviewer, so a sample of convenience is chosen within each subpopulation.

In quota sampling, specified numbers (quotas) of particular types of population units are required in the final sample. For example, to obtain a quota sample with $n = 3000$, you might specify that the sample contain 1000 white males, 1000 white females, 500 men of color, and 500 women of color, but you might give no further instructions about how these quotas are to be filled. Thus, quota sampling is not a form of probability sampling—we do not know the probabilities with which each individual is included in the sample. It is often used when probability sampling is impractical, overly costly, or considered unnecessary, or when the persons designing the sample just do not know any better.

The big drawback of quota sampling is that we do not know if the units chosen for the sample exhibit selection bias. If selection of units is totally up to the interviewer, she or he is likely to choose the most accessible members of the population—for instance, persons who are easily reached by telephone, households without menacing dogs, or areas of the forest close to the road. The most accessible members of a population are likely to differ in a systematic way from less accessible members. Thus, unlike in stratified random sampling, we cannot say that the estimate of the population total from quota sampling is unbiased over repeated sampling—one of our usual criteria of goodness in probability samples. In fact, in quota samples we cannot measure sampling error over repeated samples and we have no way of estimating the bias from the sample data. Since selection of units is up to the individual interviewer, we cannot expect that repeating the sample will give similar results. Thus, anyone drawing inferences from a quota sample must necessarily take a model-based approach.

EXAMPLE **4.11**   The 1945 survey on reading habits taken for the Book Manufacturer's Institute (Link and Hopf 1946), like many surveys in the 1940s and 1950s, used a quota sample. Some of the classifications used to define the quota classes were area, city size, age, sex, and socioeconomic status; a local supervising psychologist in each city determined the blocks of the city in which interviewers were to interview people from a specified socioeconomic group. The interviewers were then allowed to choose the specific households to be interviewed in the designated city blocks.

The quota procedure followed in the survey did not result in a sample that reflected demographic characteristics of the 1945 U.S. population. The following table compares the educational background of the survey respondents with figures from the 1940 U.S. census, adjusted to reflect the wartime changes in population:

Distribution by Educational Levels	4000 People Interviewed (%)	U.S. Census, Urban and Rural Nonfarm (%)
8th grade or less	28	48
1–3 years high school	18	19
4 years high school	25	21
1–3 years college	15	7
4 or more years college	13	5

SOURCE: Link and Hopf 1946.

The oversampling of better-educated persons casts doubt on many of the statistics given in the book. The study concluded that 31% of "active readers" (those who had read at least one book in the past month) had bought the last book they read and that

25% of all last books read by active readers cost $1 or less. Who knows whether a stratified random sample would have given the same results? ∎

In the 1948 U.S. presidential election, all major polls printed just a few days before the election predicted that Dewey would handily defeat Truman. In fact, of course, Truman won the election. According to Mosteller et al. (1949), one problem of those polls was that they all used quota sampling, not a probability-based method. This 1948 polling debacle spurred many American survey organizations to turn away from quota sampling, at least for a few years.

Many electoral polls in Britain used probability samples in the 1960s and 1970s. Probability samples, however, were much more expensive than quota samples, and quota samples used in the 1970s gave accurate predictions of the election results, so several large polling organizations went back to quota sampling (Taylor 1995). The polls that erred in predicting the winner in the 1992 British general election all used quota methods in selecting persons to interview in their homes or in the street; the primary quota classes used were sex, age, socioeconomic class, and employment status. Although we may never know exactly what went wrong in those polls (see Crewe 1992 for some other explanations), the use of quota samples may have played a part—if interviewing persons "in the street," it is certainly plausible that persons from a quota class who are accessible differ from persons who are less accessible.

Although quota sampling is not as good as probability sampling under ideal conditions, it will usually give much better results than the convenience samples that are often taken because it at least forces the inclusion of members of the different quota groups. Quota samples have the advantage of being less expensive than probability samples. The quality of the data from quota samples can be improved by allowing the interviewer less discretion in the choice of persons or households to be included in the sample. Many survey organizations use probability sampling along with quotas; they use probability sampling to select small blocks of potential respondents and then take a quota sample within each block, using variables such as age, sex, and race to define the quota classes.

A quota sample performs unfavorably compared with a stratified random sample when there is no nonresponse in the stratified random sample. When there is nonresponse, the comparison is unclear. Quota sampling can be considered as a substitution method for dealing with nonresponse, as is considered in Chapter 8: A nonrespondent is replaced by another person in the same quota class.

Because we do not know the probabilities with which units were sampled, we must take a model-based approach when analyzing data from a quota sample. The model generally adopted is that of Section 4.6—within each subclass the random variables generating the subpopulation are independent and identically distributed. Such a model implies that any selection of units from the subclass will give a representative sample; if the model holds, then quota sampling will likely give good estimates of the population quantity. If the model does not hold, then the estimates from quota sampling may be badly biased.

Deville (1991, 177) argues that quota samples may be useful for market research, when the organization requesting the survey is aware of the model being used. Persons collecting official statistics about crime, unemployment, or other matters that may be debated should use probability samples, however.

EXAMPLE 4.12   Sanzo et al. (1993) used a combination of stratified random sampling and quota sampling for estimating the prevalence of *Coxiella burnetii* infection within the Basque country in northern Spain. *Coxiella burnetii* can cause Q fever, which can lead to complications such as heart and nerve damage. Reviews of Q fever patient records from Basque hospitals showed that about three-fourths of the victims were male, about one-half were between 16 and 30 years old, and victims were disproportionately likely to be from areas with low population density.

The authors stratified the target population by population density and then randomly selected health-care centers from the three strata. In selecting persons for blood testing, however, "a probabilistic approach was rejected as we considered that the refusal rate of blood testing would be high" (p. 1185). Instead, they used quota sampling to balance the sample by age and gender; physicians asked patients who needed laboratory tests whether they would participate in the study and recruited subjects for the study until the desired sample sizes in the six quota groups were reached for each stratum.

Because a quota sample was taken instead of a probability sample, persons analyzing the data must make strong assumptions about the representativeness of the sample in order to apply the results to the general population of the Basque country. First, the assumption must be made that persons attending a health clinic for laboratory tests (the sampled population of the study) are neither more nor less likely to be infected than persons who would not be visiting the clinic. Second, one must assume that persons who are requested and agree to do the study are similar in terms of the infection to persons in the same quota class having laboratory tests that do not participate in the study. These are strong assumptions. The authors of the article argue that the assumptions are justified, but of course they cannot prove that the assumptions hold unless follow-up investigations are done.

If they had taken a probability sample of persons instead of the quota sample, they would not have had to make these strong assumptions. A probability sample of persons, however, would have been exhorbitantly expensive when compared with the quota design used, and a probability sample would also have taken longer to design and implement. With the quota sample, the authors could collect information about the public health problem; it is unclear whether the results can be generalized to the entire population, but the data do provide a great deal of quick information on the prevalence of infection that can be used in future investigation of who is likely to be infected, and why.   ∎

4.9
Exercises

1   What stratification variable(s) would you use for each of the following situations?

a   A political poll to estimate the percentage of registered voters in Arizona who approve of the job the governor is doing

b   A telephone survey of students at your university, to estimate the total amount of money students spend on textbooks

c A sample of high schools in New York City, to estimate what percentage of high schools offer one or more classes in computer programming

d A sample of public libraries in California, to study the availability of computer resources and the per capita expenditures

e A survey of anglers visiting a freshwater lake, to learn about which species of fish are preferred

f An aerial survey to estimate the number of walrus in the pack ice near Alaska between 173° east and 154° west longitude

g A sample of prime-time (7–10 P.M., Monday through Saturday; 6–10 P.M., Sunday) TV programs on CBS, to estimate the average number of promotional announcements (ads for other programming on the station) per hour of broadcast

2 The data set agstrat.dat also contains information on other variables. For each of the following quantities, plot the data and estimate the population mean for that variable along with its standard error. Give a 95% CI for each estimate. Compare your answers with those from the SRS in Exercise 11 in Chapter 2.

a Number of acres devoted to farms, 1987

b Number of farms, 1992

c Number of farms with 1000 acres or more, 1992

d Number of farms with 9 acres or fewer, 1992

3 Hard-shell clams can be sampled by using a dredge. Clams do not tend to be uniformly distributed in a body of water, however, because some areas provide better habitat than others. Thus, taking an SRS is likely to result in a large estimated variance for the number of clams in an area. Russell (1972) used stratified random sampling to estimate the total number of bushels of hard-shell clams *(Mercenaria mercenaria)* in Narragansett Bay, Rhode Island. The area of interest was divided into four strata based on preliminary surveys that identified areas in which clams were abundant. Then, n_h dredge tows were made in stratum h, for $h = 1, 2, 3, 4$. The acreage for each stratum was known, and Russell calculated that the area fished during a standard dredge tow was 0.039 acre—that is, 25.6 dredge tows would fish 1 acre.

a Here are the results from the survey taken before the commercial season. Estimate the total number of bushels of clams in the area and give the standard error of your estimate. HINT: First calculate N_h, the number of dredge tows needed to cover stratum h.

Stratum	Area (Acres)	Number of Tows Made	Average Number of Bushels per Tow	Sample Variance for Stratum
1	222.81	4	0.44	0.068
2	49.61	6	1.17	0.042
3	50.25	3	3.92	2.146
4	197.81	5	1.80	0.794

b   Another survey was performed at the end of the commercial season. In this survey, strata 1, 2, and 3 were collapsed into a single stratum, called stratum 1 below. Estimate the total number of bushels of clams (with standard error) at the end of the season.

Stratum	Area (Acres)	Number of Tows Made	Average Number of Bushels per Tow	Sample Variance for Stratum
1	322.67	8	0.63	0.083
4	197.81	5	0.40	0.046

4   Return to the hypothetical population in Example 3.4. Now, instead of using x as an auxiliary variable in ratio estimation, use it as a stratification variable: A population unit is in stratum 1 if $x \leq 5$ and in stratum 2 if $x > 5$. With this stratification, $N_1 = N_2 = 4$. The population is as follows:

Unit Number	Stratum	y
1	1	1
2	1	2
3	1	4
8	1	8
4	2	4
5	2	7
6	2	7
7	2	7

Consider the stratified sampling design in which $n_1 = n_2 = 2$.

a   Write out all possible SRSs of size 2 from stratum 1 and find the probability of each sample. Do the same for stratum 2.

b   Using your work in part (a), find the sampling distribution of $\hat{t}_{str}$.

c   Find the mean and variance of the sampling distribution of $\hat{t}_{str}$. How do these compare with the mean and variance in Examples 2.1 and 3.4?

5   Suppose a city has 90,000 dwelling units, of which 35,000 are houses, 45,000 are apartments, and 10,000 are condominiums. You believe that the mean electricity usage is about twice as much for houses as for apartments or condominiums and that the standard deviation is proportional to the mean.

a   How would you allocate a sample of 900 observations if you want to estimate the mean electricity consumption for all households in the city?

b   Now suppose that you want to estimate the overall proportion of households in which energy conservation is practiced. You have strong reason to believe that about 45% of house dwellers use some sort of energy conservation and that the corresponding percentages are 25% for apartment dwellers and 3% for condominium residents. What gain would proportional allocation offer over simple random sampling?

c Someone else has taken a small survey, using an SRS, of energy usage in houses. On the basis of the survey, each house is categorized as having electric heating or some other kind of heating. The January electricity consumption in kilowatt-hours for each house is recorded (y_i) and the results are given below:

Type of Heating	Number of Houses	Sample Mean	Sample Variance
Electric	24	972	202,396
Nonelectric	36	463	96,721
Total	60		

From other records, it is known that 16,450 of the 35,000 houses have electric heating, and 18,550 have nonelectric heating.

i Using the sample, give an estimate and its standard error of the proportion of houses with electric heating. Does your 95% CI include the true proportion?

ii Give an estimate and its standard error of the average number of kilowatt-hours used by houses in the city. What type of estimator did you use, and why did you choose that estimator?

6 A public opinion researcher has a budget of $20,000 for taking a survey. She knows that 90% of all households have telephones. Telephone interviews cost $10 per household; in-person interviews cost $30 each if all interviews are conducted in person and $40 each if only nonphone households are interviewed in person (because there will be extra travel costs). Assume that the variances in the phone and nonphone strata are similar and that the fixed costs are $c_0 = \$5000$. How many households should be interviewed in each stratum if

a All households are interviewed in person.

b Households with a phone are contacted by telephone and households without a phone are contacted in person.

7 For Example 4.3, construct a data set with 3835 observations. Include three columns: column 1 is the stratum number (from 1 to 7), column 2 contains the response variable of gender (0 for males and 1 for females), and column 3 contains the sampling weight N_h/n_h for each observation. Using columns 2 and 3 along with (4.10), calculate $\hat{p}_{str}$. Is it possible to calculate SE($\hat{p}_{str}$) by using only columns 2 and 3, with no additional information?

8 The survey in Example 4.3 collected much other data on the subjects. Another of the survey's questions asked whether the respondent agreed with the following statement: "When I look at a new issue of my discipline's major journal, I rarely find an article that interests me." The results are as follows:

Discipline	Agree (%)
Literature	37
Classics	23
Philosophy	23
History	29
Linguistics	19
Political science	43
Sociology	41

a   What is the sampled population in this survey?

b   Find an estimate of the proportion of persons in the sampled population that agree with the statement and give the standard error of your estimate.

9   Construct a small population and stratification for which $V(\hat{t}_{str})$ using proportional allocation is larger than the variance that would be obtained by taking an SRS with the same number of observations. HINT: Use (4.11).

10   In Exercise 8 of Chapter 2, data on numbers of publications were given for an SRS of 50 faculty members. Not all departments, however, were represented in the SRS. The SRS contained several faculty members from psychology and from chemistry but none from foreign languages. The following data are from a stratified sample of faculty, using the areas biological sciences, physical sciences, social sciences, and humanities as the strata. Proportional allocation was used in this sample.

Stratum	Number of Faculty Members in Stratum	Number of Faculty Members in Sample
Biological sciences	102	7
Physical sciences	310	19
Social sciences	217	13
Humanities	178	11
Total	807	50

The frequency table for number of publications in the strata is given below.

Number of Refereed Publications	Number of Faculty Members			
	Biological	Physical	Social	Humanities
0	1	10	9	8
1	2	2	0	2
2	0	0	1	0
3	1	1	0	1
4	0	2	2	0
5	2	1	0	0
6	0	1	1	0
7	1	0	0	0
8	0	2	0	0

a Estimate the total number of refereed publications by faculty members in the college and give the standard error.

b How does your result from part (a) compare with the result from the SRS in Exercise 8 of Chapter 2?

c Estimate the proportion of faculty with no refereed publications and give the standard error.

d Did stratification increase precision in this example? Explain why you think it did or did not.

11 Lydersen and Ryg (1991) used stratification techniques to estimate ringed seal populations in Svalbard fjords. The 200-km^2 study area was divided into three zones: Zone 1, outer Sassenfjorden, was covered with relatively new ice during the study period in March 1990 and had little snow cover; zone 3, Tempelfjorden, had a stable ice cover throughout the year; zone 2, inner Sassenfjorden, was intermediate between the stable zone 3 and the unstable zone 1. Ringed seals need good ice to establish territories with breeding holes, and snow cover enables females to dig out birth lairs. Thus, it was thought that the three zones would have different seal densities.

To select the sample, investigators divided the entire region into 200 1-km^2 areas; "a sampling grid covering 20% of the total area was made . . . by picking 40 numbers between one and 200 with the random number generator." In each sampled area, Imjak the Siberian husky tracked seal structures by scent; the number of breathing holes in each sampled square was recorded. A total of 199 breathing holes were located in zones 1–3. The data (reconstructed from information given in the paper) are in the file seals.dat.

The following table gives the number of plots, and the number of plots sampled, in each zone:

Zone	Number of Plots	Plots Sampled
1	68	17
2	84	12
3	48	11
Total	200	40

a Is this a stratified random sample, or a poststratified SRS? Explain.

b Estimate the total number of breathing holes in the study region, along with its standard error.

c If you were designing this survey, how would you allocate observations to strata if the goal was to estimate the total number of breathing holes? If the goal was to compare the density of breathing holes in the three zones?

12 Proportional allocation was used in the stratified sample in Example 4.1. It was noted, however, that variability was much higher in the West than in the other regions. Using the estimated variances in Example 4.1 and assuming that the sampling costs are the same in each stratum, find an optimal allocation for a stratified sample of size 300.

13 Select a stratified random sample of size 300 from the data in the file agpop.dat, using your allocation in Exercise 12. Estimate the total number of acres devoted to farming in the United States and give the standard error of your estimate. How does this standard error compare with that found in Example 4.1?

14   Burnard (1992) sent a questionnaire to a stratified sample of nursing tutors and students in Wales, to study what the tutors and students understood by the term *experiential learning*. The population size and sample size obtained for each of the four strata are given below:

Stratum	Population Size	Sample Size
General nursing tutors (GT)	150	109
Psychiatric nursing tutors (PT)	34	26
General nursing students (GS)	2680	222
Psychiatric nursing students (PS)	570	40
Total	3434	397

Respondents were asked which of the following techniques could be identified as experiential learning methods; the number of students and tutors in each group who identified the method as an experiential learning method are given below:

Method	GS	PS	PT	GT
Role play	213	38	26	104
Problem-solving activities	182	33	22	95
Simulations	95	20	22	64
Empathy-building exercises	89	25	20	54
Gestalt exercises	24	4	5	12

Estimate the overall percentage of nursing students and tutors who identify each of these techniques as *experiential learning*. Be sure to give standard errors for your estimates.

15   Kruuk et al. (1989) used a stratified sample to estimate the number of otter *(Lutra lutra)* dens along the 1400-km coastline of Shetland, UK. The coastline was divided into 242 (237 that were not predominantly buildings) 5-km sections, and each section was assigned to the stratum whose terrain type predominated. Sections were then chosen randomly from the sections in each stratum. In each section chosen, investigators counted the total number of dens in a 110-m-wide strip along the coast.

The data are in the file otters.dat. The population sizes for the strata are as follows:

Stratum	Total Sections	Sections Counted
1 Cliffs over 10 m	89	19
2 Agriculture	61	20
3 Not 1 or 2, peat	40	22
4 Not 1 or 2, nonpeat	47	21

a Estimate the total number of otter dens along the coast in Shetland, along with a standard error for your estimate.

b Discuss possible sources of bias in this study. Do you think it is possible to avoid all selection and measurement bias?

16 Marriage and divorce statistics are compiled by the National Center for Health Statistics and published in volumes of *Vital Statistics of the United States*. State and local officials provide NCHS with annual counts of marriages and divorces in each county. In addition, some states send computer tapes of additional data or microfilm copies of marriage or divorce certificates. These additional data are used to calculate statistics about age at marriage or divorce, previous marital status of marrying couples, and children involved in divorce. In 1987, if a state sent a computer tape, all records were included in the divorce statistics; if a state sent microfilm copies, a specified fraction of the divorce certificates was randomly sampled and data recorded. The sampling rates (probabilities of selection) and number of records sampled in each state in the divorce registration area for 1987 are in the file divorce.dat.

a How many divorces were there in the divorce registration area in 1987? HINT: Use the sampling weights.

b Why did NCHS use different sampling rates in different states?

c Estimate the total number of divorces granted to men aged 24 or less; to women aged 24 or less. Give 95% CIs for your estimates.

d In what proportion of all divorces is the husband between 40 and 49 years old? In what proportion is the wife between 40 and 49 years old? Give confidence intervals for your estimates.

17 Jackson et al. (1987) compared the precision of systematic and stratified sampling for estimating the average concentration of lead and copper in the soil. The 1-km^2 area was divided into 100-m squares, and a soil sample was collected at each of the resulting 121 grid intersections. Summary statistics from this systematic sample are given below:

Element	n	Average (mg kg^{-1})	Range (mg kg^{-1})	Standard Deviation (mg kg^{-1})
Lead	121	127	22–942	146
Copper	121	35	15–90	16

The investigators also poststratified the same region. Stratum A consisted of farmland away from roads, villages, and woodlands. Stratum B contained areas within 50 m of roads and was expected to have larger concentrations of lead. Stratum C contained the woodlands, which were also expected to have larger concentrations of lead because the foliage would capture airborne particles. The data on concentration of lead and copper were not used in determining the strata. The data from the grid points falling in each stratum are in the following table:

Element	Stratum	n_h	Average (mg kg^{-1})	Range (mg kg^{-1})	Standard Deviation (mg kg^{-1})
Lead	A	82	71	22–201	28
Lead	B	31	259	36–942	232
Lead	C	8	189	88–308	79
Copper	A	82	28	15–68	9
Copper	B	31	50	22–90	18
Copper	C	8	45	31–69	15

a   Calculate a 95% CI for the average concentration of lead in the area, using the systematic sample. (You may assume that this sample behaves like an SRS.) Repeat for the average concentration of copper.

b   Now use the poststratified sample and find 95% CIs for the average concentration of lead and copper. How do these compare with the confidence intervals in part (a)? Do you think that using stratification in future surveys would increase precision?

18   In Exercise 17 the sample size in each stratum was proportional to the area of the stratum. Using the sample standard deviations, what would an optimal allocation be for taking a stratified random sample with 121 observations? Is the optimal allocation the same for copper and lead?

19   Wilk et al. (1977) report data on the number and types of fish and environmental data for the area of the Atlantic continental shelf between eastern Long Island, New York, and Cape May, New Jersey. The ocean survey area was divided into strata based on depth. Sampling was done at a higher rate close to shore than farther away from shore: "In-shore strata (0–28 m) were sampled at a rate of approximately one station per 515 km^2 and off-shore strata (29–366 m) were sampled at a rate of approximately one station per 1,030 km^2" (p. 1). Thus, each record in strata 3–6 represents twice as much area as each record in strata 1 and 2. In calculating average numbers of fish caught and numbers of species, we can use a relative sampling weight of 1 for strata 1 and 2, and weight 2 for strata 3–6.

Stratum	Depth (m)	Relative Sampling Weight
1	0–19	1
2	20–28	1
3	29–55	2
4	56–100	2
5	111–183	2
6	184–366	2

The file nybight.dat contains data on the total catch for sampling stations visited in June 1974 and June 1975.

a   Construct side-by-side boxplots of the number of fish caught in the trawls in June 1974. Does there appear to be a large variation among the strata?

b Calculate estimates of the average number and average weight of fish caught per haul in June 1974, along with the standard error.

c Calculate estimates of the average number and average weight of fish caught per haul in June 1975, along with the standard error.

d Is there evidence that the average weight of fish caught per haul differs between June 1974 and June 1975? Answer using an appropriate hypothesis test.

20 In January 1995 the Office of University Evaluation at Arizona State University surveyed faculty and staff members to find out their reaction to the closure of the university during the winter break in 1994. Faculty and staff in academic units that were closed during the winter break were divided into four strata and sub-sampled:

Stratum Number	Employee Type	Population Size (N_h)	Sample Size
1	Faculty	1374	500
2	Classified staff	1960	653
3	Administrative staff	252	98
4	Academic professional	95	95

Questionnaires were sent through campus mail to persons in strata 1–4; the sample size in the above table is the number of questionnaires mailed in each stratum. We'll come back to the issue of nonresponse in this survey in Chapter 8; for now, just analyze the respondents in the stratified sample of employees in closed units; the data for the 985 survey respondents are found in the file winter.dat. For this exercise, look at the answers to the question "Would you want to have Winter Break closure again?" (variable *breakaga*).

a Not all persons in the survey responded to the question. Find the number of persons who responded to the question in each of the four strata. For this exercise, use these values as the n_h.

b Use (4.6) and (4.7) to estimate the proportion of faculty and staff that would answer yes to the question "Would you want to have Winter Break closure again?" and give the standard error.

c Create a new variable, in which persons who respond yes to the question take on the value 1, persons who respond no to the question take on the value 0, and persons who do not respond are either left blank (if you are using a spreadsheet) or assigned the missing value code (if you are using statistical software). Construct a column of sampling weights N_h/n_h for the observations in the sample. (The sampling weight will be zero or missing for nonrespondents.) Now use (4.10) to estimate the proportion of faculty and staff that would answer yes to the question "Would you want to have Winter Break closure again?"

d Using the column of 0s and 1s you constructed in the previous question, find s_h^2 for each stratum by calculating the sample variance of the observations in that stratum. Now use (4.5) to calculate the standard error of your estimate of the proportion. Why is your answer the same as you calculated in part (b)?

e Stratification is sometimes used as a method of dealing with nonresponse. Calculate the response rates (the number of persons who responded divided by the number of questionnaires mailed) for each stratum. Which stratum has the lowest response rate for this question? How does stratification treat the nonrespondents?

21 A stratified sample is being designed to estimate the prevalence p of a rare characteristic—say, the proportion of residents in Milwaukee who have Lyme disease. Stratum 1, with N_1 units, has a high prevalence of the characteristic; stratum 2, with N_2 units, has low prevalence. Assume that the cost to sample a unit (for example, the cost to select a person for the sample and determine whether he or she has Lyme disease) is the same for each stratum and that at most 2000 units are to be sampled.

   a Let p_1 and p_2 be the respective proportions in stratum 1 and stratum 2 with the rare characteristic. If $p_1 = 0.10$, $p_2 = 0.03$, and $N_1/N = 0.4$, what are n_1 and n_2 under optimal allocation?

   b If $p_1 = 0.10$, $p_2 = 0.03$, and $N_1/N = 0.4$, what is $V(\hat{p}_{str})$ under proportional allocation? Under optimal allocation? What is the variance if you take an SRS of 2000 units from the population?

   c (Use a spreadsheet for this part of the exercise.)   Now fix $p = 0.05$. Let p_1 range from 0.05 to 0.50, and N_1/N range from 0.01 to 0.50 (these two values then determine the value of p_2). For each combination of p_1 and N_1/N, find the optimal allocation and the variance under both proportional allocation and optimal allocation. Also find the variance from an SRS of 2000 units. When does the optimal allocation give a substantial increase in precision when compared to proportional allocation? When compared to an SRS?

*22 (Requires calculus.) Show that the variance of $\hat{t}_{str}$ is minimized for a fixed cost with the cost function in (4.12) when $n_h \propto N_h S_h / \sqrt{c_h}$, as in (4.13). HINT: Use Lagrange multipliers.

23 Suppose the Arizona Department of Health wishes to take a survey of 2-year-olds whose families receive medical assistance, to determine the proportion who have been immunized. The medical care is provided by several different health-care organizations, and the state has 15 counties. Table 4.6 shows the population number of 2-year-olds for each county/organization combination. The sample is to be stratified by county and organization. It is desired to select sample sizes for each combination so that

   a The margin of error for estimating percentage immunized is 0.05 or less when the data are tabulated for each county (summing over all health-care organizations).

   b The margin of error for estimating percentage immunized is 0.05 or less when the data are tabulated for each health-care organization (summing over all counties).

   c At least two children (fewer, of course, if the cell does not have two children) are selected from every cell.

Note that for this problem, as for many survey designs, many different designs would be possible.

TABLE 4.6

Table for Exercise 23

	A	B	C	D	E	Other	Total
Apache	1	13	19	0	0	94	127
Cochise	2	5	0	637	40	0	694
Coconino	1	6	0	125	0	289	421
Gila	0	2	51	151	0	0	204
Graham	0	2	0	63	0	143	208
Greenlee	0	0	0	58	0	0	58
Maricopa	118	169	0	3,732	2,675	5,105	11,799
Mohave	4	6	0	44	0	476	530
Navajo	2	5	132	124	0	0	263
Pima	62	26	0	1,097	727	1,786	3,698
Pinal	5	10	13	22	360	478	888
Santa Cruz	0	5	0	118	150	0	273
Yavapai	7	8	0	173	0	198	386
Yuma	5	5	0	837	0	0	847
LaPaz	0	1	0	89	0	0	90
Total	217	263	215	7,270	3,952	8,569	20,486

SURVEY Exercises

24 In the quest to estimate the average price a household in Stephens County is willing to pay for cable TV service, we are fortunate to know a great deal about some demographic aspects of the county, as given in the district map and tables in Appendix A. According to the SURVEY assumptions, what information might be used to stratify Stephens County in order to improve the precision of estimates? Are any other reasons for stratification relevant to Stephens County?

25 Use any considerations you like to divide Stephens County into strata. Your stratification should divide Lockhart City into approximately five strata. Why did you choose your stratification variable? Count the total number of households in each of your strata. (You may use the ADDGEN program to do this.)

The remainder of these exercises concern Lockhart City *only*.

26 Using ADDGEN, generate a stratified random sample of size 200 from Lockhart City with your stratification in Exercise 25 and proportional allocation. Find the responses using the SURVEY program. Estimate the average price a household in Lockhart City is willing to pay for cable service and the average number of TVs per household in Lockhart City. How do these estimates compare with those obtained with simple random sampling and sample mean and ratio estimates? Which estimates are the most precise?

27 Pilot studies are often used to estimate S_h. In this case we are fortunate to have a very large pilot study from the sample of size 200 used in Exercise 28 in Chapter 2.

Divide your sample from Chapter 2 into the strata you chose above and thus obtain estimates of the variances S_h^2 in each of the strata for the average price a household is willing to pay for cable TV service.

28   The sampling costs for Stephens County are given in Appendix A. Using your estimates of S_h, optimally allocate a sample of size 200 to estimate the average price a household in Lockhart City is willing to pay for cable TV service. Using that allocation, take a stratified random sample of Lockhart City and estimate the average price a household is willing to pay for cable TV service and the average number of TVs per household.

29   Under what conditions can optimal allocation be expected to perform much better than proportional allocation? Do these conditions occur in Lockhart City? Comment on the relative performance that you observed between these two allocations.

30   Using the variances estimated in Exercise 28 of Chapter 2, what sample size would be needed with simple random sampling to achieve the same precision in estimating the average price a household is willing to pay as a stratified sample of size 200 using the strata you have designed and optimal allocation? Proportional allocation?

31   Are there any deficiencies in your design? How would you correct them if you were to do this exercise a second time?

5

Cluster Sampling with Equal Probabilities

"But averages aren't real," objected Milo; "they're just imaginary."

"That may be so," he agreed, "but they're also very useful at times. For instance, if you didn't have any money at all, but you happened to be with four other people who had ten dollars apiece, then you'd each have an average of eight dollars. Isn't that right?"

"I guess so," said Milo weakly.

"Well, think how much better off you'd be, just because of averages," he explained convincingly. "And think of the poor farmer when it doesn't rain all year: if there wasn't an average yearly rainfall of 37 inches in this part of the country, all his crops would wither and die."

It all sounded terribly confusing to Milo, for he had always had trouble in school with just this subject.

"There are still other advantages," continued the child. "For instance, if one rat were cornered by nine cats, then, on the average, each cat would be 10 per cent rat and the rat would be 90 per cent cat. If you happened to be a rat, you can see how much nicer it would make things."

—Norton Juster, *The Phantom Tollbooth*

In all the sampling procedures discussed so far, we have assumed that the population is given and all we must do is reach in and take a suitable sample of units. But units are not necessarily nicely defined, even when the population is. There may be several ways of listing the units, and the unit size we choose may very well contain smaller subunits.

Suppose we want to find out how many bicycles are owned by residents in a community of 10,000 households. We could take a simple random sample (SRS) of 400 households, or we could divide the community into blocks of about 20 households each and sample every household (or subsample some of the households) in each of 20 blocks selected at random from the 500 blocks in the community. The latter plan is an example of cluster sampling. The blocks are the **primary sampling units** (psu's), or **clusters.** The households are the **secondary sampling units** (ssu's); often the ssu's are the elements in the population.

The cluster sample of 400 households is likely to give less precision than an SRS of 400 households; some blocks of the community are composed mainly of families (with more bicycles), whereas the residents of other blocks are mainly retirees (with fewer bicycles). Twenty households in the same block are not as likely to mirror the diversity of the community as well as 20 households chosen at random. Thus, cluster sampling in this situation will probably result in less information per observation than an SRS of the same size. However, if you conduct the survey in person, it is much cheaper and easier to interview all 20 households in a block than 20 households selected at random from the community, so cluster sampling may well result in more information per dollar spent.

In cluster sampling, individual elements of the population are allowed in the sample only if they belong to a cluster (primary sampling unit) that is included in the sample. The sampling unit (psu) is not the same as the observation unit (ssu), and the two sizes of experimental units must be considered when calculating standard errors from cluster samples.

Why use cluster samples?

1   Constructing a sampling frame list of observation units may be difficult, expensive, or impossible. We cannot list all honeybees in a region or all customers of a store; we may be able to construct a list of all trees in a stand of northern hardwood forest or a list of individuals in a city for which we only have a list of housing units, but constructing the list will be time-consuming and expensive.

2   The population may be widely distributed geographically or may occur in natural clusters such as households or schools. If the target population is residents of nursing homes in the United States, it is much cheaper to sample nursing homes and interview every resident in the selected homes than to interview an SRS of nursing home residents: With an SRS of residents, you might have to travel to a nursing home just to interview one resident. If taking an archaeological survey, you would examine all artifacts found in a region—you would not just choose points at random and examine only artifacts found at those isolated points.

Clusters bear a superficial resemblance to strata: A cluster, like a stratum, is a grouping of the members of the population. The selection process, though, is quite different in the two methods. Similarities and differences between cluster samples and stratified samples are illustrated in Figure 5.1.

Whereas stratification generally increases precision when compared with simple random sampling, cluster sampling generally decreases it. Members of the same cluster tend to be more similar than elements selected at random from the whole population—members of the same household tend to have similar political views; fish in the same lake tend to have similar concentrations of mercury; residents of the same nursing home tend to have similar opinions of the quality of care. These similarities usually arise because of some underlying factors that may or may not be measurable—residents of the same nursing home may have similar opinions because the care is poor, and the concentration of mercury in the fish will reflect the concentration of mercury in the lake. Thus, we do not obtain as much information about all nursing home residents in the United States by sampling two residents in the same home as by sampling two residents in different homes, because the two residents in the same home are likely to have more similar opinions. By sampling everyone in the cluster, we partially repeat the same information instead of obtaining new information, and that gives us less precision for estimates of population quantities. Cluster sampling is

FIGURE 5.1

Similarities and differences between cluster sampling and stratified sampling

Stratified Sampling	Cluster Sampling
Each element of the population is in exactly one stratum.	Each element of the population is in exactly one cluster.
Population of H strata; stratum h has n_h elements: 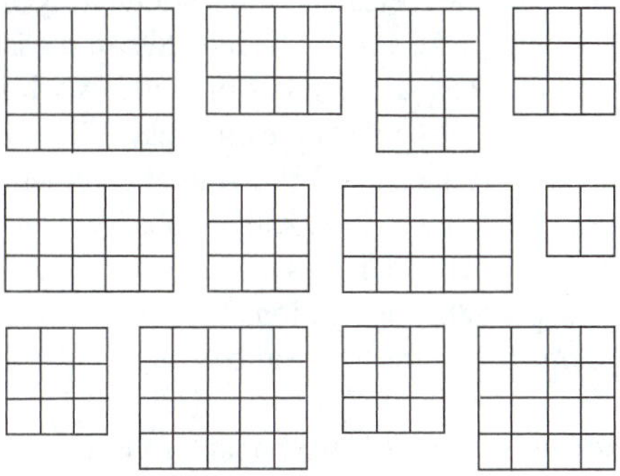	One-stage cluster sampling; population of N clusters: 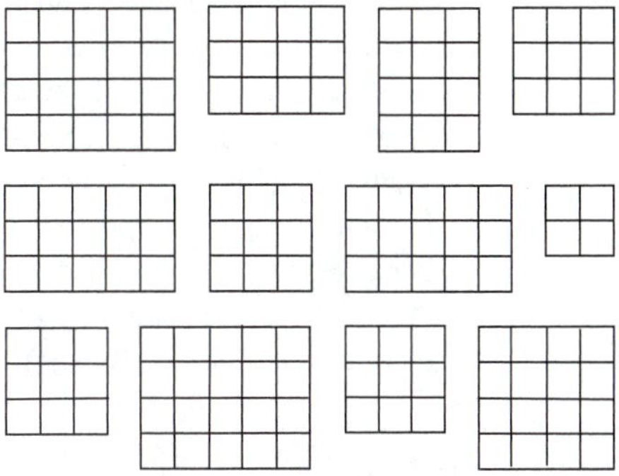

| Take an SRS from *every* stratum: 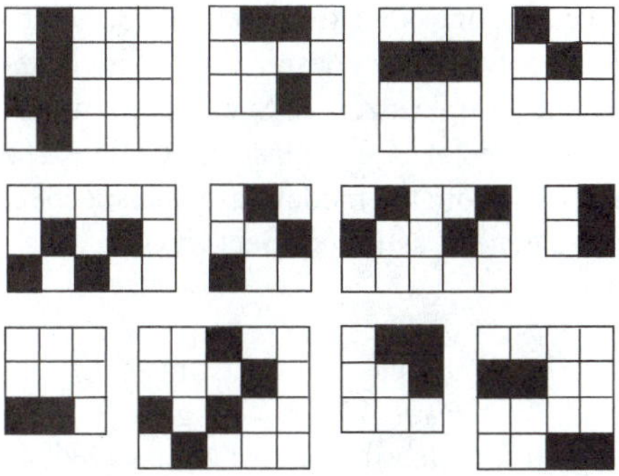 | Take an SRS of clusters; observe all elements within the clusters in the sample: 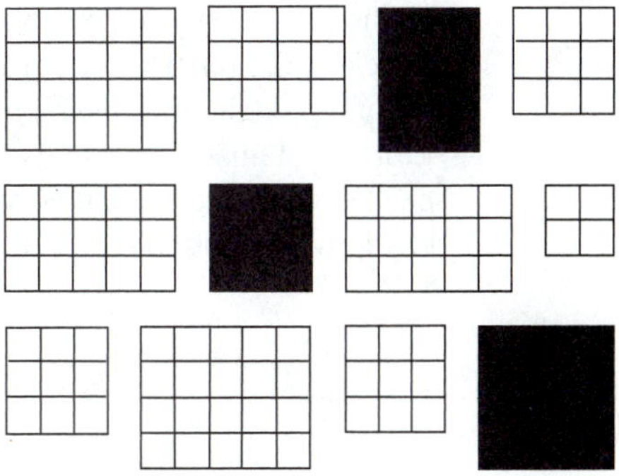 |

| Variance of the estimate of $\bar{y}_U$ depends on the variability of values *within* strata. | The cluster is the sampling unit; the more clusters we sample, the smaller the variance. The variance of the estimate of $\bar{y}_U$ depends primarily on the variability *between* cluster means. |
| For greatest precision, individual elements within each stratum should have similar values, but stratum means should differ from each other as much as possible. | For greatest precision, individual elements within each cluster should be heterogeneous, and cluster means should be similar to one another. |

used in practice because it is usually much cheaper and more convenient to sample in clusters than randomly in the population. Almost all large household surveys carried out by the U.S. government, or by commercial or academic institutions, use cluster sampling because of the cost savings.

One of the biggest mistakes made by researchers using surveys is to analyze a cluster sample as if it were an SRS. Such confusion usually results in the researchers reporting standard errors that are much smaller than they should be; this gives the impression that the survey results are much more precise than they really are.

EXAMPLE **5.1**   Basow and Silberg (1987) report results of their research on whether students evaluate female college professors differently than they evaluate male college professors. The authors matched 16 female professors with 16 male professors by subject taught, years of teaching experience, and tenure status, and then gave evaluation questionnaires to students in those professors' classes. The sample size for analyzing this study is $n = 32$, the number of faculty studied; it is not 1029, the number of students who returned questionnaires. Students' evaluations of faculty reflect the different styles of faculty teaching; students within the same class are likely to have some agreement in their rating of the professor and should not be treated as independent observations because their ratings will probably be positively correlated. If this positive correlation is ignored and the student ratings treated as independent observations, differences will be declared statistically significant far more often than they should be.   ∎

After a brief journey into "notation land" in Section 5.1, we begin by discussing **one-stage cluster sampling,** in which every element within a sampled cluster is included in the sample. We then generalize the results to **two-stage cluster sampling,** in which we subsample only some of the elements of selected clusters, in Section 5.3. In Section 5.4, we show how to use sampling weights, introduced in Section 4.3, to estimate population means and totals. In Section 5.5, we discuss design issues for cluster sampling, including selection of subsample and sample sizes. In Section 5.6, we return to systematic sampling and show that it is a special case of cluster sampling. The chapter concludes with theory of cluster sampling from the model-based perspective; we derive the design-based theory in the more general setting of Section 6.6.

5.1
Notation for Cluster Sampling

In simple random sampling, the units sampled are also the elements observed. In cluster sampling, the sampling units are the clusters, and the elements observed are the ssu's within the clusters. The universe $\mathcal{U}$ is the population of N psu's; $\mathcal{S}$ designates the sample of psu's chosen from the population of psu's, and $\mathcal{S}_i$ is the sample of ssu's chosen from the ith psu. The notation given below is used throughout this chapter and Chapter 6. The measured quantities are

$$y_{ij} = \text{measurement for } j\text{th element in } i\text{th psu.}$$

In cluster sampling, however, it is easiest to think at the psu level in terms of cluster totals. No matter how you define it, the notation for cluster sampling is messy because you need notation for both the psu and the ssu levels. The notation used in this chapter and Chapter 6 is presented in this section for easy reference. Note that in Chapters 5 and 6, N is the number of psu's, not the number of observation units.

psu Level—Population Quantities

$N =$ number of psu's in the population

$M_i =$ number of ssu's in ith psu

$$K = \sum_{i=1}^{N} M_i = \text{total number of ssu's in the population}$$

$$t_i = \sum_{j=1}^{M_i} y_{ij} = \text{total in the } i\text{th psu}$$

$$t = \sum_{i=1}^{N} t_i = \sum_{i=1}^{N} \sum_{j=1}^{M_i} y_{ij} = \text{population total}$$

$$S_t^2 = \sum_{i=1}^{N} \frac{\left(t_i - \dfrac{t}{N}\right)^2}{N-1} = \text{population variance of the psu totals}$$

ssu Level—Population Quantities

$$\bar{y}_U = \sum_{i=1}^{N} \sum_{j=1}^{M_i} \frac{y_{ij}}{K} = \text{population mean}$$

$$\bar{y}_{iU} = \sum_{j=1}^{M_i} \frac{y_{ij}}{M_i} = \frac{t_i}{M_i} = \text{population mean in the } i\text{th psu}$$

$$S^2 = \sum_{i=1}^{N} \sum_{j=1}^{M_i} \frac{(y_{ij} - \bar{y}_U)^2}{K-1} = \text{population variance (per ssu)}$$

$$S_i^2 = \sum_{j=1}^{M_i} \frac{(y_{ij} - \bar{y}_{iU})^2}{M_i - 1} = \text{population variance within the } i\text{th psu}$$

Sample Quantities

$$n = \text{number of psu's in the sample}$$

$$m_i = \text{number of elements in the sample from the } i\text{th psu}$$

$$\bar{y}_i = \sum_{j \in \mathcal{S}_i} \frac{y_{ij}}{m_i} = \text{sample mean (per ssu) for } i\text{th psu}$$

$$\hat{t}_i = \sum_{j \in \mathcal{S}_i} \frac{M_i}{m_i} y_{ij} = \text{estimated total for } i\text{th psu}$$

$$\hat{t}_{\text{unb}} = \sum_{i \in \mathcal{S}} \frac{N}{n} \hat{t}_i = \text{unbiased estimator of population total}$$

$$s_t^2 = \frac{1}{n-1} \sum_{i \in \mathcal{S}} \left(\hat{t}_i - \frac{\hat{t}_{\text{unb}}}{N}\right)^2 = \text{estimated variance of psu totals}$$

$$s_i^2 = \sum_{j \in \mathcal{S}_i} \frac{(y_{ij} - \bar{y}_i)^2}{m_i - 1} = \text{sample variance within the } i\text{th psu}$$

5.2
One-Stage Cluster Sampling

In one-stage cluster sampling, either all or none of the elements that compose a cluster (= psu) are in the sample. One-stage cluster sampling is used in many surveys in which the cost of sampling ssu's is negligible compared with the cost of sampling psu's. For education surveys, a natural psu is the classroom; all students in a selected classroom are often included as the ssu's since little extra cost is added by handing out a questionnaire to all students in the classroom rather than some.

In the population of N psu's, the ith psu contains M_i ssu's (elements). From the population, we take an SRS of n psu's and measure our variable of interest on *every* element in the chosen psu's. Thus, for one-stage cluster sampling, $M_i = m_i$.

5.2.1   Clusters of Equal Sizes: Estimation

Let's consider the simplest case in which each cluster has the same number of elements, with $M_i = m_i = M$. Most naturally occurring clusters of people do not fit into this framework, but it can occur in agricultural and industrial sampling. Estimating population means or totals is simple: We treat the cluster means or totals as the observations and simply ignore the individual elements.

Thus, we have an SRS of n observations $\{t_i, i \in \mathcal{S}\}$; t_i is the total for all the elements in psu i. Then, $\bar{t}_\mathcal{S}$ estimates the average of the cluster totals. In a household survey to estimate income in two-person households, the individual observations y_{ij} are the incomes of individual persons within the household, t_i is the total income for household i (t_i is *known* for sampled households because both persons are interviewed), $\bar{t}_U$ is the average income per household, and $\bar{y}_U$ is the average income per person. To estimate the total income t, we can use the estimator

$$\hat{t} = \frac{N}{n} \sum_{i \in \mathcal{S}} t_i. \tag{5.1}$$

The results in Sections 2.3 and 2.7 apply to $\hat{t}$ because we have an SRS of n units from a population of N units. As a result, $\hat{t}$ is an unbiased estimator of t, with variance given by

$$V(\hat{t}) = N^2 \left(1 - \frac{n}{N}\right) \frac{S_t^2}{n} \tag{5.2}$$

and with

$$SE(\hat{t}) = N \sqrt{\left(1 - \frac{n}{N}\right) \frac{s_t^2}{n}}, \tag{5.3}$$

where S_t^2 and s_t^2 are the population and sample variance, respectively, of the psu totals:

$$S_t^2 = \frac{1}{N-1} \sum_{i=1}^{N} \left(t_i - \frac{t}{N}\right)^2$$

and

$$s_t^2 = \frac{1}{n-1} \sum_{i \in S} \left(t_i - \frac{\hat{t}}{N} \right)^2.$$

To estimate $\bar{y}_U$, divide the estimated total by the number of persons, obtaining

$$\hat{\bar{y}} = \frac{\hat{t}}{NM}, \tag{5.4}$$

with

$$V(\hat{\bar{y}}) = \left(1 - \frac{n}{N} \right) \frac{S_t^2}{nM^2} \tag{5.5}$$

and

$$\mathrm{SE}(\hat{\bar{y}}) = \frac{1}{M} \sqrt{\left(1 - \frac{n}{N} \right) \frac{s_t^2}{n}}. \tag{5.6}$$

No new ideas are introduced to carry out one-stage cluster sampling; we simply use the results for simple random sampling with the cluster totals as the observations.

EXAMPLE 5.2 A student wants to estimate the average grade point average (GPA) in his dormitory. Instead of obtaining a listing of all students in the dorm and conducting an SRS, he notices that the dorm consists of 100 suites, each with four students; he chooses 5 of those suites at random and asks every person in the 5 suites what her or his GPA is. The results are as follows:

Person Number	Suite (Cluster)				
	1	2	3	4	5
1	3.08	2.36	2.00	3.00	2.68
2	2.60	3.04	2.56	2.88	1.92
3	3.44	3.28	2.52	3.44	3.28
4	3.04	2.68	1.88	3.64	3.20
Total	12.16	11.36	8.96	12.96	11.08

The psu's are the suites, so $N = 100$, $n = 5$, and $M = 4$. The estimate of the population total (the estimated sum of all the GPAs for everyone in the dorm—a meaningless quantity for this example but useful for demonstrating the procedure) is

$$\hat{t} = \frac{100}{5} (12.16 + 11.36 + 8.96 + 12.96 + 11.08) = 1130.4,$$

and

$$s_t^2 = \frac{1}{5-1} \left[(12.16 - 11.304)^2 + \cdots + (11.08 - 11.304)^2 \right] = 2.256.$$

In this example, s_t^2 is simply the usual sample variance of the 5 suite totals. Thus, using (5.4) and (5.6), $\hat{\bar{y}} = 1130.4/400 = 2.826$, and

$$\mathrm{SE}(\hat{\bar{y}}) = \sqrt{\left(1 - \frac{5}{100} \right) \frac{2.256}{(5)(4)^2}} = 0.164.$$

Note that in these calculations only the "total" row of the data table is used—the individual GPAs are only used for their contribution to the suite total. ∎

One-stage cluster sampling with an SRS of psu's produces a self-weighting sample. The weight for each observation unit is

$$w_{ij} = \frac{1}{P\{\text{ssu } j \text{ of psu } i \text{ is in sample}\}} = \frac{N}{n}.$$

For the data in Example 5.2, then,

$$\hat{t} = \sum_{i \in S} \sum_{j \in S_i} w_{ij} y_{ij}$$

$$= \frac{N}{n}(3.08 + 2.60 + \cdots + 3.28 + 3.20)$$

$$= \frac{100}{5}(56.52) = 1130.4.$$

Thus, as in stratified sampling, we can estimate a population total by summing the product of the observed values and the sampling weights.

If we had taken an SRS of nM elements, each element in the sample would have been assigned weight $(NM)/(nM) = N/n$—the same weights we obtain for cluster sampling. The precision obtained for the two types of sampling, however, can differ greatly; the difference in precision is explored in the next section.

5.2.2 Clusters of Equal Sizes: Theory

In this section we compare cluster sampling with simple random sampling: Cluster sampling almost always provides less precision for the estimators than one would obtain by taking an SRS with the same number of elements.

As in stratified sampling, let's look at the ANOVA table (Table 5.1) for the whole population. In stratified sampling, the variance of the estimator of t depends on the variability *within* the strata; Equation (4.3) and Table 4.3 imply that the variance in stratified sampling is small if SSW is small relative to SSTO, or equivalently, if the within mean square (MSW) is small relative to S^2. In stratified sampling, you have some information about *every* stratum, so you need not worry about variability due to unsampled strata. If MSB/MSW is large—that is, the variability among the stratum means is large when compared with the variability within strata—then stratified sampling increases precision.

T A B L E   5.1

Population ANOVA Table—Cluster Sampling

Source	df	Sum of Squares	Mean Square
Between psu's	$N-1$	$SSB = \sum_{i=1}^{N} \sum_{j=1}^{M} (\bar{y}_{iU} - \bar{y}_U)^2$	MSB
Within psu's	$N(M-1)$	$SSW = \sum_{i=1}^{N} \sum_{j=1}^{M} (y_{ij} - \bar{y}_{iU})^2$	MSW
Total, about $\bar{y}_U$	$NM-1$	$SSTO = \sum_{i=1}^{N} \sum_{j=1}^{M} (y_{ij} - \bar{y}_U)^2$	S^2

The opposite situation occurs in cluster sampling. In one-stage cluster sampling, the variability of the unbiased estimator of t depends entirely on the *between*-cluster part of the variability, because

$$S_t^2 = \sum_{i=1}^{N} \frac{(t_i - \bar{t}_U)^2}{N - 1} = \sum_{i=1}^{N} \frac{M^2(\bar{y}_{iU} - \bar{y}_U)^2}{N - 1} = M(\text{MSB}).$$

Thus, for cluster sampling,

$$V(\hat{t}_{\text{cluster}}) = N^2 \left(1 - \frac{n}{N}\right) \frac{M(\text{MSB})}{n}. \tag{5.7}$$

If MSB/MSW is large in cluster sampling, then cluster sampling decreases precision. In that situation, MSB is relatively large because it measures the cluster-to-cluster variability: Elements in different clusters often vary more than elements in the same cluster because different clusters have different means. If we took a cluster sample of classes and sampled all students within the selected classes, we would likely find that average reading scores varied from class to class. An excellent reading teacher might raise the reading scores for the entire class; a class of students from an area with much poverty might tend to be undernourished and not score as high at reading. Unmeasured factors, such as teaching skill or poverty, can affect the overall mean for a cluster and thus cause MSB to be large.

Within a class, too, students' reading scores vary. The MSW is the pooled value of the within-cluster variances: the variance from element to element, present for all elements of the population. If the clusters are relatively homogeneous—if, for example, students in the same class have similar scores—the MSW will be small.

Now let's compare cluster sampling to simple random sampling. If, instead of taking a cluster sample of M elements in each of n clusters, we had taken an SRS with nM observations, the variance of the estimated total would have been

$$V(\hat{t}_{\text{SRS}}) = (NM)^2 \left(1 - \frac{nM}{NM}\right) \frac{S_.^2}{nM} = N^2 \left(1 - \frac{n}{N}\right) \frac{MS_.^2}{n}.$$

Comparing this with (5.7), we see that if MSB $> S_.^2$, then cluster sampling is less efficient than simple random sampling.

The **intraclass** (sometimes called **intracluster**) **correlation coefficient** (ICC) tells us how similar elements in the same cluster are. It provides a **measure of homogeneity** within the clusters. ICC is defined to be the Pearson correlation coefficient for the $NM(M - 1)$ pairs (y_{ij}, y_{ik}) for i between 1 and N and $j \neq k$ (see Exercise 9) and can be written in terms of the population ANOVA table quantities as

$$\text{ICC} = 1 - \frac{M}{M - 1} \frac{\text{SSW}}{\text{SSTO}}. \tag{5.8}$$

Because $0 \leq \text{SSW}/\text{SSTO} \leq 1$, it follows from (5.8) that

$$-\frac{1}{M - 1} \leq \text{ICC} \leq 1.$$

If the clusters are perfectly homogeneous and hence SSW $= 0$, then ICC $= 1$. Equation (5.8) also implies that

$$\text{MSB} = \frac{NM - 1}{M(N - 1)} S^2 [1 + (M - 1)\text{ICC}].$$

How much precision do we lose by taking a cluster sample? From the above equation and (5.7),

$$\frac{V(\hat{t}_{\text{cluster}})}{V(\hat{t}_{\text{SRS}})} = \frac{\text{MSB}}{S^2} = \frac{NM-1}{M(N-1)}[1 + (M-1)\text{ICC}]. \tag{5.9}$$

If N, the number of psu's in the population, is large so that $NM - 1 \approx M(N-1)$, then the ratio of the variances in (5.9) is approximately $1 + (M-1)\text{ICC}$. So $1 + (M-1)\text{ICC}$ ssu's, taken in a one-stage cluster sample, give us approximately the same amount of information as one ssu from an SRS. If $\text{ICC} = 1/2$ and $M = 5$, then $1 + (M-1)\text{ICC} = 3$, and we would need to measure 300 elements using a cluster sample to obtain the same precision as an SRS of 100 elements. We hope, though—because it is often much cheaper and easier to collect data in a cluster sample—that we will have more precision per dollar spent in cluster sampling.

The ICC provides a measure of homogeneity for the clusters. The ICC is positive if elements within a psu tend to be similar; then, SSW will be small relative to SSTO, and the ICC relatively large. When the ICC is positive, cluster sampling is less efficient than simple random sampling of elements.

If the clusters occur naturally in the population, the ICC is usually positive. Elements within the same cluster tend to be more similar than elements selected at random from the population. This may occur because the elements in a cluster share a similar environment—we would expect wells in the same geographic cluster to have similar levels of pesticides, or we would expect one area of a city to have a different incidence of measles than another area of a city. In human populations, personal choice as well as interactions among household members or neighbors may cause the ICC to be positive—wealthy households tend to live in similar neighborhoods, and persons in the same neighborhood may share similar opinions.

The ICC is negative if elements within a cluster are dispersed *more* than a randomly chosen group would be. This forces the cluster means to be very nearly equal—because SSTO = SSW + SSB, if SSTO is held fixed and SSW is large, then SSB must be small. If $\text{ICC} < 0$, cluster sampling is more efficient than simple random sampling of elements. The ICC is rarely negative in naturally occurring clusters; negative values can occur in some systematic samples or artificial clusters, as discussed in Section 5.6.

The ICC is only defined for clusters of equal sizes. An alternative quantity that can be used as a measure of homogeneity in general populations is the adjusted R^2, called R_a^2 and defined as

$$R_a^2 = 1 - \frac{\text{MSW}}{S^2}. \tag{5.10}$$

If all clusters are of the same size, then the increase in variance due to cluster sampling is

$$\frac{\text{MSB}}{S^2} = 1 + \frac{N(M-1)}{N-1}R_a^2;$$

by comparing with (5.9), you can see that for many populations R_a^2 is close to the ICC. R_a^2 is a reasonable measure of homogeneity because of its interpretation in linear regression: It is the relative amount of variability in the population explained by the cluster means, adjusted for the number of degrees of freedom. If the clusters are homogeneous, then the cluster means are highly variable relative to the variation within clusters, and R_a^2 will be high.

EXAMPLE 5.3   Consider two artificial populations, each having three clusters with three elements per cluster.

	Population A			Population B		
Cluster 1	10	20	30	9	10	11
Cluster 2	11	20	32	17	20	20
Cluster 3	9	17	31	31	32	30

The elements are the same in the two populations, so the populations share the values $\bar{y}_U = 20$ and $S^2 = 84.5$. In population A, most of the variability occurs within clusters; in population B, most of the variability occurs between clusters.

	Population A		Population B	
	$\bar{y}_{iU}$	S_i^2	$\bar{y}_{iU}$	S_i^2
Cluster 1	20	100	10	1
Cluster 2	21	111	19	3
Cluster 3	19	124	31	1

ANOVA Table for Population A:

Source	df	SS	MS	F
Between clusters	2	6	3	0.03
Within clusters	6	670	111.67	
Total, about mean	8	676	84.5	

ANOVA Table for Population B:

Source	df	SS	MS	F
Between clusters	2	666	333	199.8
Within clusters	6	10	1.67	
Total, about mean	8	676	84.5	

$$R_a^2 = -0.3215 \quad \text{and} \quad \text{ICC} = 1 - \left(\frac{3}{2}\right)\frac{670}{676} = -0.4867 \text{ for population A.}$$

$$R_a^2 = 0.9803 \quad \text{and} \quad \text{ICC} = 1 - \left(\frac{3}{2}\right)\frac{10}{676} = 0.9778 \text{ for population B.}$$

Population A has much variation among elements within the clusters but little variation among the cluster means. This is reflected in the large negative values of the ICC and R_a^2: Elements in the same cluster are actually less similar than randomly selected elements from the whole population. For this situation, cluster sampling is more efficient than simple random sampling.

The opposite situation occurs in population B: Most of the variability occurs between clusters, and the clusters themselves are relatively homogeneous. The ICC and R_a^2 are very close to 1, indicating that little new information would be gleaned by sampling more than one element in a cluster. Here, one-stage cluster sampling is much less efficient than simple random sampling. ■

Most real-life populations fall somewhere between these two extremes. The ICC is usually positive but not overly close to 1. Thus, there is a penalty in efficiency for using cluster sampling, and that decreased efficiency should be offset by cost savings.

EXAMPLE 5.4   When all clusters are the same size, we can estimate the variance of $\hat{t}$ and the ICC from the sample ANOVA table. Here is the sample ANOVA table for the GPA data from Example 5.2:

Source	df	SS	MS
Between suites	4	2.2557	0.56392
Within suites	15	2.7756	0.18504
Total	19	5.0313	0.26480

In one-stage cluster sampling with equal cluster sizes, the mean squares for within suites and between suites are unbiased estimators of the corresponding quantities in the population ANOVA table (see Exercise 11). Thus,

$$E\big[\widehat{\text{MSB}}\big] = \text{MSB} = \frac{S_t^2}{M}$$

and, using (5.7),

$$\text{SE}(\hat{\bar{y}}) = \sqrt{\left(1 - \frac{n}{N}\right)\frac{\widehat{\text{MSB}}}{nM}} = \sqrt{\left(1 - \frac{5}{100}\right)\frac{0.56392}{(5)(4)}} = 0.164,$$

as calculated in Example 5.2.

The sample mean square total should not be used to estimate S^2 when n is small, however: These data were collected as a cluster sample and thus do not reflect enough of the cluster-to-cluster variability. Instead, multiply the unbiased estimates of MSB and MSW by the degrees of freedom from the population ANOVA table to estimate the population sums of squares in the table below. First, estimate the population quantities SSB and SSW, then add them to estimate SSTO.

For these data, because the population has 100 suites and hence 99 df for suites, $\widehat{\text{SSB}} = 99 \times 0.56392 = 55.828$. The estimates of the population sums of squares are given in the following table:

Source	df	$\widehat{\text{SS}}$ (Estimated)	MS
Between suites	99	55.828	0.56392
Within suites	300	55.512	0.18504
Total	399	111.340	0.279

Using these estimates, $\hat{S}^2 = 111.340/399 = 0.279$ (note the small difference between this estimate and the one from the sample ANOVA table, 0.265). In addition,

$$\widehat{\text{ICC}} = 1 - \left(\frac{4}{3}\right)\frac{55.512}{111.34} = 0.335$$

and

$$\hat{R}_a^2 = 1 - \frac{0.18504}{0.279} = 0.337.$$

The increase in variance for using cluster sampling is estimated to be

$$\frac{\widehat{\text{MSB}}}{\hat{S}^2} = \frac{0.56392}{0.279} = 2.02.$$

This says that we need to sample about $2.02n$ elements in a cluster sample to get the same precision as an SRS of size n. There are four persons in a cluster, so in terms of precision one cluster is worth about $4/2.02 = 1.98$ SRS persons. ∎

EXAMPLE **5.5** When is a cluster not a cluster? When it's the whole population.

Consider the situation of sampling oak trees on Santa Cruz Island, described in Example 3.5. There, the sampling unit was one tree, and an observation unit was a seedling by the tree. The population of interest was seedlings of oak trees on Santa Cruz Island. Since a random sample was taken of trees, we treated them as independent in the context of the problem; the independence was reasonable since we were only interested in generalizing to the population of oak trees on the island.

But suppose the investigator had been interested in seedling survival in all of California, had divided the regions with oak trees into equal-sized areas, and had randomly selected five of those areas to be in the study. Then the primary sampling unit is the area, and trees are subsampled in each area. If Santa Cruz Island had been selected as one of the five areas, we could no longer treat the ten trees on Santa Cruz Island as though they were part of a random sample of trees from the population; instead, those trees are part of the Santa Cruz Island cluster. We would expect all ten trees on Santa Cruz Island to experience, as a group, different environmental factors (such as weather conditions and numbers of predators) than the ten trees selected in the Santa Ynez Valley on the mainland. Thus, the ICC within each cluster (area) would likely be positive.

However, suppose we were only interested in the seedlings from tree number 10 on Santa Cruz Island. Then the population is all seedlings from tree number 10, and the primary sampling unit is the seedling. In this situation, then, the tree is not a cluster but the entire population. ■

5.2.3 Clusters of Unequal Sizes

Clusters are rarely of equal sizes in social surveys. In one of the early probability samples (Converse 1987), the Enumerative Check Census of 1937, a 2% sample of postal routes was chosen, and questionnaires were distributed to all households on each chosen postal route, with the goal of checking unemployment figures. Since postal routes had different numbers of households, the cluster sizes could vary greatly.

In a one-stage cluster sample of n of the N psu's, we know how to estimate population totals and means in two ways: using unbiased estimation and using ratio estimation.

5.2.3.1 Unbiased Estimation

An unbiased estimator of t is calculated exactly as in (5.1):

$$\hat{t}_{\text{unb}} = \frac{N}{n} \sum_{i \in \mathcal{S}} t_i. \tag{5.11}$$

By (5.3),

$$\text{SE}(\hat{t}_{\text{unb}}) = N \sqrt{\left(1 - \frac{n}{N}\right) \frac{s_t^2}{n}}. \tag{5.12}$$

The difference between unequal- and equal-sized clusters is that the variation among the individual cluster totals t_i is likely to be large when the clusters have different sizes. The investigators conducting the Enumerative Check Census of 1937 were interested in the total number of unemployed persons, and t_i would be the number of unemployed persons in postal route i. One would expect to find more persons, and hence more unemployed persons, on a postal route with a large number of households than on a postal route with a small number of households. So we would expect that t_i would be large when the cluster size M_i was large, and small when M_i was small. Often, then, s_t^2 is larger in a cluster sample when the psu's have unequal sizes than when the psu's all have the same number of ssu's.

The probability that a psu is in the sample is n/N, as an SRS of n of the N psu's is taken. Since one-stage cluster sampling is used, an ssu is included in the sample whenever its psu is included in the sample. Thus, as on page 138,

$$w_{ij} = \frac{1}{P\{\text{ssu } j \text{ of psu } i \text{ is in sample}\}} = \frac{N}{n}.$$

One-stage cluster sampling produces a self-weighting sample when the psu's are selected with equal probabilities. Using the weights, (5.11) may be written as

$$\hat{t}_{\text{unb}} = \sum_{i \in \mathcal{S}} \sum_{j \in \mathcal{S}_i} w_{ij} y_{ij}. \tag{5.13}$$

We can use (5.11) and (5.12) to derive an unbiased estimator for $\bar{y}_U$ and its variance. Define

$$K = \sum_{i=1}^{N} M_i$$

as the total number of ssu's in the population; then

$$\hat{\bar{y}}_{\text{unb}} = \frac{\hat{t}_{\text{unb}}}{K} \tag{5.14}$$

and

$$\text{SE}(\hat{\bar{y}}_{\text{unb}}) = \frac{\text{SE}(\hat{t}_{\text{unb}})}{K}. \tag{5.15}$$

To use (5.14), though, we need to know K, and we often know M_i only for the sampled clusters. In the Enumerative Check Census, for example, the number of households on a postal route would be ascertained only for the postal routes actually chosen to be in the sample.

5.2.3.2   Ratio Estimation

We usually expect t_i to be correlated with M_i; using ratio estimation, the M_i's are the auxiliary variables, taking the role of the x_i's in Chapter 3. Define:

$$\hat{\bar{y}}_r = \frac{\displaystyle\sum_{i \in \mathcal{S}} t_i}{\displaystyle\sum_{i \in \mathcal{S}} M_i}, \tag{5.16}$$

$$\hat{t}_r = K \hat{\bar{y}}_r. \tag{5.17}$$

The estimator $\hat{\bar{y}}_r$ in (5.16) is the quantity $\hat{B}$ from Chapter 3: The denominator depends on which particular psu's are included in the sample, so both numerator and denominator vary from sample to sample. From (3.7),

$$
\mathrm{SE}(\hat{\bar{y}}_r) = \sqrt{\left(1 - \frac{n}{N}\right)\frac{1}{n\bar{M}_U^2}\frac{\displaystyle\sum_{i\in\mathcal{S}}(t_i - \hat{\bar{y}}_r M_i)^2}{n-1}}
$$

$$
= \sqrt{\left(1 - \frac{n}{N}\right)\frac{1}{n\bar{M}_U^2}\frac{\displaystyle\sum_{i\in\mathcal{S}}M_i^2(\bar{y}_i - \hat{\bar{y}}_r)^2}{n-1}}
$$

(5.18)

and, consequently,

$$
\mathrm{SE}(\hat{t}_r) = N\sqrt{\left(1 - \frac{n}{N}\right)\frac{1}{n}\frac{\displaystyle\sum_{i\in\mathcal{S}}M_i^2(\bar{y}_i - \hat{\bar{y}}_r)^2}{n-1}}.
$$

(5.19)

If $\bar{M}_U \doteq K/N$, the average cluster size in the population, is unknown, one may substitute the average of the psu sizes in the sample, $\bar{M}_S$, for $\bar{M}_U$ in (5.18). Rao and Rao (1971) found that the variance estimator using $\bar{M}_S$ has less bias than the variance estimator using $\bar{M}_U$ if the variance of the y's at x_i is proportional to x_i^t for $0 \le t \le 3/2$, under certain conditions.

Note that $\hat{\bar{y}}_r$ from (5.16) may also be calculated using the weights w_{ij}, as

$$
\hat{\bar{y}}_r = \frac{\displaystyle\sum_{i\in\mathcal{S}}\sum_{j\in\mathcal{S}_i}w_{ij}y_{ij}}{\displaystyle\sum_{i\in\mathcal{S}}\sum_{j\in\mathcal{S}_i}w_{ij}}.
$$

(5.20)

The variance of the ratio estimator depends on the variability of the means per element in the clusters and can be much smaller than that of the unbiased estimator. Note, though, that $\hat{t}_r$ requires that we know the total number of elements in the population, K; the unbiased estimator in (5.11) makes no such requirement.

5.3
Two-Stage Cluster Sampling

In one-stage cluster sampling, we examine all the ssu's within the selected psu's. In many situations, though, the elements in a cluster may be so similar that examining all subunits within a psu wastes resources; alternatively, it may be expensive to measure ssu's relative to the cost of sampling psu's. In these situations, taking a subsample within each psu selected may be much cheaper. The stages within a two-stage cluster sample, when we sample the psu's and subsample the ssu's with equal probabilities, are as follows:

1 Select an SRS $\mathcal{S}$ of n psu's from the population of N psu's.

2 Select an SRS of ssu's from each selected psu. The SRS of m_i elements from the ith cluster is denoted $\mathcal{S}_i$.

FIGURE 5.2

The difference between one-stage and two-stage cluster sampling

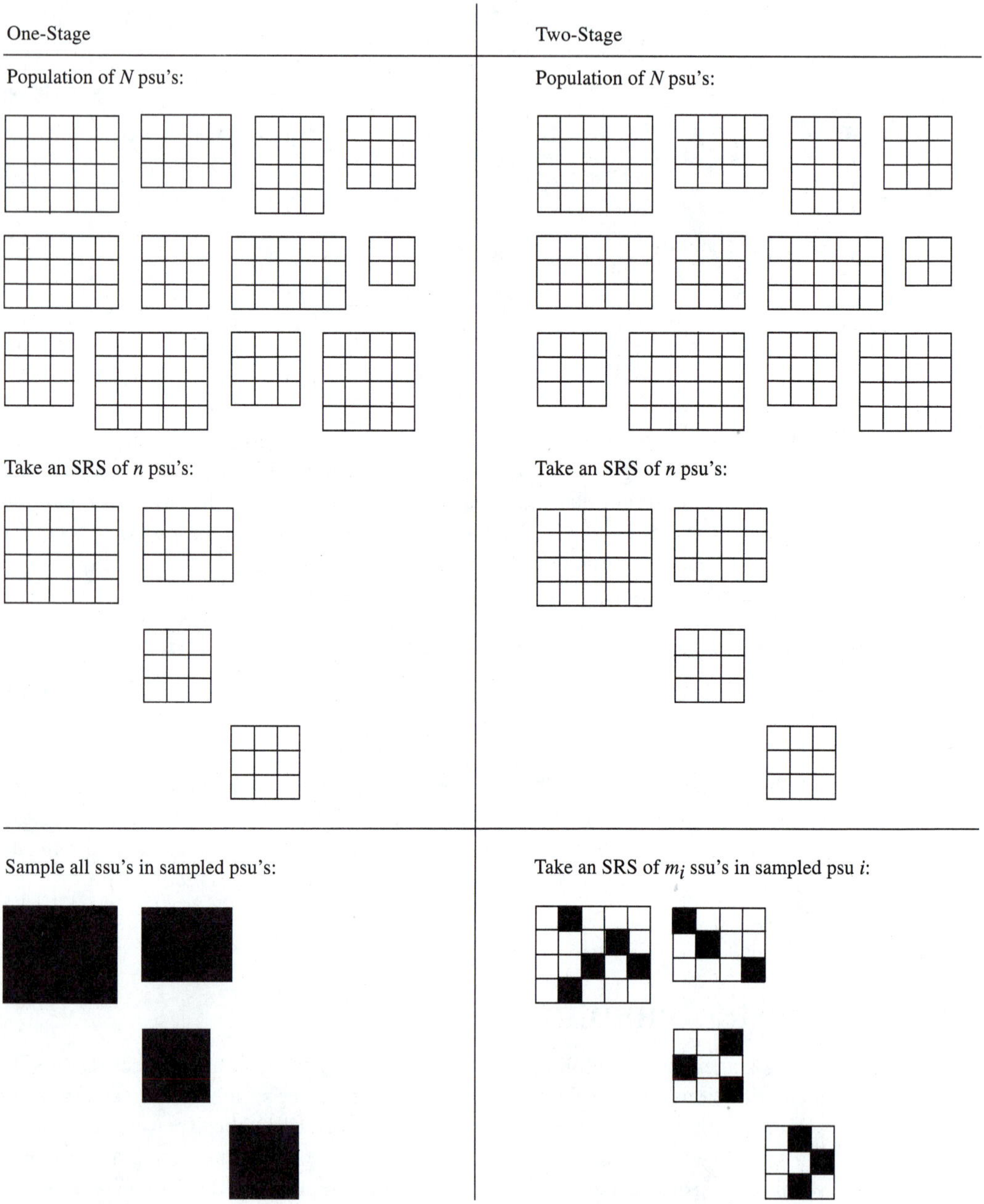

The difference between one-stage and two-stage cluster sampling is illustrated in Figure 5.2. The extra stage complicates the notation and estimators, as we need to consider variability arising from both stages of data collection. The point estimates

of t and $\bar{y}_U$ are analogous to those in one-stage cluster sampling, but the variance formulas become much messier.

In one-stage cluster sampling, we could estimate the population total by $\hat{t}_{\text{unb}} = (N/n)\sum_{i \in S} t_i$; the psu totals t_i were known because we sampled every ssu in the selected psu's. In two-stage cluster sampling, however, since we do not observe every ssu in the sampled psu's, we need to estimate the individual psu totals by

$$\hat{t}_i = \sum_{j \in S_i} \frac{M_i}{m_i} y_{ij} = M_i \bar{y}_i,$$

and an unbiased estimator of the population total is

$$\hat{t}_{\text{unb}} = \frac{N}{n} \sum_{i \in S} \hat{t}_i = \frac{N}{n} \sum_{i \in S} M_i \bar{y}_i. \tag{5.21}$$

In two-stage sampling, the $\hat{t}_i$'s are random variables. Consequently, the variance of $\hat{t}$ has two components: (1) the variability between psu's and (2) the variability of ssu's within psu's. We do not have to worry about component (2) in one-stage cluster sampling.

The variance of $\hat{t}_{\text{unb}}$ equals the variance of $\hat{t}_{\text{unb}}$ from one-stage cluster sampling plus an extra term because the $\hat{t}_i$'s estimate the cluster totals. For two-stage cluster sampling,

$$V(\hat{t}_{\text{unb}}) = N^2 \left(1 - \frac{n}{N}\right) \frac{S_t^2}{n} + \frac{N}{n} \sum_{i=1}^{N} \left(1 - \frac{m_i}{M_i}\right) M_i^2 \frac{S_i^2}{m_i}, \tag{5.22}$$

where S_t^2 is the population variance of the cluster totals and S_i^2 is the population variance among the elements within cluster i. The first term in (5.22) is the variance from one-stage cluster sampling, and the second term is the additional variance due to subsampling. To prove (5.22), we need to condition on the units included in the sample. This is more easily done in the general setting of unequal probability sampling; to avoid proving the same result twice, we will prove the general result in Section 6.6.[1]

To estimate $V(\hat{t}_{\text{unb}})$, let

$$s_t^2 = \frac{\sum_{i \in S} \left(\hat{t}_i - \frac{\hat{t}_{\text{unb}}}{N}\right)^2}{n - 1} \tag{5.23}$$

and

$$s_i^2 = \frac{\sum_{j \in S_i} (y_{ij} - \bar{y}_i)^2}{m_i - 1}. \tag{5.24}$$

As will be shown in Section 6.6, an unbiased estimator of the variance in (5.22) is given by

$$\hat{V}(\hat{t}_{\text{unb}}) = N^2 \left(1 - \frac{n}{N}\right) \frac{s_t^2}{n} + \frac{N}{n} \sum_{i \in S} \left(1 - \frac{m_i}{M_i}\right) M_i^2 \frac{s_i^2}{m_i}. \tag{5.25}$$

---

[1]Working with the additional level of abstraction will allow us to see the structure of the variance more clearly, without floundering in the notation of the special case of equal probabilities discussed in this chapter. If you prefer to see the proof before you use the variance results, read Section 6.6 now.

The standard error, $SE(\hat{t}_{unb})$, is of course the square root of (5.25). In many situations, N/n will be small relative to N^2, so the contribution of the second term in (5.25) to the variance estimator will be negligible compared with that of the first term.

If we know the total number of elements in the population, K, we can estimate the population mean by

$$\hat{\bar{y}}_{unb} = \frac{\hat{t}_{unb}}{K} \tag{5.26}$$

with standard error

$$SE(\hat{\bar{y}}_{unb}) = \frac{SE(\hat{t}_{unb})}{K}. \tag{5.27}$$

As in one-stage cluster sampling with unequal cluster sizes, the between-psu component of variance can be very large since it is affected both by variations in the unit sizes (the M_i) and by variations in the $\bar{y}_i$. If the cluster sizes are disparate, this component is large, even if the cluster means are fairly constant.

Ratio Estimation   We can also use a ratio estimator for estimating the population mean. Again, the y's of Chapter 3 are the cluster totals (now estimated) and the x's are the cluster sizes M_i:

$$\hat{\bar{y}}_r = \frac{\sum_{i \in \mathcal{S}} \hat{t}_i}{\sum_{i \in \mathcal{S}} M_i} = \frac{\sum_{i \in \mathcal{S}} M_i \bar{y}_i}{\sum_{i \in \mathcal{S}} M_i}. \tag{5.28}$$

The variance formula is based on the Taylor series approximation in (3.7) again:

$$\hat{V}(\hat{\bar{y}}_r) = \frac{1}{\bar{M}^2} \left[\left(1 - \frac{n}{N} \right) \frac{s_r^2}{n} + \frac{1}{nN} \sum_{i \in \mathcal{S}} M_i^2 \left(1 - \frac{m_i}{M_i} \right) \frac{s_i^2}{m_i} \right], \tag{5.29}$$

where the s_i^2's are defined in (5.24),

$$s_r^2 = \frac{\sum_{i \in \mathcal{S}} (M_i \bar{y}_i - M_i \hat{\bar{y}}_r)^2}{n - 1},$$

and $\bar{M}$ is the average cluster size—either the population average or sample average can be used in the estimate of the variance.

EXAMPLE **5.6**   The data in the file coots.dat come from Arnold's (1991) work on egg size and volume of American coot eggs in Minnedosa, Manitoba. In this data set, we look at volumes of a subsample of eggs in clutches (nests of eggs) with at least two eggs available for measurement.

The data are plotted in Figures 5.3–5.5. Data from a cluster sample can be plotted in many ways, and you often need to construct more than one type of plot to see features of the data. Because we have only two observations per clutch, we can plot the individual data points. If we had many observations per clutch, we could instead construct side-by-side boxplots, with one boxplot for each psu.[2] We will return to the issue of plotting data from complex surveys in Section 7.4.

---

[2] We did a similar plot in Figure 4.1 for a stratified sample, constructing a boxplot for each stratum.

FIGURE   5.3

A plot of egg-volume data. Note the wide variation in the means from clutch to clutch. This indicates that eggs within the same clutch tend to be more similar than two randomly selected eggs from different clutches and that clustering does not provide as much information per egg as would an SRS of eggs.

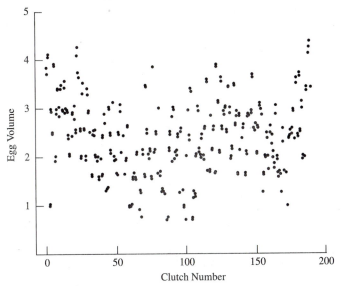

FIGURE   5.4

Another plot of egg-volume data. Here, the clutches are ordered from smallest mean to largest mean, and a line connects the two measurements of volume from the eggs in the clutch. Clutch number 88, represented by the long line in the middle of the graph, has an unusually large difference between the two eggs: One egg has volume 1.85, and the other has volume 2.84.

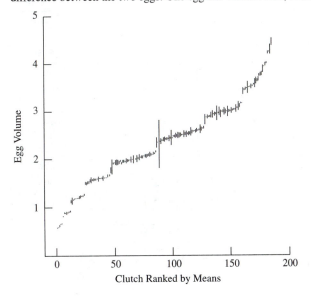

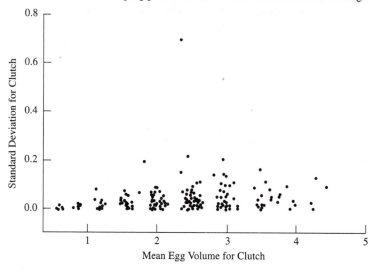

FIGURE 5.5

Yet another plot for the egg-volume data. This plot shows the relation between mean egg volume and standard deviation of egg volume within clutches. The unusual observation is from clutch 88. The clumping pattern for the means warrants further investigation.

TABLE 5.2

Spreadsheet Used for Calculations in Example 5.6

clutch	M_i	$\bar{y}_i$	s_i^2	$\hat{t}_i$	$\left(1 - \dfrac{2}{M_i}\right) M_i^2 \dfrac{s_i^2}{m_i}$	$(\hat{t}_i - M_i \hat{\bar{y}}_r)^2$
1	13	3.86	0.0094	50.23594	0.671901	318.9232
2	13	4.19	0.0009	54.52438	0.065615	490.4832
3	6	0.92	0.0005	5.49750	0.005777	89.22633
4	11	3.00	0.0008	32.98168	0.039354	31.19576
5	10	2.50	0.0002	24.95708	0.006298	0.002631
6	13	3.98	0.0003	51.79537	0.023622	377.053
7	9	1.93	0.0051	17.34362	0.159441	25.72099
8	11	2.96	0.0051	32.57679	0.253589	26.83682
9	12	3.46	0.0001	41.52695	0.006396	135.4898
10	11	2.96	0.0224	32.57679	1.108664	26.83682
⋮	⋮	⋮	⋮	⋮	⋮	⋮
180	9	1.95	0.0001	17.51918	0.002391	23.97106
181	12	3.45	0.0017	41.43934	0.102339	133.4579
182	13	4.22	0.00003	54.85854	0.002625	505.3962
183	13	4.41	0.0088	57.39262	0.630563	625.7549
184	12	3.48	0.000006	41.81168	0.000400	142.1994
sum	1757			4375.947	42.17445	11,439.58
var				149.564814		
$\hat{\bar{y}}_r =$		2.490579				

Next, we use a spreadsheet (Table 5.2) to calculate summary statistics for each clutch. The summary statistics can then be used to estimate the average egg volume and its variance. The numbers have been rounded so that they will fit on the page; in practice, of course, you should carry out all calculations to machine precision.

We use the ratio estimator to find the mean egg volume. In this case we cannot use the unbiased estimator since K, the total number of eggs in the population, is unknown. From (5.28),

$$\hat{\bar{y}}_r = \frac{\sum\limits_{i \in S} \hat{t}_i}{\sum\limits_{i \in S} M_i} = \frac{4375.947}{1757} = 2.49.$$

From the spreadsheet (Table 5.2),

$$s_r^2 = \frac{\sum\limits_{i \in S} (\hat{t}_i - M_i \hat{\bar{y}}_r)^2}{n-1} = \frac{11,439.58}{183} = 62.511$$

and $\bar{M}_S = 1757/184 = 9.549$. Using (5.29), then,

$$\hat{V}(\hat{\bar{y}}_r) = \frac{1}{9.549^2}\left[\left(1 - \frac{184}{N}\right)\frac{62.511}{184} + \left(\frac{1}{N}\right)\frac{42.17}{184}\right].$$

Now N, the total number of clutches in the population, is unknown but presumed to be large (and known to be larger than 184). Thus, we take the psu-level fpc to be 1 and note that the second term in the estimated variance will be very small relative to the first term. We then use

$$SE(\hat{\bar{y}}_r) = \frac{1}{9.549}\sqrt{\frac{62.511}{184}} = 0.061.$$

The estimated coefficient of variation for $\hat{\bar{y}}_r$ is

$$\frac{SE(\hat{\bar{y}}_r)}{\hat{\bar{y}}_r} = \frac{0.061}{2.49} = 0.0245. \quad \blacksquare$$

In Example 5.6 we could only use the ratio estimator because we know neither N nor K. The M_i's, however, did not vary widely, so the unbiased estimator would probably have had similar coefficient of variation. If all M_i's are equal, the unbiased estimator is in fact the same as the ratio estimator (see Exercise 11); if the M_i's vary, the unbiased estimator often performs poorly. The next example illustrates that the unbiased estimator of t may have large variance when the cluster sizes are highly variable.

EXAMPLE 5.7   *The Case of the Six-Legged Puppy*

Suppose we want to estimate the average number of legs on the healthy puppies in Sample City puppy homes. Sample City has two puppy homes: Puppy Palace with 30 puppies and Dog's Life with 10 puppies. Let's select one puppy home with probability $1/2$. After the home is selected, then select 2 puppies at random from the home and use $\hat{\bar{y}}_{unb}$ to estimate the average number of legs per puppy.

Suppose we select Puppy Palace. Not surprisingly, each of the 2 puppies sampled has four legs, so $\hat{t}_{PP} = 30 \times 4 = 120$. Then, using (5.21) and (5.26), an unbiased estimate for the total number of puppy legs in both homes is

$$\hat{t}_{\text{unb}} = \frac{2}{1}\hat{t}_{\text{PP}} = 240.$$

We divide the estimated total by the number of puppies to estimate the mean number of legs per puppy as $240/40 = 6$.

If we select Dog's Life instead, $\hat{t}_{\text{DL}} = 10 \times 4 = 40$, and

$$\hat{t}_{\text{unb}} = \frac{2}{1}\hat{t}_{\text{DL}} = 80.$$

If Dog's Life is selected, the unbiased estimate of the mean number of legs per puppy is $80/40 = 2$.

These are not good estimates of the number of legs per puppy. But the estimator is mathematically unbiased: $(6 + 2)/2 = 4$, so averaging over all possible samples results in the right number. The poor quality of the estimator is reflected in the very large variance of the estimate, calculated using (5.22):

$$V(\hat{t}_{\text{unb}}) = \left(1 - \frac{1}{2}\right) 2^2 \frac{S_t^2}{1} + \frac{2}{1} \sum_{i=1}^{2} \left(1 - \frac{m_i}{M_i}\right) M_i^2 \frac{S_i^2}{m_i}$$

$$= \frac{1}{2}(4)(3200) = 6400.$$

The ratio estimator, however, is right on target: If Puppy Palace is selected, $\hat{\bar{y}}_r = 120/30 = 4$; if Dog's Life is selected, $\hat{\bar{y}}_r = 40/10 = 4$. Because the estimate is the same for all possible samples, $V(\hat{\bar{y}}_r) = 0$. ∎

In general, the unbiased estimator of the population total is inefficient if the cluster sizes are unequal and t_i is roughly proportional to M_i. The variance of $\hat{t}_{\text{unb}}$ depends on the variance of t_i, and that variance may be large if the M_i's are unequal.

The ratio estimator, however, generally performs well when t_i is roughly proportional to M_i. Recall from (3.5) that the approximate mean squared error (MSE) of the estimator $\hat{B}$ is proportional to the variance of the residuals from the model: Using the notation of this chapter, the approximate MSE of $\hat{\bar{y}}_r(=\hat{B})$ is proportional to $\sum_{i=1}^{N}(t_i - \bar{y}_U M_i)^2$. When t_i (the response variable) is highly positively correlated with M_i (the auxiliary variable), the residuals are small. In Example 5.7, the total number of puppy legs in a puppy home (t_i) is exactly four times the total number of puppies in the home (M_i), so the variance of the ratio estimator is zero.

This is an important issue, since many naturally occurring clusters are of unequal sizes, and we expect that the cluster totals will often be proportional to the number of ssu's. In a cluster sample of nursing homes, we expect that a larger number of residents will be satisfied with the level of care in a home with 500 residents than in a home with 20 residents, even though the proportions of residents who are satisfied may be the same. The total of the math scores for all students in a class will be much greater for large classes than for small classes. In general, we expect to see more honeybees in a large area than in a small area. For all these situations, then, while the estimator $\hat{\bar{y}}_r$ works well, the estimator $\hat{t}_{\text{unb}}$ tends to have large variability. In Chapter 6, we will discuss an alternative design and estimator for cluster sampling that result in a much lower variance for the estimated population total when t_i is proportional to M_i.

5.4
Using Weights in Cluster Samples

For estimating overall means and totals in cluster samples, most survey statisticians use sampling weights. As we will discuss in Sections 7.2 and 7.3, weights can be used to find a point estimate of almost any quantity of interest from any probability sampling design. They are thus an extremely valuable tool for analyzing survey data.

Remember from stratified sampling that the weight of an element is the reciprocal of the probability of its selection. For cluster sampling,

$P(j$th ssu in ith psu is selected)

$$= P(i\text{th psu selected}) \times P(j\text{th ssu selected} \mid i\text{th psu selected}) \tag{5.30}$$

$$= \frac{n}{N}\frac{m_i}{M_i}.$$

Thus,

$$w_{ij} = \frac{N M_i}{n m_i}. \tag{5.31}$$

If psu's are blocks, for example, and ssu's are households, then household j in block i represents $(N M_i)/(n m_i)$ households in the population: itself, and $(N M_i)/(n m_i) - 1$ other households. Then,

$$\hat{t}_{\text{unb}} = \sum_{i \in \mathcal{S}} \sum_{j \in \mathcal{S}_i} w_{ij} y_{ij} \tag{5.32}$$

and

$$\hat{\bar{y}}_r = \frac{\hat{t}_{\text{unb}}}{\displaystyle\sum_{i \in \mathcal{S}} \sum_{j \in \mathcal{S}_i} w_{ij}}. \tag{5.33}$$

Note that $\hat{t}_{\text{unb}}$ is the same as in (5.21) and that $\hat{\bar{y}}_r$ is the same as in (5.28). The sampling weights merely provide a convenient way of calculating these estimates; they do not avoid associated shortcomings such as large variances. Also, the sampling weights give no information on how to find standard errors; either the formulas in this chapter or a method from Chapter 9 needs to be used.

In two-stage cluster sampling, a self-weighting design has each ssu representing the same number of ssu's in the population. For a self-weighting sample of persons in Illinois, we could take an SRS of counties in Illinois and then take a sample of m_i of the M_i persons from county i. To have every person in the sample represent the same number of persons in the population, m_i needs to be proportional to M_i, so m_i/M_i is approximately constant. Thus, the large counties have more persons sampled than the small counties.

EXAMPLE 5.8   In Example 5.6, the weights for the observations are

$$\frac{N}{n}\frac{M_i}{m_i} = \frac{N}{184}\frac{M_i}{2}.$$

Because N is unknown, we display the relative weights $M_i/2$ in a spreadsheet

TABLE 5.3

Spreadsheet for Egg Volume Calculations Using Relative Weights

clutch	csize	volume	relweight	wt*vol
1	13	3.795757	6.5	24.67242
1	13	3.93285	6.5	25.56352
2	13	4.215604	6.5	27.40142
2	13	4.172762	6.5	27.12295
3	6	0.931765	3	2.795294
3	6	0.900736	3	2.702209
4	11	3.018272	5.5	16.6005
4	11	2.978397	5.5	16.38118
$\vdots$	$\vdots$	$\vdots$	$\vdots$	$\vdots$
183	13	4.481221	6.5	29.12794
183	13	4.348412	6.5	28.26468
184	12	3.486132	6	20.91679
184	12	3.482482	6	20.89489
sum	3514		1757	4375.947

(Table 5.3). Column 5 is set equal to y_i times the relative weight; using (5.33), $\hat{\bar{y}}_r = 4375.947/1757 = 2.49$. The weights do not allow us to calculate the standard error, however; we still need to use (5.29) for that. ∎

5.5
Designing a Cluster Sample

Persons and organizations taking an expensive, large-scale survey need to devote a great deal of time to designing the survey; typically, large surveys administered by the Bureau of the Census take several years to design and test. Even then, the Fundamental Principle of Survey Design often holds true: You can best design the survey you should have taken after you have finished the survey. After the survey is completed, you can assess the effect of the clustering on the estimates and know where you could have allocated more resources to obtain better information.

The more you know about a population, the better you can design an efficient sampling scheme to study it. If you know the value of y_{ij} for every person in your population, then you can design a flawless (but unnecessary because you already know everything!) survey for studying the population. If you know very little about the population, chances are that you will gain information about it after collecting the survey, but you may not have the most efficient design possible for that survey. You may, however, be able to use your newly gained knowledge to make the next survey more efficient.

When designing a cluster sample, you need to decide four major issues:

1 What overall precision is needed?

2 What size should the psu's be?

3 How many ssu's should be sampled in each psu selected for the sample?

4 How many psu's should be sampled?

Question 1 must be faced in any survey design. To answer questions 2 through 4, you need to know the cost of sampling a psu for possible psu sizes, the cost of sampling an ssu, and a measure of homogeneity (R_a^2 or ICC) for the possible sizes of psu.

5.5.1 Choosing the psu Size

The psu size is often a natural unit. In Example 5.6, a clutch of eggs was an obvious cluster unit. A survey to estimate calf mortality might use farms as the psu's; a survey of sixth-grade students might use classes or schools as the psu's.

In other surveys, however, the investigator may have a wide choice for psu size. In a survey to estimate the sex and age ratios of mule deer in a region of Colorado (see Bowden et al. 1984 for more discussion of the problem), psu's might be designated areas, and ssu's might be individual deer or groups of deer in those areas. But should the size of the psu's be 1 km^2, 2 km^2, or 100 m^2?

A general principle in area surveys is that the larger the psu size, the more variability you expect to see within a psu. Hence, you expect R_a^2 and ICC to be smaller with a large psu than with a small psu. If the psu size is too large, however, you may lose the cost savings of cluster sampling.

Bellhouse (1984) gives a review of optimal design for sampling, and the theory provides useful guidance for designing your own survey. There are many ways to "try out" different psu sizes before taking your survey. One way is to postulate a model for the relationship between R_a^2 or MSW and M and to fit the model using preliminary data or information from other studies. Then use different combinations of R_a^2 and M and compare the costs. Another way is to perform an experiment and collect data on relative costs and variances with different psu sizes.

EXAMPLE 5.9 The Colorado potato beetle has long been considered a major pest by potato farmers. Zehnder et al. (1990) studied different sizes of sampling units that could be used to estimate potato beetle counts. Ten randomly selected sites were sampled from each of ten fields. The investigators visually inspected each site for small larvae, large larvae, and adults on all foliage from a single stem on each of five adjacent plants.

They then considered different psu sizes, ranging from one stem per site to five stems per site. To study the efficiency of a one-stem-per-site design, they examined data from stem 1 of each site. Similarly, the data from stems 1 and 2 of each site gave a cluster sample with two ssu's per psu, and so on. It takes about 30 minutes to walk among the sites in each field; sampling one stem requires about 10 seconds during the early part of the season. Thus, the total cost to sample all ten sites with the one-stem-per-site design is estimated to be $30 + 100/60 = 31.67$ minutes. Data for estimating the number of small larvae are given in Table 5.4.

The relative SE is calculated as 1000(SE/cost). For this example, since the cost to sample additional stems at a site is small compared with the time to traverse the field, the five-stem-per-site design is most efficient among those studied. ∎

TABLE   5.4

Relative Standard Errors in the Potato Beetle Study

Number of Stems Sampled per Site	$\hat{\bar{y}}$	$SE(\hat{\bar{y}})$	Cost to Sample One Field	Relative SE
1	1.12	0.15	31.67	4.7
2	1.01	0.10	33.33	3.0
3	0.96	0.08	35.00	2.3
4	0.91	0.07	36.67	1.9
5	0.91	0.06	38.33	1.6

5.5.2 Choosing Subsampling Sizes

The goal in designing a sample is generally to get the most information for the least cost and inconvenience. In this section, we concentrate on designing a two-stage cluster survey when all clusters have the same number, M, of ssu's; designing cluster samples will be treated more generally in Chapters 6 and 7. One approach for equal-sized clusters, discussed in Cochran (1977), is to minimize the variance in (5.22) for a fixed cost. If $M_i = M$ and $m_i = m$ for all psu's, then $V(\hat{\bar{y}}_{unb})$ may be rewritten as (see Exercise 10)

$$V(\hat{\bar{y}}_{unb}) = \left(1 - \frac{n}{N}\right)\frac{MSB}{nM} + \left(1 - \frac{m}{M}\right)\frac{MSW}{nm}, \tag{5.34}$$

where MSB and MSW are the between and within mean squares, respectively, in Table 5.1, the population ANOVA table.

If $MSW = 0$ and hence $R_a^2 = 1$, for R_a^2 defined in (5.10), then all elements within a cluster have the value of the cluster mean. In that case you may as well take $m = 1$; examining more than one element per cluster just costs extra time and money without increasing precision. For other values of R_a^2, the optimal allocation depends on the relative costs of sampling psu's and ssu's.

Consider the simple cost function

$$\text{total cost} = C = c_1 n + c_2 nm, \tag{5.35}$$

where c_1 is the cost per psu (not including the cost of measuring ssu's) and c_2 is the cost of measuring each ssu. One can easily determine, using calculus, that the values

$$n = \frac{C}{c_1 + c_2 m}$$

and

$$m = \sqrt{\frac{c_1 M(MSW)}{c_2(MSB - MSW)}} = \sqrt{\frac{c_1 M(N-1)}{c_2(NM-1)}\left(\frac{1}{R_a^2} - 1\right)}$$

minimize the variance for fixed total cost C under this cost function (see Exercises 10 and 23); often, though, a number of different values of m will work about equally well, and graphing the projected variance of the estimate will give more information than merely computing one fixed solution. A graphical approach also allows you to perform what-if analyses on the designs: What if the costs or the cost function are slightly different? Or the value of R_a^2 is changed slightly? You can also explore different cost functions with this approach.

EXAMPLE 5.10   Would subsampling have been more efficient for Example 5.2 than the one-stage cluster sample that was used? We do not know the population quantities but have information from the sample that can be used for planning future studies. Recall that $\hat{S}^2 = 0.279$, and we estimated R_a^2 as 0.337. Figures 5.6 and 5.7 show the estimated variance that would be achieved for different subsample sizes for different values of c_1 and c_2 and for different values of R_a^2.   ∎

FIGURE 5.6
Estimated variance that would be obtained for the GPA example, for different values of c_1 and c_2 and different values of m. The sample estimate R_a^2 is 0.337. The total cost C is 300, for this graph. If it takes 40 minutes per suite and 5 minutes per person, then one-stage cluster sampling should be used; if it takes 10 minutes per suite and 20 minutes per person, then only one person should be sampled per suite; if it takes 20 minutes per suite and 10 minutes per person, the minimum is reached at $m \approx 2$, although the flatness of the curve indicates that any subsampling size would be acceptable.

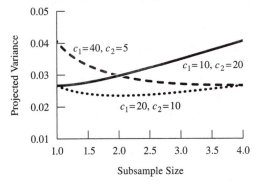

FIGURE 5.7
Estimated variance that would be obtained for the GPA example, for different values of R_a^2 and different values of m. The costs used in constructing this graph are $C = 300$, $c_1 = 20$, and $c_2 = 10$. The higher the value of R_a^2, the smaller the subsample size m should be.

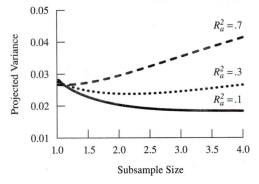

For design purposes, we need only a rough estimate of R_a^2; usually, the adjusted R^2 from the ANOVA table from sample data provides a good starting point, even though the sample value of the mean square total often underestimates S^2 when the number of psu's in the sample is small.

EXAMPLE 5.11   Here is the sample ANOVA table for the coots data, calculated using SAS.

Source	DF	Sum of Squares	Mean Square	F Value
Model	183	257.4175336	1.4066532	237.44
Error	184	1.0900782	0.0059243	
Corrected Total	367	258.5076118		

R-Square	C.V.	Root MSE	VOLUME Mean
0.995783	3.298616	0.076970	2.333394

If a future survey were planned to estimate average egg volume, one might explore subsample sizes using R_a^2's around $1 - 0.0059243/(258.5/367) = 0.99$. These data indicate a high degree of homogeneity within clutches for egg volume. For this survey, however, the marginal cost of measuring additional eggs within a clutch is very small compared with the cost of locating and accessing a clutch—it might be best to take $m_i = M_i$ despite the high degree of homogeneity, because the additional information can be used to answer other research questions concerning variability from clutch to clutch or possible effects of laying sequence. ∎

Although we discussed only designs where all M_i's are equal, we can use these methods with unequal M_i's as well: just substitute $\bar{M}$ for M in the preceding work and decide the average subsample size $\bar{m}$ to take. Then either take $\bar{m}$ observations in every cluster or allocate observations so that

$$\frac{m_i}{M_i} = \text{constant.}$$

As long as the M_i's do not vary too much, this should produce a reasonable design. If the M_i's are widely variable and the t_i's are correlated with the M_i's, a cluster sample with equal probabilities is not necessarily very efficient; an alternative design is presented in Chapter 6.

5.5.3 Choosing the Sample Size (Number of psu's)

After the psu size is determined and the subsampling fraction set, we then look at the number of psu's to sample, n. Like any survey design, design of a cluster sample is an iterative process: (1) Determine a desired precision, (2) choose the psu and subsample sizes, (3) conjecture the variance that will be achieved with that design, (4) set n to achieve the precision, and (5) repeat (adding stratification and auxiliary variables to use in ratio estimation) until the cost of the survey is within your budget.

If clusters are of equal size and we ignore the psu-level finite population correction (fpc), (5.34) implies that

$$V(\hat{\bar{y}}_{\text{unb}}) \leq \frac{1}{n}\left[\frac{\text{MSB}}{M} + \left(1 - \frac{m}{M}\right)\frac{\text{MSW}}{m}\right] = \frac{1}{n}v.$$

An approximate $100(1 - \alpha)\%$ confidence interval (CI) will be

$$\hat{\bar{y}}_{\text{unb}} \pm z_{\alpha/2}\sqrt{\frac{1}{n}\,v}.$$

Thus, to achieve a desired confidence interval half-width e, set $n = z_{\alpha/2}^2 v/e^2$. Of course, this approach presupposes that you have some knowledge of v, perhaps from a prior survey. In Section 7.5, we will examine how to determine sample sizes for any situation in which you know the efficiency of the specified design relative to an SRS design.

5.6
Systematic Sampling

Systematic sampling, discussed briefly in Chapter 2, is really a special case of cluster sampling. Suppose we want to take a sample of size 3 from a population that has 12 elements:

$$1 \quad 2 \quad 3 \quad 4 \quad 5 \quad 6 \quad 7 \quad 8 \quad 9 \quad 10 \quad 11 \quad 12$$

To take a systematic sample, choose a number randomly between 1 and 4. Draw that element and every fourth element thereafter. Thus, the population contains four psu's (they are clusters even though the elements are not contiguous):

$$\{1, 5, 9\} \quad \{2, 6, 10\} \quad \{3, 7, 11\} \quad \{4, 8, 12\}.$$

Now we take an SRS of one psu.

In a population of NM elements, there are N possible choices for the systematic sample, each of size M. We observe only the mean of the one cluster that comprises our systematic sample,

$$\bar{y}_i = \bar{y}_{iU} = \hat{\bar{y}}_{\text{sys}}.$$

Because one-stage cluster sampling with equal-sized clusters produces unbiased estimates, $E[\hat{\bar{y}}_{\text{sys}}] = \bar{y}_U$. For a simple systematic sample, we select $n = 1$ of the N clusters, so by (5.5) and (5.9), the theoretical variance is

$$V(\hat{\bar{y}}_{\text{sys}}) = \left(1 - \frac{1}{N}\right)\frac{S_t^2}{M^2}$$

$$= \left(1 - \frac{1}{N}\right)\frac{\text{MSB}}{M} \tag{5.36}$$

$$\approx \frac{S^2}{M}[1 + (M - 1)\text{ICC}].$$

With the notation for cluster sampling, M is the size of the systematic sample. Ignoring the fpc, we see that systematic sampling is more precise than an SRS of size M if the ICC is negative. Systematic sampling is more precise than simple random sampling when the variance within the possible systematic samples (clusters) is *larger* than the overall population variance—then the cluster means will be more similar. If there is little variation within the systematic samples relative to that in the population (that is, ICC > 0), then the elements in the sample all give similar information, and systematic sampling would be expected to have higher variance than an SRS.

Since $n = 1$, however, we cannot obtain an unbiased estimate of $V(\hat{\bar{y}}_{sys})$; we need to know something about the structure of the population to estimate the variance. Let's look at three different population structures.

1 *The list is in random order.* Systematic sampling is likely to produce a sample that behaves like an SRS. In many situations, the ordering of the population is unrelated to the characteristics of interest, as when the list of persons in the sampling frame is in alphabetic order. There is no reason to believe that the persons in a systematic sample will be more or less similar than a random sample of persons: We expect that ICC ≈ 0. In this situation, simple random and systematic sampling will give similar results. We can use SRS results and formulas to estimate $V(\hat{\bar{y}}_{sys})$.

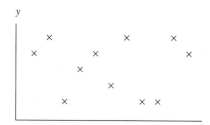

Position in Sampling Frame

2 *The sampling frame is in increasing or decreasing order.* Systematic sampling is likely to be more precise than simple random sampling. Financial records may be listed with the largest amounts first and the smallest amounts last. Such a population is said to have **positive autocorrelation:** Adjacent elements tend to be more similar than elements that are farther apart. In this case, $V(\hat{\bar{y}}_{sys})$ is less than the variance of the sample mean in an SRS of the same size since ICC < 0. A systematic sample forces the sample values to be spread out; it is possible that an SRS would consist of all low values or all high values. When the frame is in increasing or decreasing order, you may use the SRS formula for standard error, but it will likely be an overestimate and confidence intervals constructed using the SRS standard error will be too wide.

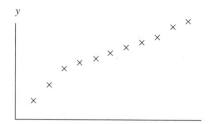

Position in Sampling Frame

Stratified sampling may work better than systematic sampling for positively autocorrelated populations: If the random start is close to either end of the sampling interval, a systematic sample will tend to give an estimate that is too low or too high.

3 *The sampling frame has a periodic pattern.* If we sample at the same interval as the periodicity, systematic sampling will be less precise than simple random sampling. Systematic sampling is most dangerous when the population is in a cyclical or periodic order, and the sampling interval coincides with a multiple of the period.

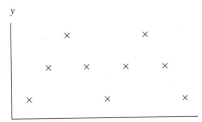

Position in Sampling Frame

Suppose the population values (in order) are

$$1 \quad 2 \quad 3 \quad 1 \quad 2 \quad 3 \quad 1 \quad 2 \quad 3 \quad 1 \quad 2 \quad 3$$

and the sampling interval is 3. Then all elements in the systematic sample will be the same; if we use the SRS formula to estimate the variance, we will have $\hat{V}(\hat{\bar{y}}_{sys}) = 0$. But the true value of $V(\hat{\bar{y}}_{sys})$ for this population is $2/3$; this sample is no more precise than a single observation chosen randomly from the population.

Systematic sampling is often used when a researcher wants a representative sample of the population but does not have the resources to construct a sampling frame in advance. It is commonly used to select elements at the bottom stage of a cluster sample. In many situations in which systematic sampling is used, the systematic sample can be treated as if it were an SRS.

EXAMPLE 5.12   *Sampling for Hazardous Waste Sites*

Many dumps and landfills in the United States contain toxic materials. These materials may have been sealed in containers when deposited but may now be suspected of leaking. But we no longer know where the materials were deposited—containers of hazardous waste may be randomly distributed throughout the landfill, or they may be concentrated in one area, or there may be none at all.

A common practice is to take a systematic sample of grid points and to take soil samples from each to look for evidence of contamination. Choose a point at random in the area, then construct a grid containing that point so that grid points are an equal distance apart. One such grid is shown in Figure 5.8. The advantages of taking a systematic sample rather than an SRS are that the systematic sample forces an even coverage of the region and is easier to implement in the field. If you are not worried about periodic patterns in the distribution of toxic materials and you have little prior knowledge where the toxic materials might be, a systematic sample is a good design.

With any grid in systematic sampling, you need to worry if the toxic materials are regularly placed so that the grid may miss all of them, as shown in Figure 5.9. If this is a concern, you would be better off taking a stratified sample. Lay out the grid but select a point at random in each square at which to take the soil sample. ▪

If periodicity is a concern in a population, one solution is to use **interpenetrating systematic samples** (Mahalanobis 1946). Instead of taking one systematic sample, take several systematic samples from the population. Then you can use the formulas for cluster samples to estimate variances; each systematic sample acts as one cluster. (This approach is explored in Exercise 22.)

FIGURE   5.8

A grid used for detecting hazardous wastes

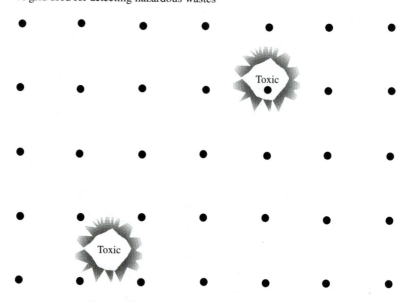

FIGURE   5.9

A grid used for detecting hazardous wastes: the worst-case scenario. Since the waste occurs in a similar pattern to the grid, the systematic sample misses every deposit of toxic waste.

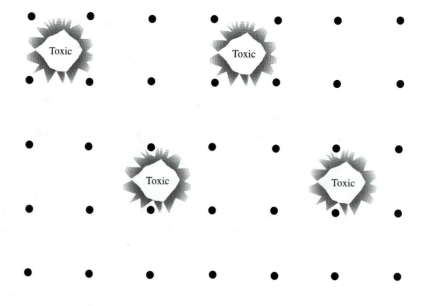

5.7
Models for Cluster Sampling*

The one-way ANOVA model with fixed effects provides a theoretical framework for stratified sampling; one possible analogous model for cluster sampling is the one-way ANOVA model with random effects (Scott and Smith 1969). Let's look at a simple version of this model:

$$\text{M1:} \quad Y_{ij} = A_i + \varepsilon_{ij}, \tag{5.37}$$

with A_i generated by a distribution with mean μ and variance σ_A^2, ε_{ij} generated by a distribution with mean 0 and variance σ^2, and all A_i's and ε_{ij}'s independent.

Model M1 implies that the expected total for a cluster increases linearly with the number of elements in the cluster, because $E_{M1}[Y_{ij}] = \mu$ and

$$E_{M1}[T_i] = E_{M1}\left[\sum_{j=1}^{M_i} Y_{ij}\right] = M_i\mu.$$

This assumption is often appropriate for cluster samples taken in practice. Suppose we are taking a two-stage cluster sample to estimate total hospital charges for delivering babies; hospitals are selected at the first stage, and birth records are selected at the second stage (twins and triplets count as one record). For illustrative purposes, assume that μ, the nationwide average cost for a hospital birth, is about \$10,000. We expect total costs billed by a hospital to be larger if the hospital delivers more babies.

The average cost per birth, however, varies from hospital to hospital—some hospitals may have higher personnel costs, and others may serve a higher-risk population or have more expensive equipment. That variation is reflected in the model by the random effects A_i: A_i is the random variable representing the average cost per birth in the ith hospital, and σ_A^2 is the population variance among the hospital means. In addition, costs vary from birth to birth within the hospitals; that variation is incorporated into the model by the term ε_{ij} with variance σ^2. These ideas are illustrated in Figure 5.10, presuming that the A_i's and ε_{ij}'s are normally distributed.

Figure 5.10 illustrates that, according to the model in (5.37), costs for births in the same hospital tend to be more similar than costs for births selected randomly across the entire population of hospital births, because the cost for a birth in a given hospital incorporates the hospital characteristics such as personnel costs or nurse/patient ratios. The intraclass correlation coefficient for model M1 is defined to be

$$\rho = \frac{\sigma_A^2}{\sigma_A^2 + \sigma^2}. \tag{5.38}$$

Note that ρ in model M1 is always nonnegative, in contrast to the ICC that can take on negative values.[3] Thus, if model M1 describes the data, cluster sampling *must* be

---

[3]Model M1, with $\rho \geq 0$, would not be appropriate if there is competition within clusters so that one member of a cluster profits at the expense of another. For example, if other environmental factors can be discounted, competition within the uterus might cause some fraternal twins to be more variable than nontwin full siblings.

FIGURE 5.10

An illustration of random effects for hospitals and births

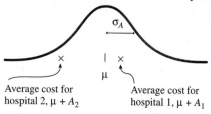

Average cost for
hospital 2, $\mu + A_2$

μ

Average cost for
hospital 1, $\mu + A_1$

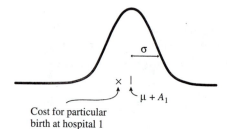

$\mu + A_1$

Cost for particular
birth at hospital 1

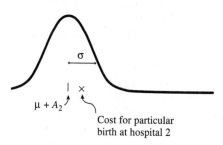

$\mu + A_2$

Cost for particular
birth at hospital 2

less efficient than an SRS of equal size. With model M1,

$$\text{Cov}_{\text{M1}}[Y_{ij}, Y_{kl}] = \begin{cases} \sigma^2 + \sigma_A^2 & \text{if } i = k \text{ and } j = l. \\ \sigma_A^2 & \text{if } i = k \text{ and } j \neq l. \\ 0 & \text{if } i \neq k. \end{cases}$$

5.7.1 Estimation Using Models

Now let's find properties of various estimates under model M1. To save some work later, we look at a general linear estimator of the form

$$\hat{T} = \sum_{i \in \mathcal{S}} \sum_{j \in \mathcal{S}_i} b_{ij} Y_{ij}$$

for b_{ij} any constants. The random variable representing the finite population total is

$$T = \sum_{i=1}^{N} \sum_{j=1}^{M_i} Y_{ij}.$$

Then, the bias is

$$E_{M1}[\hat{T} - T] = E_{M1}\left[\sum_{i \in \mathcal{S}} \sum_{j \in \mathcal{S}_i} b_{ij} Y_{ij} - \sum_{i=1}^{N} \sum_{j=1}^{M_i} Y_{ij}\right]$$

$$= \mu\left(\sum_{i \in \mathcal{S}} \sum_{j \in \mathcal{S}_i} b_{ij} - K\right).$$

Thus, $\hat{T}$ is model-unbiased when $\sum_{i \in \mathcal{S}} \sum_{j \in \mathcal{S}_i} b_{ij} = K$. The model-based (for model M1) variance of $\hat{T} - T$ is

$$V_{M1}[\hat{T} - T] = \sigma_A^2 \left[\sum_{i \in \mathcal{S}}\left(\sum_{j \in \mathcal{S}_i} b_{ij} - M_i\right)^2 + \sum_{i \notin \mathcal{S}} M_i^2\right]$$

$$+ \sigma^2 \left[\sum_{i \in \mathcal{S}} \sum_{j \in \mathcal{S}_i}(b_{ij}^2 - 2b_{ij}) + K\right].$$

(5.39)

(See Exercise 26.)

Now let's look at what happens with design-based estimators under model M1. The random variable for the design-unbiased estimator is

$$\hat{T}_{\text{unb}} = \sum_{i \in \mathcal{S}} \sum_{j \in \mathcal{S}_i} \frac{N M_i}{n m_i} Y_{ij};$$

the coefficients b_{ij} are simply the sampling weights $(N M_i)/(n m_i)$. But

$$\sum_{i \in \mathcal{S}} \sum_{j \in \mathcal{S}_i} b_{ij} = \frac{N}{n} \sum_{i \in \mathcal{S}} \sum_{j \in \mathcal{S}_i} \frac{M_i}{m_i} = \frac{N}{n} \sum_{i \in \mathcal{S}} M_i,$$

so the bias under model (5.37) is

$$\mu\left(\frac{N}{n} \sum_{i \in \mathcal{S}} M_i - K\right).$$

Note that the bias depends on which sample is taken, and the estimator is model-unbiased under (5.37) only when the average of the M_i's in the sample equals the average of the M_i's in the population, such as will occur when all M_i's are the same. This result helps explain why the design-unbiased estimator performs poorly when cluster totals are roughly proportional to cluster sizes: It is a poor estimator for a model that describes the population.

For the ratio estimator, the coefficients are $b_{ij} = K(M_i/m_i)/\sum_{k \in \mathcal{S}} M_k$ and

$$\hat{T}_r = \frac{K \sum_{i \in \mathcal{S}} \sum_{j \in \mathcal{S}_i} \frac{M_i}{m_i} Y_{ij}}{\sum_{k \in \mathcal{S}} M_k}.$$

For these b_{ij}'s,

$$\sum_{i \in \mathcal{S}} \sum_{j \in \mathcal{S}_i} b_{ij} = \sum_{i \in \mathcal{S}} \sum_{j \in \mathcal{S}_i} \frac{K M_i}{m_i \sum_{k \in \mathcal{S}} M_k} = K,$$

so the ratio estimator is model-unbiased under model M1. If model M1 describes the population, then the ratio estimator adjusts for the sizes of the particular psu's chosen for the sample; it uses M_i, a quantity that is correlated with the ith psu total, to compensate for the possibility that the sample may have a different proportion of large psu's than does the population.

The variance expression in (5.39) is complicated; if $M_i = M$ and $m_i = m$ for all i, then $\hat{T}_{unb} = \hat{T}_r$, $b_{ij} = (NM)/(nm)$, and the variance in (5.39) simplifies to

$$V_{M1}[\hat{T}_{unb} - T] = KM(N - n)\frac{\sigma_A^2}{n} + K(MN - mn)\frac{\sigma^2}{mn}. \tag{5.40}$$

EXAMPLE 5.13   Let's return to the puppy homes discussed in Example 5.7. They certainly follow model M1: All puppies have four legs, so $Y_{ij} = \mu = 4$ for all i and j. Consequently, $\sigma_A^2 = \sigma^2 = 0$. The model-based variance of the estimate $\hat{T}_{unb}$ is therefore zero, no matter which puppy home and puppies are chosen. If Puppy Palace is selected for the sample, the bias under model (5.37) is $4(2 \times 30 - 40) = 80$; if Dog's Life is selected, the bias is $4(2 \times 10 - 40) = -80$. The large variance in the design-based approach thus becomes a bias when a model-based approach is adopted. It is not surprising that $\hat{T}_{unb}$ performs poorly for the puppy homes—it is a poor estimator for a model that describes the situation well. Both bias and variance for $\hat{T}_r$, though, are zero.   ∎

The above results are only for model M1. Suppose a better model for the population is

$$\text{M2:} \quad Y_{ij} = B_i + \varepsilon_{ij}, \tag{5.41}$$

with $E[B_i] = \mu/M_i$, $V[M_i B_i] = \sigma_B^2$, $E[\varepsilon_{ij}] = 0$, $V[\varepsilon_{ij}] = \sigma^2$, and all B_i and ε_{ij} independent. Under model M2, then, the cluster totals all have expected value μ, regardless of cluster size. Examples that are described by this model are harder to come by in practice, but let's construct one based on the principle that tasks expand to fill up the allotted time. All students at Idyllic College have 100 hours available for writing term papers, but an individual student may have from one to five papers assigned. It would never occur to an Idyllic student to finish a paper quickly and relax in the extra time, so a student with one paper spends all 100 hours on the paper, a student with two papers spends 50 hours on each, and so on. Thus, the expected total amount of time spent writing term papers, $E[T_i]$, is 100 for each student, although the numbers of papers assigned (M_i) vary.

The estimator $\hat{T}_{unb}$ is unbiased under model M2:

$$E_{M2}[\hat{T}_{unb} - T] = E_{M2}\left[\sum_{i \in S}\sum_{j \in S_i}\frac{NM_i}{nm_i}Y_{ij} - \sum_{i=1}^{N}\sum_{j=1}^{M_i}Y_{ij}\right]$$

$$= \sum_{i \in S}\sum_{j \in S_i}\frac{NM_i}{nm_i}\frac{\mu}{M_i} - \sum_{i=1}^{N}\sum_{j=1}^{M_i}\frac{\mu}{M_i} = 0.$$

Thus, $\hat{T}_{unb}$ performs poorly if model (5.37) is appropriate but often quite well if model (5.41) is appropriate. Of course, these are not the only two possible models: Royall (1976a) derives results for a general class of possible models that includes both (5.37) and (5.41) and allows unequal variances for different clusters.

If you decide to use a model-based approach to analyze cluster sample data, be very careful that the model chosen is appropriate. We saw in the puppy example that the model M1 variance for $\hat{T}_{unb}$ is zero, but the bias is large; we could only evaluate the bias, however, because we knew the results for the whole population. A person who sampled only Puppy Palace and did not know the results for Dog's Life could not evaluate the bias and might conclude that puppies average six legs each! Thus, assessing the adequacy of the model is crucial in any model-based analysis. You must check the assumption that $V[\varepsilon_{ij}] = \sigma^2$ by plotting the variances of each cluster, just as you assess the equal variance assumption in ANOVA. A plot of $\hat{t}_i$ versus M_i is often useful in assessing the appropriateness of a model for the data in the sample. As always in model-based inference, we must assume that the model also holds for population elements not in the sample.

EXAMPLE 5.14 Let's fit model M1, a one-way random-effects model, to the coots data. Looking at Figures 5.4 and 5.5, it seems plausible (except for one clutch) that the within-clutch variance is the same for each clutch. Figure 5.11 shows the plot of $\hat{t}_i$ versus M_i for the coots data.

For these data, $\text{Corr}(\hat{t}_i, M_i) = 0.97$. If model M1 is appropriate for the data, we expect that $\hat{t}_i$ will increase with M_i; if model M2 is appropriate, we expect that horizontal line will fit the plotted points. For these data, $\hat{t}_i$ and M_i are clearly related, although the relationship does not appear to be a straight line.

Using SAS Proc Mixed, the estimated variance components are $\hat{\sigma}_A^2 = 0.70036$ and $\hat{\sigma}^2 = 0.00592$. Using $b_{ij} = M_i/(m_i \sum_{k \in \mathcal{S}} M_k)$, the estimated mean egg volume is 2.492196; adapting (5.39) to ignore the fpc (see Exercise 26), the estimated model-based variance is

$$\sum_{i \in \mathcal{S}} \left(\frac{M_i}{\sum_{k \in \mathcal{S}} M_k} \right)^2 \hat{\sigma}_A^2 + \sum_{i \in \mathcal{S}} \frac{1}{m_i} \left(\frac{M_i}{\sum_{k \in \mathcal{S}} M_k} \right)^2 \hat{\sigma}^2 = 0.003944 + 0.000017 = 0.00396.$$

If a different model were adopted, the estimated variance would be different. ∎

FIGURE 5.11

The plot of $\hat{t}_i$ vs. M_i, for the coots data

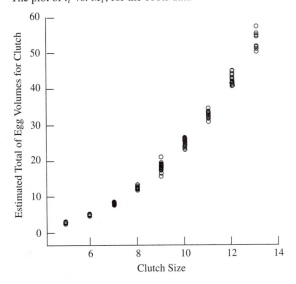

5.7.2 Design Using Models

Models are extremely useful for designing a cluster sample. Using a model for design does not mean you have to use a model for analysis of your survey data when it is collected; rather, the model provides a useful way of summarizing information you can use to make the survey more efficient. Much research has been done on using models for design; see Rao (1979b), Bellhouse (1984), and Royall (1992b) for literature reviews.

Suppose model M1 seems reasonable for your population and all psu sizes in the population are equal. Then you would like to design the survey to minimize the variance in (5.40), subject to cost constraints. Then, using the cost function in (5.35), the model-based variance is minimized when

$$m = \sqrt{\frac{c_1 \sigma^2}{c_2 \sigma_A^2}}.$$

Suppose the M_i's are unequal and model M1 holds. We can use the variance in (5.39) to determine the optimal subsampling size m_i for each cluster. This approach was used by Royall (1976a) for more general models than considered in this section. For $\hat{T}_r$, $b_{ij} = K M_i / (m_i \sum_{k \in S} M_k)$, and the variance is minimized when m_i is proportional to M_i (see Exercise 28).

5.8
Summary

Cluster sampling is commonly used in large surveys, but estimates obtained from cluster samples usually have greater variance than if we were able to measure the same number of observation units using an SRS. If it is much less expensive to sample clusters than individual elements, though, cluster sampling can provide more precision per dollar spent.

All the formulas in this chapter for cluster sampling with equal probabilities are special cases of the general results for two-stage cluster sampling with unequal cluster sizes. They can be applied to any two-stage cluster sample in which the clusters were selected with equal probability. These formulas were given in (5.21), (5.25), (5.28), and (5.29) and are repeated here, respectively:

$$\hat{t}_{\text{unb}} = \frac{N}{n} \sum_{i \in S} \hat{t}_i = \frac{N}{n} \sum_{i \in S} M_i \bar{y}_i, \tag{5.21}$$

$$\hat{V}(\hat{t}_{\text{unb}}) = N^2 \left(1 - \frac{n}{N}\right) \frac{s_t^2}{n} + \frac{N}{n} \sum_{i \in S} \left(1 - \frac{m_i}{M_i}\right) M_i^2 \frac{s_i^2}{m_i}, \tag{5.25}$$

$$\hat{\bar{y}}_r = \frac{\displaystyle\sum_{i \in S} M_i \bar{y}_i}{\displaystyle\sum_{i \in S} M_i}, \tag{5.28}$$

$$\hat{V}(\hat{\bar{y}}_r) = \left(\frac{1}{\bar{M}^2}\right)\left[\left(1 - \frac{n}{N}\right)\frac{s_r^2}{n} + \frac{1}{nN} \sum_{i \in S} \left(1 - \frac{m_i}{M_i}\right) M_i^2 \frac{s_i^2}{m_i}\right], \tag{5.29}$$

with

$$s_t^2 = \frac{\sum_{i \in S} \left(\hat{t}_i - \frac{\hat{t}_{unb}}{N} \right)^2}{n - 1}$$

and

$$s_r^2 = \frac{\sum_{i \in S} (M_i \bar{y}_i - M_i \hat{\bar{y}}_r)^2}{n - 1}.$$

For one-stage cluster sampling, $m_i = M_i$, so the second term in (5.25) and (5.29) is zero. In fact, the formulas for stratified sampling are also a special case of those in this chapter: For stratified sampling, $n = N$, and we sample m_i observations from the M_i observations in stratum i.

In practice, point estimates of the population mean and total are usually calculated using weights. You need to use the preceding formulas or a method such as the jackknife from Chapter 9 to calculate standard errors.

5.9
Exercises

1 A city council of a small city wants to know the proportion of eligible voters who oppose having an incinerator built for burning Phoenix garbage, just outside city limits. They randomly select 100 residential numbers from the city's telephone book that contains 3000 such numbers. Each selected residence is then called and asked for (a) the total number of eligible voters and (b) the number of voters opposed to the incinerator. A total of 157 voters are surveyed; of these, 23 refuse to answer the question. Of the remaining 134 voters, 112 oppose the incinerator, so the council estimates the proportion by

$$\hat{p} = \frac{112}{134} = .83582$$

with

$$\hat{V}[\hat{p}] = \frac{.83582(1 - .83582)}{134} = 0.00102.$$

Are these estimates valid? Why, or why not?

2 Senturia et al. (1994) describe a survey taken to study how many children have access to guns in their households. Questionnaires were distributed to all parents who attended selected clinics in the Chicago area during a 1-week period for well- or sick-child visits.

   a Suppose the quantity of interest is percentage of the households with guns. Describe why this is a cluster sample. What is the psu? The ssu? Is it a one-stage or two-stage cluster sample? How would you estimate the percentage of households with guns and the standard error of your estimate?

   b What is the sampling population for this study? Do you think this sampling procedure results in a representative sample of households with children? Why, or why not?

3   An accounting firm is interested in estimating the error rate in a compliance audit it is conducting. The population contains 828 claims, and the firm audits an SRS of 85 of those claims. In each of the 85 sampled claims, 215 fields are checked for errors. One claim has errors in 4 of the 215 fields, 1 claim has three errors, 4 claims have two errors, 22 claims have one error, and the remaining 57 claims have no errors. (Data courtesy of Fritz Scheuren.)

    a   Treating the claims as psu's and the observations for each field as ssu's, estimate the error rate for all 828 claims. Give a standard error for your estimate.

    b   Estimate (with SE) the total number of errors in the 828 claims.

    c   Suppose that, instead of taking a cluster sample, the firm takes an SRS of $85 \times 215 = 18{,}275$ fields from the 178,020 fields in the population. If the estimated error rate from the SRS is the same as in part (a), what will be the estimated variance $\hat{V}(\hat{p}_{SRS})$? How does this compare with the estimated variance from part (a)?

4   Survey evidence is often introduced in court cases involving trademark violation and employment discrimination. There has been controversy, however, about whether nonprobability samples are acceptable as evidence in litigation. Jacoby and Handlin (1991) selected 26 from a list of 1285 scholarly journals in the social and behavioral sciences. They examined all articles published during 1988 for the selected journals and recorded (1) the number of articles in the journal that described empirical research from a survey (they excluded articles in which the authors analyzed survey data that had been collected by someone else) and (2) the total number of articles for each journal that used probability sampling, nonprobability sampling, or for which the sampling method could not be determined. The data are in file journal.dat.

    a   Explain why this is a cluster sample.

    b   Estimate the proportion of articles in the 1285 journals that use nonprobability sampling, and give the standard error of your estimate.

    c   The authors conclude that, because "an overwhelming proportion of ... recognized scholarly and practitioner experts rely on non-probability sampling designs," courts "should have no problem admitting otherwise well-conducted non-probability surveys and according them due weight" (p. 175). Comment on this statement.

5   Use the data in the file coots.dat to estimate the average egg length, along with its standard error. Be sure to plot the data appropriately.

6   A home owner with a large library needs to estimate the purchase cost and replacement value of the book collection for insurance purposes. She has 44 shelves containing books and selects 12 shelves at random. To prepare for the second stage of sampling, she counts the books on the selected shelves. She then generates five random numbers between 1 and M_i for each selected shelf (see Table 5.5) to determine which specific books, numbered from left to right, to examine more closely. She then looks up the replacement value for the sampled books in *Books in Print*. The data are given in the file books.dat.

    a   Draw side-by-side boxplots for the replacement costs of books on each shelf. Does it appear that the means are about the same? The variances?

TABLE   5.5

Table for Exercise 6

Shelf Number	Number of Books (M_i)	Book Numbers Selected				
2	26	3	5	6	18	19
4	52	2	15	25	36	37
11	70	19	45	48	56	65
14	47	8	9	16	40	44
20	5	1	2	3	4	5
22	28	1	3	7	14	27
23	27	5	14	16	19	26
31	29	10	14	16	19	23
37	21	8	16	17	18	21
38	31	5	9	17	20	27
40	14	5	6	7	8	14
43	27	4	6	12	16	24

b Estimate the total replacement cost for the library and find the standard error of your estimate. What is the estimated coefficient of variation?

c Estimate the average replacement cost per book, along with the standard error. What is the estimated coefficient of variation?

7 Repeat Exercise 6 for the purchase cost for each book. Plot the data and estimate the total and average amount she has spent for books, along with the standard errors.

8 Construct a sample ANOVA table for the replacement cost data in Exercise 6. What is your estimate for R_a^2? Do books on the same shelf tend to have more similar replacement costs? Suppose $c_1 = 10$ and $c_2 = 4$. If all shelves had 30 books, how many books should be sampled per shelf?

*9 The ICC was defined on page 139 as the Pearson correlation coefficient for the $NM(M-1)$ pairs (y_{ij}, y_{ik}) for i between 1 and N and $j \neq k$:

$$\text{ICC} = \frac{\sum_{i=1}^{N}\sum_{j=1}^{M}\sum_{k \neq j}^{M}(y_{ij} - \bar{y}_U)(y_{ik} - \bar{y}_U)}{(NM-1)(M-1)S^2}. \tag{5.42}$$

Show that the above definition is equivalent to (5.8). HINT: First show that

$$\sum_{i=1}^{N}\sum_{j=1}^{M}\sum_{k \neq j}^{M}(y_{ij} - \bar{y}_U)(y_{ik} - \bar{y}_U) + \sum_{i=1}^{N}\sum_{j=1}^{M}(y_{ij} - \bar{y}_U)^2 = M(\text{SSB}).$$

*10 Suppose in a two-stage cluster sample that all population cluster sizes are equal ($M_i = M$ for all i) and that all sample sizes for the clusters are equal ($m_i = m$ for all i).

a Show (5.34).

b Show that $\text{MSW} = S^2(1 - R_a^2)$ and that

$$\text{MSB} = S^2\left[\frac{N(M-1)R_a^2}{N-1} + 1\right].$$

c   Using parts (a) and (b), express $V(\hat{\bar{y}})$ as a function of n, m, N, M, S^2, and R_a^2.

d   Show that if S^2 and the sample and population sizes are fixed, and if $(m-1)/m > n/N$, then $V(\hat{\bar{y}})$ is an increasing function of R_a^2.

*11   Suppose in a two-stage cluster sample that all population cluster sizes are equal ($M_i = M$ for all i) and that all sample sizes for the clusters are equal ($m_i = m$ for all i).

a   Show that $\hat{t}_{\text{unb}} = \hat{t}_r$ and, hence, that $\hat{\bar{y}}_{\text{unb}} = \hat{\bar{y}}_r$.

b   Fill in the formulas for the sums of squares in the following ANOVA table, for the sample data.

Source	df	Sum of Squares	Mean Square
Between clusters	$n-1$		$\widehat{\text{MSB}}$
Within clusters	$n(m-1)$		$\widehat{\text{MSW}}$
Total	$nm-1$		

c   Show that $E[\widehat{\text{MSW}}] = \text{MSW}$ and $E[\widehat{\text{MSB}}] = (m/M)\text{MSB} + [1-(m/M)]\text{MSW}$, where MSB and MSW are the between and within mean squares, respectively, from the *population* ANOVA table.

d   Show, using (5.25) or (5.29), that

$$\hat{V}(\hat{\bar{y}}_{\text{unb}}) = \left(1 - \frac{n}{N}\right)\frac{\widehat{\text{MSB}}}{nm} + \frac{1}{N}\left(1 - \frac{m}{M}\right)\frac{\widehat{\text{MSW}}}{m}.$$

12   An inspector samples cans from a truckload of canned creamed corn to estimate the average number of worm fragments per can. The truck has 580 cases; each case contains 24 cans. The inspector samples 12 cases at random and subsamples 3 cans randomly from each selected case.

						Case						
	1	2	3	4	5	6	7	8	9	10	11	12
Can 1	1	4	0	3	4	0	5	3	7	3	4	0
Can 2	5	2	1	6	9	7	5	0	3	1	7	0
Can 3	7	4	2	6	8	3	1	2	5	4	9	0

a   Estimate the mean number of worm fragments per can, along with the standard error of your estimate. (You may use the result of Exercise 11 to calculate the SE.)

b   Suppose a new truckload is to be inspected and is thought to be similar to this one. It takes 10 minutes to locate and open a case, and 8 minutes to locate and examine each specified can within a case. How many cans should be examined per case?

13   The new candy Green Globules is being test-marketed in an area of upstate New York. The market research firm decides to sample 6 of the 45 cities in the area and then to sample supermarkets within those cities, wanting to know the number of cases of Green Globules sold.

City	Number of Supermarkets	Number of Cases Sold
1	52	146, 180, 251, 152, 72, 181, 171, 361, 73, 186
2	19	99, 101, 52, 121
3	37	199, 179, 98, 63, 126, 87, 62
4	39	226, 129, 57, 46, 86, 43, 85, 165
5	8	12, 23
6	14	87, 43, 59

Use any statistical package to obtain summary statistics for each cluster. Plot the data, and estimate the total number of cases sold and the average number sold per supermarket, along with the standard errors of your estimates.

14 The Arizona Health Care Cost Containment System (AHCCCS) provides medical assistance to low-income households in Arizona. Each county determines whether households are eligible for assistance. Sometimes, however, households are certified to be eligible when they really are not. The Arizona Statutes, Section 36-2905.01, mandate the collection of a "statistically valid quality control sample of the eligibility certifications made by each county." The certification error rate for each county is to be determined "by dividing the number of members in the sample who were erroneously certified by the total number of members in the sample." Quality control audits are done by sampling household records, however; once a household record is selected and audited, it costs the same amount to evaluate one person in the household as it does to evaluate all persons in the household.

a Explain how to use cluster sampling to estimate the certification error rate for a county.

b Suppose a county certified 1572 households to be eligible for medical assistance in 1995. In past years, the certification error rate per household has been about 10%. How many households should be included in your sample so that the half-width of a 95% CI for estimating the per-person certification error rate is less than 0.03? What assumptions did you need to make to arrive at your sample size?

15 A researcher wants to study the prevalence of smoking and other hight-risk behaviors among female high school students in a region with 35 high schools.

Number of Students	Number of Schools
0–499	3
500–999	7
1000–1499	18
1500–2000	5

She intends to drive to n of the schools and then interview some or all female students in the selected schools. She has conducted a similar study with 4 schools out of 29 in another region. The results were as follows:

School	Number of Students	Number of Female Students	Number of Females Interviewed	Number of Smokers
1	1471	792	25	10
2	890	447	15	3
3	1021	511	20	6
4	1587	800	40	27

a   Estimate the percentage of female students who smoke, from the study of the 4 schools.

b   Using information from the previous study, propose a design for the new one. Suppose it takes about 50 hours per school to make contact with school officials, obtain permission, obtain a list of female students, and travel back and forth. Although interviews themselves are only about 10 minutes, it takes about 30 minutes per interview obtained to allow for additional scheduling of no-shows, obtaining parental permission, and other administrative tasks. The investigator would like to spend 300 hours or less on the data collection.

16   Gnap (1995) conducted a survey to estimate the teacher workload in Maricopa County, Arizona, public school districts. Her target population was all first- through sixth-grade, full-time public school teachers with at least 1 year of experience. In 1994 Maricopa County had 46 school districts with 311 elementary schools and 15,086 teachers. Gnap stratified the schools by size of school district; the large stratum, consisting of schools in districts with more than 5000 students, is considered in this exercise. The stratum contained 245 schools; 23 participated in the survey. All teachers in the selected schools were asked to fill out the questionnaire. Due to nonresponse, however, some questionnaires were not returned. (We will examine possible effects of nonresponse in Exercise 17 of Chapter 8.) The data are found in the file teachers.dat, with psu information in teachmi.dat.

a   Why would a cluster sample be a better design than an SRS for this study? Consider issues such as cost, ease of collecting data, and confidentiality for respondents. What are some disadvantages of using a cluster sample?

b   Calculate the mean and standard deviation of the variable *hrwork* for each school in the "large" stratum. Construct a graph of the means for each school and a separate graph of the standard deviations. Does there seem to be more variation within a school, or does more of the variability occur between different schools? How did you deal with the missing values (coded as -9)?

c   Construct a scatterplot of the standard deviations versus the means for the schools for the variable *hrwork*. Is there more variability in schools with higher workloads? Less? No apparent relation?

d   Estimate the average of *hrwork* in the large stratum in Maricopa County, along with its standard error. Use *popteach* in the file teachmi.dat for the M_i's.

17   The file measles.dat contains data consistent with that obtained in a survey of parents whose children had not been immunized for measles during a recent campaign to immunize all children between the ages of 11 and 15. During the campaign, 7633 children from the 46 schools in the area were immunized; 9962 children whose records showed no previous immunization were not immunized. In a follow-up survey to explore why the children had not been immunized during the campaign, Roberts et al. (1995) sent questionnaires to the parents of a cluster sample of the 9962 children. Ten schools were randomly selected, then nonimmunized children from each school were selected, and the parents of those children were sent a questionnaire. Not all parents responded to the questionnaire (you will examine the effects of nonresponse in Exercise 18 of Chapter 8).

School	Number of Students Not Immunized (M_i)
1	78
2	238
3	261
4	174
5	236
6	188
7	113
8	170
9	296
10	207

a   Using the data from the returned questionnaires, estimate, separately for each school, the percentage of parents who returned a consent form. For this exercise, ignore the "no answer" responses.

b   Estimate the overall percentage of parents who returned a consent form, and give a 95% CI for your estimate.

c   How do your estimate and interval in part (b) compare with the results you would have obtained if you had ignored the clustering and analyzed the data as an SRS? Find the ratio:

$$\frac{\text{estimated variance from part (b)}}{\text{estimated variance if the data were analyzed as an SRS}}.$$

What is the effect of clustering?

18   Repeat Exercise 17, for estimating the percentage of children who had previously had measles.

19   Refer to Example 5.9. Later in the potato-growing season, it takes more time to inspect stems. Suppose it takes 2 minutes to inspect each stem. Which psu size is most efficient?

20   a   For the SRS from the Census of Agriculture in the file agsrs.dat (discussed in Example 2.4), find the sample ANOVA table of *acres92*, using *state* as the cluster variable. What is R_a^2 for this sample? Is there a clustering effect?

b   Suppose $c_1 = 15c_2$, where c_1 is the cost to sample a state and c_2 is the cost to sample a county within a state. Using $\bar{M}$ as the cluster size, what should $\bar{m}$ be, if it is desired to sample a total of 300 counties? How many states would be sampled (that is, what is n)?

21   Using the value of n determined in Exercise 20, draw a self-weighting cluster sample of 300 counties from the file agpop.dat. Plot the data using side-by-side boxplots. Estimate the total number of acres devoted to farms in the United States, along with the standard error, using both the unbiased estimate and the ratio estimate. How do these values compare, and how do they compare with the SRS and stratified samples from Examples 2.4 and 4.1?

22   The file ozone.dat contains hourly ozone readings from Eskdalemuir, Scotland, for 1994 and 1995.

a   Construct a histogram of the population values. Find the mean, standard deviation, and median of the population.

b   Take a systematic sample with period 24. To do this, select a random integer k between 1 and 24 and select the column containing the observations with GMT k. Construct a histogram of the sample values.

c   Now suppose you treat your systematic sample as though it were an SRS. Find the sample mean, standard deviation, and median. Construct an interval estimate of the population mean using the procedure in Section 2.4. Does your interval contain the true value of the population mean from part (a)?

d   Take four independent systematic samples, each with period 96. Now use formulas from cluster sampling to estimate the population mean and construct a 95% CI for the mean.

*23   (Requires calculus.) Show that if $M_i = M$ and $m_i = m$ for all i and if the cost function is $C = c_1 n + c_2 nm$, then

$$m = \sqrt{\frac{c_1 M(N-1)(1 - R_a^2)}{c_2(NM-1)R_a^2}}$$

minimizes the variance of $\hat{\bar{y}}_{\text{unb}}$ for fixed total cost C. HINT: Use Exercise 10.

*24   (Requires knowledge of trigonometry.) In Example 5.12, a systematic sampling scheme was proposed for detecting hazardous wastes in landfills. How far apart should sampling points be placed? Suppose there is a leakage and it spreads to a circular region with radius R. Let $2D$ be the distance between adjacent sampling points in the same row or column.

a   Calculate the probability with which a contaminant will be detected. HINT: Consider three cases, with $R < D$, $D \le R \le \sqrt{2}D$, and $R > \sqrt{2}D$.

b   Propose a sampling design that gives a higher probability that a contaminant will be detected than the square grid, but does not increase the number of sampling points.

*25   (Requires knowledge of random-effects models.) Under model M1 in (5.37), a one-way random-effects model, ρ can be estimated by

$$\hat{\rho} = \frac{\hat{\sigma}_A^2}{\hat{\sigma}_A^2 + \hat{\sigma}^2},$$

where $\hat{\sigma}_A^2$ and $\hat{\sigma}^2$ estimate the variance components σ_A^2 and σ^2. The method-of-moment estimators for one-stage cluster sampling when all clusters are of the same size are $\hat{\sigma}^2 = $ MSW and $\hat{\sigma}_A^2 = ($MSB $-$ MSW$)/M$.

a What is $\hat{\rho}$ in Example 5.4? How does it compare with $\widehat{\text{ICC}}$?

b Calculate $\hat{\rho}$ for populations A and B in Example 5.3. Why do these differ from the ICC?

*26 (Requires knowledge of random-effects models.)

a Suppose we ignore the fpc of a model-based estimator. Find

$$V_{\text{M1}} \left(\sum_{i \in \mathcal{S}} \sum_{j \in \mathcal{S}_i} b_{ij} Y_{ij} \right).$$

b Prove (5.39). HINT: Let

$$c_{ij} = \begin{cases} b_{ij} - 1 & \text{if } i \in \mathcal{S} \text{ and } j \in \mathcal{S}_i. \\ -1 & \text{otherwise.} \end{cases}$$

Then, $\hat{T} - T = \sum_{i=1}^{N} \sum_{j=1}^{M_i} c_{ij} Y_{ij}$.

*27 (Requires linear algebra and calculus.) Although $\hat{T}_r$ is unbiased for model M1, constructing an estimator with smaller variance is possible. Let

$$c_k = \frac{m_k}{1 + \rho(m_k - 1)}$$

and

$$\hat{T}_{\text{opt}} = \sum_{i \in \mathcal{S}} \sum_{j \in \mathcal{S}_i} \frac{c_i}{m_i} \left[\rho M_i + \frac{K - \rho \displaystyle\sum_{k \in \mathcal{S}} c_k M_k}{\displaystyle\sum_{k \in \mathcal{S}} c_k} \right] Y_{ij}.$$

Show that $\hat{T}_{\text{opt}}$ is unbiased and minimizes the variance in (5.39) among all unbiased estimators for model (5.37).

*28 (Requires calculus.) Suppose the M_i's are unequal and model M1 holds. The budget allows you to take a total of L measurements on subunits. Show that the variance in (5.39) is minimized for $\hat{T}_r$ when m_i is proportional to M_i. HINT: Use Lagrange multipliers, with the constraint $\sum_{i \in \mathcal{S}} m_i = L$.

29 The January 1994 issue of *The Nation* ranked 22 columnists by how much they used the words *I*, *me*, and *myself*. Select your favorite newspaper columnist. Randomly select five of the columnist's columns that appeared in the past year and use one-stage cluster sampling to estimate the proportion of total words taken up by *I*, *me*, and *myself*. What is your psu? Your ssu?

SURVEY Exercises

30 We would like to see if a cluster sample from the rural areas of Stephens County can improve on the precision of an SRS of size 100 while costing the same. To do this, we need to know the cost of sampling 100 houses randomly in districts 1 through 43. Use ADDGEN to generate ten different SRSs of size 100 from the rural districts; calculate how much each of those different samples would cost, and average the costs to get an estimate of the cost of sampling 100 houses randomly in districts 1 through 43.

31   Design a two-stage cluster sampling scheme for the rural areas (districts 1–43) of Stephens County. Your design should (a) choose between 25 and 50% of the districts (clusters) with equal probability, (b) subsample within each chosen district with sample size proportional to district size (number of houses), and (c) cost about the same amount as an SRS of size 100.

32   Using your sample from Exercise 31, estimate the average price a rural household is willing to pay for cable TV, using both an unbiased estimate and a ratio estimate. Be sure to give standard errors and to plot the data appropriately.

6

Sampling with Unequal Probabilities

'Personally I never care for fiction or storybooks. What I like to read about are facts and statistics of any kind. If they are only facts about the raising of radishes, they interest me. Just now, for instance, before you came in'—he pointed to an encyclopædia on the shelves—'I was reading an article about "Mathematics." Perfectly pure mathematics.

'My own knowledge of mathematics stops at "twelve times twelve," but I enjoyed that article immensely. I didn't understand a word of it; but facts, or what a man believes to be facts, are always delightful. That mathematical fellow believed in his facts. So do I. Get your facts first, and'—the voice dies away to an almost inaudible drone—'then you can distort 'em as much as you please.'

—Mark Twain, quoted in Rudyard Kipling, *From Sea to Sea*

Up to now, we have only discussed sampling schemes in which the probabilities of choosing sampling units are equal. Equal probabilities give schemes that are often easy to design and explain. Such schemes are not, however, always possible or, if practicable, as efficient as schemes using unequal probabilities. We saw in Example 5.7 that a cluster sample with equal probabilities may result in a large variance for the design-unbiased estimator of the population mean and total.

EXAMPLE 6.1 O'Brien et al. (1995) took a sample of nursing home residents in the Philadelphia area, with the objective of determining residents' preferences on life-sustaining treatments. Do they wish to have cardiopulmonary resuscitation (CPR) if the heart stops beating, or to be transferred to a hospital if a serious illness develops, or to be fed through an enteral tube if no longer able to eat? The target population was all residents of licensed nursing homes in the Philadelphia area. There were 294 such homes, with a total of 37,652 beds (before sampling, they only knew the number of beds, not the number of residents).

Because the survey was to be done in person, cluster sampling was essential for keeping survey costs manageable. Had the researchers chosen to use cluster sampling with equal probabilities of selection, they would have taken a simple random sample (SRS) of nursing homes, then another SRS of residents within each selected home.

In a cluster sample with equal probabilities, however, a nursing home with 20 beds is as likely to be chosen for the sample as a nursing home with 1000 beds. The sample is only self-weighting if the subsample size for each home is proportional to the number of beds in the home. Each bed sampled represents the same number of beds in the population if one-stage cluster sampling is used, or if 10% (or any other percentage) of beds are sampled in each selected home.

Sampling homes with equal probabilities would result in a mathematically valid estimator, but it has three major shortcomings. First, you would expect that the total number of patients in a home who desire CPR (t_i) would be proportional to the number of beds in the home (M_i), so estimators from Chapter 5 may have large variance. Second, a self-weighting equal-probability sample may be cumbersome to administer. It may require driving out to a nursing home just to interview one or two residents, and equalizing workloads of interviewers may be difficult. Third, the cost of the sample is unknown in advance—a random sample of 40 homes may consist primarily of large nursing homes, which would lead to greater expense than anticipated.

Instead of taking a cluster sample of homes with equal probabilities, the investigators randomly drew a sample of 57 nursing homes with probabilities proportional to the number of beds. They then took an SRS of 30 beds (and their occupants) from a list of all beds within the nursing home. If the number of residents equals the number of beds and if a home has the same number of beds when visited as are listed in the sampling frame, then the sampling design results in every resident having the same probability of being included in the sample. The cost is known before selecting the sample, the same number of interviews are taken at each home, and the estimator of a population total will likely have a smaller variance than estimators in Chapter 5.

Since this sample is self-weighting, you can easily obtain point estimates (but *not* standard errors) of desired quantities by usual methods. You can estimate the median age of the nursing home residents by finding the sample median of the residents in the sample, or the 70th percentile by finding the 70th percentile of the sample. If a sample is not self-weighting, point estimates are still easily calculated using weights. A warning, though: Always consider the cluster design when calculating the precision of your estimates. ■

In Chapter 4 we noted that sometimes stratified sampling is used to sample different units with different probabilities. In a survey to estimate total business expenditures on advertising, we might want to stratify by company sales or income. The largest companies such as IBM would be in one stratum, medium-sized companies would be in a number of different strata, and very small companies such as Robin's Tailor Shop would be in yet another stratum. An optimal allocation scheme would sample a very high fraction (perhaps 100%) in the stratum with the largest companies and a small fraction of companies in the stratum with the smallest companies; the variance from company to company will be much higher among IBM, AT&T, and Phillip Morris than among Robin's Tailor Shop, Pat's Shoe Repair, and Flowers by Leslie. The variance is larger in the large companies just because the amounts of money involved are so much larger. Thus, the sampling variance is decreased by assigning unequal probabilities to sampling units in different strata.

To estimate the total spent on advertising using this stratified sample, we assign higher weights to companies with lower probabilities of selection. As discussed in Section 4.3, the probability that a company in stratum h will be included in the sample is n_h/N_h; the sampling weight for that company is N_h/n_h. Each company sampled in stratum h represents N_h/n_h companies in the population, and $\hat{t}_{str} = \sum_{h=1}^{H} \sum_{j \in \mathcal{S}_h} (N_h/n_h) y_{hj}$.

We can also use unequal probability of selection to decrease variances without explicitly stratifying. When sampling with unequal probabilities, we deliberately vary the probabilities that we will select different psu's for the sample and compensate by providing suitable weights in the estimation. The key is that we *know* the probabilities[1] with which we will select a given unit:

$$P(\text{unit } i \text{ selected on first draw}) = \psi_i. \qquad \text{(6.1)}$$

$$P(\text{unit } i \text{ in sample}) = \pi_i. \qquad \text{(6.2)}$$

The deliberate selection of psu's with known but unequal probabilities differs greatly from the selection bias discussed in Chapter 1. Many surveys with selection bias do sample with unequal probabilities, but the probabilities of selection are unknown and unestimable, so the survey takers cannot compensate for the unequal probabilities in the weighting. If you take a survey of students by asking students who walk by the library to participate, you certainly are sampling with unequal probabilities— students who use the library frequently are more likely to be asked to participate in the survey, while other students never go by the library at all. But you have no idea how many students in the population are represented by a participant in your survey and no way of correcting for the unequal probabilities of selection in the estimation.

When first presented with the idea of unequal-probability sampling, some people think of it as "unnatural" or "contrived." On the contrary, for many populations with clustering, unequal-probability sampling at the psu level produces a sample that mirrors the population better than an equal-probability sample. Examples of unequal-probability samples are given in Section 6.5. To understand these examples and to design your own samples, it is essential that you have an understanding of probability. We will consider with-replacement sampling first, starting with the simple design of selecting only one primary sampling unit (psu). In Section 6.4, we consider unequal-probability sampling without replacement. Notation used in this chapter is defined in Section 5.1.

6.1
Sampling One Primary Sampling Unit

As a special case, suppose we select just one ($n = 1$) of the N psu's to be in the sample. The total for psu i is denoted by t_i, and we want to estimate the population total, t. Sampling one psu will demonstrate the ideas of unequal-probability sampling without introducing the complications.

Let's start out by looking at what happens for a situation in which we know the whole population. A town has four supermarkets, ranging in size from 100 square meters (m^2) to 1000 m^2. We want to estimate the total amount of sales in the four stores for last month by sampling just one of the stores. (Of course, this is just an illustration—if we really had only four supermarkets we would probably take a census.) You might expect that a larger store would have more sales than a smaller

---

[1]We consider two different probabilities in this chapter because, when sampling with unequal probabilities without replacement (see Section 6.4), selecting a unit on the first draw can affect the selection probabilities for other units.

store and that the variability in total sales among several 1000-m² stores will be greater than the variability in total sales among several 100-m² stores.

Since we sample only one store, we have that the probability that a store is selected on the first draw (ψ_i) is the same as the probability that the store is included in the sample (π_i). For this example, take

$$\pi_i = \psi_i = P(\text{store } i \text{ selected})$$

proportional to the size of the store. Since store A accounts for 1/16 of the total floor area of the four stores, it is sampled with probability 1/16. For illustrative purposes, we know the values of t_i for the whole population:

Store	Size (m²)	ψ_i	t_i (in Thousands)
A	100	$\dfrac{1}{16}$	11
B	200	$\dfrac{2}{16}$	20
C	300	$\dfrac{3}{16}$	24
D	1000	$\dfrac{10}{16}$	245
Total	1600	1	300

We could select a probability sample of size 1 with the probabilities given above by shuffling cards numbered 1 through 16 and choosing one card. If the card's number is 1, choose store A; if 2 or 3, choose B; if 4, 5, or 6, choose C; and if 7 through 16, choose D. Or we could spin once on a spinner like this:

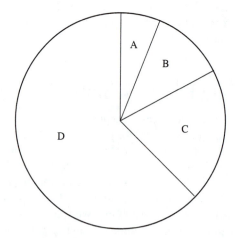

We compensate for the unequal probabilities of selection by also using ψ_i in the estimator. We have already seen such compensation for unequal probabilities of selection in stratified sampling: If we select 10% of the units in stratum 1 and 20% of the units in stratum 2, the sampling weight is 10 for each unit in stratum 1 and 5 for each unit in stratum 2. Here, we select store A with probability 1/16, so store

A's sampling weight is 16. If the size of the store is roughly proportional to the total sales for that store, we would expect that store A also has about 1/16 of the total sales and that multiplying store A's sales by 16 would estimate the total sales for all four stores. As always, the sampling weight of unit i is the reciprocal of the probability of selection:

$$w_i = \frac{1}{P(\text{unit } i \text{ in sample})} = \frac{1}{\psi_i}.$$

Thus, our estimator of the population total from an unequal probability sample of size 1 is

$$\hat{t}_\psi = \sum_{i \in S} w_i t_i = \sum_{i \in S} \frac{t_i}{\psi_i}.$$

Four samples of size 1 are possible from this simple population:

Sample	ψ_i	t_i	$\hat{t}_\psi$	$(\hat{t}_\psi - t)^2$
{A}	$\dfrac{1}{16}$	11	176	15,376
{B}	$\dfrac{2}{16}$	20	160	19,600
{C}	$\dfrac{3}{16}$	24	128	29,584
{D}	$\dfrac{10}{16}$	245	392	8,464

As defined in Chapter 2,

$$E[\hat{t}_\psi] = \sum_{\substack{\text{possible} \\ \text{samples } S}} P(S) \hat{t}_{\psi S}$$

$$= \frac{1}{16}(176) + \frac{2}{16}(160) + \frac{3}{16}(128) + \frac{10}{16}(392) = 300.$$

Of course, $\hat{t}_\psi$ will always be unbiased because, in general,

$$E[\hat{t}_\psi] = \sum_{i=1}^{N} \psi_i \frac{t_i}{\psi_i} = t. \tag{6.3}$$

The variance of $\hat{t}_\psi$ is

$$V[\hat{t}_\psi] = E[(\hat{t}_\psi - t)^2]$$

$$= \sum_{\substack{\text{possible} \\ \text{samples } S}} P(S)(\hat{t}_{\psi S} - t)^2$$

$$= \sum_{i=1}^{N} \psi_i \left(\frac{t_i}{\psi_i} - t \right)^2. \tag{6.4}$$

For this example,

$$V[\hat{t}_\psi] = \frac{1}{16}(15,376) + \frac{2}{16}(19,600) + \frac{3}{16}(29,584) + \frac{10}{16}(8,464) = 14,248.$$

Compare these results to those from an SRS of size 1, in which the probability of selecting each unit is $\psi_i = 1/4$, so $1/\psi_i = 4 = N$. Note that if all of the probabilities of selection are equal, as in simple random sampling, $1/\psi_i$ always equals N.

Sample	ψ_i	t_i	$\hat{t}_\psi$	$(\hat{t}_\psi - t)^2$
{A}	$\dfrac{1}{4}$	11	44	65,536
{B}	$\dfrac{1}{4}$	20	80	48,400
{C}	$\dfrac{1}{4}$	24	96	41,616
{D}	$\dfrac{1}{4}$	245	980	462,400

As always, $\hat{t}_{\text{SRS}}$ is unbiased and thus has expectation 300, but for this example the SRS variance is much larger than the variance from the unequal-probability sampling scheme:

$$V[\hat{t}_\psi] = \frac{1}{4}(65,536) + \frac{1}{4}(48,400) + \frac{1}{4}(41,616) + \frac{1}{4}(462,400) = 154,488.$$

The variance from the unequal-probability scheme, 14,248, is much smaller because it uses auxiliary information: We expect the store size to be related to the sales, and we use that information in designing the sampling scheme.

We believe that t_i is correlated to the size of the store, which is known. Since store D accounts for 10/16 of the total floor area of supermarkets, it is reasonable to believe that store D will account for about 10/16 of the total sales as well. Thus, if store D is chosen and is believed to account for about 10/16 of the total sales, we would have a good estimate of total sales by multiplying store D's sales by 16/10.

What if store D accounts for only 4/16 of the total sales? Then the unequal-probability estimator $\hat{t}_\psi$ will still be unbiased over repeated sampling, but it will have a large variance (see Exercise 5). The method still works mathematically but is not as efficient as if t_i is roughly proportional to ψ_i.

Sampling only one psu is not as unusual as you might think. Many large, complex surveys are so highly stratified that each stratum contains only a few psu's. A large number of strata is used to increase the precision of the survey estimates. In such a survey, it may be perfectly reasonable to want to select only one psu from each stratum. But, with only one psu per stratum in the sample, we do not have an estimate of the variability between psu's within a stratum. When large survey organizations sample only one psu per stratum, they often split the psu selected in some way to estimate the stratum variance; this method is discussed in Chapter 9.

6.2
One-Stage Sampling with Replacement

Now suppose $n > 1$, and we sample *with* replacement. Sampling with replacement means that the selection probabilities do not change after we have drawn the first unit. Let

$$\psi_i = P(\text{select unit } i \text{ on first draw}).$$

If we sample with replacement, then ψ_i is also the probability that unit i is selected on the second draw, or the third draw, or any other given draw. The overall probability that unit i is in the sample at least once is

$$\pi_i = 1 - P(\text{unit } i \text{ is not in sample}) = 1 - (1 - \psi_i)^n.$$

If $n = 1$, then $\pi_i = \psi_i$.

The idea behind unequal-probability sampling is simple. Draw n psu's with replacement. Then estimate the population total, using the estimator from the previous section, separately for each psu drawn. Some psu's may be drawn more than once—the estimated population total, calculated using a given psu, is included as many times as the psu is drawn. Since the psu's are drawn with replacement, we have n independent estimates of the population total. We then estimate the population total t by averaging those n independent estimates of t. The estimated variance is the sample variance of the n independent estimates of t, divided by n.

6.2.1 Selecting Primary Sampling Units

6.2.1.1 The Cumulative-Size Method

There are several ways to sample psu's with unequal probabilities. All require that you have a measure of size for all psu's in the population. The cumulative-size method extends the method used in the previous section, in which random numbers are generated, and psu's corresponding to those numbers are included in the sample. For the supermarkets, we drew cards from a deck with cards numbered 1 through 16. If the card's number is 1, choose store A; if 2 or 3, choose B; if 4, 5, or 6, choose C; and if 7 through 16, choose D. To sample with replacement, put the card back after selecting a psu and draw again.

EXAMPLE 6.2 Consider the population of introductory statistics classes at a college shown in Table 6.1. The college has 15 such classes; class i has M_i students, for a total of 647 students in introductory statistics courses. We decide to sample 5 classes with replacement, with probability proportional to M_i, and then collect a questionnaire from each student in the sampled classes. For this example then, $\psi_i = M_i/647$.

To select the sample, generate five random integers with replacement between 1 and 647. Then the psu's to be chosen for the sample are those whose range in the cumulative M_i includes the randomly generated numbers. The set of five random numbers {487, 369, 221, 326, 282} results in the sample of units {13, 9, 6, 8, 7}. The cumulative-size method allows the same unit to appear more than once: The five random numbers {553, 082, 245, 594, 150} leads to the sample {14, 3, 6, 14, 5}—psu 14 is then included twice in the data. ∎

Of course, we can take an unequal-probability sample when the ψ_i's are not proportional to the M_i's: Simply form a cumulative ψ_i range instead, and sample uniform random numbers between 0 and 1. This variation of the method is discussed in Exercise 4.

Systematic sampling is often used to select psu's in large, complex samples, rather than generating random numbers with replacement. Systematic sampling really gives a sample without replacement, but in large populations sampling without replacement and sampling with replacement are very similar, as the probability that a unit will be selected twice is small. To sample psu's systematically, list the population elements

TABLE   6.1

Population of Introductory Statistics Classes

Class Number	M_i	ψ_i	Cumulative M_i Range	
1	44	0.068006	1	44
2	33	0.051005	45	77
3	26	0.040185	78	103
4	22	0.034003	104	125
5	76	0.117465	126	201
6	63	0.097372	202	264
7	20	0.030912	265	284
8	44	0.068006	285	328
9	54	0.083462	329	382
10	34	0.052550	383	416
11	46	0.071097	417	462
12	24	0.037094	463	486
13	46	0.071097	487	532
14	100	0.154560	533	632
15	15	0.023184	633	647
Total	647	1		

for the first psu in the sample, followed by the elements for the second psu, and so on. Then take a systematic sample of the elements. The psu's to be included in the sample are those in which at least one element is in the systematic sample of elements. The larger the psu, the higher the probability it will be in the sample.

The statistics classes have a total of 647 students. To take a (roughly, because 647 is not a multiple of 5) systematic sample, choose a random number k between 1 and 129 and select the psu containing student k, the psu containing student $129 + k$, the psu containing student $2(129) + k$, and so on. Suppose the random number we select as a start value is 112. Then the systematic sample of elements results in the following psu's being chosen:

Number in Systematic Sample	psu Chosen
112	4
241	6
370	9
499	13
628	14

Larger classes (psu's) have a higher chance of being in the sample because it is more likely that a multiple of the random number chosen will be one of the numbered elements in a large psu. Systematic sampling does not give us a true random sample with replacement, though, because it is impossible for classes with 129 or fewer students to occur in the sample more than once, and classes with more than 129 students are sampled with probability 1. In many populations, however, it is much easier to implement than methods that give a random sample. If the psu's are arranged geographically, taking a systematic sample may force the selected psu's to be spread out over more of the region and may give better results than a random sample with replacement.

6.2.1.2   Lahiri's Method

Lahiri's (1951) method may be more tractable than the cumulative-size method when the number of psu's is large. It is an example of a *rejective* method, because you generate pairs of random numbers to select psu's and then reject some of them if the psu size is too small. Let N = number of psu's in population and $\max\{M_i\}$ = maximum psu size. You will show that Lahiri's method produces a with-replacement sample with the desired probabilities in Exercise 14.

1 Draw a random number between 1 and N. This indicates which psu you are considering.

2 Draw a random number between 1 and $\max\{M_i\}$; if the random number is less than or equal to M_i, then include psu i in the sample; otherwise, go back to step 1.

3 Repeat until the desired sample size is obtained.

EXAMPLE 6.3   Let's use Lahiri's method for the classes in Example 6.2. For Lahiri's method, we only need to know M_i for each psu. The largest class has $\max\{M_i\} = 100$ students, so we generate pairs of random integers, the first between 1 and 15, the second between 1 and 100, until the sample has five psu's (Table 6.2). The psu's to be sampled are {12, 14, 14, 5, 1}.   ■

6.2.2   Theory of Estimation

Because we are sampling with replacement, the sample may contain the same unit more than once. To allow us to keep track of which psu's occur multiple times in the sample, define the random variable Q_i by

$$Q_i = \text{number of times unit } i \text{ occurs in the sample.}$$

TABLE 6.2
Lahiri's Method, for Example 6.3

First Random Number (psu i)	Second Random Number	M_i	Action
12	6	24	6 < 24; include psu 12 in sample
14	24	100	Include in sample
1	65	44	65 > 44; discard pair of numbers and try again
7	84	20	84 > 20; try again
10	49	34	Try again
14	47	100	Include
15	43	15	Try again
5	24	76	Include
11	87	46	Try again
1	36	44	Include

Then, $\hat{t}_\psi$ is the average of all t_i / ψ_i for units chosen to be in the sample:

$$\hat{t}_\psi = \frac{1}{n} \sum_{i=1}^{N} Q_i \frac{t_i}{\psi_i}. \tag{6.5}$$

If a unit appears k times in the sample, it is counted k times in the estimator. Note that $\sum_{i=1}^{N} Q_i = n$ and $E[Q_i] = n\psi_i$, so $\hat{t}_\psi$ is unbiased for estimating t.

To calculate the variance, note that the estimator in (6.5) is the average of n independent observations, each with variance $\sum_{i=1}^{N} \psi_i (t_i / \psi_i - t)^2$ [from (6.4)], so

$$V[\hat{t}_\psi] = \frac{1}{n} \sum_{i=1}^{N} \psi_i \left(\frac{t_i}{\psi_i} - t \right)^2. \tag{6.6}$$

To estimate $V[\hat{t}_\psi]$ from a sample, you might think we could use a formula of the same form as (6.6), but that will not work. Equation (6.6) involves a weighted average of the $(t_i / \psi_i - t)^2$, weighted by the unequal probabilities of selection. But in taking the sample, we have already used the unequal probabilities—they appear in the random variables Q_i in (6.5). If we included the ψ_i's again as multipliers in estimating the sample variance, we would be using the unequal probabilities twice. Instead, to estimate the variance, use

$$\hat{V}(\hat{t}_\psi) = \frac{1}{n} \sum_{i=1}^{N} Q_i \frac{\left(\dfrac{t_i}{\psi_i} - \hat{t}_\psi \right)^2}{n-1}. \tag{6.7}$$

Note that (6.7) is just a variation of the formula s^2/n you used in introductory statistics: The sum is simply the sample variance of the numbers t_i / ψ_i for the sampled psu's. Equation (6.7) is an unbiased estimator of the variance in (6.6) because

$$E[\hat{V}(\hat{t}_\psi)] = \frac{1}{n(n-1)} \sum_{i=1}^{N} E\left[Q_i \left(\frac{t_i}{\psi_i} - \hat{t}_\psi \right)^2 \right]$$

$$= \frac{1}{n(n-1)} \sum_{i=1}^{N} E\left[Q_i \left(\frac{t_i}{\psi_i} - t \right)^2 - Q_i \left(\hat{t}_\psi - t \right)^2 \right]$$

$$= \frac{1}{n(n-1)} \left[\sum_{i=1}^{N} n\psi_i \left(\frac{t_i}{\psi_i} - t \right)^2 - n V(\hat{t}_\psi) \right]$$

$$= V(\hat{t}_\psi).$$

We are sampling with replacement, so unit i will occur in the sample with approximate frequency $n\psi_i$. One caution: If N is small or some of the ψ_i's are unusually large, it is possible that the sample will consist of one psu sampled n times. In that case, the estimated variance is zero; it is better to use sampling without replacement (see Section 6.4) if this may occur.

EXAMPLE 6.4   For the situation in Example 6.3, suppose we sample the psu's selected by Lahiri's method, {12, 14, 14, 5, 1}. The response t_i is the total number of hours all students in class i spent studying statistics last week, with the following data:

Class	ψ_i	t_i	t_i/ψ_i
12	$\dfrac{24}{647}$	75	2021.875
14	$\dfrac{100}{647}$	203	1313.410
14	$\dfrac{100}{647}$	203	1313.410
5	$\dfrac{76}{647}$	191	1626.013
1	$\dfrac{44}{647}$	168	2470.364

The numbers in the last column of the table are the estimates of t that would be obtained if that psu were the only one selected in a sample of size 1. The population total is estimated by averaging the five values of t_i/ψ_i:

$$\hat{t}_\psi = \frac{2021.875 + 1313.410 + 1313.410 + 1626.013 + 2470.364}{5} = 1749.014.$$

The standard error (SE) of $\hat{t}_\psi$ is simply $s/\sqrt{n}$, where s is the sample standard deviation of the five numbers in the rightmost column of the table:

$$\text{SE}[\hat{t}_\psi] = \frac{1}{\sqrt{5}}\sqrt{\frac{(2021.875 - 1749.014)^2 + \cdots + (2470.364 - 1749.014)^2}{4}}$$

$$= 222.42.$$

The average amount of time a student spent studying statistics is

$$\hat{\bar{y}}_\psi = \frac{1749.014}{647} = 2.70$$

hours with $\text{SE}(\hat{\bar{y}}_\psi) = 222.42/647 = 0.34$ hour. ∎

6.2.3 Designing the Selection Probabilities

We would like to choose the ψ_i's so that the variances of the estimates are as small as possible. Ideally, we would use $\psi_i = t_i/t$ (then $\hat{t}_\psi = t$ for all samples and $V[\hat{t}_\psi] = 0$), so if t_i was the annual income of the ith household, ψ_i would be the proportion of total income in the population that came from the ith household. But of course, the t_i's are unknown until sampled. Even if the income were known before the survey was taken, we are often interested in more than one quantity; using income for designing the probabilities of selection may not work well for estimating other quantities.

Because many totals in a psu are related to the number of elements in a psu, we often take ψ_i to be the relative proportion of elements in psu i or the relative size of psu i. Then, a large psu has a greater chance of being in the sample than a small psu. With M_i the number of elements in the ith psu and K the number of elements in the population, we take $\psi_i = M_i/K$. With this choice of the probabilities ψ_i, we have **probability proportional to size** (pps) sampling. We used pps sampling in Example 6.2.

Then, for one-stage pps sampling, $t_i/\psi_i = K\bar{y}_i$, so

$$\hat{t}_\psi = \frac{K}{n} \sum_{i=1}^{N} Q_i \bar{y}_i,$$

$$\hat{\bar{y}}_\psi = \frac{1}{n} \sum_{i=1}^{N} Q_i \bar{y}_i,$$

$$\hat{V}(\hat{t}_\psi) = \frac{1}{n} \sum_{i=1}^{N} Q_i \frac{\left(\dfrac{t_i}{\psi_i} - \hat{t}_\psi\right)^2}{n-1} = \frac{K^2}{n} \sum_{i=1}^{N} Q_i \frac{(\bar{y}_i - \hat{\bar{y}}_\psi)^2}{n-1}$$

$$\hat{V}(\hat{\bar{y}}_\psi) = \frac{1}{n} \sum_{i=1}^{N} \frac{Q_i(\bar{y}_i - \hat{\bar{y}}_\psi)^2}{n-1}.$$

The sum in the variance estimates is simply the sample variance of the psu means $\bar{y}_i$.

All the work in pps sampling has been done in the sampling design itself. The pps estimates can be calculated simply by treating the $\bar{y}_i$'s as individual observations and finding their mean and sample variance. In practice, however, there are usually some deviations from a strict pps scheme, so you should use (6.5) and (6.7) for estimating the population total and its estimated variance.

EXAMPLE **6.5**   The file statepop.dat contains data from an unequal-probability sample of 100 counties in the United States. Counties were chosen using the cumulative-size method from the listings in the *City and County Data Book, 1994,* with probabilities proportional to their populations. Sampling was done with replacement, so very large counties occur multiple times in the sample: Los Angeles County, with the largest population in the United States, occurs four times.

One of the quantities recorded for each county was the number of physicians in the county. You would expect larger counties to have more physicians, so pps sampling should work well for estimating the total number of physicians in the United States.

You must be careful in plotting data from an unequal-probability sample, as you need to consider the unequal probabilities when interpreting the plots. A plot of t_i versus ψ_i (Figure 6.1a) tells the efficiency of the unequal-probability design: The closer the plot is to a straight line, the better unequal-probability sampling works. A histogram of t_i in a pps sample will not give a representative view of the population of psu's, as psu's with large ψ_i's are overrepresented in the sample. A histogram of t_i/ψ_i, however, may give an idea of the spread involved in the population estimates, and may help you identify unusual psu's (Figure 6.1b).

The sample was chosen using the cumulative-size method; Table 6.3 shows the sampled counties arranged alphabetically by state. The ψ_i's were calculated as $M_i/255{,}077{,}536$.

FIGURE 6.1

Selected plots for pps sample estimating the total number of physicians in the United States. (a) Plot of t_i vs. ψ_i; there is a strong linear relationship between the variables, which indicates that pps sampling increases efficiency. The unusual observation is New York County, New York. (b) Histogram of the 100 values of t_i/ψ_i. Each value estimates t.

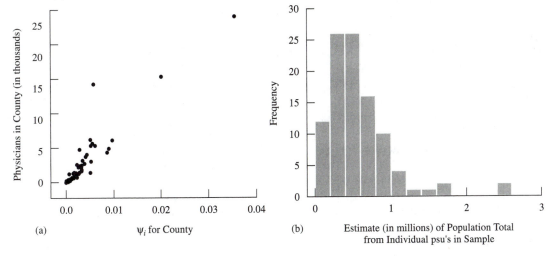

(a) ψ_i for County

(b) Estimate (in millions) of Population Total from Individual psu's in Sample

TABLE 6.3

Sampled Counties in Example 6.5

State	County	Population Size, M_i	ψ_i	Number of Physicians, t_i	t_i/ψ_i
AL	Wilcox	13,672	0.00005360	4	74,627.72
AZ	Maricopa	2,209,567	0.00866233	4320	498,710.81
AZ	Maricopa	2,209,567	0.00866233	4320	498,710.81
AZ	Pinal	120,786	0.00047353	61	128,820.64
AR	Garland	76,100	0.00029834	131	439,095.36
AR	Mississippi	55,060	0.00021586	48	222,370.54
CA	Contra Costa	840,585	0.00329541	1761	534,379.68
⋮	⋮	⋮	⋮	⋮	⋮
VA	Chesterfield	225,225	0.00088297	181	204,990.72
WA	King	1,557,537	0.00610613	5280	864,704.59
WI	Lincoln	27,822	0.00010907	28	256,709.47
WI	Waukesha	320,306	0.00125572	687	547,096.42
		average			570,304.30
		std. dev.			414,012.30

The average of the t_i/ψ_i column is 570,304.3, the estimated total number of physicians in the United States. The standard error of the estimate is $414,012.3/\sqrt{100} = 41,401.23$. For comparison, the *City and County Data Book* lists a total of 532,638 physicians in the United States, a value that is less than 1 SE away from our estimate. ■

6.3
Two-Stage Sampling with Replacement

The estimators for two-stage unequal-probability sampling with replacement are almost the same as those for one-stage sampling. Take a sample of psu's with replacement, choosing the ith psu with known probability ψ_i. As in one-stage sampling with replacement, Q_i is the number of times psu i occurs in the sample. Then take a probability sample of m_i subunits in the ith psu. Simple random sampling without replacement or systematic sampling is often used to select the subsample, although any probability sampling method may be used.

The only difference between two-stage sampling with replacement and one-stage sampling with replacement is that in two-stage sampling, we must estimate t_i. If psu i is in the sample more than once, there are Q_i estimates of the total for psu i: $\hat{t}_{i1}, \hat{t}_{i2}, \ldots, \hat{t}_{iQ_i}$.

The subsampling procedure needs to meet two requirements:

1   Whenever psu i is selected to be in the sample, the same subsampling design is used to select secondary sampling units (ssu's) from that psu. Different subsamples from the same psu, though, must be sampled independently. Thus, if you decide before sampling that you will take an SRS of size 5 from psu 42 if it is selected, every time psu 42 appears in the sample you must generate a different set of random numbers to select 5 of the ssu's in psu 42. WARNING: If you just take one subsample of size 5 and use it more than once for psu 42, you do not have independent subsamples, and (6.9) will not be an unbiased estimator of the variance.

2   The jth subsample taken from psu i (for $j = 1, \ldots, Q_i$) is selected in such a way that $E[\hat{t}_{ij}] = t_i$. As the same procedure is used each time psu i is selected for the sample, we can define $V[\hat{t}_{ij}] = V_i$ for all j.

The estimators from one-stage unequal sampling with replacement are modified slightly to allow for different subsamples in psu's that are selected more than once:

$$\hat{t}_\psi = \frac{1}{n} \sum_{i=1}^{N} \sum_{j=1}^{Q_i} \frac{\hat{t}_{ij}}{\psi_i}. \tag{6.8}$$

$$\hat{V}(\hat{t}_\psi) = \frac{1}{n} \sum_{i=1}^{N} \sum_{j=1}^{Q_i} \frac{\left(\dfrac{\hat{t}_{ij}}{\psi_i} - \hat{t}_\psi\right)^2}{n-1}. \tag{6.9}$$

In Exercise 15 you will show that (6.9) is an unbiased estimator of the variance $V(\hat{t}_\psi)$, given in (6.27). Because sampling is with replacement, and hence it is possible to have more than one subsample from a given psu, the variance estimator captures both parts of the variance: the part due to the variability among psu's and the part that arises because t_i is estimated from a subsample rather than observed. The population mean is estimated by $\hat{\bar{y}}_\psi = \hat{t}_\psi / K$, with estimated variance $\hat{V}(\hat{\bar{y}}_\psi) = \hat{V}(\hat{t}_\psi)/K^2$.

In a pps sample, in which the ith psu is selected with probability $\psi_i = M_i / K$, the estimators again simplify. Then, $\hat{\bar{y}}_\psi$ is simply the average of the estimated psu means for psu's in the sample, and $\hat{V}(\hat{\bar{y}}_\psi)$ is the sample variance of those estimated psu means divided by n. The subsample sizes do not appear in the estimates.

In summary, here are the steps for taking a two-stage unequal-probability sample with replacement:

1 Determine the probabilities of selection ψ_i, the number n of psu's to be sampled, and the subsampling procedure to be used within each psu. With any method of selecting the psu's, we take a probability sample of ssu's within the psu's: Often in two-stage cluster sampling, we take an SRS without replacement of elements within the chosen psu's.

2 Select n psu's with probabilities ψ_i and with replacement. Either the cumulative-size method or Lahiri's method may be used to select the psu's for the sample.

3 Use the procedure determined in step 1 to select subsamples from the psu's chosen. If a psu occurs in the sample more than once, independent subsamples are used for each replicate.

4 Estimate the population total t from each psu in the sample as though it were the only one selected. The result is n estimates of the form $\hat{t}_{ij}/\psi_i$.

5 $\hat{t}_\psi$ is the average of the n estimates in step 4.

6 $\text{SE}(\hat{t}_\psi) = (1/\sqrt{n})$ (sample standard deviation of the n estimates in step 4).

EXAMPLE 6.6   Let's return to the situation in Example 6.4. Now suppose we subsample five students in each class rather than observing t_i. We will see that the estimation process is almost the same as in Example 6.4. Here, the response y_{ij} is the total number of hours student j in class i spent studying statistics last week (Table 6.4). Note that class 14 appears twice in the sample; each time it appears, a different subsample is collected.

Thus, $\hat{t}_\psi = 1617.5$ and $\text{SE}(\hat{t}_\psi) = 521.628/\sqrt{5} = 233.28$. From this sample, the average amount of time a student spent studying statistics is

$$\hat{\bar{y}}_\psi = \frac{1617.5}{647} = 2.5$$

hours with $\text{SE}(\hat{\bar{y}}_\psi) = 233.28/647 = 0.36$ hour.   ∎

TABLE   6.4
Spreadsheet for Calculations in Example 6.6

Class	M_i	ψ_i	y_{ij}	$\bar{y}_i$	$\hat{t}_i$	$\hat{t}_i/\psi_i$
12	24	0.0371	2, 3, 2.5, 3, 1.5	2.4	57.6	1552.8
14	100	0.1546	2.5, 2, 3, 0, 0.5	1.6	160.0	1035.2
14	100	0.1546	3, 0.5, 1.5, 2, 3	2.0	200.0	1294.0
5	76	0.1175	1, 2.5, 3, 5, 2.5	2.8	212.8	1811.6
1	44	0.0680	4, 4.5, 3, 2, 5	3.7	162.8	2393.9
			average			1617.5
			std. dev.			521.628

In Example 6.6, classes were selected with probability proportional to number of students in the class, so $\psi_i = M_i/K$. Subsampling the same number of students in each class resulted in a self-weighting sample. Under pps sampling with replacement with simple random sampling at the second stage, the sampling weight for an element sampled from psu i is, from (6.8),

$$w_i = \frac{1}{n} \frac{M_i}{m_i} \frac{1}{\psi_i}.$$

In pps sampling, with $\psi_i = M_i/K$, we have that $w_i = K/(nm_i)$; the sample is self-weighting if all m_i's are equal. For Example 6.6, the sampling weight is $647/(5 \times 5) = 25.88$ for each observation. The population total is equivalently estimated as

$$\frac{647}{25}(2 + 3 + 2.5 + \cdots + 3 + 2 + 5) = 1617.5.$$

EXAMPLE 6.7   Let's see what happens if we use unequal-probability sampling on the puppy homes considered in Example 5.7. Take ψ_i proportional to the number of puppies in the home, so that Puppy Palace with 30 puppies is sampled with probability 3/4 and Dog's Life with 10 puppies is sampled with probability 1/4. As before, once a puppy home is chosen, take an SRS of 2 puppies in the home. Then if Puppy Palace is selected, $\hat{t}_\psi = \hat{t}_{PP}/(3/4) = (30)(4)/(3/4) = 160$. If Dog's Life is chosen, $\hat{t}_\psi = \hat{t}_{DL}/(1/4) = (10)(4)/(1/4) = 160$. Thus, either possible sample results in an estimated average of $\hat{\bar{y}}_\psi = 160/40 = 4$ legs per puppy, and the variance of the estimator is zero.   ∎

Sampling with replacement has the advantage that it is very easy to select the sample and to obtain estimates of the population total and its variance. If N is small, however, as occurs in many highly stratified complex surveys with few clusters in each stratum, sampling with replacement is less efficient than many designs for sampling without replacement. In the next section, we discuss advantages and challenges of sampling without replacement.

6.4
Unequal-Probability Sampling Without Replacement

Generally, sampling with replacement is less efficient than sampling without replacement; with-replacement sampling is used because of the ease in selecting and analyzing samples. Nevertheless, in large surveys with many small strata, the inefficiencies may wipe out the gains in convenience. Much research has been done on unequal-probability sampling without replacement; the theory is more complicated because the probability that a unit is selected is different for the first unit chosen than for the second, third, and subsequent units. When you understand the probabilistic arguments involved, however, you can find the properties of any sampling scheme.

EXAMPLE 6.8   The supermarket example from Section 6.1 can be used to illustrate some of the features of unequal-probability sampling with replacement. Here is the population again:

Store	Size (m²)	t_i (in Thousands)
A	100	11
B	200	20
C	300	24
D	1000	245
Total	1600	300

Let's select two psu's without replacement and with unequal probabilities. As in Sections 6.1 to 6.3, let

$$\psi_i = P(\text{select unit } i \text{ on first draw}).$$

Since we are sampling without replacement, though, the probability that unit j is selected on the second draw depends on which unit was selected on the first draw.

One way to select the units with unequal probabilities is to use ψ_i as the probability of selecting unit i on the first draw, and then adjust the probabilities of selecting the other stores on the second draw. If store A was chosen on the first draw, then for selecting the second store we would spin the wheel while blocking out the section for store A, or shuffle the deck and redeal without card 1. Thus,

$$P(\text{store A chosen on first draw}) = \psi_A = \frac{1}{16}$$

and

$$P(\text{B chosen on second draw} \mid \text{A chosen on first draw}) = \frac{\frac{2}{16}}{1 - \frac{1}{16}} = \frac{\psi_B}{1 - \psi_A}.$$

The denominator is the sum of the ψ_i for stores B, C, and D. In general,

$P(\text{unit } i \text{ chosen first, unit } j \text{ chosen second})$

$$= P(\text{unit } i \text{ chosen first}) \, P(\text{unit } j \text{ chosen second} \mid \text{unit } i \text{ chosen first})$$

$$= \psi_i \frac{\psi_j}{1 - \psi_i}.$$

Similarly,

$$P(\text{unit } j \text{ chosen first, unit } i \text{ chosen second}) = \psi_j \frac{\psi_i}{1 - \psi_j}.$$

Note that $P(\text{unit } i \text{ chosen first, unit } j \text{ chosen second})$ is not the same as $P(\text{unit } j$ chosen first, unit i chosen second): The order of selection makes a difference! By adding the probabilities of the two choices, though, we can find the probability that a sample of size 2 consists of psu's i and j:

$$\text{For } n = 2, \quad P(\text{units } i \text{ and } j \text{ in sample}) = \pi_{ij} = \psi_i \frac{\psi_j}{1 - \psi_i} + \psi_j \frac{\psi_i}{1 - \psi_j}.$$

For a sample of size 2, the probability that psu i is in the sample is then the sum over

j of the probabilities that psu's i and j are both in the sample:

$$\text{For } n = 2, \quad P(\text{unit } i \text{ in sample}) = \pi_i = \sum_{\substack{j=1 \\ j \neq i}}^{N} \pi_{ij}.$$

The following table gives the π_i and π_{ij} for the supermarkets. The entries of the table are π_{ij} for each pair of stores (rounded to four decimal places); the margins give the π_i for the four stores.

		Store j				
		A	B	C	D	π_i
	A	—	.0173	.0269	.1458	.1900
Store i	B	.0173	—	.0556	.2976	.3705
	C	.0269	.0556	—	.4567	.5393
	D	.1458	.2976	.4567	—	.9002
	π_j	.1900	.3705	.5393	.9002	2.0000

6.4.1   The Horvitz–Thompson Estimator

In without-replacement sampling, π_i is the *inclusion probability,* the probability that the ith unit is in the sample; π_{ij} is the probability that units i and j are both in the sample. The inclusion probability π_i can be calculated as the sum of the probabilities of all samples containing the ith unit and has the property that

$$\sum_{i=1}^{N} \pi_i = n. \tag{6.10}$$

For the π_{ij}'s, as will be shown in Theorem 6.1 of Section 6.6,

$$\sum_{\substack{j=1 \\ j \neq i}}^{N} \pi_{ij} = (n-1)\pi_i. \tag{6.11}$$

For the supermarkets, the resulting π_i's are not proportional to the sizes of the stores—in fact, they cannot be proportional to the store sizes, as store D accounts for more than half of the total floor area but cannot be sampled with a probability greater than 1. The π_i's that result from this draw-by-draw method due to Yates and Grundy (1953) may or may not be the desired probabilities of inclusion in the sample; you may need to adjust the ψ_i's to obtain a prespecified set of π_i's.

In Example 6.8, $\pi_A = P(\text{store A in sample}) = 0.19$, and the π_i's sum to 2. Thus, π_i / n is the *average probability* that a unit will be selected on one of the draws: It is the probability we would assign to the ith unit's being selected on draw k $(k = 1, \ldots, n)$ if we did not know the true probabilities.

Recall that for sampling with replacement, $\hat{t}_\psi$ is the average of $\hat{t}_{ij}/\psi_i$ for psu's in the sample. But when samples are drawn without replacement, the probabilities of selection depend on what was drawn before. Instead of dividing the estimated total for psu i by ψ_i, we divide by the *average* probability of selecting that unit in a draw, π_i / n. We then have the **Horvitz–Thompson (HT) estimator** of the population total

(Horvitz and Thompson 1952):

$$\hat{t}_{HT} = \frac{1}{n} \sum_{i \in S} \frac{\hat{t}_i}{\pi_i/n} = \sum_{i \in S} \frac{\hat{t}_i}{\pi_i} = \sum_{i=1}^{N} Z_i \frac{\hat{t}_i}{\pi_i},$$ (6.12)

where $Z_i = 1$ if psu i is in the sample, and 0 otherwise.

The Horvitz–Thompson estimator is easily shown to be unbiased for t by using Theorem 6.2, to be proven in Section 6.6. Here, $P(Z_i = 1) = \pi_i$, so by (6.19),

$$E[\hat{t}_{HT}] = \sum_{i=1}^{N} \pi_i \frac{t_i}{\pi_i} = t.$$

Using (6.20) through (6.22), the variance of the Horvitz–Thompson estimator is

$$V(\hat{t}_{HT}) = \sum_{i=1}^{N} \frac{1 - \pi_i}{\pi_i} t_i^2 + \sum_{i=1}^{N} \sum_{k \neq i}^{N} \frac{\pi_{ik} - \pi_i \pi_k}{\pi_i \pi_k} t_i t_k + \sum_{i=1}^{N} \frac{V(\hat{t}_i)}{\pi_i}$$

$$= \sum_{i=1}^{N} \sum_{k>i}^{N} (\pi_i \pi_k - \pi_{ik}) \left(\frac{t_i}{\pi_i} - \frac{t_k}{\pi_k} \right)^2 + \sum_{i=1}^{N} \frac{V(\hat{t}_i)}{\pi_i}.$$ (6.13)

The second expression in (6.13) is the Sen-Yates-Grundy form (Sen 1953; Yates and Grundy 1953).

Theorem 6.3 in Section 6.6 implies that

$$\hat{V}_1[\hat{t}_{HT}] = \sum_{i \in S} (1 - \pi_i) \frac{\hat{t}_i^2}{\pi_i^2} + \sum_{i \in S} \sum_{\substack{k \in S \\ k \neq i}} \frac{\pi_{ik} - \pi_i \pi_k}{\pi_{ik}} \frac{\hat{t}_i}{\pi_i} \frac{\hat{t}_k}{\pi_k} + \sum_{i \in S} \frac{\hat{V}(\hat{t}_i)}{\pi_i}$$ (6.14)

and the Sen–Yates–Grundy form,

$$\hat{V}_2[\hat{t}_{HT}] = \sum_{i \in S} \sum_{\substack{k \in S \\ k>i}} \frac{\pi_i \pi_k - \pi_{ik}}{\pi_{ik}} \left(\frac{\hat{t}_i}{\pi_i} - \frac{\hat{t}_k}{\pi_k} \right)^2 + \sum_{i \in S} \frac{\hat{V}(\hat{t}_i)}{\pi_i}$$ (6.15)

are both unbiased estimators of the variance in (6.13). A problem can arise in estimating the variance of $\hat{t}_{HT}$, however: The unbiased estimators in (6.14) or (6.15) can result in a negative estimate of the variance in some unequal-probability designs! The stability can sometimes be improved by careful choice of the sampling design, but in general the calculations are cumbersome.

An alternative, which avoids some of the potential instability and computational complexity, is to use the with-replacement variance estimator in (6.9) rather than (6.14) or (6.15). This was suggested by Durbin (1953). If without-replacement sampling is more efficient than with-replacement sampling, the with-replacement variance estimator in (6.9) is expected to overestimate the variance and result in conservative confidence intervals, but in many instances the bias is small. The commonly used computer-intensive methods described in Chapter 9 calculate the with-replacement variance.

Note that to use the Horvitz–Thompson estimator when $n > 1$, we must know the inclusion probability π_i for each psu. The draw-by-draw procedure used in the supermarket example—finding the probability of any pair of psu's being in the sample and then finding the overall probability that the ith psu would be in the sample—becomes somewhat tedious for large populations and sample sizes larger than 2. Systematic sampling can be used to draw a sample without replacement and is relatively simple to implement (hence its widespread use), but many of the π_{ij}'s for the population are zero. Brewer and Hanif (1983) present over 50 methods for selecting without-replacement unequal-probability samples. Most of these methods are for $n = 2$. Some methods are easier to compute, some are more suitable for specific applications, and some give a more stable estimate of the variance of the Horvitz–Thompson estimator of t.

6.4.2   Weights in Unequal-Probability Samples

All without-replacement sampling schemes discussed so far can be considered as special cases of two-stage cluster sampling with (possibly) unequal probabilities.

In the Horvitz–Thompson estimator, the sampling weight for the ith psu is

$$w_i = \frac{1}{\pi_i}.$$

Thus, the Horvitz–Thompson estimator for the population total is

$$\hat{t}_{\text{HT}} = \sum_{i \in \mathcal{S}} w_i \hat{t}_i.$$

For a without-replacement probability sample of ssu's within psu's, we can define, using the notation of Särndal et al. (1992),

$$\pi_{j|i} = P(\,j\text{th ssu in }i\text{th psu included in sample} \mid i\text{th psu is in the sample}).$$

Then,

$$\hat{t}_i = \sum_{j \in \mathcal{S}_i} \frac{y_{ij}}{\pi_{j|i}}.$$

The overall probability that the (i, j)th element is selected is $\pi_{j|i} \pi_i$. Thus, we can define the sampling weight for the (i, j)th element as

$$w_{ij} = \frac{1}{\pi_{j|i} \pi_i} \tag{6.16}$$

and the Horvitz–Thompson estimator of the population total as

$$\hat{t}_{\text{HT}} = \sum_{i \in \mathcal{S}} \sum_{j \in \mathcal{S}_i} w_{ij} y_{ij}. \tag{6.17}$$

The population mean is estimated as

$$\hat{\bar{y}}_{\text{HT}} = \frac{\displaystyle\sum_{i \in \mathcal{S}} \sum_{j \in \mathcal{S}_i} w_{ij} y_{ij}}{\displaystyle\sum_{i \in \mathcal{S}} \sum_{j \in \mathcal{S}_i} w_{ij}}. \tag{6.18}$$

6.4.3   The Horvitz–Thompson Estimator for General Without-Replacement Designs

We noted in Section 5.8 that the formulas for stratified sampling were a special case of those for two-stage cluster sampling. In fact, all formulas for unbiased estimation of totals in without-replacement sampling in Chapters 2, 4, 5, and 6 are special cases of (6.12) through (6.15).

In simple random sampling, for example, a psu is an individual element and $t_i = y_i$. We show in Appendix B that

$$\pi_i = P(Z_i = 1) = \frac{n}{N}$$

and

$$\pi_{ij} = P(Z_i = 1, Z_j = 1) = \frac{n(n-1)}{N(N-1)}.$$

Thus, for simple random sampling,

$$\hat{t}_{HT} = \sum_{i=1}^{N} Z_i \frac{N}{n} y_i = N\bar{y},$$

and

$$V(\hat{t}_{HT}) = \sum_{i=1}^{N} \frac{1-\pi_i}{\pi_i} y_i^2 + \sum_{i=1}^{N} \sum_{k \neq i}^{N} \frac{\pi_{ik} - \pi_i \pi_k}{\pi_i \pi_k} y_i y_k$$

$$= \sum_{i=1}^{N} \frac{1 - \frac{n}{N}}{\frac{n}{N}} y_i^2 + \sum_{i=1}^{N} \sum_{k \neq i}^{N} \frac{\frac{n-1}{N-1} - \frac{n}{N}}{\frac{n}{N}} y_i y_k$$

$$= \cdots = N^2 \left(1 - \frac{n}{N}\right) \frac{S^2}{n} = V(N\bar{y}).$$

In Exercise 18 you will show that the formulas for stratified sampling are a special case of Horvitz–Thompson estimation.

6.5
Examples of Unequal-Probability Samples

Many sampling situations are well suited for unequal-probability samples. This section gives three examples of sampling designs in common use.

EXAMPLE 6.9   *Random Digit Dialing*

In telephone surveys, it is important to have a well-defined and efficient procedure by which telephone numbers to appear in the sample are generated. In the early days of telephone surveys, many organizations simply took numbers from the telephone directory. That approach leads to selection bias, however, because unlisted telephone numbers do not appear in the directory and a directory does not contain telephone numbers added since its publication. Modifications of sampling from the directory have been suggested to allow inclusion of unlisted numbers, but most have some difficulties with undercoverage.

Random Digit Dialing Element Sampling   Generating telephone numbers at random from the frame of all possible telephone numbers avoids undercoverage of unlisted numbers. In the United States, telephone numbers consist of

| area code | + | prefix (or exchange) | + | suffix. |
| (3 digits) | | (3 digits) | | (4 digits) |

Thus, a random sample of telephone numbers in the United States can be chosen by randomly selecting an area code and prefix combination known to be in use and appending a four-digit number chosen randomly from 0000 to 9999. If the random number chosen does not belong to a household, the number is discarded and a new ten-digit number tried.

This method is simple to understand and explain and, assuming no nonresponse, produces an SRS of telephone numbers from the frame of all possible telephone numbers. The method is self-weighting because we expect to dial residential numbers from a prefix at a rate proportional to the relative frequency of residential numbers beginning with the prefix. In practice, the method can be expensive: Lepkowski (1988) reports that fewer than 25% of all potential telephone numbers generated by this method belong to a household. Multiple calls to a number may be needed to ascertain whether or not the number is residential.

The Mitofsky–Waksberg Method   Mitofsky (1970) and Waksberg (1978) developed a cluster-sampling method for sampling residential telephone numbers. The following description is of the "sampler's utopia" procedure in which everyone answers the phone; Lavrakas (1993) and Potthoff (1994) give suggestions for how to use the Mitofsky–Waksberg method when residents are not as cooperative.

1   Construct a frame of all area codes and prefixes in the area of interest.

2   Draw a random sample of ten-digit telephone numbers from the set of telephone numbers with area code and prefix in the frame and suffix between 0000 and 9999. After step 2, you have a sample of telephone numbers exactly as in random digit dialing element sampling.

3   Dial each number selected in step 2. If the selected number is residential, interview the household and choose its psu to be in the sample; the associated psu is the block of 100 telephone numbers that have the same first eight digits as the selected number. For example, if the randomly selected telephone number (202)456-1414 is determined to be residential, then the psu of all numbers of the form (202)456-14xx is included in the sample. The telephone numbers kept are an SRS of residential households in the region. If the selected number is not residential, discard it and its psu. Continue sampling at the first stage until the desired number of psu's, n, is selected.

4   For the second stage of sampling, randomly select additional telephone numbers without replacement from each psu in the sample until the desired sample size for each psu is attained.

The Mitofsky–Waksberg method dramatically increases the percentage of calls made that reach residential households. Lepkowski (1988) found that 60% of telephone numbers chosen at stage two reached households, as compared with 25% for random digit element sampling. The method works because the psu's of 100 telephone numbers are clustered—some psu's are unassigned, some tend to be assigned to commercial establishments, and some are largely residential. The two-stage procedure eliminates sampling unassigned psu's at the second stage and reduces the probability of selecting psu's with few residential telephone numbers.

Under ideal conditions, the Mitofsky–Waksberg procedure samples psu's with probabilities proportional to the number of residential telephone numbers in the psu's.

If the second stage prescribes selecting an additional $(k - 1)$ residential telephone numbers in each sampled psu and if all psu's in the sample have at least k residential telephone numbers, then the Mitofsky–Waksberg procedure gives each residential telephone number the same probability of being selected in the sample—the result is a self-weighting sample of residential telephone numbers.

Our procedures for pps sampling require that the probabilities of selection be known for all psu's in the sample. In the Mitofsky–Waksberg procedure, the probabilities are unknown before sampling, but we can still calculate weights. Let M_i be the number of residential telephone numbers in the ith psu, and let k be the number of residential telephone numbers in each psu that are selected to be in the sample. Then,

$$P(\text{number selected}) = P(i\text{th psu selected}) \, P(\text{number selected} \mid \text{psu selected})$$
$$\propto \frac{M_i}{K} \frac{k}{M_i} = \frac{k}{K}.$$

To estimate a population total, you would need to know K, the total number of residential telephone numbers in the population, and use sampling weights $w_{ij} = K/k$.

To estimate an average or proportion, the typical goal of telephone surveys, you do not need to know K. You only need to know a "relative weight" w_{ij} for each response y_{ij} in the sample, and you can estimate the population mean as

$$\hat{\bar{y}} = \frac{\displaystyle\sum_{i \in S} \sum_{j \in S_i} w_{ij} y_{ij}}{\displaystyle\sum_{i \in S} \sum_{j \in S_i} w_{ij}}.$$

Here, with a self-weighting sample, you can use relative weights of $w_{ij} = 1$.

Note that although under ideal conditions the Mitofsky–Waksberg method leads to a self-weighting sample of residential telephone numbers, it does *not* give a self-weighting sample of households—some households may have more than one telephone number; others may not have a telephone. In practice, someone using the Mitofsky–Waksberg method would adjust the weights to compensate for multiple telephone lines and nonresponse, as will be discussed in Chapter 8. ■

EXAMPLE **6.10**   *3-P Sampling*

Probability **P**roportional to **P**rediction (3-P) sampling, described by Schreuder et al. (1968), is commonly recommended as a sampling scheme in forestry. Suppose an investigator wants to estimate the total volume of timber in an area. Several options are available: (1) Estimate the volume for each tree in the area. There may be thousands of trees, however, and this can be very time-consuming. (2) Use a cluster sample in which plots of equal areas are selected and the volume of every tree in the selected plots measured. (3) Use an unequal-probability sampling scheme in which points in the area are selected at random, and the trees closest to the points are included in the sample. In this design, a tree is selected with probability proportional to the area of the region that is closer to that tree than to any other tree. (4) Estimate the volume of each tree by eye and then select trees with probability proportional to the estimated volume. When done in one pass, with trees selected as the volume is estimated, this is 3-P sampling—the prediction P stands for the predicted (estimated) volume used in determining the π_i's.

As a form of unequal-probability sampling in which the probability of selecting a psu is unknown in advance of taking the sample, 3-P sampling is a special case of **Poisson sampling.** The largest trees tend to produce the most timber and contribute most to the variability of the estimate of total volume. Thus, unequal-probability sampling can be expected to lead to less sampling effort. Theoretically, you could estimate the volume of each of the N trees in the forest by eye, obtaining a value x_i for tree i. Then, you could revisit trees randomly selected with probabilities proportional to x_i and carefully measure the volume t_i. Such a procedure, however, requires two trips through the forest and adds much work to the sampling process. In 3-P sampling, only one trip is made through the forest, and trees are selected for the sample at the same time the x_i's are measured. The procedure is as follows:

1   Estimate or guess what the maximum value of x_i for the trees is likely to be. Define a value L that is larger than your estimated maximum value of x_i.

2   Proceed to a tree in the forest and determine x_i for that tree. Generate a random number u_i in $[0, L]$. If $u_i \le x_i$, then measure the volume y_i on that tree; otherwise, go on to the next tree.

3   Repeat step 2 on every tree in the forest.

The unequal-probability sampling in this case essentially gives every board-foot of timber an equal chance of being selected for the sample. Note that the size of the unequal-probability sample is unknown until sampling is completed. The probability that tree i is included in the sample is $\pi_i = x_i/L$. The Horvitz–Thompson estimator is

$$\hat{t}_{\text{HT}} = \sum_{i \in S} \frac{y_i}{\pi_i} = L \sum_{i \in S} \frac{y_i}{x_i} = \sum_{i=1}^{N} Z_i \frac{y_i}{\pi_i},$$

where $Z_i = 1$ if tree i is in the sample, and 0 otherwise. Then, the sample size is the random variable $\sum_{i=1}^{N} Z_i$ with expected value $\sum_{i=1}^{N} x_i/L$.

Because the sample size is variable rather than fixed, Poisson sampling provides a different method of unequal-probability sampling than those discussed in Sections 6.1 through 6.4. Brewer and Hanif (1983) give additional theory and references for Poisson sampling.   ■

In natural-resource sampling, 3-P sampling is one example of the use of unequal probabilities. A number of other examples are given in Overton and Stehman (1995).

EXAMPLE **6.11**   *Dollar Unit Sampling*

An accountant auditing the accounts receivable for a company often takes a sample to estimate the true total accounts receivable balance. The book value x_i is known for each account in the population; the audited value t_i will be known only for accounts in the sample. In Section 3.2 we saw how the auxiliary information x_i could be used in difference estimation to improve the precision from an SRS of accounts. Ratio or regression estimation could be used similarly.

Instead of being used in the analysis, the book values could be used in the design of the sample. You could stratify the accounts by the value of x_i, or you could take an unequal-probability sample with selection probabilities proportional to x_i. (Or you could do both: First stratify, then sample with unequal probabilities within each stratum.) If you sample accounts with probabilities proportional to x_i, then each individual dollar in the book values has the same probability of being selected in the sample (hence the name **dollar unit sampling**). With each dollar equally likely to be included in the sample, an account with book value $10,000 is ten times as likely to be in the sample as an account with book value $1000.

Consider a client with 87 accounts receivable, with a book balance of $612,824. The auditor has decided that a sample of size 25 will be sufficient for estimating the error in accounts receivable and takes a random sample with replacement of the 612,824 dollars in the book value population. As individual dollars can only be audited as part of the whole account, each dollar selected serves as a "hook" to snag the whole account for audit. The cumulative-size method is used to select psu's (accounts) for this example; often, in practice, auditors take a systematic sample of dollars and their accompanying psu's. A systematic sample guarantees that accounts with book values greater than the sampling interval will be included in the sample. Table 6.5 shows the first few lines of the account selection; the full table is in file audit.dat. Here, accounts 3 and 13 are included once, and account 9 is included twice (but only needs to be audited once since this is a one-stage cluster sample). This is thus an example of one-stage pps sampling with replacement, as discussed in Section 6.2.

The selected accounts are audited, and the audit values recorded in Table 6.6. Using the results from Section 6.2, the total overstatement is estimated to be $4334 with standard error $13,547/\sqrt{25} = \$2709$. In many auditing situations, however, most of the audited values agree with the book values, so most of the differences are zeros. A confidence interval based on a normal approximation does not perform well in this situation, so auditors typically use confidence bounds based on the Poisson or multinomial distribution (see Neter et al. 1978) rather than a confidence interval of the form (average $\pm$ 1.96 SE).

TABLE 6.5
Account Selection for Audit Sample

Account (Audit Unit)	Book Value	Cumulative Book Value	Random Number	
1	2,459	2,459		
2	2,343	4,802		
3	6,842	11,644	11,016	
4	4,179	15,823		
5	750	16,573		
6	2,708	19,281		
7	3,073	22,354		
8	4,742	27,096		
9	16,350	43,446	31,056	38,500
10	5,424	48,870		
11	9,539	58,409		
12	3,108	61,517		
13	3,935	65,452	63,047	
14	900	66,352		

TABLE   **6.6**

Results of the Audit on Accounts in the Sample

Account (Audit Unit)	Book Value (BV)	ψ_i	Audit Value (AV)	BV − AV Difference	$\dfrac{\text{Diff}}{\psi_i}$	Difference per Dollar
3	6,842	0.0111647	6,842	0	0	0.00000
9	16,350	0.0266798	16,350	0	0	0.00000
9	16,350	0.0266798	16,350	0	0	0.00000
13	3,935	0.0064211	3,935	0	0	0.00000
24	7,090	0.0115694	7,050	40	3,457	0.00564
29	5,533	0.0090287	5,533	0	0	0.00000
34	2,163	0.0035296	2,163	0	0	0.00000
36	2,399	0.0039147	2,149	250	63,862	0.10421
43	8,941	0.0145898	8,941	0	0	0.00000
44	3,716	0.0060637	3,716	0	0	0.00000
45	8,663	0.0141362	8,663	0	0	0.00000
46	69,540	0.1134747	69,000	540	4,759	0.00777
46	69,540	0.1134747	69,000	540	4,759	0.00777
46	69,540	0.1134747	69,000	540	4,759	0.00777
49	6,881	0.0112283	6,881	0	0	0.00000
55	70,100	0.1143885	70,100	0	0	0.00000
55	70,100	0.1143885	70,100	0	0	0.00000
55	70,100	0.1143885	70,100	0	0	0.00000
56	6,467	0.0105528	6,467	0	0	0.00000
61	21,000	0.0342676	21,000	0	0	0.00000
70	3,847	0.0062775	3,847	0	0	0.00000
74	2,422	0.0039522	2,422	0	0	0.00000
75	2,291	0.0037384	2,191	100	26,749	0.04365
79	4,667	0.0076156	4,667	0	0	0.00000
81	31,257	0.0510049	31,257	0	0	0.00000
		average			4,334	0.007071874
		std. dev.			13,547	0.02210527

Another way of looking at the unequal-probability estimate is to find the overstatement for each individual dollar in the sample. Account 24, for example, has a book value of $7090 and an error of $40. The error is prorated to every dollar in the book value, leading to an overstatement of $0.00564 for each of the 7090 dollars. The average overstatement for the individual dollars in the sample is $0.007071874, so the total overstatement for the population is estimated as $(0.007071874)(612,824) = 4334$. ∎

6.6
Randomization Theory Results and Proofs*

In two-stage cluster sampling, we always select the psu's first and then select subunits within the sampled psu's. One approach to calculate a theoretical variance for any estimator in multistage sampling is to condition on which psu's are included in the sample. To do this, we need to use Properties 4 (successive conditioning) and 5 (calculating variances conditionally) of conditional expectation, stated in Section B.4.

In this section, we state and prove Theorem 6.2, the Horvitz–Thompson theorem (Horvitz and Thompson 1952), which gives the properties of the Horvitz–Thompson estimator in (6.12). In Theorem 6.3, we find unbiased estimators of the variance. We then show that the variance for cluster sampling with equal probabilities in (5.22) follows as a special case of these theorems. First, however, we prove (6.10) and (6.11).

THEOREM 6.1

For a without-replacement probability sample of n units, let

$$
Z_i = \begin{cases} 1 & \text{if psu } i \text{ is in the sample} \\ 0 & \text{if psu } i \text{ is not in the sample} \end{cases}
$$

and define

$$
P(Z_i = 1) = \pi_i
$$

and

$$
P(Z_i = 1 \text{ and } Z_k = 1) = \pi_{ik}.
$$

Then

$$
\sum_{i=1}^{N} \pi_i = n
$$

and

$$
\sum_{\substack{k=1 \\ k \neq i}}^{N} \pi_{ik} = (n-1)\pi_i.
$$

Proof   Since the sample size is n, $\sum_{i=1}^{N} Z_i = n$. Also,

$$
E[Z_i] = E[Z_i^2] = \pi_i
$$

because $P(Z_i = 1) = \pi_i$. Consequently,

$$
n = E\left[\sum_{i=1}^{N} Z_i \right] = \sum_{i=1}^{N} \pi_i.
$$

Also,

$$
\sum_{\substack{k=1 \\ k \neq i}}^{N} \pi_{ik} = \sum_{\substack{k=1 \\ k \neq i}}^{N} E[Z_i Z_k] = E[Z_i(n - Z_i)] = \pi_i(n-1),
$$

which completes the proof.   ∎

THEOREM 6.2   *Horvitz–Thompson*

Let Z_i, π_i, and π_{ik} be as in Theorem 6.1. Suppose that sampling is done at the second stage so that sampling in any psu is done independently of the sampling in any other psu, and that $\hat{t}_i$ is independent of $(Z_1, \ldots, Z_N)$ with $E[\hat{t}_i] = E[\hat{t}_i \mid Z_1, \ldots, Z_N] = t_i$. Then

$$
E\left[\sum_{i=1}^{N} Z_i \frac{\hat{t}_i}{\pi_i} \right] = \sum_{i=1}^{N} \pi_i \frac{t_i}{\pi_i} = t \tag{6.19}
$$

and

$$V\left[\sum_{i=1}^{N} Z_i \frac{\hat{t}_i}{\pi_i}\right] = V_{psu} + V_{ssu} \tag{6.20}$$

where

$$V_{psu} = V\left[\sum_{i=1}^{N} Z_i \frac{t_i}{\pi_i}\right] = \sum_{i=1}^{N}(1-\pi_i)\frac{t_i^2}{\pi_i} + \sum_{i=1}^{N}\sum_{\substack{k=1 \\ k \neq i}}^{N}(\pi_{ik} - \pi_i\pi_k)\frac{t_i}{\pi_i}\frac{t_k}{\pi_k} \tag{6.21}$$

and

$$V_{ssu} = \sum_{i=1}^{N} \frac{V(\hat{t}_i)}{\pi_i}. \tag{6.22}$$

Proof   First note that

$$\text{Cov}(Z_i, Z_k) = \begin{cases} \pi_i(1-\pi_i) & \text{if } i = k. \\ \pi_{ik} - \pi_i\pi_k & \text{if } i \neq k. \end{cases}$$

We use successive conditioning to show (6.19):

$$E\left[\sum_{i=1}^{N} Z_i \frac{\hat{t}_i}{\pi_i}\right] = E\left\{E\left[\sum_{i=1}^{N} Z_i \frac{\hat{t}_i}{\pi_i}\bigg| Z_1, \ldots, Z_N\right]\right\}$$

$$= E\left[\sum_{i=1}^{N} Z_i \frac{t_i}{\pi_i}\right]$$

$$= \sum_{i=1}^{N} \pi_i \frac{t_i}{\pi_i}$$

$$= t.$$

The first step simply applies successive conditioning; in the second step, we use the independence of $\hat{t}_i$ and $(Z_1, \ldots, Z_N)$.

To find the variance, use the expression for calculating the variance conditionally in Property 5 of Section B.4, and again use the independence of $\hat{t}_i$ and $(Z_1, \ldots, Z_N)$:

$$V\left[\sum_{i=1}^{N} Z_i \frac{\hat{t}_i}{\pi_i}\right] = V\left[E\left(\sum_{i=1}^{N} Z_i \frac{\hat{t}_i}{\pi_i}\bigg| Z_1, \ldots, Z_N\right)\right]$$

$$+ E\left[V\left(\sum_{i=1}^{N} Z_i \frac{\hat{t}_i}{\pi_i}\bigg| Z_1, \ldots, Z_N\right)\right]$$

$$= V\left[\sum_{i=1}^{N} Z_i \frac{t_i}{\pi_i}\right] + E\left[\sum_{i=1}^{N} Z_i^2 \frac{V(\hat{t}_i)}{\pi_i^2}\right]$$

$$= \sum_{i=1}^{N}\sum_{k=1}^{N} \frac{t_i}{\pi_i}\frac{t_k}{\pi_k} \text{Cov}(Z_i, Z_k) + \sum_{i=1}^{N} \pi_i \frac{V(\hat{t}_i)}{\pi_i^2}$$

$$= \sum_{i=1}^{N}\pi_i(1-\pi_i)\frac{t_i^2}{\pi_i^2} + \sum_{i=1}^{N}\sum_{\substack{k=1 \\ k \neq i}}^{N}(\pi_{ik} - \pi_i\pi_k)\frac{t_i}{\pi_i}\frac{t_k}{\pi_k} + \sum_{i=1}^{N} \frac{V(\hat{t}_i)}{\pi_i}. \quad \blacksquare$$

Equation (6.19) establishes that the Horvitz–Thompson estimator is unbiased, and Equations (6.20) through (6.22) show that (6.13) is the variance of the Horvitz–Thompson estimator.

Theorem 6.3 gives an unbiased estimator for the variance in (6.13).

THEOREM 6.3

Suppose the conditions of Theorem 6.2 hold and that $\hat{V}(\hat{t}_i)$ is an unbiased estimator of $V(\hat{t}_i)$. Then,

$$E\left[Z_i \frac{\hat{V}(\hat{t}_i)}{\pi_i^2}\right] = \frac{V(\hat{t}_i)}{\pi_i}, \tag{6.23}$$

$$E\left[\sum_{i=1}^{N} Z_i \frac{\hat{V}(\hat{t}_i)}{\pi_i^2}\right] = V_{\text{ssu}}, \tag{6.24}$$

and

$$E\left[\sum_{i=1}^{N} Z_i(1-\pi_i)\frac{\hat{t}_i^2}{\pi_i^2} + \sum_{i=1}^{N}\sum_{\substack{k=1\\k\neq i}}^{N} Z_i Z_k \frac{\pi_{ik} - \pi_i\pi_k}{\pi_{ik}}\frac{\hat{t}_i}{\pi_i}\frac{\hat{t}_k}{\pi_k}\right]$$

$$= E\left[\sum_{i=1}^{N}\sum_{k=i+1}^{N} Z_i Z_k \frac{\pi_i\pi_k - \pi_{ik}}{\pi_{ik}}\left(\frac{\hat{t}_i}{\pi_i} - \frac{\hat{t}_k}{\pi_k}\right)^2\right] \tag{6.25}$$

$$= V_{\text{psu}} + \sum_{i=1}^{N}(1-\pi_i)\frac{V(\hat{t}_i)}{\pi_i}.$$

Proof We prove (6.23) and (6.24) by again using successive conditioning:

$$E\left[Z_i \frac{\hat{V}(\hat{t}_i)}{\pi_i^2}\right] = E\left[E\left(Z_i \frac{\hat{V}(\hat{t}_i)}{\pi_i^2}\,\bigg|\, Z_1, \ldots, Z_N\right)\right] = E\left[Z_i \frac{V(\hat{t}_i)}{\pi_i^2}\right] = \frac{V(\hat{t}_i)}{\pi_i}.$$

Result (6.24) follows immediately.

To prove (6.25), note that because $\hat{t}_i$ and $(Z_1, \ldots, Z_N)$ are independent,

$$E[\hat{t}_i^2 \mid Z_1, \ldots, Z_N] = E[\hat{t}_i^2] = t_i^2 + V(\hat{t}_i).$$

Thus,

$$E\left[\sum_{i=1}^{N} Z_i(1-\pi_i)\frac{\hat{t}_i^2}{\pi_i^2}\right] = E\left[E\left(\sum_{i=1}^{N} Z_i(1-\pi_i)\frac{\hat{t}_i^2}{\pi_i^2}\,\bigg|\, Z_1, \ldots, Z_N\right)\right]$$

$$= E\left[\sum_{i=1}^{N} Z_i\frac{1-\pi_i}{\pi_i^2}\{t_i^2 + V(\hat{t}_i)\}\right]$$

$$= \sum_{i=1}^{N} \frac{1-\pi_i}{\pi_i}[t_i^2 + V(\hat{t}_i)].$$

Because subsampling is done independently in different clusters, $E[\hat{t}_i \hat{t}_k] = t_i t_k$ for $k \neq i$, so

$$
E\left[\sum_{i=1}^{N} \sum_{\substack{k=1 \\ k\neq i}}^{N} Z_i Z_k \frac{\pi_{ik} - \pi_i \pi_k}{\pi_{ik}} \frac{\hat{t}_i}{\pi_i} \frac{\hat{t}_k}{\pi_k}\right]
$$

$$
= E\left[E\left(\sum_{i=1}^{N} \sum_{\substack{k=1 \\ k\neq i}}^{N} Z_i Z_k \frac{\pi_{ik} - \pi_i \pi_k}{\pi_{ik}} \frac{\hat{t}_i}{\pi_i} \frac{\hat{t}_k}{\pi_k}\,\middle|\, Z_1, \ldots, Z_N\right)\right]
$$

$$
= E\left[\sum_{i=1}^{N} \sum_{\substack{k=1 \\ k\neq i}}^{N} Z_i Z_k \frac{\pi_{ik} - \pi_i \pi_k}{\pi_{ik}} \frac{t_i}{\pi_i} \frac{t_k}{\pi_k}\right]
$$

$$
= \sum_{i=1}^{N} \sum_{\substack{k=1 \\ k\neq i}}^{N} (\pi_{ik} - \pi_i \pi_k) \frac{t_i}{\pi_i} \frac{t_k}{\pi_k}.
$$

Combining the two results, we see that

$$
E\left[\sum_{i=1}^{N} Z_i(1 - \pi_i)\frac{\hat{t}_i^2}{\pi_i^2} + \sum_{i=1}^{N} \sum_{\substack{k=1 \\ k\neq i}}^{N} Z_i Z_k \frac{\pi_{ik} - \pi_i \pi_k}{\pi_{ik}} \frac{\hat{t}_i}{\pi_i} \frac{\hat{t}_k}{\pi_k}\right]
$$

$$
= V_{\text{psu}} + \sum_{i=1}^{N} \frac{1 - \pi_i}{\pi_i} V(\hat{t}_i),
$$

which proves the first part of (6.25). We show the second part of (6.25) similarly:

$$
E\left[\sum_{i=1}^{N} \sum_{k=i+1}^{N} Z_i Z_k \frac{\pi_i \pi_k - \pi_{ik}}{\pi_{ik}} \left(\frac{\hat{t}_i}{\pi_i} - \frac{\hat{t}_k}{\pi_k}\right)^2\right]
$$

$$
= E\left\{E\left[\sum_{i=1}^{N} \sum_{k=i+1}^{N} Z_i Z_k \frac{\pi_i \pi_k - \pi_{ik}}{\pi_{ik}} \left(\frac{\hat{t}_i}{\pi_i} - \frac{\hat{t}_k}{\pi_k}\right)^2\,\middle|\, Z_1, \ldots, Z_N\right]\right\}
$$

$$
= E\left[\sum_{i=1}^{N} \sum_{k=i+1}^{N} Z_i Z_k \frac{\pi_i \pi_k - \pi_{ik}}{\pi_{ik}} \left(\frac{t_i^2 + V(\hat{t}_i)}{\pi_i^2} - 2\frac{t_i}{\pi_i} \frac{t_k}{\pi_k} + \frac{t_k^2 + V(\hat{t}_k)}{\pi_k^2}\right)\right]
$$

$$
= \frac{1}{2} \sum_{i=1}^{N} \sum_{\substack{k=1 \\ k\neq i}}^{N} (\pi_i \pi_k - \pi_{ik}) \left(\frac{t_i^2 + V(\hat{t}_i)}{\pi_i^2} - 2\frac{t_i}{\pi_i} \frac{t_k}{\pi_k} + \frac{t_k^2 + V(\hat{t}_k)}{\pi_k^2}\right)
$$

$$
= \sum_{i=1}^{N} [\pi_i(n - \pi_i) - (n - 1)\pi_i] \left(\frac{t_i^2 + V(\hat{t}_i)}{\pi_i^2}\right) + \sum_{i=1}^{N} \sum_{\substack{k=1 \\ k\neq i}}^{N} (\pi_{ik} - \pi_i \pi_k) \frac{t_i}{\pi_i} \frac{t_k}{\pi_k}.
$$

The last step follows from Theorem 6.1, and the last expression is easily seen to be equivalent to $V_{\text{psu}} + \sum_{i=1}^{N}(1 - \pi_i)V(\hat{t}_i)/\pi_i$. This completes the proof of Theorem 6.3. ∎

Theorem 6.3 implies that (6.14) and (6.15) are unbiased estimators of the variance of the Horvitz–Thompson estimator.

If psu's are selected with equal probabilities, as in Chapter 5, then

$$P(Z_i = 1) = \pi_i = \frac{n}{N},$$

$$P(Z_i = 1 \text{ and } Z_j = 1) = \pi_{ij} = \frac{n}{N}\frac{n-1}{N-1},$$

$$\hat{t}_{\text{unb}} = \sum_{i \in S} \frac{N}{n}\hat{t}_i = \sum_{i=1}^{N} Z_i \frac{N}{n}\hat{t}_i,$$

so we can apply Theorem 6.2 with $\pi_i = n/N$. Then,

$$E[\hat{t}_{\text{unb}}] = \sum_{i=1}^{N} \frac{n}{N}\frac{N}{n}t_i = t,$$

and, from (6.21),

$$V_{\text{psu}}[\hat{t}_{\text{unb}}] = \sum_{i=1}^{N}\left(1 - \frac{n}{N}\right)\left(\frac{N}{n}\right)t_i^2 + \sum_{i=1}^{N}\sum_{\substack{k=1\\k \neq i}}^{N}\left[\frac{n}{N}\frac{n-1}{N-1} - \left(\frac{n}{N}\right)^2\right]\left(\frac{N}{n}\right)^2 t_i t_k$$

$$= \frac{N}{n}\left(1 - \frac{n}{N}\right)\left[\sum_{i=1}^{N} t_i^2 - \frac{1}{N-1}\sum_{i=1}^{N}\sum_{\substack{k=1\\k \neq i}}^{N} t_i t_k\right]$$

$$= \frac{N}{n(N-1)}\left(1 - \frac{n}{N}\right)\left[(N-1)\sum_{i=1}^{N} t_i^2 - \sum_{i=1}^{N}\sum_{k=1}^{N} t_i t_k + \sum_{i=1}^{N} t_i^2\right]$$

$$= \frac{N}{n(N-1)}\left(1 - \frac{n}{N}\right)\left[N\sum_{i=1}^{N} t_i^2 - t^2\right]$$

$$= N^2\left(1 - \frac{n}{N}\right)\frac{S_t^2}{n}.$$

By result (2.7) from SRS theory,

$$V(\hat{t}_i) = M_i^2\left(1 - \frac{m_i}{M_i}\right)\frac{S_i^2}{m_i},$$

so, using (6.22),

$$V_{\text{ssu}} = \sum_{i=1}^{N} \frac{N}{n}M_i^2\left(1 - \frac{m_i}{M_i}\right)\frac{S_i^2}{m_i}.$$

This completes the proof of (5.22).

After a bit of algebra, it may be shown that when taking a cluster sample with equal probabilities,

$$\sum_{i=1}^{N} Z_i(1 - \pi_i)\left(\frac{N}{n}\right)^2 \hat{t}_i^2 + \sum_{i=1}^{N}\sum_{\substack{k=1 \\ k \neq i}}^{N} Z_i Z_k \frac{\pi_{ik} - \pi_i \pi_k}{\pi_{ik}}\left(\frac{N}{n}\right)^2 \hat{t}_i \hat{t}_k = N^2\left(1 - \frac{n}{N}\right)\frac{s_t^2}{n}.$$

Thus, by Theorem 6.3,

$$E\left[N^2\left(1 - \frac{n}{N}\right)\frac{s_t^2}{n}\right] = N^2\left(1 - \frac{n}{N}\right)\frac{S_t^2}{n} + \frac{N}{n}\left(1 - \frac{n}{N}\right)\sum_{i=1}^{N} V(\hat{t}_i); \qquad \textbf{(6.26)}$$

consequently,

$$E[s_t^2] = S_t^2 + \frac{1}{N}\sum_{i=1}^{N} V(\hat{t}_i).$$

Note that the expected value of s_t^2 is larger than S_t^2: It includes the variation from psu total to psu total, plus variation from not knowing the psu total.

Because

$$\hat{V}(\hat{t}_i) = \left(1 - \frac{m_i}{M_i}\right)M_i^2 \frac{s_i^2}{m_i}$$

is an unbiased estimator of $V(\hat{t}_i)$, Theorem 6.3 implies that

$$E\left[\sum_{i=1}^{N} Z_i \left(\frac{N}{n}\right)^2 \hat{V}(\hat{t}_i)\right] = E\left[\sum_{i \in \mathcal{S}} \left(\frac{N}{n}\right)^2 \hat{V}(\hat{t}_i)\right] = V_{\text{ssu}}.$$

Using (6.26), then,

$$E\left[N^2\left(1 - \frac{n}{N}\right)\frac{s_t^2}{n} + \frac{N}{n}\sum_{i \in \mathcal{S}} \hat{V}(\hat{t}_i)\right]$$

$$= N^2\left(1 - \frac{n}{N}\right)\frac{S_t^2}{n} + \frac{N}{n}\left(1 - \frac{n}{N}\right)\sum_{i=1}^{N} V(\hat{t}_i) + \sum_{i=1}^{N} V(\hat{t}_i)$$

$$= N^2\left(1 - \frac{n}{N}\right)\frac{S_t^2}{n} + \frac{N}{n}\sum_{i=1}^{N} V(\hat{t}_i)$$

Thus, (5.25) is an unbiased estimator of (5.22).

The methods used in these proofs can be applied to any number of levels of clustering. You may want to sample schools, then classes within schools, then students within classes. Exercise 28 asks you to find an expression for the variance in three-stage cluster sampling. Rao (1979a) presents an alternative and elegant approach, relying on properties of nonnegative definite matrices, for deriving mean squared errors and variance estimators for linear estimators of population totals.

6.7
Models and Unequal-Probability Sampling*

In general, data from a good sampling design should produce reasonable inferences from either a model-based or randomization approach. Let's see how the pps estimator performs for model M1 from (5.37). The model is

$$\text{M1:} \quad Y_{ij} = A_i + \varepsilon_{ij},$$

with the A_i's generated by a distribution with mean μ and variance σ_A^2, the ε_{ij}'s generated by a distribution with mean 0 and variance 1, and all A_i's and ε_{ij}'s independent.

As we did for the estimators in Chapter 5, we can write the pps estimator as a linear combination of the random variables Y_{ij}. For a pps design, $\psi_i = M_i/K$, so

$$\hat{T}_P = \sum_{i \in \mathcal{S}} \frac{K}{nM_i} \hat{T}_i = \sum_{i \in \mathcal{S}} K \frac{\bar{Y}_{\mathcal{S}_i}}{n} = \sum_{i \in \mathcal{S}} \sum_{j \in \mathcal{S}_i} \frac{K}{nm_i} Y_{ij}.$$

Note that $\sum_{i \in \mathcal{S}} \sum_{j \in \mathcal{S}_i} K/(nm_i) = K$, so $\hat{T}_P$ is unbiased under model M1 in (5.37). In addition, from (5.39),

$$V_{\text{M1}}[\hat{T}_P - T] = \sigma_A^2 \left[\sum_{i \in \mathcal{S}} \left(\sum_{j \in \mathcal{S}_i} \frac{K}{nm_i} - M_i \right)^2 + \sum_{i \notin \mathcal{S}} M_i^2 \right]$$

$$+ \sigma^2 \left[\sum_{i \in \mathcal{S}} \sum_{j \in \mathcal{S}_i} \left\{ \left(\frac{K}{nm_i} \right)^2 - 2 \frac{K}{nm_i} \right\} + K \right]$$

$$= \sigma_A^2 \left[\frac{K^2}{n} - 2 \frac{K}{n} \sum_{i \in \mathcal{S}} M_i + \sum_{i=1}^{N} M_i^2 \right] + \sigma^2 \left[\sum_{i \in \mathcal{S}} \frac{K^2}{n^2 m_i} - K \right].$$

The model-based variance for $\hat{T}_P$ has implications for design. Suppose a sample is desired that will minimize $V_{\text{M1}}[\hat{T}_P - T]$. The psu sizes M_i for the sample units appear only in the term $-2\sigma_A^2 (K/n) \sum_{i \in \mathcal{S}} M_i$, so for fixed n the variance is smallest when the n units with largest M_i's are included in the sample. If, in addition, a constraint is placed on the number of subunits that can be examined, $\sum_{i \in \mathcal{S}} (1/m_i)$ is smallest when all m_i's are equal.

Inference in the model-based approach does not depend on the sampling design. As long as model M1 holds for the population, $\hat{T}_P$ is model-unbiased with the variance given above. In a model-based approach, an investigator with complete faith in the model can simply select the psu's with the largest values of M_i to be the sample. In practice, however, this would not be done—no one has complete faith in a model, especially before data collection. Royall and Eberhardt (1975) suggest using balanced sampling, in which the sample is selected in such a way that inferences are robust to certain forms of model misspecification.

As described in Section 6.2, pps sampling can be thought of as a way of introducing randomness into the optimal design for model M1 and estimator $\hat{T}_P$. The self-weighting design of taking all m_i's to be equal also minimizes the variance in the model-based approach. Thus, if model M1 is thought to describe the data, pps sampling and estimation should perform well in practice. Särndal (1978) and Thompson (1997) discuss differences between design- and model-based inference in survey samples.

We conclude our discussion with a widely quoted example from Basu, often used to demonstrate that Horvitz–Thompson estimates can be as silly as any other statistical procedures improperly applied.

> The circus owner is planning to ship his 50 adult elephants and so he needs a rough estimate of the total weight of the elephants. As weighing an elephant is a cumbersome process, the owner wants to estimate the total weight by weighing just one elephant. Which elephant should he weigh? So the owner looks back on his records and discovers a list of the elephants' weights taken 3 years ago. He finds that 3 years ago Sambo the middle-sized elephant was the average (in weight) elephant in his herd. He checks with the elephant trainer who reassures him (the owner) that Sambo may still be considered to be the average elephant in the herd. Therefore, the owner plans to weigh Sambo and take 50 y (where y is the present weight of Sambo) as an estimate of the total weight $Y = Y_1 + Y_2 + \cdots + Y_{50}$ of the 50 elephants. But the circus statistician is horrified when he learns of the owner's purposive sampling plan. "How can you get an unbiased estimate of Y this way?" protests the statistician. So, together they work out a compromise sampling plan. With the help of a table of random numbers they devise a plan that allots a selection probability of 99/100 to Sambo and equal selection probabilities of 1/4900 to each of the other 49 elephants. Naturally, Sambo is selected and the owner is happy. "How are you going to estimate Y?", asks the statistician. "Why? The estimate ought to be 50y of course," says the owner. "Oh! No! That cannot possibly be right," says the statistician, "I recently read an article in the *Annals of Mathematical Statistics* where it is proved that the Horvitz–Thompson estimator is the unique hyperadmissible estimator in the class of all generalized polynomial unbiased estimators." "What is the Horvitz–Thompson estimate in this case?" asks the owner, duly impressed. "Since the selection probability for Sambo in our plan was 99/100," says the statistician, "the proper estimate of Y is 100y/99 and not 50y." "And how would you have estimated Y," inquires the incredulous owner, "if our sampling plan made us select, say, the big elephant Jumbo?" "According to what I understand of the Horvitz–Thompson estimation method," says the unhappy statistician, "the proper estimate of Y would then have been 4900y, where y is Jumbo's weight." That is how the statistician lost his circus job (and perhaps became a teacher of statistics!). (1971, 212–213)

Should the circus statistician have been fired? A statistician desiring to use a model in analyzing survey data would say yes: The circus statistician is using the model $y_i \propto 99/100$ for Sambo and $y_i \propto 1/4900$ for all other elephants in the herd—certainly not a model that fits the data well. A randomization-inference statistician would also say yes: Even though models are not used explicitly in the Horvitz–Thompson theory, the estimator is most efficient (has the smallest variance) when the psu total (here, y_i) is proportional to the probability of selection. The silly design used by the circus statistician leads to a huge variance for the Horvitz–Thompson estimator. If that were not reason enough, the statistician proposes a sample of size 1—he can neither check the validity of the model in a model-based approach nor estimate the variance of the Horvitz–Thompson estimator!

6.8
Exercises

1 For each of the following situations, which unit might be used as the psu? Do you believe there would be a strong clustering effect? Would you sample psu's with equal or unequal probabilities?

    a You want to estimate the percentage of patients who wear contact lenses from the population of all patients of U.S. Air Force optometrists and ophthalmologists.

    b Human taeniasis is acquired by ingesting larvae of the pork tapeworm in inadequately cooked pork. You have been asked to design a survey to estimate the percentage of inhabitants of a village who have taeniasis. A medical examination is required to diagnose the condition.

    c You wish to estimate the total number of cows and heifers on all Ontario dairy farms; in addition, you would like to find estimates of the birth rate and stillbirth rate.

    d You want to estimate the percentages of undergraduate students at U.S. universities who are registered to vote, and who are affiliated with each political party.

    e A fisheries agency is interested in the distribution of carapace width of snow crabs. A trap hauled from a fishing boat has a limit of 30 crabs.

    f You wish to conduct a customer satisfaction survey of persons who have taken guided bus tours of the Grand Canyon rim area. Tour groups range in size from 8 to 44 persons.

2 Historians wanting to use data from U.S. censuses collected in the precomputer age faced the daunting task of poring over reels of handwritten records on microfilm, arranged in geographic order. The Public Use Microdata Samples (PUMS) were constructed by taking samples of the records and typing those records into the computer. Ruggles describes the PUMS construction for the 1940 census:

> The population schedules of the 1940 census are preserved on 4,576 microfilm reels. Each census page contains information on forty individuals. Two lines on each page were designated as "sample lines" by the Census Bureau: the individuals falling on those lines—5 percent of the population—were asked a set of supplemental questions that appear at the bottom of the census page.
>
>   Two of every five census pages were systematically selected for examination. On each selected census page, one of the two designated sample lines was then randomly selected. Data-entry personnel then counted the size of the sample unit containing the targeted sample line. Units size six or smaller were included in the sample in inverse proportion to their size. Thus, every one-person unit was included in the sample, every second two-person unit, every third three-person unit, and so on. Units with seven or more persons were included with a probability of 1-in-7: every seventh household of size seven or more was selected for the sample. (1995, 44)

    a Explain why this is a cluster sample. What are the psu's? The ssu's?

    b What effect do you think the clustering will have on estimates of race? Age? Occupation?

c Construct a table for the probability of selection for persons in one-person units, two-person units, and so on.

d What happens if you estimate the mean age of the population by the average age of all persons in the sample? What estimator should you use?

e Do you think that taking a systematic sample was a good idea for this sample? Why, or why not?

f Does this method provide a representative sample of households? Why, or why not?

g What type of sample is taken of the individuals with supplementary information? Explain.

3 Ruggles also describes the 1950 PUMS:

The 1950 census schedules are contained on 6,278 microfilm reels. Each census page contains information on thirty individuals. Every fifth line on the census page was designated as a sample line, and additional questions for the sample-line individuals appear at the bottom of the form. For the last sample-line individual on each page, there was a block of additional supplemental questions. Thus, 20 percent of individuals were asked a basic set of supplemental questions, and 3.33 percent of individuals were asked a full set of supplemental questions.

One-in-eleven pages within enumeration districts was selected randomly. On each selected census page, the sixth sample-line individual (the one with the full set of questions) was selected for inclusion in the sample. Any other members of the sample unit containing the selected individual were also included. (1995, 45)

For the 1950 PUMS, answer the same questions from Exercise 2.

4 An investigator wants to take an unequal-probability sample of 10 of the 25 psu's in the population listed below and wishes to sample units with replacement.

psu	ψ_i	psu	ψ_i
1	0.000110	14	0.014804
2	0.018556	15	0.005577
3	0.062999	16	0.070784
4	0.078216	17	0.069635
5	0.075245	18	0.034650
6	0.073983	19	0.069492
7	0.076580	20	0.036590
8	0.038981	21	0.033853
9	0.040772	22	0.016959
10	0.022876	23	0.009066
11	0.003721	24	0.021795
12	0.024917	25	0.059185
13	0.040654		

a Adapt the cumulative-size method to draw a sample of size 10 with replacement with probabilities ψ_i.

b Adapt Lahiri's method to draw a sample of size 10 with replacement with probabilities ψ_i.

5 For the supermarket example in Section 6.1, suppose the ψ_i's are as given but that each store has $t_i = 75$. What is $E[\hat{t}_\psi]$? $V[\hat{t}_\psi]$?

6 For the supermarket example in Section 6.1, suppose the ψ_i's are 7/16 for store A and 3/16 for each of stores B, C, and D. Show that $\hat{t}_\psi$ is unbiased and find its variance. Do you think that the sampling scheme with these ψ_i's is a good one?

7 Return to the supermarket example of Section 6.1. Now let's select two supermarkets with replacement. List the 16 possible samples (A, A), (A, B), etc., and find the probability with which each sample would be selected. Calculate $\hat{t}_\psi$ for each sample. What is $E[\hat{t}_\psi]$? $V[\hat{t}_\psi]$?

8 The file statepps.dat lists the number of counties, land area, and 1992 population for the 50 states plus the District of Columbia.

   a Use the cumulative-size method to draw a sample of size 10 with replacement, with probabilities proportional to land area. What is ψ_i for each state in your sample?

   b Use the cumulative-size method to draw a sample of size 10 with replacement, with probabilities proportional to population. What is ψ_i for each state in your sample?

   c How do the two samples differ? Which states tend to be in each sample?

9 Use your sample of states drawn with probability proportional to population, from Exercise 8, for this problem.

   a Using the sample, estimate the total number of counties of the United States and find the standard error of your estimate. How does your estimate compare with the true value of total number of counties (which you can calculate, since the file statepps.dat contains the data for the whole population)?

   b Now suppose that your friend Tom finds the ten values of numbers of counties in your sample but does not know that you selected these states with probabilities proportional to population. Tom then estimates the total land area using formulas for an SRS. What values for the estimated total and its standard error are calculated by Tom? How do these values differ from yours? Is Tom's estimate unbiased for the population total?

10 In Example 2.4, we took an SRS to estimate the total acreage devoted to farming in the United States in 1992. In Example 3.2, we used ratio estimation, with auxiliary variable the number of acres of farms in 1987, to increase the precision of the estimate. Now, use the sample of states drawn with probability proportional to land area in Exercise 8 and then subsample five counties randomly from each state using file agpop.dat. Estimate the total acreage devoted to farming in 1992, along with its standard error.

11 The file statepop.dat, used in Example 6.5, also contains information on total number of farms, number of veterans, and other items.

   a Plot the total number of farms versus the probabilities of selection ψ_i. Does your plot indicate that unequal-probability sampling will be helpful here?

   b Estimate the total number of farms in the United States, along with its standard error.

12   Use the file statepop.dat for this problem.

    a   Plot the total number of veterans versus the probabilities of selection ψ_i. Does your plot indicate that unequal-probability sampling will be helpful here?

    b   Estimate the total number of veterans in the United States and find the standard error for your estimate.

    c   Estimate the total number of Vietnam veterans in the United States and find the standard error for your estimate.

13   Let's return to the situation in Exercise 8 of Chapter 2, in which we took an SRS to estimate the average and total numbers of refereed publications of faculty and research associates. Now, consider a pps sample of faculty: The 27 academic units range in size from 2 to 92. We used Lahiri's method to choose ten psu's with probabilities proportional to size and with replacement and took an SRS of four (or fewer, if $M_i < 4$) members from each psu. Note that academic unit 14 appears three times in the sample; each time it appears, a different subsample was collected.

Academic Unit	M_i	ψ_i	y_{ij}
14	65	0.0805452	3, 0, 0, 4
23	25	0.0309789	2, 1, 2, 0
9	48	0.0594796	0, 0, 1, 0
14	65	0.0805452	2, 0, 1, 0
16	2	0.0024783	2, 0,
6	62	0.0768278	0, 2, 2, 5
14	65	0.0805452	1, 0, 0, 3
19	62	0.0768278	4, 1, 0, 0
21	61	0.0755886	2, 2, 3, 1
11	41	0.0508055	2, 5, 12, 3

Find the estimated total number of publications, along with its standard error.

*14   (Requires probability.)

    a   Prove that Lahiri's method results in a pps sample with replacement.

    b   Suppose the population has N psu's, with sizes $M_1, M_2, \ldots, M_N$. Let X represent the number of pairs of random numbers that must be generated to obtain a sample of size n. Find $E[X]$.

*15   (Requires probability.) In Section 6.3, note that the random variables $Q_1, \ldots, Q_N$ have a joint multinomial distribution with probabilities $\psi_1, \psi_2, \ldots, \psi_N$. Use properties of the multinomial distribution to show that $\hat{t}_\psi$ in (6.8) is an unbiased estimator of t with variance given by

$$V(\hat{t}_\psi) = \frac{1}{n} \sum_{i=1}^{N} \psi_i \left(\frac{t_i}{\psi_i} - t \right)^2 + \frac{1}{n} \sum_{i=1}^{N} \frac{V_i}{\psi_i}. \tag{6.27}$$

Also show that (6.9) is an unbiased estimator of the variance in (6.27). HINT: Use properties of conditional expectation in Appendix B and write

$$V(\hat{t}_\psi) = V(E[\hat{t}_\psi \mid Q_1, \ldots, Q_N]) + E(V[\hat{t}_\psi \mid Q_1, \ldots, Q_N]).$$

16 Show that the two expressions for the variance in (6.13) are equivalent. HINT: Use (6.10) and (6.11).

17 Show that (6.14) and (6.15) are equivalent when psu's and ssu's are selected with equal probabilities, as in Chapter 5. Are they equal if psu's are selected with unequal probabilities?

18 Show that the formulas for stratified sampling in (4.3) and (4.5) follow from the formulas for the Horvitz–Thompson estimator.

19 Use the population in Example 3.4 for this exercise. Let ψ_i be proportional to x_i.

    a Using the draw-by-draw method illustrated in Example 6.8, calculate π_i for each unit and π_{ij} for each pair of units, for a without-replacement sample of size 2.

    b What is $V(\hat{t}_{\mathrm{HT}})$? How does it compare with the with-replacement variance using (6.27)?

*20 (Requires probability.) *Brewer's* (1963, 1975) *procedure for without-replacement unequal-probability sampling.* For a sample of size $n = 2$, let π_i be the desired probability of inclusion for psu i, with the usual constraint that $\sum_{i=1}^{N} \pi_i = n$. Let $\psi_i = \pi_i/2$ and

$$a_i = \frac{\psi_i(1 - \psi_i)}{1 - \pi_i}.$$

Draw the first psu with probability $a_i / \sum_{k=1}^{N} a_k$ of selecting psu i. Supposing psu i is selected at the first draw, select the second psu from the remaining $N - 1$ psu's with probabilities $\psi_j/(1 - \psi_i)$.

    a Show that

$$\pi_{ij} = \frac{\psi_i \psi_j}{\displaystyle\sum_{k=1}^{N} a_k} \left(\frac{1}{1 - \pi_i} + \frac{1}{1 - \pi_j} \right).$$

    b Show that $P(\text{psu } i \text{ selected in sample}) = \pi_i$. HINT: First show that

$$2 \sum_{k=1}^{N} a_k = 1 + \sum_{k=1}^{N} \frac{\psi_k}{1 - \pi_k}.$$

    c The Sen–Yates–Grundy (SYG) estimator of the variance in (6.15) for one-stage sampling is

$$\hat{V}(\hat{t}_{\mathrm{HT}}) = \sum_{i \in \mathcal{S}} \sum_{\substack{j \in \mathcal{S} \\ j > i}} \frac{\pi_i \pi_j - \pi_{ij}}{\pi_{ij}} \left(\frac{t_i}{\pi_i} - \frac{t_j}{\pi_j} \right)^2.$$

    Show that $\pi_i \pi_j - \pi_{ij} \geq 0$ for Brewer's method, so that the SYG estimator of the variance is always nonnegative.

21   The following table gives population values for a small population of clusters:

psu, i	M_i	Values, y_{ij}	t_i
1	5	3, 5, 4, 6, 2,	20
2	4	7, 4, 7, 7,	25
3	8	7, 2, 9, 4, 5, 3, 2, 6	38
4	5	2, 5, 3, 6, 8,	24
5	3	9, 7, 5	21

You wish to select two psu's without replacement with probabilities of inclusion proportional to M_i. Using Brewer's method from Exercise 20, construct a table of π_{ij} for the possible samples. What is the variance of the Horvitz–Thompson estimator?

*22   (Requires probability.) Rao (1963) discusses the following rejective method for selecting a pps sample without replacement: Select n psu's with probabilities ψ_i and with replacement. If any psu appears more than once in the sample, reject the whole sample and select another n psu's with replacement. Repeat until you obtain a sample of n psu's with no duplicates.

Find π_{ij} and π_i for this procedure, for $n = 2$.

*23   (Requires probability.) *The Rao–Hartley–Cochran (1962) method for selecting psu's with unequal probabilities.* To take a sample of size n, divide the population into n random groups of psu's, $U_1, U_2, \ldots, U_n$. Then select one psu from each group (independently) with probability proportional to size. If psu i is in group k, it is selected with probability $x_{ki} = M_i / \sum_{j \in U_k} M_j$. Let $\alpha(k)$ be the label of the psu selected from group k. Then, conditionally on the groups, $\hat{t}_{\alpha(k)}/x_{k,\alpha(k)}$ estimates the total in group k. The estimator of the population total is

$$\hat{t}_{\text{RHC}} = \sum_{k=1}^{n} \frac{t_{\alpha(k)}}{x_{k,\alpha(k)}}.$$

Show that $\hat{t}_{\text{RHC}}$ is unbiased for t and find its variance. HINT: Use two sets of indicator variables. Let $I_{ki} = 1$ if psu i is in group k, and 0 otherwise; let $Z_i = 1$ if psu i is selected to be in the sample. Then, $\hat{t}_{\text{RHC}} = \sum_{k=1}^{n} \sum_{i=1}^{N} I_{ki} Z_i t_i / x_{ki}$.

*24   (Requires calculus.) Suppose in (6.27) that the variance of the estimator of the total in psu i is $V_i = M_i^2 S_i^2 / m_i$. If you can only subsample an expected total of $C = E\left[\sum_{i \in \mathcal{S}} m_i\right]$ ssu's, what values of m_i minimize (6.27)?

25   In Example 6.9, it was shown that the Mitofsky–Waksberg method produces a self-weighting sample if any psu in the sample has at least k residential telephone numbers. Suppose a psu in the sample has $x < k$ residential numbers. What is the relative weight for a telephone number in that psu?

26   One drawback of the Mitofsky–Waksberg method as described in Example 6.9 is that the sequential sampling procedure of selecting numbers in the psu until one has a total of k residential numbers can be cumbersome to implement. Suppose in the second stage you dial an additional $(k - 1)$ numbers whether they are residential or not and let x be the number of residential lines among the $(k-1)$ numbers. What are the relative weights for the residential telephone numbers?

27 The Mitofsky–Waksberg method, described in Example 6.9, gives a self-weighting sample of telephone numbers under ideal circumstances. Does it give a self-weighting sample of adults? Why, or why not? If not, what relative weights should be used?

*28 (Requires probability.) Suppose a three-stage cluster sample is taken from a population with N psu's, M_i ssu's in the ith psu, and L_{ij} tsu's (tertiary sampling units) in the jth ssu of the ith psu. To draw the sample, n psu's are randomly selected, then m_i ssu's from the selected psu's, then l_{ij} tsu's from the selected ssu's.

a Show that the sample weights are

$$w_{ijk} = \frac{N}{n} \frac{M_i}{m_i} \frac{L_{ij}}{l_{ij}}.$$

b Let

$$\hat{t} = \sum_{i \in \mathcal{S}} \sum_{j \in \mathcal{S}_i} \sum_{k \in \mathcal{S}_{ij}} w_{ijk} y_{ijk}.$$

Show that

$$E[\hat{t}] = t = \sum_{i=1}^{N} \sum_{j=1}^{M_i} \sum_{k=1}^{L_{ij}} y_{ijk}.$$

c Using the properties of conditional expectation in Section B.4, find an expression for $V(\hat{t})$.

*29 (Model based.) Suppose the entire population is observed in the sample so that $n = N$ and $m_i = M_i$. Examine the three estimators $\hat{T}_{\text{unb}}$, $\hat{T}_r$ (from Section 5.7), and $\hat{T}_P$ (from Section 6.7). If the entire population is observed, which of these estimators equal $T = \sum_{i=1}^{N} \sum_{j=1}^{M_i} Y_{ij}$?

SURVEY Exercises

30 Design a self-weighting sampling scheme for districts 1–43, with districts (clusters) chosen with probability proportional to size and with replacement. Your design should have the same number of clusters and cost about the same amount as the sample in Chapter 5. For this sample, estimate the average price a rural household is willing to pay for cable TV, along with the standard error of your estimate.

31 Comment on the relative performance of the three estimates:

a Cluster sample with equal probabilities, unbiased estimate

b Cluster sample with equal probabilities, ratio estimate

c Cluster sample with probabilities proportional to district size [You calculated the estimates for parts (a) and (b) in Exercise 32, Chapter 5.]

Which estimate is most precise? Explain.

7

Complex Surveys

There is no more effective medicine to apply to feverish public sentiment than figures. To be sure, they must be properly prepared, must cover the case, not confine themselves to a quarter of it, and they must be gathered for their own sake, not for the sake of a theory. Such preparation we get in a national census.

—Ida Tarbell, *The Ways of Woman* (1915)

Most large surveys involve several of the ideas we have discussed: A survey may be stratified with several stages of clustering and rely on ratio and regression estimates to adjust for other variables. The formulas for estimating standard errors can become horrendous, especially if there are several stages of clustering without replacement. Sampling weights and design effects are commonly used in complex surveys to simplify matters. These, and plots for complex survey data, are discussed in this chapter. The chapter concludes with a description of the National Crime Victimization Survey design, and with parallels between survey samples and designed experiments.

7.1
Assembling Design Components

We have seen most of the components of a complex survey: random sampling, ratio estimation, stratification, and clustering. Now, let's see how to assemble them into one sampling design. Although in practice weights (Section 7.2) are often used to find point estimates and computer-intensive methods (Chapter 9) are used to calculate variances of the estimates, understanding the basic principles of how the components work together is important. Here are the concepts you already know, in a modular form ready for assembly.

7.1.1 Building Blocks for Surveys

1 *Cluster sampling with replacement.* Select a sample of n clusters with replacement; cluster i is selected with probability ψ_i. Estimate the total for cluster i by an unbiased estimate $\hat{t}_i$. Then treat the n values (the sample is with replacement, so some of the values in the set may be from the same psu's) of $u_i = \hat{t}_i/\psi_i$ as observations: Estimate the population total by $\bar{u}$ and estimate the variance of the estimated total by s_u^2/n.

2   Cluster sampling without replacement. Select a sample of n clusters without replacement; the probability that cluster i is selected for the sample is π_i. Estimate the total for cluster i by an unbiased estimate $\hat{t}_i$ and calculate an unbiased estimate of the variance of $\hat{t}_i$, $\hat{V}(\hat{t}_i)$. Then estimate the population total with the Horvitz–Thompson estimator[1] from Equation (6.12):

$$\hat{t}_{\text{HT}} = \sum_{i \in \mathcal{S}} \frac{\hat{t}_i}{\pi_i}.$$

Use an exact formula from Chapters 5 or 6 or a method from Chapter 9 to estimate the variance.

3   Stratification. Let $\hat{t}_1, \ldots, \hat{t}_H$ be unbiased estimators of the strata totals $t_1, \ldots, t_H$ and let $\hat{V}(\hat{t}_1), \ldots, \hat{V}(\hat{t}_H)$ be unbiased estimators of the variances. Then estimate the total by

$$\hat{t} = \sum_{h=1}^{H} \hat{t}_h$$

and its variance by

$$\hat{V}(\hat{t}) = \sum_{h=1}^{H} \hat{V}(\hat{t}_h).$$

4   Ratio estimation. Let $\hat{t}_y$ and $\hat{t}_x$ be unbiased estimators of t_y and t_x, respectively. Then, the ratio is estimated by

$$\hat{B} = \frac{\hat{t}_y}{\hat{t}_x}$$

and its variance by

$$\hat{V}(\hat{B}) = \frac{\hat{B}^2}{t_x^2} \hat{V}(\hat{t}_x) + \frac{1}{t_x^2} \hat{V}(\hat{t}_y) - 2\frac{\hat{B}}{t_x^2} \widehat{\text{Cov}}(\hat{t}_x, \hat{t}_y), \tag{7.1}$$

as will be shown in Section 9.1. The ratio estimator of the total is $\hat{B}t_x$ with estimated variance $t_x^2 \hat{V}(\hat{B})$.

We often use ratios for estimating means, letting the auxiliary variable x_i be 1 if unit i is in the sample, and 0 otherwise. Then, $\hat{t}_x$ estimates the population size, and the ratio divides the estimated population total by the estimated population size.

Stratification usually forms the coarsest classification: Strata may be, for example, areas of the country, different area codes, or types of habitat. Clusters (sometimes several stages of clusters) are sampled from each stratum in the design, and additional stratification may occur within clusters. With several stages of clustering and

---

[1]Recall that the Horvitz–Thompson estimator encompasses the other without-replacement, unbiased estimators of the total as special cases, as discussed in Section 6.4.

stratification, it helps to draw a diagram or construct a table of the survey design, as illustrated in the following example.

EXAMPLE **7.1**   Malaria is a serious health problem in The Gambia. Malaria morbidity can be reduced by using bed nets that are impregnated with insecticide, but this is only effective if the bed nets are in widespread use. In 1991 a nationwide survey was designed to estimate the prevalence of bed net use in rural areas. The survey is described and results reported in D'Alessandro et al. (1994).

The sampling frame consisted of all rural villages of fewer than 3000 people in The Gambia. The villages were stratified by three geographic regions (eastern, central, and western) and by whether the village had a public health clinic (PHC) or not. In each region five districts were chosen with probability proportional to the district population as estimated in the 1983 national census. In each district four villages were chosen, again with probability proportional to census population: two PHC villages and two non-PHC villages. Finally, six compounds were chosen more or less randomly from each village, and a researcher recorded the number of beds and nets, along with other information, for each compound.

In summary, the sample design is the following:

Stage	Sampling Unit	Stratification
1	District	Region
2	Village	PHC/non-PHC
3	Compound	

To calculate estimates or standard errors using formulas from the previous chapters, you would start at stage 3 and work up. The following are steps you would use to estimate the total number of bed nets (without using ratio estimation):

1   Record the total number of nets for each compound.

2   Estimate the total number of nets for each village by (number of compounds in the village) × (average number of nets per compound). Find the estimated variance of the total number of nets, for each village.

3   Estimate the total number of nets for the PHC villages in each district. Villages were sampled from the district with probabilities proportional to population, so formulas from Chapter 6 need to be used to estimate the total and the variance of the estimated total. Repeat for the non-PHC villages in each district.

4   Add the estimates from the two strata (PHC and non-PHC) to estimate the number of nets in each district; sum the estimated variances from the two strata to estimate the variance for the district.

5   At this point you have the estimated total number of nets and the estimated variance, for each district. Now use two-stage cluster-sampling formulas to estimate the total number of nets for each region.

6   Finally, add the estimated totals for each region to estimate the total number of bed nets in The Gambia. Add the region variances as called for in stratified sampling.

Sounds a little complicated, doesn't it? And we have not even included ratio estimation, which would almost certainly be incorporated here because we know approximate population numbers for the numbers of beds at each stage. Fortunately, we do not always have to go to this much trouble in complex surveys. As we will see later in this chapter and in Chapter 9, we can use sampling weights and computer-intensive methods to avoid much of this effort. ∎

7.1.2   Ratio Estimation in Complex Surveys

Ratio estimation is part of the analysis, not the design, and does not appear in a diagram of the design. Ratio estimation may be used at almost any level of the survey, although it is usually used near the top.

One quantity of interest in the bed net survey was the proportion of beds that have nets. The ratio used for the proportions could be calculated at almost any level of the survey; for simplicity, assume we are only interested in the PHC villages. In the following, x refers to beds and y refers to nets.

1   *Compound level.* Calculate the proportion of beds in the compound that have nets and use these proportions as the observations. Then, the estimate at the village level would be the average of the six compound proportions, the estimate at the district level would be calculated from the five village estimates, and so on. This is similar to the mean-of-ratios estimator from Exercise 22 of Chapter 3.

2   *Village level.* For each village, calculate (total number of nets)/(total number of beds). The estimated variance at the village level will be calculated from (7.1). Then, at the district level, average the ratios obtained for the villages in the district.

3   *District level.* This is similar to the village level, except ratios are formed for each district.

4   *Region level.* Use the pps formulas to estimate the total number of beds and total number of nets for the regions C (central), E (eastern), and W (western). The result is six estimates of totals—$\hat{t}_{xC}, \hat{t}_{xE}, \hat{t}_{xW}, \hat{t}_{yC}, \hat{t}_{yE}, \hat{t}_{yW}$—and estimates of the variances and covariances associated with the estimated totals. Now calculate the three ratios $\hat{t}_{yC}/\hat{t}_{xC}$, $\hat{t}_{yE}/\hat{t}_{xE}$, and $\hat{t}_{yW}/\hat{t}_{xW}$ and use the ratio estimate formula to estimate the variance of each ratio. Then combine the three ratio estimates by using stratification:

$$\hat{B} = \sum_{h=1}^{H} \left(\frac{N_h}{N} \right) \frac{\hat{t}_{yh}}{\hat{t}_{xh}}.$$

$$\hat{V}(\hat{B}) = \sum_{h=1}^{H} \left(\frac{N_h}{N} \right)^2 \hat{V}\left(\frac{\hat{t}_{yh}}{\hat{t}_{xh}} \right).$$

5   *Above the region level.* Use the stratification to estimate $\hat{t}_y$ and $\hat{t}_x$ for the whole population, along with the estimated variances and covariance. Now estimate the ratio $\hat{t}_y/\hat{t}_x$ and use (7.1) to estimate the variance.

Recall from Chapter 3 that the ratio estimator is biased and the bias can be serious with small sample sizes. The sample size is small for many of the levels, so you need to be very careful with the estimator: Only six compounds are sampled per village, and five villages per district, so bias is a concern at those levels. Cassel et al. (1977, ch. 7) compare several strategies involving ratio estimators.

At the region level, a comparable estimate of the population total is the **separate ratio estimator:**

$$\sum_{h=1}^{H} \frac{t_{xh}\hat{t}_{yh}}{\hat{t}_{xh}}.$$

Ratio estimation, done separately in each stratum, can improve efficiency if $\hat{t}_{yh}/\hat{t}_{xh}$'s vary from stratum to stratum. It should *not* be used when strata sample sizes are small because each ratio is biased and the bias can propagate through the strata.

Above the region level, the **combined ratio estimator** $t_x\hat{t}_y/\hat{t}_x$ provides a comparable estimate of the population total. The combined estimator has less bias when few psu's are sampled per stratum. When the ratios vary greatly from stratum to stratum, however, the combined estimator does not take advantage of the extra efficiency afforded by stratification, as does the separate ratio estimator.

7.1.3   Simplicity in Survey Design

All these design components have been shown to increase efficiency in survey after survey. Sometimes, though, an inexperienced survey designer is tempted to use a complex sampling design simply because it is there or has been used in the past, not because it has been demonstrated to be more efficient. Make sure you know from pretests or previous research that a complex design really is more efficient and practical. A simpler design giving the same amount of information per dollar spent is almost always to be preferred to a more complicated design: It is often easier to administer and easier to analyze, and data from the survey are less likely to be analyzed incorrectly by subsequent analysts. A complex design should be efficient for estimating *all* quantities of primary interest—an optimal allocation in stratified sampling for estimating the total amount U.S. businesses spend on health-care benefits may be very inefficient for estimating the percentage of businesses that declare bankruptcy in a year.

7.2
Sampling Weights

7.2.1   Constructing Sampling Weights

In many large sample surveys, weights are used to deal with the effects of stratification and clustering on point estimates. We have already seen how sampling weights are used in stratified sampling and in cluster sampling. The sampling weight for an observation unit is always the reciprocal of the probability that the observation unit is selected to be in the sample.

Recall that for stratified sampling,

$$\hat{t}_{\text{str}} = \sum_{h=1}^{H} \sum_{j \in \mathcal{S}_h} w_{hj} y_{hj},$$

where the sampling weight $w_{hj} = (N_h/n_h)$ can be thought of as the number of observations in the population represented by the sample observation y_{hj}. The probability of selecting the jth unit in the hth stratum to be in the sample is $\pi_{hj} = n_h/N_h$, so the sampling weight is simply the inverse of the probability of selection: $w_{hj} = 1/\pi_{hj}$.

The sum of the sampling weights in stratified sampling equals the population size N; each sampled unit "represents" a certain number of units in the population, so the whole sample "represents" the whole population. The stratified-sampling estimate of $\bar{y}_U$ is

$$\bar{y}_{\text{str}} = \frac{\displaystyle\sum_{h=1}^{H} \sum_{j \in \mathcal{S}_h} w_{hj} y_{hj}}{\displaystyle\sum_{h=1}^{H} \sum_{j \in \mathcal{S}_h} w_{hj}}.$$

The same forms of the estimators were used in cluster sampling in Section 5.4, and the general form of weighted estimators was given in Section 6.4. In cluster sampling with equal probabilities,

$$w_{ij} = \frac{NM_i}{nm_i} = \frac{1}{\text{probability that the } j\text{th ssu in the } i\text{th psu is in the sample}}.$$

Again,

$$\hat{t} = \sum_{i \in \mathcal{S}} \sum_{j \in \mathcal{S}_i} w_{ij} y_{ij},$$

and the estimate of the population mean is

$$\frac{\hat{t}}{\displaystyle\sum_{i \in \mathcal{S}} \sum_{j \in \mathcal{S}_i} w_{ij}}.$$

For cluster sampling with unequal probabilities, when π_i is the probability that the ith psu is in the sample and $\pi_{j|i}$ is the probability that the jth ssu is in the sample given that the ith psu is in the sample, the sampling weights are $w_{ij} = 1/(\pi_i \pi_{j|i})$.

For three-stage cluster sampling, the principle extends: Let w_p be the weight for the psu, $w_{s|p}$ be the weight for the ssu, and $w_{t|s,p}$ be the weight associated with the tsu (tertiary sampling unit). Then, the overall sampling weight for an observation unit is

$$w = w_p \times w_{s|p} \times w_{t|s,p}.$$

All the information needed to construct point estimates is contained in the sampling weights; when computing point estimates, the sometimes cumbersome probabilities with which psu's, ssu's, and tsu's are selected appear only through the weights. But the sampling weights give no information on how to find standard errors of the estimates, and thus knowing the sampling weights alone will not allow you to do inferential statistics. Variances of estimates depend on the probabilities that any pair of units is selected to be in the sample and requires more knowledge of the sampling design than given by weights alone.

Very large weights are often truncated, so that no single observation has a very large contribution to the overall estimate. While this biases the estimators, it can reduce the mean squared error (MSE). Truncation is often used when weights are used to adjust for nonresponse, as described in Chapter 8.

Since we will be considering stratified multistage designs in the remainder of this book, from now on we will adopt a unified notation for estimates of population totals. We will consider y_i to be a measurement on observation unit i and w_i to be the sampling weight of observation unit i. Thus, for a stratified sample, y_i is an observation unit within a particular stratum, and $w_i = N_h/n_h$, where unit i is in stratum h. This allows us to write the general estimator of the population total as

$$\hat{t}_y = \sum_{i \in S} w_i y_i, \tag{7.2}$$

where all measurements are at the observation unit level. The general estimator of the population mean is

$$\hat{\bar{y}} = \frac{\hat{t}_y}{\displaystyle\sum_{i \in S} w_i}; \tag{7.3}$$

$\sum_{i \in S} w_i$ estimates the number of observation units, N, in the population.

EXAMPLE 7.2   The Gambia bed net survey in Example 7.1 was designed so that within each region each compound would have almost the same probability of being included in the survey; probabilities varied only because different districts had different numbers of persons in PHC villages and because number of compounds might not always be exactly proportional to village population. For the central region PHC villages, for example, the probability that a given compound would be included in the survey was

$$P(\text{district selected}) \times P(\text{village selected} \mid \text{district selected})$$

$$\times\, P(\text{compound selected} \mid \text{district and village selected})$$

$$\propto \frac{D1}{R} \times \frac{V}{D2} \times \frac{1}{C},$$

where

C = number of compounds in the village

V = number of people in the village

$D1$ = number of people in the district

$D2$ = number of people in the district in PHC villages

R = number of people in PHC villages in all central districts

Since the number of compounds in a village will be roughly proportional to the number of people in a village, V/C should be approximately the same for all compounds. R is also the same for all compounds within a region. The weights for each region, the reciprocals of the inclusion probabilities, differ largely because of the variability in $D1/D2$. As R varies from stratum to stratum, though, compounds in more populous strata have higher weights than those in less populous strata. ■

7.2.2   Self-Weighting and Non-Self-Weighting Samples

Sampling weights for all observation units are equal in self-weighting surveys. Self-weighting samples can, in the absence of nonsampling errors, be considered representative of the population because each observed unit represents the same number of unobserved units in the population. Standard statistical methods may then be applied to the sample to obtain point estimates. A histogram of the sample values displays the approximate frequencies of occurrence in the population; the sample mean, median, and other sample statistics estimate the corresponding population quantities. In addition, self-weighting samples often yield smaller variances, and sample statistics are more robust (Kish 1992).

Most large self-weighting samples used in practice are not simple random samples (SRSs), however. Stratification is used to reduce variances and obtain separate estimates for domains of interest; clustering, usually with pps, is used to reduce costs. Standard statistical software—software written to analyze data fulfilling the usual statistical assumption that observations are independent and identically distributed—gives correct estimates for the mean, percentiles, and other quantities in a self-weighting complex survey. Standard errors, hypothesis-test statistics, and confidence intervals constructed by standard software are wrong, however, as mentioned above. When you read a paper or book in which the authors analyze data from a complex survey, see whether they accounted for the data structure in the analysis or whether they simply ran the raw data through a standard SAS or SPSS procedure and reported the results. If the latter, their inferential results must be viewed with suspicion; it is possible that they only find statistical significance because they fail to account for the survey design in the standard errors.

Many surveys, of course, purposely sample observation units with different probabilities. The disproportionate sampling probabilities often occur in the stratification: A higher sampling fraction is used for a stratum of large businesses than for a stratum of small businesses. The U.S. National Health and Nutrition Examination Survey (NHANES) purposely oversamples areas containing large black and Mexican American populations (Ezzati-Rice and Murphy 1995); oversampling these populations allows comparison of the health of racial and ethnic minorities.

7.2.3   Weights and a Model-Based Analysis of Survey Data

You might think that a statistician taking a model-based perspective could ignore the weights altogether. After all, to a model-based survey statistician, the sample design is irrelevant and the important part of the analysis is finding a model that summarizes the population structure; as sampling weights are functions of the probabilities of selection in the design, perhaps they too are irrelevant.

The model-based and randomization-based approaches, however, are not as far apart as some of the literature debating the issue would have you believe. Remember, a statistician designing a survey to be analyzed using weights implicitly visualizes a model for the data; NHANES is stratified and subpopulations oversampled precisely because researchers believe there will be a difference among the subpopulations. Such differences also need to be included in the model. If you ignore the weights in analyzing data from NHANES, for example, you implicitly assume that whites, blacks, and Mexican Americans are largely interchangeable in health status. Ignoring the clustering in the inference assumes that observations in the same cluster are uncorrelated, which is not generally true. A data analyst who ignores stratification variables and dependence among observations is not fitting a good model to the data but is simply being lazy. A good analysis of survey data using models is difficult and requires extensive validation of the model. The book by Skinner et al. (1989) contains several chapters on modeling data from complex surveys.

Many researchers have found that sampling weights contain information that can be used in a model-based analysis. Little (1991) develops a class of models that result in estimators that behave like estimators obtained using survey weights. Pfeffermann (1993) describes a framework for deciding on whether to use sampling weights in regression models of survey data.

7.3
Estimating a Distribution Function

So far, we have concentrated on estimating population means, totals, and ratios. Historically, sampling theory was developed primarily to find these basic statistics and to answer questions such as "What percentage of adult males are unemployed?" or "What is the total amount of money spent on health care in the United States?" or "What is the ratio of the numbers of exotic to native birds in an area?"

But statistics other than means or totals may be of interest. You may want to estimate the median income in Canada, find the 95th percentile of test scores, or construct a histogram to show the distribution of fish lengths. An insurance company may set reimbursements for a medical procedure using the 75th percentile of charges for the procedure. We can estimate any of these quantities (but not their standard errors, however) with sampling weights. The sampling weights allow us to construct an empirical distribution for the population.

Suppose the values for the entire population of N units are known. Then any quantity of interest may be calculated from the **probability mass function,**

$$f(y) = \frac{\text{number of units whose value is } y}{N},$$

or the **distribution function,**

$$F(y) = \frac{\text{number of units with value} \leq y}{N} = \sum_{x \leq y} f(x).$$

In probability theory, these are the probability mass function and distribution function for the random variable Y, where Y is the value obtained from a random sample of size 1 from the population. Then $f(y) = P\{Y = y\}$ and $F(y) = P\{Y \leq y\}$. Of course, $\sum f(y) = F(\infty) = 1$.

Any population quantity can be calculated from the probability mass function or distribution function. The population mean is

$$\bar{y}_U = \sum yf(y).$$

A population median is any value m that satisfies $F(m) \geq 1/2$ and $P(Y \geq m) \geq 1/2$; in general, x is a $100r$th percentile if $F(x) \geq r$ and $P(Y \geq x) \geq 1-r$. The population variance, too, can be written using the probability mass function:

$$S^2 = \frac{1}{N-1} \sum_{i=1}^{N}(y_i - \bar{y}_U)^2$$

$$= \frac{N}{N-1} \sum_y f(y) \left[y - \sum_x xf(x) \right]^2$$

$$= \frac{N}{N-1} \left[\sum_y y^2 f(y) - \left\{ \sum_y yf(y) \right\}^2 \right].$$

EXAMPLE 7.3   Consider an artificial population of 1000 men and 1000 women in file htpop.dat. Each person's height is measured to the nearest centimeter (cm). The frequency table (Table 7.1) gives the probability mass function and distribution function for the 2000 persons in the population. Figures 7.1 and 7.2 show the graphs of $F(y)$ and $f(y)$. The population mean is $\sum yf(y) = 168.6$.

Now let's take an SRS of size 200 from the population (file htsrs.dat). An SRS is self-weighting; each person in the sample represents 10 persons in the population. Hence, the histogram of the sample should resemble $f(y)$ from the population; Figure 7.3 shows that it does.

But suppose a stratified sample of 160 women and 40 men (file htstrat.dat) is taken instead of a self-weighting sample. A histogram of the raw data will distort the population distribution, as illustrated in Figure 7.4. The sample mean and median are too low because men are underrepresented in the sample.   ■

Sampling weights allow us to construct empirical probability mass and distribution functions for the data. Any statistics can then be calculated. Define the **empirical probability mass function** (epmf) to be the sum of the weights for all observations taking on the value y, divided by the sum of all the weights:

$$\hat{f}(y) = \frac{\sum\limits_{i \in S: y_i = y} w_i}{\sum\limits_{i \in S} w_i}.$$

The **empirical distribution function** $\hat{F}(y)$ is the sum of all weights for observations with values $\leq y$, divided by the sum of all weights:

$$\hat{F}(y) = \sum_{x \leq y} \hat{f}(x).$$

TABLE 7.1
Frequency Table for Population in Example 7.3

Value, y	Frequency	f(y)	F(y)	Value, y	Frequency	f(y)	F(y)
136	1	0.0005	0.0005	172	57	0.0285	0.6540
140	1	0.0005	0.0010	173	45	0.0225	0.6765
141	2	0.0010	0.0020	174	52	0.0260	0.7025
142	1	0.0005	0.0025	175	57	0.0285	0.7310
143	6	0.0030	0.0055	176	49	0.0245	0.7555
144	3	0.0015	0.0070	177	54	0.0270	0.7825
145	4	0.0020	0.0090	178	57	0.0285	0.8110
146	3	0.0015	0.0105	179	40	0.0200	0.8310
147	14	0.0070	0.0175	180	35	0.0175	0.8485
148	11	0.0055	0.0230	181	43	0.0215	0.8700
149	13	0.0065	0.0295	182	29	0.0145	0.8845
150	20	0.0100	0.0395	183	26	0.0130	0.8975
151	15	0.0075	0.0470	184	29	0.0145	0.9120
152	18	0.0090	0.0560	185	23	0.0115	0.9235
153	28	0.0140	0.0700	186	21	0.0105	0.9340
154	38	0.0190	0.0890	187	19	0.0095	0.9435
155	38	0.0190	0.1080	188	17	0.0085	0.9520
156	57	0.0285	0.1365	189	15	0.0075	0.9595
157	53	0.0265	0.1630	190	10	0.0050	0.9645
158	49	0.0245	0.1875	191	14	0.0070	0.9715
159	55	0.0275	0.2150	192	10	0.0050	0.9765
160	77	0.0385	0.2535	193	9	0.0045	0.9810
161	72	0.0360	0.2895	194	7	0.0035	0.9845
162	66	0.0330	0.3225	195	2	0.0010	0.9855
163	62	0.0310	0.3535	196	7	0.0035	0.9890
164	61	0.0305	0.3840	197	8	0.0040	0.9930
165	60	0.0300	0.4140	198	4	0.0020	0.9950
166	75	0.0375	0.4515	199	2	0.0010	0.9960
167	79	0.0395	0.4910	200	4	0.0020	0.9980
168	62	0.0310	0.5220	201	1	0.0005	0.9985
169	79	0.0395	0.5615	204	1	0.0005	0.9990
170	72	0.0360	0.5975	206	2	0.0010	1.0000
171	56	0.0280	0.6255				

FIGURE 7.1
The function F(y) for the population of heights

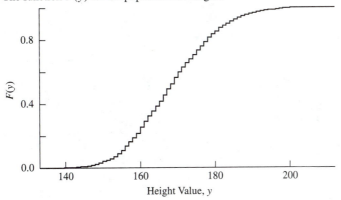

FIGURE 7.2

The function $f(y)$ for the population of heights

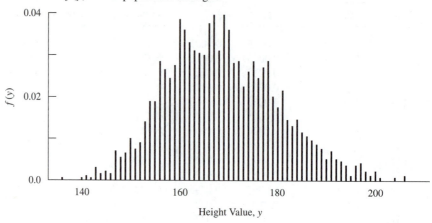

FIGURE 7.3

A histogram of raw data from an SRS of size 200. The general shape is similar to that of $f(y)$ for the population because the sample is self-weighting.

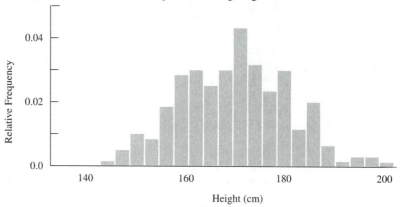

FIGURE 7.4

A histogram of raw data from a stratified sample of 160 women and 40 men. Tall persons are underrepresented in the sample.

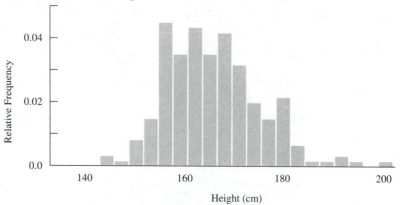

FIGURE 7.5

The estimates $\hat{f}(y)$ and $\hat{F}(y)$ for the stratified sample of 160 women and 40 men

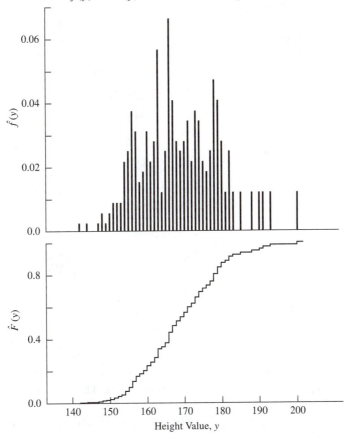

For a self-weighting sample, $\hat{f}(y)$ reduces to the relative frequency of y in the sample. For a non-self-weighting sample, $\hat{f}(y)$ and $\hat{F}(y)$ are attempts to reconstruct the population functions f and F from the sample. The weight w_i is the number of population units represented by unit i, so $\sum_{i \in S: y_i = y} w_i$ estimates the total number of units in the population that have value y.

Each woman in the stratified sample has sampling weight 6.25; each man has sampling weight 25. The empirical probability mass and distribution functions from the stratified sample are in Figure 7.5. The weights correct the underrepresentation of taller people found in the histogram in Figure 7.4. The scarcity of men in the sample, however, demands a price: The right tail of $\hat{f}(y)$ has a few spikes of size 25/2000, rather than a number of values tapering off.

The epmf $\hat{f}(y)$ can be used to find estimates of population quantities. First, express the population characteristic in terms of $f(y)$: $\bar{y}_U = \sum yf(y)$ or

$$S^2 = \frac{N}{N-1} \left\{ \sum_y f(y) \left[y - \sum_x xf(x) \right]^2 \right\} = \frac{N}{N-1} \left\{ \sum_y y^2 f(y) - \left[\sum_y yf(y) \right]^2 \right\}.$$

TABLE 7.2

Estimates from samples in Example 7.3

Quantity	Population	SRS	Stratified, No Weights	Stratified, with Weights
Mean	168.6	168.9	164.6	169.0
Median	168	169	163	168
25th percentile	160	160	157	161
90th percentile	184	184	178	182
Variance	124.5	122.6	93.4	116.8

Then, substitute $\hat{f}(y)$ for every appearance of $f(y)$ to obtain an estimate of the population characteristic. Using this method, then,

$$\hat{\bar{y}} = \sum_y y \hat{f}(y) = \frac{\sum_{i \in \mathcal{S}} y_i w_i}{\sum_{i \in \mathcal{S}} w_i}$$

and

$$\hat{S}^2 = \frac{N}{N-1} \left\{ \sum_y y^2 \hat{f}(y) - \left[\sum_y y \hat{f}(y) \right]^2 \right\}. \qquad (7.4)$$

Table 7.2 shows the difference in the estimates when weights for the stratified sample are incorporated through the function $\hat{f}(y)$. The statistics calculated using weights are much closer to the population quantities.

This simple example involved only stratification, but the method is the same for any survey design. You need to know only the sampling weights to estimate almost anything through the empirical distribution function. If desired, you can smooth the empirical distribution function before estimating quantiles; see Silverman (1986), Scott (1992), or Venables and Ripley (1994, sec. 5.5). Gill et al. (1988) show that this empirical distribution function is uniformly asymptotically consistent; in small samples, though, the tails of the epmf $\hat{f}(y)$ are often too short, whether or not the sample is self-weighting, because extreme values may not be included in the sample.[2] Nusser et al. (1996) use a semiparametric approach for estimating daily dietary intakes of various nutrients from the Continuing Survey of Food Intakes by Individuals, a stratified multistage survey.

Although the weights may be used to find point estimates through the empirical distribution function, calculating standard errors is much more complicated and requires knowledge of the sampling design. Variances of statistics calculated from the empirical distribution function will be discussed in Chapter 9.

---

[2] Additional problems may occur in estimating distribution functions because respondents may round their answers. For example, some respondents may round their height to 165 or 170 cm, causing spikes at those values. If you smooth the epmf, you may want to choose a bandwidth to increase the amount of smoothing, or you may want to adopt a model for the effect of rounding by the respondent.

7.4
Plotting Data from a Complex Survey

Simple plots reveal much information about data from a small SRS or representative systematic sample. Histograms or smoothed density estimates display the shape of the data; scatterplots and scatterplot matrices show relationships between variables; other plots discussed in Chambers et al. (1983) and Cleveland (1994) emphasize other features of the data. In a complex sampling design, however, a single plot will not display the richness of the data. As seen in Figure 7.4, plots commonly used for SRSs can mislead when applied to raw data from non-self-weighting samples. Clustering causes numerous difficulties in plotting data from a complex survey, as noted in Example 5.6, because the clustering structure as well as possible unequal weighting must be displayed in the graphs; the problems are compounded because data sets from surveys are often very large and involve several layers of clustering.

Data should be plotted both with and without weights to see the effect of the weights. In addition, data should be plotted separately for each stratum and for each psu, if possible, to examine variability in the responses. You already know how to plot the raw data without weights; in this section we provide some examples of incorporating the weights into the graphics.

EXAMPLE 7.4 The 1987 Survey of Youth in Custody (Beck et al. 1988; U.S. Department of Justice 1989) sampled juveniles and young adults in long-term, state-operated juvenile institutions. Residents of facilities at the end of 1987 were interviewed about family background, previous criminal history, and drug and alcohol use. Selected variables from the survey are in the file syc.dat.

The facilities form a natural cluster unit for an in-person survey; the sampling frame of 206 facilities was constructed from the 1985 Children in Custody (CIC) Census. The psu's (facilities) were divided into 16 strata by number of residents in the 1985 CIC. Each of the 11 facilities with 360 or more youth formed its own stratum (strata 6–16); each of these facilities was included in the sample, and residents of the 11 facilities were subsampled. In strata 1–5, facilities were sampled with probability proportional to size from the 195 remaining facilities; residents were subsampled with predetermined sampling fractions. Table 7.3 contains information about the strata.

TABLE   7.3

Survey of Youth in Custody Stratum Information

Stratum	CIC Size (Number of Residents)	Number of psu's in Frame	Number of Residents in CIC	Number of Eligible psu's in Sample
1	1–59	99	2881	11
2	60–119	39	3525	7
3	120–179	30	4355	7
4	180–239	13	2594	7
5	240–359	14	4129	7

The stratum boundaries were chosen so that the number of residents in each stratum would be comparable. It was originally intended that each resident have probability 1/8 of inclusion in the sample, which would result in a self-weighting sample with constant weight 8. The facilities in strata 14 and 16, however, had experienced a great deal of growth between 1985 and 1987, so the sampling fractions in those strata were changed to 1/11 and 1/12, respectively. In strata 1–5, weights varied from about 5 to about 15, depending on the facility's probability of selection and the predetermined sampling fraction in that facility. The weights were further adjusted for nonresponse and to match the sample counts with the 1987 census count of youths in long-term, state-operated facilities. After all weighting adjustments were made, weights ranged from 5 (in stratum 4) to 58 (for some youths in states that required parental permission and hence had lower response rates).

Let's look at some plots of the age of residents. Some youths are over age 18 because California Youth Authority facilities were included in the sample. As the survey aimed to be approximately self-weighting, the histogram of the unweighted data in Figure 7.6 and the epmf incorporating weights in Figure 7.7 are overall similar in shape. Some discrepancies appear on closer examination, though—the weights indicate that youths aged 15 were somewhat undersampled due to unequal selection probabilities and nonresponse, while youths aged 17 were somewhat oversampled.

If we were only interested in the distribution for the entire population, we could concentrate on plots such as those in Figures 7.6 and 7.7, and similar plots informative about univariate distributions such as quantile-quantile plots (see Chambers et al. 1983). But we would also like to explore stratum-to-stratum differences in age distribution. Figure 7.8 incorporates weights into boxplots of the data.

As the response variable *age* is discrete, we can show even more detail for each stratum. Figure 7.9 displays the sum of the weights for each age within each stratum. The estimated relative frequency of youths with that age in each stratum is indicated by a circle whose area is proportional to the sum of the weights.

F I G U R E   7.6

A histogram of all data, not incorporating weights. The histogram shows the distribution of ages in the sample.

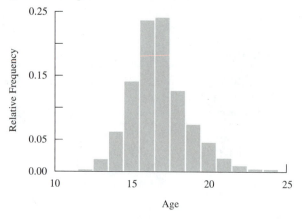

FIGURE   7.7

An estimated probability mass function for age, $\hat{f}(y)$. The shape is similar to that of the histogram of the raw data, but there are relatively more 15-year-olds and relatively fewer 17-year-olds.

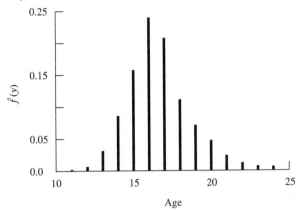

FIGURE   7.8

A boxplot of age distributions for each stratum, incorporating the weights. Note the wide variability from stratum to stratum.

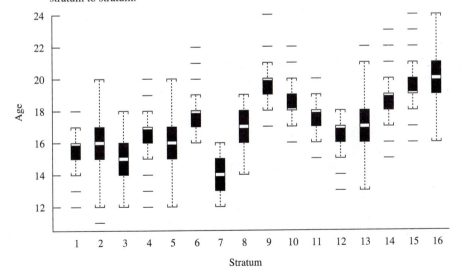

We may also be interested in the facility-to-facility variability. Figures 7.10 and 7.11 show similar plots for the psu's in stratum 5. These plots could be drawn for each stratum to show differences in psu variability among the strata.   ■

FIGURE   **7.9**

The age distribution for each stratum. The area of each circle is proportional to the sum of the weights for sample observations in that stratum and age class. The highest number of youths under age 18 are in strata 1–5.

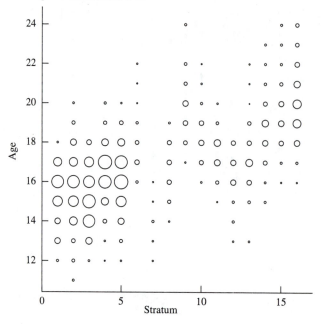

FIGURE   **7.10**

A boxplot of ages, incorporating weights, for the psu's in stratum 5. The width of each boxplot is proportional to the number of sample observations in that facility.

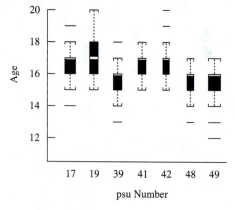

FIGURE   7.11

The age distribution for each psu in stratum 5. The area of each circle is proportional to the sum of the weights for sample observations in that psu and age class.

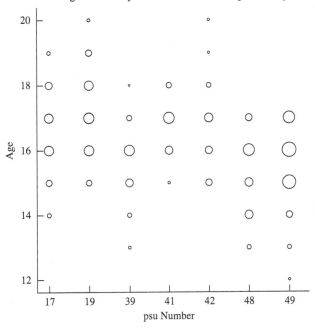

7.5
Design Effects

Cornfield (1951) suggested measuring the efficiency of a sampling plan by the ratio of the variance that would be obtained from an SRS of k observation units to the variance obtained from the complex sampling plan with k observation units. Kish (1965) named the reciprocal of Cornfield's ratio the **design effect** (abbreviated deff) of a sampling plan and estimator and used it to summarize the effect of the design on the variance of the estimate:

$$\text{deff(plan, statistic)} = \frac{V(\text{estimate from sampling plan})}{V(\text{estimate from an SRS with same number of observation units})} \quad (7.5)$$

For estimating a mean from a sample with n observation units,

$$\text{deff(plan, } \hat{\bar{y}}) = \frac{V(\hat{\bar{y}})}{\left(1 - \frac{n}{N}\right) \frac{S^2}{n}}.$$

The design effect provides a measure of the precision gained or lost by use of the more complex design instead of an SRS. Although it is a useful concept, it is not a way to avoid calculating variances: You need an estimate of the variance from the complex design to find the design effect. Of course, different quantities in the same survey may have different design effects. Kish shows how the design effect allows you to use prior knowledge for the survey design.

The SRS variance is generally easier to obtain than $V(\hat{\bar{y}})$: If estimating a proportion, the SRS variance is approximately $p(1-p)/n$; if estimating another type of mean, the SRS variance is approximately S^2/n. So if the design effect is approximately known, the variance from the complex sample can be estimated by (deff $\times$ SRS variance). We can estimate the variance of an estimated proportion $\hat{p}$ by

$$\hat{V}[\hat{p}] = \text{deff} \times \frac{\hat{p}(1-\hat{p})}{n}.$$

We have seen design effects for several sampling plans. In Section 4.4 the design effect for stratified sampling with proportional allocation was shown to be approximately

$$\frac{V_{\text{prop}}}{V_{\text{SRS}}} \approx \frac{\sum\limits_{h=1}^{H} \dfrac{N_h}{N} S_h^2}{S^2}$$

$$\approx \frac{\sum\limits_{h=1}^{H} \dfrac{N_h}{N} S_h^2}{\sum\limits_{h=1}^{H} \dfrac{N_h}{N} \left[S_h^2 + (\bar{y}_{hU} - \bar{y}_U)^2 \right]}. \tag{7.6}$$

Unless all the stratum means are equal, the design effect for a stratified sample will usually be less than 1—stratification generally gives more precision per observation unit than an SRS.

We also looked extensively at design effects in cluster sampling, particularly in Section 5.2.2. From (5.9), the design effect for single-stage cluster sampling when all psu's have M ssu's is approximately

$$1 + (M-1)\text{ICC}.$$

The intraclass correlation coefficient (ICC) is usually positive in cluster sampling, so the design effect is usually larger than 1; cluster samples usually give less precision per observation unit than an SRS.

In surveys with both stratification and clustering, we cannot say before calculating variances for our sample whether the design effect for a given quantity will be less than 1 or greater than 1. Stratification tends to increase precision and clustering tends to decrease it, so the overall design effect depends on whether more precision is lost by clustering than gained by stratification.

EXAMPLE **7.5**   For the bed net survey discussed in Example 7.1, the design effect for the proportion of beds with nets was calculated to be 5.89. This means that about six times as many observations are needed with the complex sampling design used in the survey to obtain the same precision that would have been achieved with an SRS. The high design effect in this survey is due to the clustering: Villages tend to be homogeneous in bed net use. If you ignored the clustering and analyzed the sample as though it were an SRS, the estimated standard errors would be much too low, and you would think you had much more precision than really existed.   ■

7.5.1   Design Effects and Confidence Intervals

If the design effect for each statistic is known, one can use it in conjunction with standard software to obtain confidence intervals (CIs) for means and totals. If n observation units are sampled from a population of N possible observation units and if $\hat{p}$ is the survey estimate of the proportion of interest, an approximate 95% CI for p is (assuming the finite population correction is close to 1):

$$\hat{p} \pm 1.96 \sqrt{\text{deff}} \sqrt{\frac{\hat{p}(1-\hat{p})}{n}}. \tag{7.7}$$

When estimating a mean rather than a proportion, if the sample is large enough to apply a central limit theorem, an approximate 95% CI is

$$\hat{\bar{y}} \pm 1.96 \sqrt{\text{deff}} \sqrt{\frac{\hat{S}^2}{n}},$$

where $\hat{S}^2$ may be calculated using (7.4).

Kish (1995) and other authors sometimes now use design effect to refer to the quantity[3]

$$\text{deft}(\bar{y}) = \frac{\text{SE}(\bar{y}_{\text{plan}})}{\frac{s}{\sqrt{n}}},$$

so that deft will be an appropriate multiplier for a standard error or confidence interval half-width. In practice, as Kish points out, choice of deff or deft makes little difference, but you need to pay attention to which definition a survey uses.

7.5.2   Design Effects and Sample Sizes

Design effects are extremely useful for estimating the sample size needed for a survey. That is the purpose for which it was introduced by Cornfield (1951), who used it to estimate the sample size that would be needed if the sampling unit in a survey estimating the prevalence of tuberculosis was a census tract or block rather than an individual. The maximum allowable error was specified to be 20% of the true prevalence, or $0.2 \times p$. If the prevalence of tuberculosis was $p = 0.01$, the sample size for an SRS would need to be

$$n = \frac{1.96^2 p(1-p)}{(0.2p)^2} = 9508.$$

Cornfield recommended increasing the sample size for an SRS to 20,000, to allow more precision in separate estimates for subpopulations. He estimated the design effect for sampling census tracts rather than individuals to be 7.4 and concluded that if census tracts, which averaged 4600 individuals, were used as a sampling unit, a sample size of 148,000 adults, rather than 20,000 adults, would be needed.

If you know the design effect for a similar survey, you need to estimate only the sample size you would take using an SRS. Then multiply that sample size by

---

[3] The term *deft* is due to Tukey (1968).

deff to obtain the number of observation units you need to observe with the complex design. For sample-size purposes, you may wish to use separate design effects for each stratum.

7.6
The National Crime Victimization Survey

Most crime statistics given in U.S. newspapers come from the Uniform Crime Reports, compiled by the FBI from reports submitted by law enforcement agencies. But the Uniform Crime Reports underestimate the amount of crime in the United States, largely because not all crimes are reported to the police.

The National Crime Victimization Survey (NCVS)[4] is a large national survey administered by the Bureau of Justice Statistics, with interviews conducted by the Bureau of the Census. Like the Current Population Survey (CPS), the NCVS follows a stratified, multistage cluster design. Information on the design of the CPS is found in Hanson (1978) and in McGuiness (1994); additional information on the NCVS is in documentation released by the Bureau of Justice Statistics. The NCVS surveys households from across the United States and asks household members 12 years old and older about their experiences as victims of crime in the past 6 months. The NCVS is the only national source of information about victims of crime.

The CPS and the NCVS once used similar designs—in fact, the 1970-based NCVS design used a subset of the CPS's primary sampling units. This was done to save on administrative and interviewer costs. For the 1980 and 1990 NCVS sample designs, overlap with the previous NCVS design was maximized as much as possible. We describe the 1980-based design here as used to produce the 1990 NCVS estimates; the basic features of the design are the same for the post-1990 NCVS. We return to the NCVS in Chapter 8, to show how weights are adjusted for nonresponse and undercoverage in this large complex survey.

A psu in the NCVS is a county, a group of adjacent counties, or a metropolitan statistical area (MSA). An MSA is a large city together with adjacent communities that are economically and socially integrated with the city. Examples are the Montgomery, Alabama MSA (included in both the 1980 and 1990 sample designs), which includes Autauga, Elmore, and Montgomery counties; the Columbus, Ohio MSA, including Delaware, Fairfield, Franklin, Madison, Pickaway, and Union counties; and the Albany–Schenectady–Troy, New York MSA, including Albany, Greene, Montgomery, Rensselaer, Saratoga, and Schenectady counties.

Any psu with population about 550,000 or more (according to the 1980 census) is automatically included in the sample. Such a psu is said to be *self-representing* (SR) because it does not represent any psu's other than itself. The probability this psu will be selected is 1.

All other psu's are grouped into strata so that each stratum group has a population of about 650,000. In the NCVS, psu's are grouped into strata based on geographic location, demographic information available from the 1980 census, and on Uniform Crime Report crime rates. One psu is selected from each of these strata, with

---

[4]The survey was previously called the National Crime Survey. We use NCVS to refer to both names.

probability proportional to population size; this psu is called *non-self-representing* (NSR) because it is supposed to represent not just itself but all psu's in that stratum. Within a stratum, a psu with 100,000 population is twice as likely to be selected for the sample as a psu with population 50,000. For the 1990 NCVS, there were 84 SR psu's and 153 NSR psu's. As victimization rates vary regionally, the large number of strata in the NCVS increases the precision of the estimates.

The second stage of sampling involves selecting enumeration districts (EDs), geographic areas used in the 1980 decennial census; an ED typically contains about 300 to 400 households, but EDs vary considerably in population and land area.[5] The EDs are selected with probability proportional to their 1980 population size; the number of EDs selected within a psu is determined so that the sample of EDs will be approximately self-weighting. In the census listing, EDs are arranged by geographic location; EDs are selected using systematic sampling, as described in Section 6.2, so that the sampled EDs will be distributed geographically over the selected psu. If the overall sampling rate is $1/x$, in SR psu's the sampling interval is x. If using census records for the sampling frame, the addresses are numbered from 1 to the number of households in the psu. A random number k is chosen between 1 and x, and the EDs chosen to be in the sample are the ones containing addresses $k, k + x, k + 2x$, and so on. In NSR psu's, the sampling interval is (probability psu is selected) $\times$ (x).

In the third stage of sampling, each selected ED is divided into clusters of approximately four housing units each. The census lists housing units within an ED in geographic order, and when possible that listing is used. A sample of those clusters is taken, and each housing unit in a selected cluster of about four housing units is included in the sample. All persons aged 12 and over in the housing unit are to be interviewed for the survey.

In some regions *area sampling* is used. If the census listing of housing units were the only one used throughout that decade, there would be substantial undercoverage of the population, since no newly built housing units would be included in the sample. To allow new housing units to be included in the sample, the NCVS uses a sample of building permits for residential units and samples those. In area sampling, a field representative lists all housing units or other living quarters within a selected area of an ED, and that listing then serves as the sampling frame for that area.

In summary, the stages for the 1990 NCVS are shown in Table 7.4.

Interviews for the NCVS with persons aged 12 and over are taken every month, with the housing units selected for the sample covered in a 6-month period—this allows the interviewing workload to be distributed evenly throughout the year. To allow for longitudinal analyses of the data and to be relatively certain that crimes reported for a 6-month period occurred during those 6 months and not during an earlier time, the residents of each housing unit are interviewed every 6 months over a 3-year period, for a total of seven interviews. The first interview is not used for estimating victimization rates but only for bounding—*bounding* establishes a time frame for reported victimizations so that a victimization reported in two successive interview periods is only counted once. Being a victim of a crime is a memorable experience for most people—so memorable, in fact, that it is easy to remember a victimization as more recent than it really was and to *telescope* an earlier victimization into the

---

[5]For the 1990 census, EDs were renamed as address register areas, ARAs.

T A B L E   7.4

Sampling Stages for the 1990 NCVS

Stage	Sampling Unit	Stratification
1	psu (county, set of adjacent counties, or MSA)	Location, demographic information, and crime-related characteristics
2	Enumeration district	
3	Cluster of four housing units	
4	Household	
5	Person within household	

6-month reference period. Using a panel study allows the Bureau of Justice Statistics to bound each interview by the previous one; a respondent is questioned in further detail if an incident appears to have been repeated from the last interview.

For 1990 about 62,600 housing units (including group quarters) were in the sample. Of those, 56,800 housing units received the main questionnaire (occupants of the remaining housing units were given a new questionnaire being phased in). About 8200 of the 56,800 selected housing units were ineligible for the NCVS because they were vacant, demolished, or no longer used as residences. No interviews were completed in about 1600 of the housing units, however, because the residents could not be reached or refused to participate in the survey. The NCVS for 1990 had a household nonresponse rate of 1600/48,600, or about 3.3%. Altogether, about 95,000 persons gave responses to the questionnaire.

Clearly, this is a complex survey design, and weights are used to calculate estimates of victimization rates and total numbers of crimes. The survey is designed to be approximately self-weighting, so initially each individual is assigned the same base weight of (1/probability of housing-unit selection). For the NCVS in the late 1980s, each person represents approximately 1658 other persons in the United States, so the base weight is 1658.

The NCVS is designed to be self-weighting, but sometimes a selected cluster within an ED has more housing units than originally thought; for example, an apartment building might have been erected in place of detached housing units. Then only housing units in a subsample of the cluster are interviewed. If subsampling is used, the units subsampled are assigned a weighting-control factor (WCF). If only one-third are sampled, for instance, the sampled units are assigned a WCF of 3 because they will represent three times as many units. If a housing unit is in a cluster in which subsampling is not needed, it is assigned a WCF of 1. At this level, a sampled housing unit represents

$$\text{base weight} \times \text{WCF}$$

housing units in the population. This is the sampling weight for a housing unit sampled in the NCVS; as the survey attempts to interview all persons aged 12 and older in the sampled housing units, the sampling weight for a person in the sample is set equal to the weight for the housing unit.

All other weighting adjustments in the NCVS adjust for nonresponse or are used in poststratification. Some persons selected to be in the sample are not interviewed because they are absent or refuse to participate. The interviewer gathers demographic information on the nonrespondents, and that demographic information is used to adjust the weights in an attempt to counteract the nonresponse. (This is an example of weighting-class adjustments for nonresponse, as discussed in Section 8.5.) Two different weighting adjustments for nonresponse are used: the within-household noninterview adjustment factor (WHHNAF) and the household noninterview adjustment factor (HHNAF). In each adjustment factor, the goal is to increase weights of interviewed units that are most similar to units that cannot be interviewed.

The WHHNAF is used to compensate for individual nonrespondents in households in which at least one member responded to the survey. It is computed separately for each of the regions (Northeast, Midwest, South, and West) of the United States. Within each region, the persons from households in which there was at least one respondent are classified into 24 cells, using the race of the person designated as reference person; the age and sex of the nonresponding household member; and the nonrespondent's relationship to the reference person. Any of the 24 cells that contain fewer than 30 interviewed cases or that produce a WHHNAF of 2 or more are combined with similar cells; the collapsing of cells prevents some individuals from having weights that are too large. Then,

$$\text{WHHNAF} = \frac{\text{sum of weights of all persons in cell}}{\text{sum of weights of all interviewed persons in cell}}.$$

The weights used to calculate the WHHNAF are the weights assigned to this point in the weighting procedure—that is, (base weight) $\times$ (WCF). Thus, the weights of respondents in a cell are increased so that they represent the nonrespondents and the persons in the population that the nonrespondent would represent, in addition to their original representation. After applying the WHHNAF, the weight for an individual is

$$\text{base weight} \times \text{WCF} \times \text{WHHNAF}.$$

Some of the nonresponse is due to nonresponding individuals in responding households; other nonresponse occurs because the entire household is nonrespondent. About 3 to 4% of households are eligible for the survey but cannot be reached or refuse to respond; the HHNAF is used to attempt to compensate for nonresponse at the household level. For the HHNAF, households are grouped into cells by MSA status, urban/rural, and race of reference person. Then,

$$\text{HHNAF} = \frac{\text{sum of weights of all persons in cell}}{\text{sum of weights of all interviewed persons in cell}}.$$

As with the WHHNAF, the weights used in calculating the HHNAF are the weights calculated so far: (base weight) $\times$ (WCF) $\times$ (WHHNAF). Cells are combined until the HHNAF is less than 2.

At this point in the construction of the weights, the weight assigned to an individual is

$$\text{base weight} \times \text{WCF} \times \text{WHHNAF} \times \text{HHNAF}.$$

The sampling weights for responding individuals have been increased so that they also represent nonrespondents who are demographically similar.

Because the NCVS is a sample, the demographic information in the sample usually differs from that of the U.S. population as a whole. Two stages of ratio estimation are used to adjust the sample values so that they agree better with updated census information. This adjustment is expected to reduce the variance of estimates of victimization rates.

The first stage of ratio estimation is used in NSR psu's only and is intended to reduce the variability that results from using one psu to represent the stratum. Ratio estimation is used to assign different weights to cells stratified by region, MSA status, and race. The first-stage factor,

$$\text{FSF} = \frac{\text{independent count of number of persons in cell}}{\text{sample estimate (sum of weights) of the number of persons in cell}},$$

adjusts for differences between census characteristics of sampled NSR psu's and characteristics of the full set of NSR psu's. The FSF equals 1 for SR psu's and is truncated at 1.3 for NSR psu's.

The second-stage factor (SSF) of ratio estimation is applied to everyone in the sample. The persons in the sample are classified into 72 groups on the basis of their age, race, and sex. Cells need to have a count of at least 30 interviewed persons, and the SSF needs to be between 0.5 and 2.0; cells are collapsed until these conditions are met.

$$\text{SSF} = \frac{\text{independent count of number of persons in cell}}{\text{sample estimate (sum of weights) of the number of persons in cell}}.$$

The SSF is a form of poststratification: It is intended to adjust the sample distribution of age, race, and sex so that the cross-classification agrees with independently taken counts that are thought to be more accurate. If the sum of weights of elderly white women in the sample is larger than the current "best" estimate of the number of elderly white women in the population from updated census information, then the SSF will be less than 1 for all elderly white women in the sample.

After all the adjustments, the final weight for person i is

$$w_i = \text{base weight} \times \text{WCF} \times \text{WHHNAF} \times \text{HHNAF} \times \text{FSF} \times \text{SSF}.$$

The weight w_i is used as though there were actually w_i persons in the population exactly like the one to which the weight is attached. In the 1990 NCVS, the person weights range from 1100 to 9000, with most weights between 1500 and 2500. Figure 7.12 gives boxplots for the weights for persons interviewed between July and December 1990. The weights are included on the public-use tapes of the NCVS: To use them to estimate the total number of aggravated assaults reported by white females, you would define

$$y_i = \begin{cases} k & \text{if person } i \text{ is a white female who reported } k \text{ aggravated assaults} \\ 0 & \text{otherwise.} \end{cases}$$

and use $\sum_{i \in S} w_i y_i$ as your estimate.

Even though the nonresponse is relatively low in the NCVS, the weights make a difference in calculating victimization rates. Estimates of victimization rates are generally higher when weights are used than when they are not used. Young black male respondents to the survey are disproportionately likely to be victims of crime, and undercoverage and nonresponse among black males is high.

Since the sampling design and the weighting scheme are so complicated in the NCVS, finding design effects requires much work. Variances are now calculated by

FIGURE   7.12

Boxplots of weights for the 1990 NCVS, for all persons, white males, white females, nonwhite males, nonwhite females, persons under age 25, persons over age 25, and victims of violent crime. The horizontal lines represent the maximum, 95th percentile, 75th percentile, median, 25th percentile, 5th percentile, and minimum. Note that the weights are much higher for nonwhite males, indicating the higher nonresponse and undercoverage in that group.

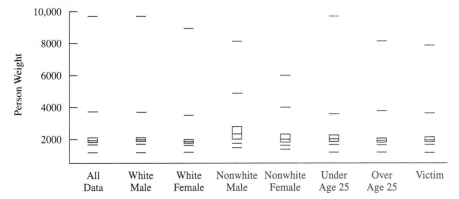

replication methods, described in Chapter 9. The sampling design affects variance estimates at several different levels:

1   In NSR strata, only one psu is selected out of several psu's in the stratum, so there is between-psu variance in those strata.

2   Within an ED, a cluster of approximately four housing units is selected to be in the sample—these housing units are likely to be positively correlated.

3   All persons within sampled households are interviewed—this gives a clustering effect for persons.

4   Systematic sampling is used to choose the EDs instead of simple random sampling. The effect of systematic sampling on the variance is difficult to determine, although it is conjectured that often systematic sampling results in a lower variance than simple random sampling, because the sampled units in a systematic sample are forced to be spread out in the sampling frame.

Weighting adjustments, especially the SSF, also affect the variance of the estimates. The SSF is thought to decrease the variance of the estimates, as would be expected since the adjustment is really a form of poststratification. The overall design effect for the NCVS, and for similar U.S. government surveys, is about 2.

7.7
Sampling and Experiment Design*

Numerous parallels between sample surveys and designed experiments are discussed in Fienberg and Tanur (1987) and Yates (1981). Some of these parallels are noted in this section.

Simple random sampling, in which the universe $\mathcal{U}$ has N units, is similar to the randomization approach to the comparison of two treatments using a total of N experimental units. To test the hypothesis $H_0 : \mu_1 = \mu_2$, randomly assign n of the N units to treatment 1 and the remaining $N - n$ units to treatment 2. The observed value of the test statistic is compared with the reference distribution based on all $\binom{N}{n}$ possible assignments of experimental units to treatments. The p-value comes from the randomization distribution. Using randomization for inference dates back to Fisher (1925), and the theory is developed in Kempthorne (1952).

Randomization serves similar purposes in sampling and in experiment design. In sampling, the goal is to generalize our results to the population, and we hope that randomization gives us a representative sample. When we design an experiment, we attempt to "randomize out" all other possible influences, and we hope that we can separate the differences due to the treatments from random error. In both cases, we can quantify how often we expect to have a sample or a design that gives us a "bad" result. This quantification appears in confidence intervals: It is expected that 95% of possible samples or possible replications of an experiment will yield a 95% CI that contains the true value.

The purpose of stratification is to increase the precision of our estimates by grouping similar items together. The same purpose is met in design of experiments with blocking. Cluster samples also group similar items together, but the purpose is convenience, not precision. An analog in experiment design is a split-plot design, which generally gives greater precision in the subplot estimate than in the whole-plot estimate.

The structural similarity between surveys and designed experiments was exploited by using ANOVA tables to develop the theory of stratification and cluster sampling. We used a fixed-effects one-way ANOVA for a model-based approach to stratification and a random-effects one-way ANOVA for a model-based approach to cluster sampling. Much of the theory in cluster sampling is similar to the theory of random-effects models; in the models in Chapters 5 and 6, we relied on variance components to explain the dependence in the data.

Poststratification and ratio and regression estimation in sampling allow us to increase the precision of our estimates by taking advantage of the relationship between the variable of interest and other classification variables; the same goal in designed experiments is met by using covariate adjustment, as in analysis of covariance.

Both experiment design and sampling are involved in similar debates between using a randomization theory approach or using a model-based approach. We have touched on the different philosophical approaches for estimating functions of totals in Sections 2.8, 3.4, 4.6, 5.7, and 6.7, but much more has been said. I encourage the interested reader to start with the discussion papers by Smith (1994) and Hansen et al. (1983) and the book by Thompson (1997). Royall (1992a) succinctly summarizes a model-based approach to sampling.

Finally, in both sample surveys and designed experiments, it is crucial that adequate effort be spent on the design of the study. No amount of statistical analysis, however sophisticated, can compensate for a poor design. Chapter 1 presented examples of disastrous results from selection bias resulting from poor survey design or execution. A call-in poll is not only useless for generalizing to a population but also harmful, as people may believe its statistics are accurate. Similarly, little can be concluded about the efficacy of treatments A and B for a medical condition if the most ill patients are assigned to treatment A; if the mean duration of symptoms is significantly less for treatment B than for treatment A, is the difference due to the treatment or to the difference in the patients?

Of course, adjusting for an imperfect design in the analysis is sometimes possible. If a measure of the severity of the illness at the beginning of the study is available, it could be used as a covariate in comparing the two treatments, although there will still be worries about confounding with other, unmeasured quantities. Values for missing cells in a two-way ANOVA design can be estimated by a model. Similarly, available information about nonrespondents can be used to improve estimation in the presence of nonresponse, as discussed in the next chapter.

7.8
Exercises

1 Obtain one of the papers listed below, or another paper employing a complex survey design, and write a short critique. Your critique should include:

a A brief summary of the design and analysis.

b A discussion of the effectiveness of the design and the appropriateness of the analysis.

c Your recommendations for future studies of this type.

Stewart, R. E., and H. A. Kantrud. 1973. Ecological distribution of breeding waterfowl populations in North Dakota. *Journal of Wildlife Management* 37 (1): 39–50.

Matson, R. G., and W. D. Lipe. 1975. Regional sampling: A case study of Cedar Mesa, Utah. In *Sampling in archaeology*, 124–143. Edited by J. W. Mueller. Tucson: University of Arizona Press.

U.S. Veterans Administration. 1980. *Study of former prisoners of war.* Washington, D.C.: Government Printing Office. The sampling design is discussed on pages 16–21.

Carra, J. S. 1984. Lead levels in blood of children around smelter sites in Dallas. In *Environmental sampling for hazardous wastes*. Edited by E. G. Schweitzer and J. A. Santolucito, ACS Symposium Series 267. Washington, D.C.: American Chemical Society.

Gerbert, B., B. T. Maguire, and T. J. Coates. 1990. Are patients talking to their physicians about AIDS? *American Journal of Public Health* 80:467–468.

Langley, G. R., D. L. Tritchler, H. A. Llewellyn-Thomas, and J. E. Till. 1991. Use of written cases to study factors associated with regional variations in referral rates. *Journal of Clinical Epidemiology* 44 (4/5): 391–402.

Oppliger, R. A., G. L. Landry, S. W. Foster, and A. C. Lambrecht. 1993. Bulimic behaviors among interscholastic wrestlers: A statewide survey. *Pediatrics* 91 (4): 826–831.

Tanfer, K. 1993. National Survey of Men: Design and execution. *Family Planning Perspectives* 25:83–86.

Wadsworth, J., J. Field, A. M. Johnson, S. Bradshaw, and K. Wellings. 1993. Methodology of the National Survey of Sexual Attitudes and Lifestyles. *Journal of the Royal Statistical Society, Ser. A*, 156:407–421.

Benson, V., and M. A. Marano. 1994. Current estimates from the National Health Interview Survey. *Vital and Health Statistics* 10 (189). The survey design is described in Appendix I, starting on page 132.

Guyon, A. B., A. Barman, J. U. Ahmed, A. U. Ahmed, and M. S. Alam. 1994. A baseline survey on use of drugs at the primary health care level in Bangladesh. *Bulletin of the World Health Organization* 72 (2): 265–271.

Heneman, H. G., D. L. Huett, R. J. Lavigna, and D. Ogsten. 1995. Assessing managers' satisfaction with staffing services. *Personnel Psychology* 48: 163–172.

Kellermann, A. L., L. Westphal, L. Fischer, and B. Harvard. 1995. Weapon involvement in home invasion crimes. *Journal of the American Medical Association* 273 (22): 1759–1762.

Tielsch, J. M., J. Katz, H. A. Quigley, J. C. Javitt, and A. Sommer. 1995. Diabetes, intraocular pressure, and primary open-angle glaucoma in the Baltimore Eye Survey. *Ophthalmology* 102 (1): 48–54.

2 Many government statistical organizations and other collectors of survey data now have Web sites where they provide information on the survey design. Some Internet addresses are given in Table 7.5 (these are subject to change, but you should be able to find the organization through a search). The first site listed, www.fedstats.gov, provides links to U.S. government agencies that spend at least $500,000 per year on statistical activities. Many of these agencies conduct surveys. The Web site www.lib.umich.edu/libhome/Documents.center/stats.html provides links to information about surveys on a wide variety of topics, from finance to agriculture.

TABLE 7.5
Web Sites with Information on Large Surveys

Organization	Address
Federal Interagency Council on Statistical Policy	www.fedstats.gov
U.S. Bureau of the Census	www.census.gov
Statistics Canada	www.statcan.ca
Statistics Norway	www.ssb.no
Statistics Sweden	www.scb.se
UK Office for National Statistics	www.ons.gov.uk
Australian Bureau of Statistics	www.statistics.gov.au
Statistics New Zealand	www.stats.govt.nz
Statistics Netherlands	www.cbs.nl
Gallup Organization	www.gallup.com
Nielsen Media Research	www.nielsenmedia.com
National Opinion Research Center	www.norc.uchicago.edu
Inter-University Consortium for Political and Social Research	www.icpsr.umich.edu

Look up a site on the Internet describing a complex survey. Write a summary of the purpose, design, and method used for analysis. Do you think that the design used could be improved upon? If so, how?

3 You are asked to design a survey to estimate the total number of cars without permits that park in handicapped parking places on your campus. What variables (if any) would you consider for stratification? For clustering? What information do you need to aid in the design of the survey? Describe a survey design that you think would work well for this situation.

4 Repeat Exercise 3 for a survey to estimate the total number of books in a library that need rebinding.

5 Repeat Exercise 3 for a survey to estimate the percentage of persons in your city who speak more than one language.

6 Repeat Exercise 3 for a survey to estimate the distribution of number of eggs laid by Canada geese.

7 Show that in a stratified sample $\sum y \hat{f}(y)$ produces the estimator in (4.2).

8 What is $\hat{S}^2$ in (7.4) for an SRS? How does it compare with the sample variance s^2?

9 In a two-stage cluster sample of rural and urban areas in Nepal, Rothenberg et al. (1985) found that the design effect for common contagious diseases was much higher than for rare contagious diseases. In the urban areas, measles—with an estimated incidence of 123.9 cases per 1000 children per year—had a design effect of 7.8; diphtheria—with an estimated incidence of 2.1 cases per 1000 children per year—had a design effect of 1.9.

Explain why one would expect this disparity in the design effects. HINT: Suppose a sample of 1000 children is taken, in 50 clusters of 20 children each. Also suppose that the disease is as aggregated as possible, so if the estimated incidence were 40 per 1000, all children in two clusters would have the disease, and no children in the remaining 38 clusters would have the disease. Now calculate deff for incidences varying from 1 per 1000 to 200 per 1000.

10 Using the data in the file nybight.dat (see Exercise 19 of Chapter 4), find the epmf of number of species caught per trawl in 1974. Be sure to use the sampling weights.

11 Using the data in the file teachers.dat (see Exercise 16 of Chapter 5), use the sampling weights to find the epmf of the number of hours worked. What is the design effect?

12 Using the data in the file measles.dat (see Exercise 17 of Chapter 5), what is the design effect for percentage of parents who returned a consent form? For the percentage of children who had previously had measles?

13 The Survey of Youth in Custody sampled youth who were residents of long-term facilities at the end of 1987. Is the sample representative of youth who have been in long-term facilities in 1987? Why, or why not?

14   The file syc.dat contains other information from the 1987 Survey of Youth in Custody. Plot data for the age of the youth at first arrest. What is the average age of first arrest? The median? The 25th percentile? (Use the "final weight" to estimate these quantities. Do *not* calculate standard errors for now.) How do your estimates compare to estimates obtained without using weights?

15   Using the file syc.dat and the final weights, estimate the proportion of youths who

a   Are age 14 or younger.

b   Are held for a violent offense.

c   Lived with both parents when growing up.

d   Are male.

e   Are Hispanic.

f   Grew up primarily in a single-parent family.

g   Have used illegal drugs.

16   The file ncvs.dat includes selected variables for victimization incidents reported between July and December 1989 in the NCVS. The incident weights are the person weights divided by the number of victims involved in the incident. Using the data, find estimates of the percentage of

a   Victimization incidents that are violent.

b   Violent crime victimizations that involve injury.

c   Violent crime victimizations that are reported to the police.

Do your calculations both with and without weights. Do the weights appear to make a difference? (Do *not* find standard errors, as you are not given enough information to do so.)

17   The British Crime Survey is also a stratified, multistage survey (Aye Maung 1995). In contrast to the NCVS, the BCS is not designed to be approximately self-weighting, as inner-city areas are sampled at about twice the rate of non-inner-city areas. In the BCS, households are selected using probability sampling, but only one adult (selected at random) is interviewed in each responding household. Set the relative sampling weight for an inner-city household to be 1.

a   Consider the BCS as a sample of households. What is the relative sampling weight for a non-inner-city household?

b   Consider the BCS as a sample of adults. Construct a table of relative sampling weights for the sample of adults.

Number of Adults	Inner City	Non–Inner City
1		
2		
3		
4		
5		

*18   (Requires probability.)   *Combined ratio estimators.* In a stratified sample, the combined ratio estimator of the population total is defined to be $\hat{t}_{y\text{comb}} = t_x \hat{t}_y / \hat{t}_x$, where

$\hat{t}_y = \sum_{h=1}^{H} \hat{t}_{yh}$, $\hat{t}_{yh}$ is an unbiased estimator of the population total for y in stratum h, $\hat{t}_x = \sum_{h=1}^{H} \hat{t}_{xh}$, $\hat{t}_{xh}$ is an unbiased estimator of the population total for x in stratum h, and t_x is the population total for x.

a   Show that

$$\frac{|\text{Bias}[\hat{t}_{y\text{comb}}]|}{\sqrt{V(\hat{t}_{y\text{comb}})}} \leq \text{CV}(\hat{t}_x).$$

HINT: See page 66.

b   In a stratified random sample, find the approximate bias and MSE of $\hat{t}_{y\text{comb}}$.

*19   (Requires probability.) *Separate ratio estimators.* In a stratified sample, the separate ratio estimator of the population total is defined to be

$$\hat{t}_{y\text{sep}} = \sum_{h=1}^{H} \frac{t_{xh}\hat{t}_{yh}}{\hat{t}_{xh}},$$

where $\hat{t}_{yh}$ is an unbiased estimator of the population total for y in stratum h, $\hat{t}_{xh}$ is an unbiased estimator of the population total for x in stratum h, and t_{xh} is the population total for x in stratum h.

Using results from Section 3.1, find the bias and an approximation to the MSE of $\hat{t}_{y\text{sep}}$ in a stratified random sample. Allow different ratios, B_h, in each stratum. When will the bias be small?

SURVEY Exercises

20   Design a stratified cluster survey for Stephens County. Stratify on two variables: urban/rural and assessed valuation. Then, within each stratum, select two districts with probability proportional to population and sample an equal number of households within each district selected. Construct the sampling weights for each household in your sample.

21   Execute the sample and estimate the average price a household is willing to pay for cable TV. Be sure to give standard errors.

22   Compare your results with those from an SRS with the same number of households. What is the estimated design effect for your survey? How do the costs compare?

8

Nonresponse

Miss Schuster-Slatt said she thought English husbands were lovely, and that she was preparing a questionnaire to be circulated to the young men of the United Kingdom, with a view to finding out their matrimonial preferences.

"But English people won't fill up questionnaires," said Harriet.

"Won't fill up questionnaires?" cried Miss Schuster-Slatt, taken aback.

"No," said Harriet, "they won't. As a nation we are not questionnaire-conscious."

— Dorothy Sayers, *Gaudy Night*

The best way to deal with nonresponse is to prevent it. After nonresponse has occurred, it is sometimes possible to model the missing data, but predicting the missing observations is never as good as observing them in the first place. Nonrespondents often differ in critical ways from respondents; if the nonresponse rate is not negligible, inference based upon only the respondents may be seriously flawed.

We discuss two types of nonresponse in this chapter: **unit nonresponse,** in which the entire observation unit is missing, and **item nonresponse,** in which some measurements are present for the observation unit but at least one item is missing. In a survey of persons, unit nonresponse means that the person provides no information for the survey; item nonresponse means that the person does not respond to a particular item on the questionnaire. In the Current Population Survey and the National Crime Victimization Survey (NCVS), unit nonresponse can arise for a variety of reasons: The interviewer may not be able to contact the household; the person may be ill and cannot respond to the survey; the person may refuse to participate in the survey. In these surveys, the interviewer tries to get demographic information about the nonrespondent, such as age, sex, and race, as well as characteristics of the dwelling unit, such as urban/rural status; this information can be used later to adjust for the nonresponse. Item nonresponse occurs largely because of refusals: A household may decline to give information about income, for example.

In agriculture or wildlife surveys, the term *missing data* is generally used instead of *nonresponse,* but the concepts and remedies are similar. In a survey of breeding ducks, for example, some birds will not be found by the researchers; they are, in a sense, nonrespondents. The nest may be raided by predators before the investigator can determine how many eggs were laid; this is comparable to item nonresponse.

In this chapter, we discuss four approaches to dealing with nonresponse:

1   Prevent it. Design the survey so that nonresponse is low. This is by far the best method.

2   Take a representative subsample of the nonrespondents; use that subsample to make inferences about the other nonrespondents.

3   Use a model to predict values for the nonrespondents. Weights implicitly use a model to adjust for unit nonresponse. Imputation often adjusts for item nonresponse, and parametric models may be used for either type of nonresponse.

4   Ignore the nonresponse (not recommended, but unfortunately common in practice).

8.1
Effects of Ignoring Nonresponse

EXAMPLE **8.1**   Thomsen and Siring (1983) report results from a 1969 survey on voting behavior carried out by the Central Bureau of Statistics in Norway. In this survey, three calls were followed by a mail survey. The final nonresponse rate was 9.9%, which is often considered to be a small nonresponse rate. Did the nonrespondents differ from the respondents?

In the Norwegian voting register, it was possible to find out whether a person voted in the election. The percentage of persons who voted could then be compared for respondents and nonrespondents; Table 8.1 shows the results. The selected sample is all persons selected to be in the sample, including data from the Norwegian voting register for both respondents and nonrespondents.

The difference in voting rate between the nonrespondents and the selected sample was largest in the younger age groups. Among the nonrespondents, the voting rate varied with the type of nonresponse. The overall voting rate for the persons who refused to participate in the survey was 81%, the voting rate for the not-at-homes was 65%, and the voting rate for the mentally and physically ill was 55%, implying that absence or illness were the primary causes of nonresponse bias.   ■

TABLE **8.1**

Percentage of Persons Who Voted

	All	20–24	25–29	Age 30–49	50–69	70–79
Nonrespondents	71	59	56	72	78	74
Selected sample	88	81	84	90	91	84

SOURCE: Adapted from table 8 in Thomsen and Siring 1983.

It has been demonstrated repeatedly that nonresponse can have large effects on the results of a survey—in Example 8.1, a nonresponse rate of less than 10% led to an over-estimate of the voting rate in Norway. Holt and Elliot discuss the results of a series of studies done on nonresponse in the United Kingdom, indicating that "lower response rates are associated with the following characteristics: London residents; households with no car; single people; childless couples; older people; divorced/widowed people; new Commonwealth origin; lower educational attainment; self-employed" (1991, 334).

Moreover, increasing the sample size without targeting nonresponse does nothing to reduce nonresponse bias; a larger sample size merely provides more observations from the class of persons that would respond to the survey. Increasing the sample size may actually worsen the nonresponse bias, as the larger sample size may divert resources that could have been used to reduce or remedy the nonresponse or it may result in less care in the data collection. Recall that the infamous *Literary Digest Survey* of 1936 (discussed on p. 7) had 2.4 million respondents but a response rate of less than 25%. The U.S. decennial census itself does not include the entire population, and the undercoverage rate varies for different demographic groups. In the early 1990s, the nonresponse and undercoverage in the U.S. census prompted a lawsuit from certain cities to force the Census Bureau to adjust for the nonresponse, and the debate about census adjustment continues.

Most small surveys ignore any nonresponse that remains after callbacks and follow-ups, and report results based on complete records only. Hite (1987) did so in the survey discussed in Chapter 1, and much of the criticism of her results was based on her low response rate. Nonresponse is also ignored for many surveys reported in newspapers, both local and national.

An analysis of complete records has the underlying assumptions that the nonrespondents are similar to the respondents and that units with missing items are similar to units that have responses for every question. Much evidence indicates that this assumption does not hold true in practice. If nonresponse is ignored in the NCVS, for example, victimization rates are underestimated. Biderman and Cantor (1984) find lower victimization rates for persons who respond in three consecutive interviews than for persons who are nonrespondents in at least one of those interviews or who move before the panel study is completed.

Results reported from an analysis of only complete records should be taken as representative of the population of persons who would respond to the survey, which is rarely the same as the target population. If you insist on estimating population means and totals using only the complete records and making no adjustment for nonrespondents, at the very least you should report the rate of nonresponse.

The main problem caused by nonresponse is potential bias of population estimates. Think of the population as being divided into two somewhat artificial strata of respondents and nonrespondents. The population respondents are the units that would respond if they were chosen to be in the sample; the number of population respondents, N_R, is unknown. Similarly, the N_M (M for missing) population nonrespondents are the units that would not respond. We then have the following population quantities:

Stratum	Size	Total	Mean	Variance
Respondents	N_R	t_R	$\bar{y}_{RU}$	S_R^2
Nonrespondents	N_M	t_M	$\bar{y}_{MU}$	S_M^2
Entire population	N	t	$\bar{y}_U$	S^2

The population as a whole has variance $S^2 = \sum_{i=1}^{N}(y_i - \bar{y}_U)^2/(N-1)$, mean $\bar{y}_U$, and total t. A probability sample from the population will likely contain some respondents and some nonrespondents. But, of course, on the first call we do not observe y_i for any of the units in the nonrespondent stratum. If the population mean in the nonrespondent stratum differs from that in the respondent stratum, estimating the population mean using only the respondents will produce bias.[1]

Let $\bar{y}_R$ be an approximately unbiased estimator of the mean in the respondent stratum, using only the respondents. Because

$$\bar{y}_U = \frac{N_R}{N}\bar{y}_{RU} + \frac{N_M}{N}\bar{y}_{MU},$$

the bias is approximately

$$E[\bar{y}_R] - \bar{y}_U \approx \frac{N_M}{N}(\bar{y}_{RU} - \bar{y}_{MU}).$$

The bias is small if either (1) the mean for the nonrespondents is close to the mean for the respondents or (2) N_M/N is small—there is little nonresponse. But we can never be assured of (1), as we generally have no data for the nonrespondents. Minimizing the nonresponse rate is the only sure way to control nonresponse bias.

8.2
Designing Surveys to Reduce Nonsampling Errors

A common feature of poor surveys is a lack of time spent on the design and nonresponse follow-up in the survey. Many persons new to surveys (and some, unfortunately, not new) simply jump in and start collecting data without considering potential problems in the data-collection process; they mail questionnaires to everyone in the target population and analyze those that are returned. It is not surprising that such surveys have poor response rates. Many surveys reported in academic journals on purchasing, for example, have response rates between 10 and 15%. It is difficult to see how anything can be concluded about the population in such a survey.

A researcher who knows the target population well will be able to anticipate some of the reasons for nonresponse and prevent some of it. Most investigators, however, do not know as much about reasons for nonresponse as they think they do. They need to discover why the nonresponse occurs and resolve as many of the problems as possible before commencing the survey.

---

[1] The variance is often too low as well. In income surveys, for example, the rich and the poor are more likely to be nonrespondents on the income questions. In that case, S_R^2, for the respondent stratum, is smaller than S^2. The point estimate of the mean may be biased, and the variance estimate may be biased, too.

These reasons can be discovered through designed experiments and application of quality-improvement methods to the data collection and processing. You do not know why previous surveys related to yours have a low response rate? Design an experiment to find out. You think errors are introduced in the data recording and processing? Use a nested design to find the sources of errors. Any book on quality control or designed experiments will tell you how to collect your data.

And, of course, you can rely on previous researchers' experiments to help you minimize nonsampling errors. The references on experiment design and quality control at the end of the book are a good place to start; Hidoroglou et al. (1993) give a general framework for nonresponse.

EXAMPLE 8.2 The 1990 U.S. decennial census attempted to survey each of the over 100 million households in the United States. The response rate for the mail survey was 65%; households that did not mail in the survey needed to be contacted in person, adding millions of dollars to the cost of the census. Increasing the mail response rate for future censuses would result in tremendous savings.

Dillman et al. (1995a) report results of a factorial experiment employed in the 1992 Census Implementation Test, designed to explore the individual effects and interactions of three experimental factors on response rates. The three factors were (1) a prenotice letter alerting the household to the impending arrival of the census form, (2) a stamped return envelope included with the census form, and (3) a reminder postcard sent a few days after the census form. The results were dramatic, as shown in Figure 8.1. The experiment established that, although all three factors influenced the response rate, the letter and postcard led to greater gains in response rate than the stamped return envelope. ■

Nonresponse can have many different causes; as a result, no single method can be recommended for every survey. Platek (1977) classifies sources of nonresponse

FIGURE 8.1

Response rates achieved for each combination of the factors *letter, envelope,* and *postcard.* The observed response rate was 64.3% when all three aids were used and only 50% when none were used.

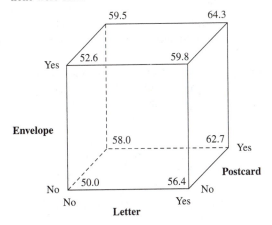

FIGURE   8.2

Factors affecting nonresponse

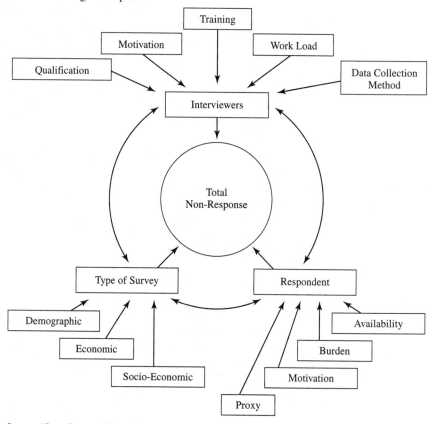

SOURCE: "Some Factors Affecting Non-Response," by R. Platek, 1977, *Survey Methodology, 3,* 191–214. Copyright ©
1977 Survey Methodology. Reprinted with permission.

as related to (1) survey content, (2) methods of data collection, and (3) respondent
characteristics, and illustrates various sources using the diagram in Figure 8.2. Groves
(1989) and Dillman (1978) discuss additional sources of nonresponse. The following
are some factors that may influence response rate and data accuracy.

- *Survey content.* A survey on drug use or financial matters may have a large
number of refusals. Sometimes the response rate can be increased for sensitive items
by careful ordering of the questions or by using a randomized response technique
(see Section 12.5).

- *Time of survey.* Some calling periods or seasons of the year may yield higher
response rates than others. The vacation month of August, for example, would be a
bad time to take a one-time household survey in Germany.

- *Interviewers.* Gower (1979) found a large variability in response rates achieved
by different interviewers, with about 15% of interviewers reporting almost no non-

response. Some field investigators in a bird survey may be better at spotting and identifying birds than others. Standard quality-improvement methods can be applied to increase the response rate and accuracy for interviewers. The same methods can be applied to the data-coding process.

■ *Data-collection method.* Generally, telephone and mail surveys have a lower response rate than in-person surveys (they also have lower costs, however). Computer Assisted Telephone Interviewing (CATI) has been demonstrated to improve accuracy of data collected in telephone surveys; with CATI, all questions are displayed on a computer, and the interviewer codes the responses in the computer as questions are asked. CATI is especially helpful in surveys in which a respondent's answer to one question determines which question is asked next (Catlin and Ingram 1988).

Mail, fax, and Internet surveys often have low response rates. Possible reasons for nonresponse in a mail survey should be explored before the questionnaire is mailed: Is the survey sent to the wrong address? Do recipients discard the envelope as junk mail even before opening it? Will the survey reach the intended recipient? Will the recipient believe that filling out the survey is worth the time?

■ *Questionnaire design.* We have already seen that question wording has a large effect on the responses received; it can also affect whether a person responds to an item on the questionnaire. The volume edited by Tanur (1993) explores some recent research on application of cognitive research on question design. In a mail survey, a well-designed form for the respondent may increase data accuracy.

■ *Respondent burden.* Persons who respond to a survey are doing you an immense favor, and the survey should be as nonintrusive as possible. A shorter questionnaire, requiring less detail, may reduce the burden to the respondent. Respondent burden is a special concern in panel surveys such as the NCVS, in which sampled households are interviewed every six months for $3\frac{1}{2}$ years. DeVries et al. (1996) discuss methods used in reducing respondent burden in the Netherlands. Techniques such as stratification can reduce respondent burden because a smaller sample suffices to give the required precision.

■ *Survey introduction.* The survey introduction provides the first contact between the interviewer and potential respondent; a good introduction, giving the recipient motivation to respond, can increase response rates dramatically. Nielsen Media Research emphasizes to households in its selected sample that their participation in the Nielsen ratings affects which television shows are aired. The respondent should be told for what purpose the data will be used (unscrupulous persons often pretend to be taking a survey when they are really trying to attract customers or converts) and assured confidentiality.

■ *Incentives and disincentives.* Incentives, financial or otherwise, may increase the response rate. Disincentives may work as well: Physicians who refused to be assessed by peers after selection in a stratified sample from the College of Physicians and Surgeons of Ontario registry had their medical licenses suspended. Not surprisingly, nonresponse was low (McAuley et al. 1990).

■ *Follow-up.* The initial contact of the sample is usually less costly per unit than follow-ups of the initial nonrespondents. If the initial survey is by mail, a reminder may increase the response rate. Not everyone responds to follow-up calls, though; some persons will refuse to respond to the survey no matter how often they are contacted. You need to decide how many follow-up calls to make before the marginal returns do not justify the money spent.

You should try to obtain at least some information about nonrespondents that can be used later to adjust for the nonresponse, and include surrogate items that can be used for item nonresponse. True, there is no complete compensation for not having the data, but partial information may be better than none. Information about the race, sex, or age of a nonrespondent may be used later to adjust for nonresponse. Questions about income may well lead to refusals, but questions about cars, employment, or education may be answered and can be used to predict income. If the pretests of the survey indicate a nonresponse problem that you do not know how to prevent, try to design the survey so that at least some information is collected for each observation unit.

The quality of survey data is largely determined at the design stage. Fisher's (1938) words about experiments apply equally well to the design of sample surveys: "To call in the statistician after the experiment is done may be no more than asking him to perform a postmortem examination: he may be able to say what the experiment died of." Any survey budget needs to allocate sufficient resources for survey design and for nonresponse follow-up. Do not scrimp on the survey design; every hour spent on design may save weeks of remorse later.

8.3
Callbacks and Two-Phase Sampling

Virtually all good surveys rely on callbacks to obtain responses from persons not at home for the first try. Analysis of callback data can provide some information about the biases that can be expected from the remaining nonrespondents.

EXAMPLE **8.3**   Traugott (1987) analyzed callback data from two 1984 Michigan polls on preference for presidential candidates. The overall response rates for the surveys were about 65%, typical for large political polls. About 21% of the interviewed sample responded on the first call; up to 30 attempts were made to reach persons who did not respond on the first call. Traugott found that later respondents were more likely to be male, older, and Republican than early respondents; while 48% of the respondents who answered the first call supported Reagan and 45% supported Mondale, 59% of the entire sample supported Reagan as opposed to 39% for Mondale. Differing procedures for nonresponse follow-up and persistence in callback may explain some of the inconsistencies among political polls.

If nonrespondents resemble late respondents, one might speculate that nonrespondents were more likely to favor Reagan. But nonrespondents do not necessarily resemble the hard-to-reach; persons who absolutely refuse to participate may differ greatly from persons who could not be contacted immediately, and nonrespondents may be more likely to have illnesses or other circumstances preventing participation. We also do not know how likely it is that nonrespondents to the surveys will vote in the election; even if we speculate that they were more likely to favor Reagan, they are not necessarily more likely to vote for Reagan. ∎

Often, when the survey is designed so that callbacks will be used, the initial contact is by mail survey; the follow-up calls use a more expensive method such as a personal interview.

Hansen and Hurwitz (1946) proposed subsampling the nonrespondents and using **two-phase sampling** (also called **double sampling**) for stratification to estimate the population mean or total. The population is divided into two strata, as described in Section 8.1; the two strata are respondents and initial nonrespondents, persons who do not respond to the first call. We will develop the theory of two-phase sampling for general survey designs in Section 12.1; here, we illustrate how it can be used for nonresponse.

In the simplest form of two-phase sampling, randomly select n units in the population. Of these, n_R respond and n_M do not respond. The values n_R and n_M, though, are random variables; they will change if a different simple random sample (SRS) is selected. Then, make a second call on a random subsample of $100v\%$ of the n_M nonrespondents in the sample, where the subsampling fraction v does not depend on the data collected.

Suppose that through some superhuman effort all the targeted nonrespondents are reached. Let $\bar{y}_R$ be the sample average of the original respondents and $\bar{y}_M$ (M stands for "missing") be the average of the subsampled nonrespondents. The two-phase sampling estimates of the population mean and total are

$$\hat{\bar{y}} = \frac{n_R}{n}\bar{y}_R + \frac{n_M}{n}\bar{y}_M \tag{8.1}$$

and

$$\hat{t} = N\hat{\bar{y}} = \frac{N}{n}\sum_{i \in \mathcal{S}_R} y_i + \frac{N}{n}\frac{1}{v}\sum_{i \in \mathcal{S}_M} y_i, \tag{8.2}$$

where $\mathcal{S}_R$ represents the sampled units in the respondent stratum and $\mathcal{S}_M$ represents the sampled units in the nonrespondent stratum. Note that $\hat{t}$ is a weighted sum of the observed units; the weights are N/n for the respondents and $N/(nv)$ for the subsampled nonrespondents. Because only a subsample was taken in the nonrespondent stratum, each subsampled unit in that stratum represents more units in the population than does a unit in the respondent stratum.

The expected value and variance of these estimators are found in Section 12.1. Because $\hat{t}$ is an appropriately weighted unequal-probability estimator, Theorem 6.2 implies that $E[\hat{t}] = t$. From (12.5), if the finite population corrections can be ignored, we can estimate the variance by

$$\hat{V}(\hat{\bar{y}}) = \frac{n_R - 1}{n - 1}\frac{s_R^2}{n} + \frac{n_M - 1}{n - 1}\frac{s_M^2}{vn} + \frac{1}{n - 1}\left[\frac{n_R}{n}(\bar{y}_R - \hat{\bar{y}})^2 + \frac{n_M}{n}(\bar{y}_M - \hat{\bar{y}})^2\right].$$

If everyone responds in the subsample, two-phase sampling not only removes the nonresponse bias but also accounts for the original nonresponse in the estimated variance.

8.4
Mechanisms for Nonresponse

Most surveys have some residual nonresponse even after careful design and follow-up of nonresponse. All methods for fixing up nonresponse are necessarily model-based. If we are to make any inferences about the nonrespondents, we must assume that they are related to respondents in some way. A good nontechnical reference for methods of dealing with nonresponse is Groves (1989); the three-volume set edited by Madow et al. (1983) contains much information on the statistical research on nonresponse up to that date.

Dividing population members into two fixed strata of would-be respondents and would-be nonrespondents is fine for thinking about potential nonresponse bias and for two-phase methods. To adjust for nonresponse that remains after all other measures have been taken, we need a more elaborate setup, letting the response or nonresponse of unit i be a random variable. Define the random variable

$$R_i = \begin{cases} 1 & \text{if unit } i \text{ responds.} \\ 0 & \text{if unit } i \text{ does not respond.} \end{cases}$$

After sampling, the realizations of the response indicator variable are known for the units selected in the sample. A value for y_i is recorded if r_i, the realization of R_i, is 1. The probability that a unit selected for the sample will respond,

$$\phi_i = P(R_i = 1),$$

is of course unknown but assumed positive. Rosenbaum and Rubin (1983) call ϕ_i the **propensity score** for the ith unit.

Suppose that y_i is a response of interest and that $\mathbf{x}_i$ is a vector of information known about unit i in the sample. Information used in the survey design is included in $\mathbf{x}_i$. We consider three types of missing data, using the Little and Rubin (1987) terminology of nonresponse classification.

Missing Completely at Random   If ϕ_i does not depend on $\mathbf{x}_i$, y_i, or the survey design, the missing data are **missing completely at random** (MCAR). Such a situation occurs if, for example, someone at the laboratory drops a test tube containing the blood sample of one of the survey participants—there is no reason to think that the dropping of the test tube had anything to do with the white blood cell count.[2] If data are MCAR, the respondents are representative of the selected sample.

Missing data in the NCVS would be MCAR if the probability of nonresponse is completely unrelated to region of the United States, race, sex, age, or any other variable measured for the sample *and* if the probability of nonresponse is unrelated to any variables about victimization status. Nonrespondents would be essentially selected at random from the sample.

---

[2]Even here, though, the suspicious mind can create a scenario in which the nonresponse might be related to quantities of interest: Perhaps laboratory workers are less likely to drop test tubes that they believe contain HIV.

If the response probabilities ϕ_i are all equal and the events $\{R_i = 1\}$ are conditionally independent of each other and of the sample-selection process given n_R, then the data are MCAR. If an SRS of size n is taken, then under this mechanism the respondents will be a simple random subsample of variable size n_R. The sample mean of the respondents, $\bar{y}_R$, is approximately unbiased for the population mean. The MCAR mechanism is implicitly adopted when nonresponse is ignored.

Missing at Random Given Covariates, or Ignorable Nonresponse   If ϕ_i depends on $\mathbf{x}_i$ but not on y_i, the data are **missing at random** (MAR); the nonresponse depends only on observed variables. We can successfully model the nonresponse, since we know the values of $\mathbf{x}_i$ for all sample units. Persons in the NCVS would be missing at random if the probability of responding to the survey depends on race, sex, and age—all known quantities—but does not vary with victimization experience within each age/race/sex class. This is sometimes termed **ignorable nonresponse**: Ignorable means that a model can explain the nonresponse mechanism and that the nonresponse can be ignored after the model accounts for it, not that the nonresponse can be completely ignored and complete-data methods used.

Nonignorable Nonresponse   If the probability of nonresponse depends on the value of a response variable and cannot be completely explained by values of the $\mathbf{x}$'s, then the nonresponse is **nonignorable.** This is likely the situation for the NCVS: It is suspected that a person who has been victimized by crime is less likely to respond to the survey than a nonvictim, even if they share the values of all known variables such as race, age, and sex. Crime victims may be more likely to move after a victimization and thus not be included in subsequent NCVS interviews. Models can help in this situation, because the nonresponse probability may also depend on known variables but cannot completely adjust for the nonresponse.

The probabilities of responding, ϕ_i, are useful for thinking about the type of nonresponse. Unfortunately, they are unknown, so we do not know for sure which type of nonresponse is present. We can sometimes distinguish between MCAR and MAR by fitting a model attempting to predict the observed probabilities of response for subgroups from known covariates; if the coefficients in a logistic regression model are significantly different from zero, the missing data are likely not MCAR. Distinguishing between MAR and nonignorable nonresponse is more difficult. In the next section, we discuss a method for estimating the ϕ_i's.

8.5
Weighting Methods for Nonresponse

In previous chapters we have seen how weights can be used in calculating estimates for various sampling schemes (see Sections 4.3, 5.4, and 7.2). The sampling weights are the reciprocals of the probabilities of selection, so an estimate of the population total is $\sum_{i \in \mathcal{S}} w_i y_i$. For stratification, the weights are $w_i = N_h / n_h$ if unit i is in stratum h; for sampling elements with unequal probabilities, $w_i = 1/\pi_i$.

Weights can also be used to adjust for nonresponse. Let Z_i be the indicator variable for presence in the selected sample, with $P(Z_i = 1) = \pi_i$. If R_i is independent of

Z_i, then the probability that unit i will be measured is

$$P(\text{unit } i \text{ selected in sample and responds}) = \pi_i \phi_i.$$

The probability of responding, ϕ_i, is estimated for each unit in the sample, using auxiliary information that is known for all units in the selected sample. The final weight for a respondent is then $1/(\pi_i \hat{\phi}_i)$. Weighting methods assume that the response probabilities can be estimated from variables known for all units; they assume MAR data. References for more information on weighting are Oh and Scheuren (1983) and Holt and Elliot (1991).

8.5.1 Weighting-Class Adjustment

Sampling weights w_i have been interpreted as the number of units in the population represented by unit i of the sample. Weighting-class methods extend this approach to compensate for nonsampling errors: Variables known for all units in the selected sample are used to form weighting-adjustment classes, and it is hoped that respondents and nonrespondents in the same weighting-adjustment class are similar. Weights of respondents in the weighting-adjustment class are increased so that the respondents represent the nonrespondents' share of the population as well as their own.

EXAMPLE **8.4**   Suppose the age is known for every member of the selected sample and that person i in the selected sample has sampling weight $w_i = 1/\pi_i$. Then weighting classes can be formed by dividing the selected sample among different age classes, as Table 8.2 shows.

We estimate the response probability for each class by

$$\hat{\phi}_c = \frac{\text{sum of weights for respondents in class } c}{\text{sum of weights for selected sample in class } c}.$$

Then the sampling weight for each respondent in class c is multiplied by $1/\hat{\phi}_c$, the weight factor in Table 8.2. The weight of each respondent with age between 15 and 24, for example, is multiplied by 1.622. Since there was no nonresponse in the over-65 group, their weights are unchanged.   ■

TABLE   **8.2**

Illustration of Weighting-Class Adjustment Factors

| | \multicolumn{5}{c}{Age} | |
	15–24	25–34	35–44	45–64	65+	Total
Sample size	202	220	180	195	203	1,000
Respondents	124	187	162	187	203	863
Sum of weights for sample	30,322	33,013	27,046	29,272	30,451	150,104
Sum of weights for respondents	18,693	28,143	24,371	28,138	30,451	
$\hat{\phi}_c$	0.6165	0.8525	0.9011	0.9613	1.0000	
Weight factor	1.622	1.173	1.110	1.040	1.000	

The probability of response is assumed to be the same within each weighting class, with the implication that within a weighting class, the probability of response does not depend on y. As mentioned earlier, weighting-class methods assume MAR data. The weight for a respondent in weighting class c is $1/(\pi_i \hat{\phi}_c)$.

To estimate the population total using weighting-class adjustments, let $x_{ci} = 1$ if unit i is in class c, and 0 otherwise. Then let the new weight for respondent i be

$$\tilde{w}_i = \sum_c \frac{w_i x_{ci}}{\hat{\phi}_c},$$

where w_i is the sampling weight for unit i; $\tilde{w}_i = w_i/\hat{\phi}_c$ if unit i is in class c. Assign $\tilde{w}_i = 0$ if unit i is a nonrespondent. Then,

$$\hat{t}_{wc} = \sum_{i \in S} \tilde{w}_i y_i$$

and

$$\hat{\bar{y}}_{wc} = \frac{\hat{t}_{wc}}{\sum_{i \in S} \tilde{w}_i}.$$

In an SRS, for example, if n_c is the number of sample units in class c, n_{cR} is the number of respondents in class c, and $\bar{y}_{cR}$ is the average for the respondents in class c, then $\hat{\phi}_c = n_{cR}/n_c$ and

$$\hat{t}_{wc} = \sum_{i \in S} \sum_c \frac{N}{n} \frac{n_c}{n_{cR}} x_{ci} y_i = N \sum_c \frac{n_c}{n} \bar{y}_{cR}.$$

EXAMPLE 8.5   *The National Crime Victimization Survey*

To adjust for individual nonresponse in the NCVS, the within-household noninterview adjustment factor (WHHNAF) of Chapter 7 is used. NCVS interviewers gather demographic information on the nonrespondents, and this information is used to classify all persons into 24 weighting-adjustment cells. The cells depend on the age of the person, the relation of the person to the reference person (head of household), and the race of the reference person.

For any cell, let W_R be the sum of the weights for the respondents and W_M be the sum of the weights for the nonrespondents. Then the new weight for a respondent in a cell will be the previous weight multiplied by the weighting-adjustment factor $(W_M + W_R)/W_R$. Thus, the weights that would be assigned to nonrespondents are reallocated among respondents with similar (we hope) characteristics.

A problem occurs if $(W_M + W_R)/W_R$ is too large. If $(W_M + W_R)/W_R > 2$, the cell contains more nonrespondents than respondents. In this case, the variance of the estimate increases; if the number of respondents in the cell is small, the weight may not be stable. The Census Bureau collapses cells to obtain weighting-adjustment factors of 2 or less. If there are fewer than 30 interviewed persons in a cell or if the weighting-adjustment factor is greater than 2, the cell is combined (collapsed) with neighboring cells until the collapsed cell has more than 30 observations and a weighting-adjustment factor of 2 or less. ■

Construction of Weighting Classes   Weighting-adjustment classes should be constructed as though they were strata; as shown in the next section, weighting adjustment is similar to poststratification. The classes should be formed so that units within each class are as similar as possible with respect to the major variables of interest and so that the response rates vary from class to class.

Little (1986) suggests estimating the response probabilities ϕ_i as a function of the known variables (perhaps using logistic regression) and grouping observations into classes based on $\hat{\phi}_i$. This approach is preferable to simply using the estimated values of ϕ_i in individual case weights, as the estimated response probabilities may be extremely variable and might cause the final estimates to be unstable.

8.5.2   Poststratification

Poststratification is similar to weighting-class adjustment, except that population counts are used to adjust the weights. Suppose an SRS is taken. After the sample is collected, units are grouped into H different poststrata, usually based on demographic variables such as race or sex. The population has N_h units in poststratum h; of these, n_h were selected for the sample and n_{hR} responded. The poststratified estimator for $\bar{y}_U$ is

$$\bar{y}_{\text{post}} = \sum_{h=1}^{H} \frac{N_h}{N} \bar{y}_{hR};$$

the weighting-class estimator for $\bar{y}_U$, if the weighting classes are the poststrata, is

$$\bar{y}_{wc} = \sum_{h=1}^{H} \frac{n_h}{n} \bar{y}_{hR}.$$

The two estimators are similar in form; the only difference is that in poststratification the N_h are known, whereas in weighting-class adjustments the N_h are unknown and estimated by $N n_h / n$.

For the poststratified estimator, often the conditional variance given the n_{hR} is used. For an SRS,

$$V(\bar{y}_{\text{post}} \mid n_{hR}, h = 1, \ldots, H) = \sum_{h=1}^{H} \left(\frac{N_h}{N}\right)^2 \left(1 - \frac{n_{hR}}{N_h}\right)\left(\frac{S_h^2}{n_{hR}}\right). \tag{8.3}$$

The unconditional variance of $\bar{y}_{\text{post}}$ is slightly larger, with additional terms of order $1/n_{hR}^2$, as given in Oh and Scheuren (1983). A variance estimator for poststratification will be given in Exercise 5 of Chapter 9.

8.5.2.1   Poststratification Using Weights

In a general survey design, the sum of the weights in subgroup h is supposed to estimate the population count N_h for that subgroup. Poststratification uses the ratio estimator within each subgroup to adjust by the true population count.

Let $x_{hi} = 1$ if unit i is a respondent in poststratum h, and 0 otherwise. Then let

$$w_i^* = \sum_{h=1}^{H} w_i x_{hi} \frac{N_h}{\displaystyle\sum_{j \in \mathcal{S}} w_j x_{hj}}.$$

Using the modified weights,

$$\sum_{i \in \mathcal{S}} w_i^* x_{hi} = N_h,$$

and the poststratified estimator of the population total is

$$\hat{t}_{\text{post}} = \sum_{i \in \mathcal{S}} w_i^* y_i.$$

Poststratification can adjust for undercoverage as well as nonresponse if the population count N_h includes individuals not in the sampling frame for the survey.

EXAMPLE **8.6**   The second-stage factor in the NCVS (see Section 7.6) uses poststratification to adjust the weights. After all other weighting adjustments have been done, including the weighting-class adjustments for nonresponse, poststratification is used to make the sample counts agree with estimates of the population counts from the Bureau of the Census. Each person is assigned to one of 72 poststrata based on the person's age, race, and sex. The number of persons in the population falling in that poststratum, N_h, is known from other sources. Then, the weight for a person in poststratum h is multiplied by

$$\frac{N_h}{\text{sum of weights for all respondents in poststratum } h}.$$

With weighting classes, the weighting factor to adjust for unit nonresponse is always at least 1. With poststratification, because weights are adjusted so that they sum to a known population total, the weighting factor can be any positive number, although weighting factors of 2 or less are desirable. ∎

Poststratification assumes that (1) within each poststratum each unit selected to be in the sample has the same probability of being a respondent, (2) the response or nonresponse of a unit is independent of the behavior of all other units, and (3) nonrespondents in a poststratum are like the respondents. The data are MCAR within each poststratum. These are big assumptions; to make them seem a little more plausible, survey researchers often use many poststrata. But a large number of poststrata may create additional problems, in that few respondents in some poststrata may result in unstable estimates, and may preclude the application of the central limit theorem. If faced with poststrata with few observations, most practitioners collapse the poststrata with others that have similar means in key variables until they have a reasonable number of observations in each poststratum. For the Current Population Survey, a "reasonable" number means that each group has at least 20 observations and that the response rate for each group is at least 50%.

8.5.2.2   Raking Adjustments

Raking is a poststratification method that can be used when poststrata are formed using more than one variable, but only the marginal population totals are known.

Raking was first used in the 1940 census to ensure that the complete census data and samples taken from it gave consistent results and was introduced in Deming and Stephan (1940); Brackstone and Rao (1976) further developed the theory. Oh and Scheuren (1983) describe raking ratio estimates for nonresponse.

Consider the following table of sums of weights from a sample; each entry in the table is the sum of the sampling weights for persons in the sample falling in that classification (for example, the sum of the sampling weights for black females is 300).

	Black	White	Asian	Native American	Other	Sum of Weights
Female	300	1200	60	30	30	1620
Male	150	1080	90	30	30	1380
Sum of Weights	450	2280	150	60	60	3000

Now suppose we know the true population counts for the marginal totals: We know that the population has 1510 women and 1490 men, 600 blacks, 2120 whites, 150 Asians, 100 Native Americans, and 30 persons in the "Other" category. The population counts for each cell in the table, however, are unknown; we do not know the number of black females in this population and cannot assume independence. Raking allows us to adjust the weights so that the sums of weights in the margins equal the population counts.

First, adjust the rows. Multiply each entry by (true row population)/(estimated row population). Multiplying the cells in the female row by 1510/1620 and the cells in the male row by 1490/1380 results in the following table:

	Black	White	Asian	Native American	Other	Sum of Weights
Female	279.63	1118.52	55.93	27.96	27.96	1510
Male	161.96	1166.09	97.17	32.39	32.39	1490
Total	441.59	2284.61	153.10	60.35	60.35	3000

The row totals are fine now, but the column totals do not yet equal the population totals. Repeat the same procedure with the columns in the new table. The entries in the first column are each multiplied by 600/441.59. The following table results:

	Black	White	Asian	Native American	Other	Sum of Weights
Female	379.94	1037.93	54.79	46.33	13.90	1532.90
Male	220.06	1082.07	95.21	53.67	16.10	1467.10
Total	600.00	2120.00	150.00	100.00	30.00	3000.00

But this has thrown the row totals off again. Repeat the procedure until both row and column totals equal the population counts. The procedure converges as long as all cell counts are positive. In this example, the final table of adjusted counts is

	Black	White	Asian	Native American	Other	Sum of Weights
Female	375.59	1021.47	53.72	45.56	13.67	1510
Male	224.41	1098.53	96.28	54.44	16.33	1490
Total	600.00	2120.00	150.00	100.00	30.00	3000

The entries in the last table may be better estimates of the cell populations (that is, with smaller variance) than the original weighted estimates, simply because they use more information about the population. The weighting-adjustment factor for each white male in the sample is 1098.53/1080; the weight of each white male is increased a little to adjust for nonresponse and undercoverage. Likewise, the weights of white females are decreased because they are overrepresented in the sample.

The assumptions for raking are the same as for poststratification, with the additional assumption that the response probabilities depend only on the row and column and not on the particular cell. If the sample sizes in each cell are large enough, the raking estimator is approximately unbiased.

Raking has some difficulties—the algorithm may not converge if some of the cell estimates are zero. There is also a danger of "overadjustment"—if there is little relation between the extra dimension in raking and the cell means, raking can increase the variance rather than decrease.it.

8.5.3 Estimating the Probability of Response: Other Methods

Some weighting-class methods use weights that are the reciprocal of the estimated probability of response. A famous example is the Politz–Simmons method for adjusting for nonavailability of sample members.

Suppose all calls are made during Monday through Friday evenings. Each respondent is asked whether he or she was at home, at the time of the interview, on each of the four preceding weeknights. The respondent replies that she was home k of the four nights. It is then assumed that the probability of response is proportional to the number of nights at home during interviewing hours, so the probability of response is estimated by $\hat{\phi}_i = (k_i + 1)/5$. The sampling weight w_i for each respondent is then multiplied by $5/(k_i + 1)$. The respondents with $k = 0$ were home on only one of the five nights and are assigned to represent their share of the population plus the share of four persons in the sample who were called on one of their "unavailable" nights. The respondents most likely to be home have $k = 4$; it is presumed that all persons in the sample who were home every night were reached, so their weights are unchanged. The estimate of the population mean is

$$\hat{\bar{y}} = \frac{\displaystyle\sum_{i \in \mathcal{S}} \frac{5 w_i y_i}{k_i + 1}}{\displaystyle\sum_{i \in \mathcal{S}} \frac{5 w_i}{k_i + 1}}.$$

This method of weighting—described by Hartley (1946) and Politz and Simmons (1949)—is based on the premise that the most accessible persons will tend to be overrepresented in the survey data. The method is easy to use, theoretically appealing, and can be used in conjunction with callbacks. But it still misses people who were not at home on any of the five nights or who refused to participate in the survey. Because nonresponse is due largely to refusals in some telephone surveys, the Politz–Simmons method may not be helpful in dealing with all nonresponse. Values of k may also be in error, because people may err when recalling how many evenings they were home.

Potthoff et al. (1993) modified and extended the Politz–Simmons method to determine weights based on the number of callbacks needed, assuming that the ϕ_i's follow a beta distribution.

8.5.4   A Caution About Weights

The models for weighting adjustments for nonresponse are strong: In each weighting cell, the respondents and nonrespondents are assumed to be similar. Each individual in a weighting class is assumed equally likely to respond to the survey, regardless of the value of the response. These models never exactly describe the true state of affairs, and you should always consider their plausibility and implications. It is an unfortunate tendency of many survey practitioners to treat the weighting adjustment as a complete remedy and to then act as though there was no nonresponse. Weights may improve many of the estimates, but they rarely eliminate all nonresponse bias. If weighting adjustments are made (and remember, making no adjustments is itself a model about the nature of the nonresponse), practitioners should always state the assumed response model and give evidence to justify it. Weighting adjustments are usually used for unit nonresponse, not for item nonresponse (which would require a different weight for each item).

8.6
Imputation

Missing items may occur in surveys for several reasons: An interviewer may fail to ask a question; a respondent may refuse to answer the question or cannot provide the information; a clerk entering the data may skip the value. Sometimes, items with responses are changed to missing when the data set is edited or cleaned—a data editor may not be able to resolve the discrepancies for an individual 3-year-old who voted in the last election and may set both values to missing.

Imputation is commonly used to assign values to the missing items. A replacement value, often from another person in the survey who is similar to the item nonrespondent on other variables, is imputed for the missing value. When imputation is used, an additional variable that indicates whether the response was measured or imputed should be created for the data set.

Imputation procedures are used not only to reduce the nonresponse bias but to produce a "clean," rectangular data set—one without holes for the missing values. We may want to look at tables for subgroups of the population, and imputation allows us to do that without considering the item nonresponse separately each time we construct a table. Some references for imputation include Sande (1983) and Kalton and Kasprzyk (1982; 1986).

EXAMPLE 8.7 The Current Population Survey (CPS) has an overall high household response rate (typically well above 90%), but some households refuse to answer certain questions. The nonresponse rate is about 20% on many income questions. This nonresponse would create a substantial bias in any analysis unless some corrective action were taken: Various studies suggest that the item nonresponse for the income items is highest for low-income and high-income households. Imputation for the missing data makes it possible to use standard statistical techniques such as regression without the analyst having to treat the nonresponse by using specially developed methods. For surveys such as the CPS, if imputation is to be done, the agency collecting the data has more information to guide it in filling in the missing values than does an independent analyst, because identifying information is not released on the public-use tapes.

The CPS uses weighting for noninterview adjustment and hot-deck imputation for item nonresponse. The sample is divided into classes using variables sex, age, race, and other demographic characteristics. If an item is missing, a corresponding item from another unit in that class is substituted. Usually, hot-deck imputation is done by taking the value of the missing item from a household that is similar to the household with the missing item in some other explanatory variable such as family size. ∎

We use the small data set in Table 8.3 to illustrate some of the different methods for imputation. This artificial data set is only used for illustration; in practice, a much larger data set is needed for imputation. A "1" means the respondent answered yes to the question.

TABLE 8.3

Small Data Set Used to Illustrate Imputation Methods

Person	Age	Sex	Years of Education	Crime Victim?	Violent-Crime Victim?
1	47	M	16	0	0
2	45	F	?	1	1
3	19	M	11	0	0
4	21	F	?	1	1
5	24	M	12	1	1
6	41	F	?	0	0
7	36	M	20	1	?
8	50	M	12	0	0
9	53	F	13	0	?
10	17	M	10	?	?
11	53	F	12	0	0
12	21	F	12	0	0
13	18	F	11	1	?
14	34	M	16	1	0
15	44	M	14	0	0
16	45	M	11	0	0
17	54	F	14	0	0
18	55	F	10	0	0
19	29	F	12	?	0
20	32	F	10	0	0

8.6.1   Deductive Imputation

Some values may be imputed in the data editing, using logical relations among the variables. In Table 8.3, person 9 is missing the response for whether she was a victim of violent crime. But she had responded that she was not a victim of any crime, so the violent-crime response should be changed to 0.

Deductive imputation may sometimes be used in longitudinal surveys. If a woman has two children in year 1 and two children in year 3, but is missing the value for year 2, the logical value to impute would be 2.

8.6.2   Cell Mean Imputation

Respondents are divided into classes (cells) based on known variables, as in weighting-class adjustments. Then, the average of the values for the responding units in cell c, $\bar{y}_{cR}$, is substituted for each missing value. *Cell mean imputation* assumes that missing items are missing completely at random within the cells.

EXAMPLE 8.8   The four cells for our example are constructed using the variables age and sex. (In practice, of course, you would want to have many more individuals in each cell.)

		Age	
		≤ 34	≥ 35
	M	Persons 3, 5, 10, 14	Persons 1, 7, 8, 15, 16
Sex			
	F	Persons 4, 12, 13, 19, 20	Persons 2, 6, 9, 11, 17, 18

Persons 2 and 6, missing the value for years of education, would be assigned the mean value for the four women aged 35 or older who responded to the question: 12.25. The mean for each cell after imputation is the same as the mean of the respondents. The imputed value, however, is not one of the possible responses to the question about education. ∎

Mean imputation gives the same point estimates for means, totals, and proportions as the weighting-class adjustments. Mean imputation methods fail to reflect the variability of the nonrespondents, however—all missing observations in a class are given the same imputed value. The distribution of y will be distorted because of a "spike" at the value of the sample mean of the respondents. As a consequence, the estimated variance in the subclass will be too small.

To avoid the spike, a stochastic cell mean imputation could be used. If the response variable were approximately normally distributed, the missing values could be imputed with a randomly generated value from a normal distribution with mean $\bar{y}_{cR}$ and standard deviation s_{cR}.

Mean imputation, stochastic or otherwise, distorts relationships among different variables because imputation is done separately for each missing item. Sample correlations and other statistics are changed. Jinn and Sedransk (1989a; 1989b) discuss the effect of different imputation methods on secondary data analysis—for instance, for estimating a regression slope.

8.6.3   Hot-Deck Imputation

In *hot-deck imputation,* as in cell mean imputation and weighting-adjustment methods, the sample units are divided into classes. The value of one of the responding units in the class is substituted for each missing response. Often, the values for a set of related missing items are taken from the same donor, to preserve some of the multivariate relationships. The name *hot deck* is from the days when computer programs and data sets were punched on cards—the deck of cards containing the data set being analyzed was warmed by the card reader, so the term *hot deck* was used to refer to imputations made using the same data set. Fellegi and Holt (1976) discuss methods for data editing and hot-deck imputation with large surveys.

How is the donor unit to be chosen? Several methods are possible.

Sequential Hot-Deck Imputation   Some hot-deck imputation procedures impute the value in the same subgroup that was last read by the computer. This is partly a carryover from the card days of computers (imputation could be done in one pass) and partly a belief that, if the data are arranged in some geographic order, adjacent units in the same subgroup will tend to be more similar than randomly chosen units in the subgroup. One problem with using the value on the previous "card" is that often nonrespondents also tend to occur in clusters, so one person may be a donor multiple times, in a way that the sampler cannot control. One of the other hot-deck imputation methods is usually used today for most surveys.

In our example, person 19 is missing the response for crime victimization. Person 13 had the last response recorded in her subclass, so the value 1 is imputed.

Random Hot-Deck Imputation   A donor is randomly chosen from the persons in the cell with information on all missing items. To preserve multivariate relationships, usually values from the same donor are used for all missing items of a person.

In our small data set, person 10 is missing both variables for victimization. Persons 3, 5, and 14 in his cell have responses for both crime questions, so one of the three is chosen randomly as the donor. In this case, person 14 is chosen, and his values are imputed for both missing variables.

Nearest-Neighbor Hot-Deck Imputation   Define a distance measure between observations, and impute the value of a respondent who is "closest" to the person with the missing item, where closeness is defined using the distance function.

If age and sex are used for the distance function, so that the person of closest age with the same sex is selected to be the donor, the victimization responses of person 3 will be imputed for person 10.

8.6.4   Regression Imputation

Regression imputation predicts the missing value by using a regression of the item of interest on variables observed for all cases. A variation is *stochastic regression imputation,* in which the missing value is replaced by the predicted value from the regression model, plus a randomly generated error term.

We only have 18 complete observations for the response crime victimization (not really enough for fitting a model to our data set), but a logistic regression of the response with explanatory variable age gives the following model for predicted probability of victimization, $\hat{p}$:

$$\log \frac{\hat{p}}{1 - \hat{p}} = 2.5643 - 0.0896 \times \text{age}.$$

The predicted probability of being a crime victim for a 17-year-old is 0.74; because that is greater than a predetermined cutoff of 0.5, the value 1 is imputed for person 10.

EXAMPLE **8.9** Paulin and Ferraro (1994) discuss regression models for imputing income in the U.S. Consumer Expenditure Survey. Households selected for the interview component of the survey are interviewed each quarter for five consecutive quarters; in each interview, they are asked to recall expenditures for the previous 3 months. The data are used to relate consumer expenditures to characteristics such as family size and income; they are the source of reports that expenditures exceed income in certain income classes.

The Consumer Expenditure Survey conducts about 5000 interviews each year, as opposed to about 60,000 for the NCVS. This sample size is too small for hot-deck imputation methods, as it is less likely that suitable donors will be found for nonrespondents in a smaller sample. If imputation is to be done at all, a parametric model needs to be adopted. Paulin and Ferraro used multiple regression models to predict the log of family income (logarithms are used because the distribution of income is skewed) from explanatory variables including total expenditures and demographic variables. These models assume that income items are MAR, given the covariates. ∎

8.6.5 Cold-Deck Imputation

In *cold-deck imputation,* the imputed values are from a previous survey or other information, such as from historical data. (Since the data set serving as the source for the imputation is not the one currently running through the computer, the deck is "cold.") Little theory exists for the method. As with hot-deck imputation, cold-deck imputation is not guaranteed to eliminate selection bias.

8.6.6 Substitution

Substitution methods are similar to cold-deck imputation. Sometimes interviewers are allowed to choose a substitute while in the field; if the household selected for the sample is not at home, they try next door. Substitution may help reduce some nonresponse bias, as the household next door may be more similar to the nonresponding household than would be a household selected at random from the population. But the household next door is still a respondent; if the nonresponse is related to the characteristics of interest, there will still be nonresponse bias. An additional problem is that, since the interviewer is given discretion about which household to choose, the sample no longer has known probabilities of selection.

The 1975 Michigan Survey of Substance Abuse was taken to estimate the number of persons that used 16 types of substances in the previous year. The sample design was a stratified multistage sample with 2100 households. Three calls were made at a dwelling; then the house to the right was tried, then the house to the left. From the data, evidence shows that the substance-use rate increases as the required number of calls increases.

Some surveys select designated substitutes at the same time the sample units are selected. If a unit does not respond, then one of the designated substitutes is randomly selected. The National Longitudinal Study (see National Center of Educational Statistics 1977) used this method. This stratified, multistage sample of the high school graduating class of 1972 was intended to provide data on the educational experiences, plans, and attitudes of high school seniors. Four high schools were randomly selected from each of 600 strata. Two were designated for the sample, and the other two were saved as backups in case of nonresponse. Of the 1200 schools designated for the sample, 948 participated, 21 had no graduating seniors, and 231 either refused or were unable to participate. Investigators chose 122 schools from the backup group to substitute for the nonresponding schools. Follow-up studies showed a consistent 5% bias in a number of estimated totals, which was attributed to the use of substitute schools and to nonresponse.

Substitution has the added danger that efforts to contact the designated units may not be as great as if no "easy way out" was provided. If substitution is used, it should be reported in the results.

8.6.7 Multiple Imputation

In *multiple imputation,* each missing value is imputed $m(\geq 2)$ different times. Typically, the same stochastic model is used for each imputation. These create m different "data" sets with no missing values. Each of the m data sets is analyzed as if no imputation had been done; the different results give the analyst a measure of the additional variance due to the imputation. Multiple imputation with different models for nonresponse can give an idea of the sensitivity of the results to particular nonresponse models. See Rubin (1987; 1996) for details on implementing multiple imputation.

8.6.8 Advantages and Disadvantages of Imputation

Imputation creates a "clean," rectangular data set that can be analyzed by standard software. Analyses of different subsets of the data will produce consistent results. If the nonresponse is missing at random given the covariates used in the imputation procedure, imputation substantially reduces the bias due to item nonresponse. If parts of the data are confidential, the data collector can perform the imputation. The data collector has more information about the sample and population than is released to the public (for example, the collector may know the exact address for each sample member) and can often perform a better imputation using that information.

The foremost danger of using imputation is that future data analysts will not distinguish between the original and the imputed values. Ideally, the imputer should record which observations are imputed, how many times each nonimputed record is used as a donor, and which donor was used for a specific response imputed to a recipient. The imputed values may be good guesses, but they are not real data.

Variances computed using the data together with the imputed values are always too small, partly because of the artificial increase in the sample size and partly because the imputed values are treated as though they were really obtained in the data collection. The true variance will be larger than that estimated from a standard software package. Rao (1996) and Fay (1996) discuss methods for estimating the variances after imputation.

8.7
Parametric Models for Nonresponse*

Most of the methods for dealing with nonresponse assume that the nonresponse is *ignorable*—that is, conditionally on measured covariates, nonresponse is independent of the variables of interest. In this situation, rather than simply dividing units among different subclasses and adjusting weights, one can fit a superpopulation model. From the model, then, one predicts the values of the y's not in the sample. The model fitting is often iterative.

In a completely model-based approach, we develop a model for the complete data and add components to the model to account for the proposed nonresponse mechanism. Such an approach has many advantages over other methods: The modeling approach is flexible and can be used to include any knowledge about the nonresponse mechanism, the modeler is forced to state the assumptions about nonresponse explicitly in the model, and some of these assumptions can be evaluated. In addition, variance estimates that result from fitting the model account for the nonresponse, if the model is a good one.

EXAMPLE 8.10   Many people believe that spotted owls in Washington, Oregon, and California are threatened with extinction because timber harvesting in mature coniferous forests reduces their available habitat. Good estimates of the size of the spotted owl population are needed for reasoned debate on the issue.

In the sampling plan described by Azuma et al. (1990), a region of interest is divided into N sampling regions (psu's), and an SRS of n psu's is selected. Let $Y_i = 1$ if psu i is occupied by a pair of owls, and 0 otherwise. Assume that the Y_i's are independent and that $P(Y_i = 1) = p$, the true proportion of occupied psu's. If occupancy could be definitively determined for each psu, the proportion of psu's occupied could be estimated by the sample proportion $\bar{y}$. While a fixed number of visits can establish that a psu is occupied, however, a determination that a psu is unoccupied may be wrong—some owl pairs are "nonrespondents," and ignoring the nonresponse will likely result in a too-low estimate of percentage occupancy.

Azuma et al. (1990) propose using a geometric distribution for the number of visits required to discover the owls in an occupied unit, thus modeling the nonresponse. The assumptions for the model are (1) the probability of determining occupancy on the first visit, η, is the same for all psu's, (2) each visit to a psu is independent, and (3) visits can continue until an owl is sighted. A geometric distribution is commonly used for number of callbacks needed in surveys of people (see Potthoff et al. 1993).

Let X_i be the number of visits required to determine whether psu i is occupied or not. Under the geometric model,

$$P(X_i = x \mid Y_i = 1) = \eta(1 - \eta)^{x-1} \qquad \text{for } x = 1, 2, 3, \dots.$$

The budget of the U.S. Forest Service, however, does not allow for an infinite number of visits. Suppose a maximum of s visits are to be made to each psu. The random variable Y_i cannot be observed; the observable random variables are

$$V_i = \begin{cases} k & \text{if } Y_i = 1, X_i = k, \text{ and } X_i \le s. \\ 0 & \text{otherwise.} \end{cases}$$

$$U_i = \begin{cases} 1 & \text{if } Y_i = 1 \text{ and } X_i \le s. \\ 0 & \text{otherwise.} \end{cases}$$

Here, $\sum_{i \in S} U_i$ counts the number of psu's observed to be occupied, and $\sum_{i \in S} V_i$ counts the total number of visits made to occupied units. Using the geometric model, the probability that an owl is first observed in psu i on visit $k(\le s)$ is

$$P(V_i = k) = \eta(1 - \eta)^{k-1} p,$$

and the probability that an owl is observed on one of the s visits to psu i is

$$P(U_i = 1) = E[U_i] = [1 - (1 - \eta)^s] p.$$

Thus, the expected value of the sample proportion of occupied units, $E[\bar{U}]$, is $[1 - (1 - \eta)^s] p$ and is less than the proportion of interest p if $\eta < 1$. The geometric model agrees with the intuition that owls are missed in the s visits.

We find the maximum likelihood estimates of p and η under the assumption that all psu's are independent. The likelihood function

$$(\eta p)^{\sum_i u_i} (1 - \eta)^{\sum_i (v_i - u_i)} \left[1 - p + p(1 - \eta)^s\right]^{n - \sum_i u_i}$$

is maximized when

$$\hat{p} = \frac{\bar{u}}{1 - (1 - \hat{\eta})^s}$$

and when $\hat{\eta}$ solves

$$\frac{\bar{v}}{\bar{u}} = \frac{1}{\eta} - \frac{s(1 - \eta)^s}{1 - (1 - \eta)^s};$$

numerical methods are needed to calculate $\hat{\eta}$. Maximum likelihood theory also allows calculation of the asymptotic covariance matrix of the parameter estimates.

An SRS of 240 habitat psu's in California had the following results:

Visit number	1	2	3	4	5	6
Number of occupied psu's	33	17	12	7	7	5

A total of 81 psu's were observed to be occupied in six visits, so $\bar{u} = 81/240 = 0.3375$. The average number of visits made to occupied units was $\bar{v}/\bar{u} = 196/81 = 2.42$. Thus, the maximum likelihood estimates are $\hat{\eta} = 0.334$ and $\hat{p} = 0.370$; using the asymptotic covariance matrix from maximum likelihood theory, we estimate the variance of $\hat{p}$ by 0.00137. Thus, an approximate 95% confidence interval for the proportions of units that are occupied is 0.370 ± 0.072.

Incorporating the geometric model for number of visits gave a larger estimate of the proportion of units occupied. If the model does not describe the data, however, the estimate $\hat{p}$ will still be biased; if the model is poor, $\hat{p}$ may be a worse estimate of the occupancy rate than $\bar{u}$. If, for example, field investigators were more likely to find owls on later visits because they accumulate additional information on where to look, the geometric model would be inappropriate.

We need to check whether the geometric model adequately describes the number of visits needed to determine occupancy. Unfortunately, we cannot determine whether the model would describe the situation for units in which owls are not detected in six visits, as the data are missing. We can, however, use a χ^2 goodness-of-fit test to see whether data from the six visits made are fit by the model. Under the model, we expect $n\eta(1 - \eta)^{k-1}p$ of the psu's to have owls observed on visit k, and we plug in our estimates of p and η to calculate expected counts:

Visit	Observed Count	Expected Count
1	33	29.66
2	17	19.74
3	12	13.14
4	7	8.75
5, 6	12	9.71
Total	81	80.99

Visits 5 and 6 were combined into one category so that the expected cell count would be greater than 5. The χ^2 test statistic is 1.75, with p-value >0.05. There is no indication that the model is inadequate for the data we have. We cannot check its adequacy for the missing data, however. The geometric model assumes observations are independent and that an occupied psu would eventually be determined to be occupied if enough visits were made. We cannot check whether that assumption of the model is reasonable or not: If some wily owls will never be detected in any number of visits, $\hat{p}$ will still be too small. ■

To use models with nonresponse, you need (1) a thorough knowledge of mathematical statistics, (2) a powerful computer, and (3) knowledge of numerical methods for optimization. Commonly, maximum likelihood methods are used to estimate parameters, and the likelihood equations rarely have closed-form solutions. Calculation of estimates required numerical methods even for the simple model adopted for the owls, and that was an SRS with a simple geometric model for the response mechanism that allowed us to easily write down the likelihood function. Likelihood functions for more complex sampling designs or nonresponse mechanisms are much more difficult to construct (particularly if observations in the same cluster are considered dependent), and calculating estimates often requires intensive computations. Little and Rubin (1987) discuss likelihood-based methods for missing data in general. Stasny (1991) gives an example of using models to account for nonresponse.

8.8
What Is an Acceptable Response Rate?

Often an investigator will say, "I expect to get a 60% response rate in my survey. Is that acceptable, and will the survey give me valid results?" As we have seen in this chapter, the answer to that question depends on the nature of the nonresponse: If the nonrespondents are MCAR, then we can largely ignore the nonresponse and use the respondents as a representative sample of the population. If the nonrespondents tend to differ from the respondents, then the biases in the results from using only the respondents may make the entire survey worthless.

Many references give advice on cutoffs for acceptability of response rates. Babbie, for example, says: "I feel that a response rate of at least 50 percent is *adequate* for analysis and reporting. A response of at least 60 percent is *good*. And a response rate of 70 percent is *very good*" (1973, 165). I believe that giving such absolute guidelines for acceptable response rates is dangerous and has led many survey investigators to unfounded complacency about nonresponse; many examples exist of surveys with a 70% response rate whose results are flawed. The NCVS needs corrections for nonresponse bias even with a response rate of about 95%.

Be aware that response rates can be manipulated by defining them differently. Researchers often do not say how the response rate was calculated or may use an estimate of response rate that is smaller than it should be. Many surveys inflate the response rate by eliminating units that could not be located from the denominator. Very different results for response rate accrue, depending on which definition of response rate is used; all of the following have been used in surveys:

$$\frac{\text{number of completed interviews}}{\text{number of units in sample}}$$

$$\frac{\text{number of completed interviews}}{\text{number of units contacted}}$$

$$\frac{\text{completed interviews} + \text{ineligible units}}{\text{contacted units}}$$

$$\frac{\text{completed interviews}}{\text{contacted units} - \text{(ineligible units)}}$$

$$\frac{\text{completed interviews}}{\text{contacted units} - \text{(ineligible units)} - \text{refusals}}$$

Note that a "response rate" calculated using the last formula will be much higher than one calculated using the first formula because the denominator is smaller.

The guidelines for reporting response rates in Statistics Canada (1993) and Hidiroglou et al. (1993) provide a sensible solution for reporting response rates. They define *in-scope units* as those that belong to the target population, and *resolved units* as those units for which it is known whether or not they belong to the target population.[3] They suggest reporting a number of different response rates for a survey,

---

[3] If, for example, the target population is residential telephone numbers, it may be impossible to tell whether or not a telephone that rings but is not answered belongs to the target population; such a number would be an *unresolved unit*.

including the following:

- Out-of-scope rate: the ratio of the number of out-of-scope units to the number of resolved units
- No-contact rate: the ratio of the number of no-contacts and unresolved units to the number of in-scope and unresolved units
- Refusal rate: the ratio of number of refusals to the number of in-scope units
- Nonresponse rate: the ratio of number of nonrespondent and unresolved units to the number of in-scope and unresolved units

Different measures of response rates may be appropriate for different surveys, and I hesitate to recommend one "fits-all" definition of response rate. The quantities used in calculating response rate, however, should be defined for every survey. The following recommendations from the U.S. Office of Management and Budget's Federal Committee on Statistical Methodology, reported in González et al. (1994), are helpful:

Recommendation 1. Survey staffs should compute response rates in a uniform fashion over time and document response rate components on each edition of a survey.

Recommendation 2. Survey staffs for repeated surveys should monitor response rate components (such as refusals, not-at-homes, out-of-scopes, address not locatable, postmaster returns, etc.) over time, in conjunction with routine documentation of cost and design changes.

Recommendation 3. Response rate components should be published in survey reports; readers should be given definitions of response rates used, including actual counts, and commentary on the relevance of response rates to the quality of the survey data.

Recommendation 4. Some research on nonresponse can have real payoffs. It should be encouraged by survey administrators as a way to improve the effectiveness of data collection operations.

8.9
Exercises

1   Ryan et al. (1991) report results from the Ross Laboratories Mothers' Survey, a national mail survey investigating infant feeding in the United States. Questionnaires asking mothers about the type of milk fed to their infants during each of the first 6 months and about socioeconomic variables were mailed to a sample of mothers of 6-month-old infants. The authors state that the number of questionnaires mailed increased from 1984 to 1989: "In 1984, 56,894 questionnaires were mailed and 30,694 were returned. In 1989, 196,000 questionnaires were mailed and 89,640 were returned." Low-income families were oversampled in the survey design because they had the lowest response rates. Respondents were divided into subclasses defined by region, ethnic background, age, and education; weights were computed using information from the Bureau of the Census.

a   Which was used: weighting-class adjustments or poststratification?

b Oversampling the low-income families is a form of substitution. What are the advantages and drawbacks of using substitution in this survey?

c Weighted counts are "comparable with those published by the U.S. Bureau of the Census and the National Center for Health Statistics" on ethnicity, maternal age, income, education, employment, birth weight, region, and participation in the Women, Infants, and Children supplemental food program. Using the weighted counts, the investigators estimated that about 53% of mothers had one child, whereas the government data indicated that about 43% of mothers had one child. Does the agreement of weighted counts with official statistics indicate that the weighting corrects the nonresponse bias? Explain.

d Discuss the use of weighting in this survey. Can you think of any improvements?

2 Investigators selected an SRS of 200 high school seniors from a population of 2000 for a survey of TV-viewing habits, with an overall response rate of 75%. By checking school records, they were able to find the grade point average (GPA) for the nonrespondents and classify the sample accordingly:

GPA	Sample Size	Number of Respondents	Hours of TV $\bar{y}$	s_y
3.00–4.00	75	66	32	15
2.00–2.99	72	58	41	19
Below 2.00	53	26	54	25
Total	200	150		

a What is the estimate for the average number of hours of TV watched per week if only respondents are analyzed? What is the standard error of the estimate?

b Perform a χ^2 test for the null hypothesis that the three GPA groups have the same response rates. What do you conclude? What do your results say about the type of missing data: Do you think the data are MCAR? MAR? Nonignorable?

c Perform a one-way ANOVA to test the null hypothesis that the three GPA groups have the same mean level of TV viewing. What do you conclude? Does your ANOVA indicate that GPA would be a good variable for constructing weighting cells? Why, or why not?

d Use the GPA classification to adjust the weights of the respondents in the sample. What is the weighting-class estimate of the average viewing time?

e The population counts are 700 students with a GPA between 3 and 4; 800 students with a GPA between 2 and 3; and 500 students with a GPA less than 2. Use these population counts to construct a poststratified estimate of the mean viewing time.

f What other methods might you use to adjust for the nonresponse?

g What other variables might be collected that could be used in nonresponse models?

3 The following description and assessment of nonresponse is from a study of Hamilton, Ontario, home owners' attitudes on composting toilets:

The survey was carried out by means of a self-administered mail questionnaire. Twelve hundred questionnaires were sent to a randomly selected sample of house-dwellers.

Follow-up thank you notes were sent a week later. In total, 329 questionnaires were returned, representing a response rate of 27%. This was deemed satisfactory since many mail surveyors consider a 15 to 20% response rate to be a good return. (Wynia et al. 1993, 362)

Do you agree that the response rate of 27% is satisfactory? Suppose the investigators came to you for statistical advice on analyzing these data and designing a follow-up survey. What would you tell them?

4   Kosmin and Lachman (1993) had a question on religious affiliation included in 56 consecutive weekly household surveys; the subject of household surveys varied from week to week from cable TV use, to preference for consumer items, to political issues. After four callbacks, the unit nonresponse rate was 50%; an additional 2.3% refused to answer the religion question. The authors say:

> Nationally, the sheer number of interviews and careful research design resulted in a high level of precision ... Standard error estimates for our overall national sample show that we can be 95% confident that the figures we have obtained have an error margin, plus or minus, of less than 0.2%. This means, for example, that we are more than 95% certain that the figure for Catholics is in the range of 25.0% to 26.4% for the U.S. population. (p. 286)

a   Critique the preceding statement.

b   If you anticipated item nonresponse, do you think it would be better to insert the question of interest in different surveys each week, as was done here, or to use the same set of additional questions in each survey? Explain your answer. How would you design an experiment to test your conjecture?

5   Find an example of a survey in a popular newspaper or magazine. Is the nonresponse rate given? If so, how was it calculated? How do you think the nonresponse might have affected the conclusions of the survey? Give suggestions for how the journalist could deal with nonresponse problems in the article.

6   Find an example of a survey in a scholarly journal. How did the authors calculate the nonresponse rate? How did the survey deal with nonresponse? How do you think the nonresponse might have affected the conclusions of the study? Do you think the authors adequately account for potential nonresponse biases? What suggestions do you have for future studies?

7   The issue of nonresponse in the Winter Break Closure Survey (in the file winter.dat) was briefly mentioned in Exercise 20 of Chapter 4. What model is adopted for nonresponse when the formulas from stratified sampling are used to estimate the proportion of university employees who would answer yes to the question "Would you want to have Winter Break Closure again?" Do you think this is a reasonable model? How else might you model the effects of nonresponse in this survey? What additional information could be collected to adjust for unit nonresponse?

8   One issue in the U.S. statistical community in recent years is whether the American Statistical Association (ASA) should offer a certification process for its members so that statisticians meeting the qualifications could be designated as "Certified Statisticians." In 1994 the ASA surveyed its membership about this issue (data are in the file certify.dat). The survey was sent to all 18,609 members, and 5001 responses were

obtained. Results from the survey were reported in the October 1994 issue of *Amstat News.*

Assume that in 1994, the ASA membership had the following characteristics: Fifty-five percent have Ph.D.s and 38% have master's degrees; 29% work in industry, 34% work in academia, and 11% work in government. The cross-classification between education and workplace was unavailable.

a What are the response rates for the various subclasses of ASA membership? Are the nonrespondents MCAR? Do you think they are MAR?

b Use raking to adjust the weights for the six cells defined by education (Ph.D. or non-Ph.D.) and workplace (industry, academia, or other). Start with an initial weight of 18,609/5001 for each respondent. What assumptions must you make to use raking?

Estimate the proportion of ASA members who respond to each of categories 0 through 5 (variable *certify*), both with and without the raking weights. For this exercise, you may want to classify missing values in the "non-Ph.D." or the "other workplace" category.

c Do you think that opponents of certification are justified in using results from this survey to claim that a majority of the ASA membership opposes certification? Why, or why not?

9 The ACLS survey in Example 4.3 had nonresponse. Calculate the response rate in each stratum for the survey. What model was adopted for the nonresponse in Example 4.3? Is there evidence that the nonresponse rate varies among the strata, or that it is related to the percentage female membership?

10 Weights are used in the Survey of Youth in Custody (discussed in Example 7.4) to adjust for unit nonresponse. Use a hot-deck procedure to impute values for the variable measuring with whom the youth lived when growing up. What variables will you use to group the data into classes?

11 Repeat Exercise 10, using a regression imputation model.

12 Repeat Exercise 10, for the variable *have used illegal drugs*.

13 Repeat Exercise 11, for the variable *have used illegal drugs*.

14 The U.S. National Science Foundation Division of Science Resources Studies published results from the 1995 Survey of Doctorate Recipients in "Characteristics of Doctoral Scientists and Engineers in the United States: 1995."[4] How does this survey deal with nonresponse? Do you think that nonresponse bias is a problem for this survey?

15 How did the survey you critiqued in Exercise 1 of Chapter 7 deal with nonresponse? In your opinion, did the investigators adequately address the problems of nonresponse? What suggestions do you have for improvement?

16 Answer the questions in Exercise 15 for the survey you examined in Exercise 2 of Chapter 7.

---

[4]NSF Publication 97-319. Single copies are available free of charge from the Division of Science Resources Studies, National Science Foundation, Arlington, VA 22230; by e-mail from pubs@nsf.gov; or through the Internet (www.nsf.gov/sbe/srs).

17   Gnap (1995) conducted a survey on teacher workload, which was used in Exercise 16 of Chapter 5.

    a   The original survey was intended as a one-stage cluster sample. What was the overall response rate?

    b   Would you expect nonresponse bias in this study? If so, in which direction would you expect the bias to be? Which teachers do you think would be less likely to respond to the survey?

    c   Gnap also collected data on a random subsample of the nonrespondents in the "large" stratum, in the file teachnr.dat. How do the respondents and nonrespondents differ?

    d   Is there evidence of nonresponse bias when you compare the subsample of nonrespondents to the respondents in the original survey?

18   Not all of the parents surveyed in the study discussed in Exercise 17 of Chapter 5 returned the questionnaire. In the original sampling design, 50 questionnaires were mailed to parents of children in each school, for a total planned sample size of 500. We know that of the 9962 children who were not immunized during the campaign, the consent form had not been returned for 6698 of the children, the consent form had been returned but immunization refused for 2061 of the children, and 1203 children whose parents had consented were absent on immunization day.

    a   Calculate the response rate for each cluster. What is the correlation of the response rate and the percentage of respondents in the school who returned the consent form? Of the response rate and the percentage of respondents in each school who refused consent?

    b   Overall, about 67% (6698/9962) of the parents in the target population did not return the consent form. Using the data from the respondents, calculate a 95% confidence interval for the proportion of parents in the sample who did not return the consent form. Calculate two additional interval estimates for this quantity: one assuming that the missing values are all 0s and one assuming that the missing values are all 1s. What is the relation between your estimates and the population quantity?

    c   Repeat part (b), examining the percentage of parents who returned the form but refused to have their children immunized.

    d   Do you think nonresponse bias is a problem for this survey?

SURVEY Exercises

When running SURVEY, you may have noticed the prompt

```
ENTER DESIRED THREE NONRESPONSE RATES:
NOT-AT-HOMES, REFUSALS, RANDOM ANSWERS
```

If you enter

```
.3 0 0
```

in response, about 30% of the households in Stephens County will "not be home." If you enter

```
0 .3 0
```

about 30% of the households in Stephens County will refuse to say how much they would be willing to pay to subscribe to cable TV. If you enter

$$0 \quad 0 \quad .3$$

about 30% of the households in Stephens County will give random answers to certain questions.

19   Generate 200 random addresses for an SRS of the households in Stephens County. You will use this same list of addresses for all exercises in this chapter. Draw the full sample of size 200 specified by those addresses with no nonresponse. This sample gives the values you would have if all households responded. Estimate the means for the assessed value of the house and for each of questions 1 through 9 in Figure A.3 on page 417.

20   Using the list of addresses from Exercise 19, draw an SRS of size 200 with 30% unit nonresponse rate. You will find that about 30% of the households have the information on district, household number, and assessed value, but the words "NOT AT HOME" instead of answers to questions 1 through 9. Find the means for the assessed value of the house and for questions 1 through 9 for just the responding households. How do these compare with the results from the full SRS? Is there evidence of nonresponse bias?

21   Apply two-phase sampling to the nonrespondents, taking a random subsample of 30% of the nonrespondents. (Assume that all households respond to the second call.) Now estimate for the price a household is willing to pay for cable TV and the number of TVs, along with their standard errors. How do these estimates compare with those in Exercise 20?

22   Poststratify your sample from Exercise 20, using the strata you constructed in Chapter 4. Now calculate the poststratified estimates for the price a household is willing to pay for cable TV and the number of TVs. Are these closer to the values from Exercise 19? What are you assuming about the nature of the nonresponse when you use this weighting scheme? Do you think these assumptions are justified?

23   For the respondents, fit the linear regression model $y = a + bx$, where $y =$ price household is willing to pay for cable and $x =$ assessed value of the house. Now, for the nonrespondents, impute the predicted value from this regression model for the missing y values and use the "completed" data set to estimate the average price a household is willing to pay for cable. Compare this estimate to the previous one and to the estimate from the full data set. Is the standard error given by your statistical package correct here? Why, or why not?

24   Generate another set of data from the same address list, this time with a 30% item nonresponse rate. (The nonresponse parameters are 0, .3, 0.) What is the average price the respondents are willing to pay for cable? Using the respondents, develop a regression model for cable price based on the other variables. Impute the predicted values from this model for your missing observations and recalculate your estimate.

25   Perform another imputation on the data, this time using a sequential hot-deck procedure. Impute the value of the household immediately preceding the one with the missing item (if that one also has missing data, move up through the previous households until you find one that has the data and then impute that value). How does the value using this imputation scheme differ from the estimate in Exercise 24?

9

Variance Estimation in Complex Surveys*

Rejoice that under cloud and star
   The planet's more than Maine or Texas.
Bless the delightful fact there are
   Twelve months, nine muses, and two sexes;
And infinite in earth's dominions
Arts, climates, wonders, and opinions.

— Phyllis McGinley, "*In Praise of Diversity*" [1]

Population means and totals are easily estimated using weights. Estimating variances is more intricate: In Chapter 7 we noted that in a complex survey with several levels of stratification and clustering, variances for estimated means and totals are calculated at each level and then combined as the survey design is ascended. Poststratification and nonresponse adjustments also affect the variance.

In previous chapters, we have presented and derived variance formulas for a variety of sampling plans. Some of the variance formulas, such as those for simple random samples (SRSs), are relatively simple. Other formulas, such as $\hat{V}(\hat{t})$ from a two-stage cluster sample without replacement, are more complicated. All work for estimating variances of estimated totals. But we often want to estimate other quantities from survey data for which we have presented no variance formula. For example, in Chapter 3 we derived an approximate variance for a ratio of two means when an SRS is taken. What if you want to estimate a ratio, but the survey is not an SRS? How would you estimate the variance?

This chapter describes several methods for estimating variances of estimated totals and other statistics from complex surveys. Section 9.1 describes the commonly used linearization method for calculating variances of nonlinear statistics. Sections 9.2 and 9.3 present random group and resampling methods for calculating variances of linear and nonlinear statistics. Section 9.4 describes the calculation of generalized variance functions, and Section 9.5 describes constructing confidence intervals. These methods are described in more detail by Wolter (1985) and Rao (1988); Rao (1997) and Rust and Rao (1996) summarize recent work.

9.1
Linearization (Taylor Series) Methods

Most of the variance formulas in Chapters 2 through 6 were for estimates of means and totals. Those formulas can be used to find variances for any linear combination of estimated means and totals. If $\hat{t}_1, \ldots, \hat{t}_k$ are unbiased estimates of k totals in the population, then

$$V\left(\sum_{i=1}^{k} a_i \hat{t}_i\right) = \sum_{i=1}^{k} a_i^2 V(\hat{t}_i) + 2 \sum_{i=1}^{k} \sum_{j=i+1}^{k} a_i a_j \operatorname{Cov}(\hat{t}_i, \hat{t}_j). \tag{9.1}$$

The result can be expressed equivalently using unbiased estimates of k means in the population:

$$V\left(\sum_{i=1}^{k} a_i \hat{\bar{y}}_i\right) = \sum_{i=1}^{k} a_i^2 V(\hat{\bar{y}}_i) + 2 \sum_{i=1}^{k} \sum_{j=i+1}^{k} a_i a_j \operatorname{Cov}(\hat{\bar{y}}_i, \hat{\bar{y}}_j).$$

Thus, if t_1 is the total number of dollars robbery victims reported stolen, t_2 is the number of days of work robbery victims missed because of the crime, and t_3 is the total medical expenses incurred by robbery victims, one measure of financial consequences of robbery (assuming \$150 per day of work lost) might be $\hat{t}_1 + 150\hat{t}_2 + \hat{t}_3$. By (9.1), the variance is

$$V(\hat{t}_1 + 150\hat{t}_2 + \hat{t}_3) = V(\hat{t}_1) + 150^2 V(\hat{t}_2) + V(\hat{t}_3)$$
$$+ 300 \operatorname{Cov}(\hat{t}_1, \hat{t}_2) + 2 \operatorname{Cov}(\hat{t}_1, \hat{t}_3) + 300 \operatorname{Cov}(\hat{t}_2, \hat{t}_3).$$

This expression requires calculation of six variances and covariances; it is easier computationally to define a new variable at the observation unit level,

$$q_i = y_{i1} + 150 y_{i2} + y_{i3},$$

and find $V(\hat{t}_q) = V\left(\sum_{i \in S} w_i q_i\right)$ directly.

Suppose, though, that we are interested in the proportion of total loss accounted for by the stolen property, t_1/t_q. This is not a linear statistic, as t_1/t_q cannot be expressed in the form $a_1 t_1 + a_2 t_q$ for constants a_i. But Taylor's theorem from calculus allows us to **linearize** a smooth nonlinear function $h(t_1, t_2, \ldots, t_k)$ of the population totals; Taylor's theorem gives the constants $a_0, a_1, \ldots, a_k$ so that

$$h(t_1, \ldots, t_k) \approx a_0 + \sum_{i=1}^{k} a_i t_i.$$

Then $V[h(\hat{t}_1, \ldots, \hat{t}_k)]$ may be approximated by $V\left(\sum_{i=1}^{k} a_i \hat{t}_i\right)$, which we know how to calculate using (9.1).

Taylor series approximations have long been used in statistics to calculate approximate variances. Woodruff (1971) illustrates their use in complex surveys. Binder

FIGURE **9.1**

The function $h(x) = x(1 - x)$, along with the tangent to the function at point p. If $\hat{p}$ is close to p, then $h(\hat{p})$ will be close to the tangent line. The slope of the tangent line is $h'(p) = 1 - 2p$.

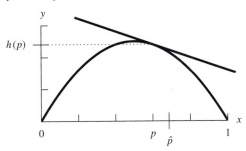

(1983) gives a more rigorous treatment of Taylor series methods for complex surveys and tells how to use linearization when the parameter of interest θ solves $h(\theta, t_1, \ldots, t_k) = 0$, but θ is not necessarily expressed as an explicit function of $t_1, \ldots, t_k$.

EXAMPLE **9.1**   The quantity $\theta = p(1 - p)$, where p is a population proportion, may be estimated by $\hat{\theta} = \hat{p}(1 - \hat{p})$. Assume that $\hat{p}$ is an unbiased estimator of p and that $V(\hat{p})$ is known. Let $h(x) = x(1 - x)$, so $\theta = h(p)$ and $\hat{\theta} = h(\hat{p})$. Now h is a nonlinear function of x, but the function can be approximated at any nearby point a by the tangent line to the function; the slope of the tangent line is given by the derivative, as illustrated in Figure 9.1.

The first-order version of Taylor's theorem states that if the second derivative of h is continuous, then

$$h(x) = h(a) + h'(a)(x - a) + \int_a^x (x - t)h''(t)dt;$$

under conditions commonly satisfied in statistics, the last term is small relative to the first two, and we use the approximation

$$h(\hat{p}) \approx h(p) + h'(p)(\hat{p} - p)$$
$$= p(1 - p) + (1 - 2p)(\hat{p} - p).$$

Then,

$$V[h(\hat{p})] \approx (1 - 2p)^2 V(\hat{p} - p),$$

and $V(\hat{p})$ is known, so the approximate variance of $h(\hat{p})$ can be calculated.   ∎

The following are the basic steps for constructing a linearization estimator of the variance of a nonlinear function of means or totals:

1   Express the quantity of interest as a function of means or totals of variables measured or computed in the sample. In general, $\theta = h(t_1, t_2, \ldots, t_k)$ or $\theta = h(\bar{y}_{1U}, \ldots, \bar{y}_{kU})$. In Example 9.1, $\theta = h(\bar{y}_U) = h(p) = p(1 - p)$.

2   Find the partial derivatives of h with respect to each argument. The partial derivatives, evaluated at the population quantities, form the linearizing constants a_i.

3 Apply Taylor's theorem to linearize the estimate:

$$h(\hat{t}_1, \hat{t}_2, \ldots, \hat{t}_k) \approx h(t_1, t_2, \ldots, t_k) + \sum_{j=1}^{k} a_j(\hat{t}_j - t_j),$$

where

$$a_j = \left. \frac{\partial h(c_1, c_2, \ldots, c_k)}{\partial c_j} \right|_{t_1, t_2, \ldots, t_k}.$$

4 Define the new variable q by

$$q_i = \sum_{j=1}^{k} a_j y_{ij}.$$

Now find the estimated variance of $\hat{t}_q = \sum_{i \in S} w_i q_i$. This will generally approximate the variance of $h(\hat{t}_1, \ldots, \hat{t}_k)$.

EXAMPLE 9.2 We used linearization methods to approximate the variance of the ratio and regression estimators in Chapter 3. In Chapter 3, we used an SRS, estimator $\hat{B} = \bar{y}/\bar{x} = \hat{t}_y/\hat{t}_x$, and the approximation

$$\hat{B} - B = \frac{\bar{y} - B\bar{x}}{\bar{x}} \approx \frac{\bar{y} - B\bar{x}}{\bar{x}_U} = \sum_{i \in S} \frac{y_i - Bx_i}{n\bar{x}_U}.$$

The resulting approximation to the variance was

$$V[\hat{B} - B] \approx \frac{1}{n^2 \bar{x}_U^2} V \left[\sum_{i \in S} (y_i - Bx_i) \right].$$

Essentially, we used Taylor's theorem to obtain this approximation. The steps below give the same result.

1 Express B as a function of the population totals. Let $h(c, d) = d/c$, so

$$B = h(t_x, t_y) = \frac{t_y}{t_x} \quad \text{and} \quad \hat{B} = h(\hat{t}_x, \hat{t}_y) = \frac{\hat{t}_y}{\hat{t}_x}.$$

Assume that the sample estimates $\hat{t}_x$ and $\hat{t}_y$ are unbiased.

2 The partial derivatives are

$$\frac{\partial h(c, d)}{\partial c} = \frac{-d}{c^2} \quad \text{and} \quad \frac{\partial h(c, d)}{\partial d} = \frac{1}{c};$$

evaluated at $c = t_x$ and $d = t_y$, these are $-t_y/t_x^2$ and $1/t_x$.

3 By Taylor's theorem,

$$\hat{B} = h(\hat{t}_x, \hat{t}_y)$$

$$\approx h(t_x, t_y) + \left. \frac{\partial h(c, d)}{\partial c} \right|_{t_x, t_y} (\hat{t}_x - t_x) + \left. \frac{\partial h(c, d)}{\partial d} \right|_{t_x, t_y} (\hat{t}_y - t_y).$$

Using the partial derivatives from step 2,

$$\hat{B} - B \approx -\frac{t_y}{t_x^2} (\hat{t}_x - t_x) + \frac{1}{t_x} (\hat{t}_y - t_y).$$

4 The approximate mean squared error of $\hat{B}$ is

$$E[(\hat{B} - B)^2] \approx E\left[\left\{-\frac{t_y}{t_x^2}(\hat{t}_x - t_x) + \frac{1}{t_x}(\hat{t}_y - t_y)\right\}^2\right]$$

$$= \frac{t_y^2}{t_x^4}V(\hat{t}_x) + \frac{1}{t_x^2}V(\hat{t}_y) - 2\frac{t_y}{t_x^3}\,\text{Cov}(\hat{t}_x, \hat{t}_y) \qquad \textbf{(9.2)}$$

$$= \frac{1}{t_x^2}\{B^2 V(\hat{t}_x) + V(\hat{t}_y) - 2B\,\text{Cov}(\hat{t}_x, \hat{t}_y)\}.$$

We can substitute estimated values for B, for the variances and covariance, and possibly for t_x from the particular sampling scheme used into (9.2). Alternatively, we would define

$$q_i = \frac{1}{t_x}[y_i - \hat{B}x_i]$$

and find $\hat{V}(\hat{t}_q)$.

If the sampling design is an SRS of size n, then $V(\hat{t}_x) = N^2(1 - n/N)S_x^2/n$, $V(\hat{t}_y) = N^2(1 - n/N)S_y^2/n$, and $\text{Cov}(\hat{t}_x, \hat{t}_y) = N^2(1 - n/N)RS_xS_y/n$. ∎

Advantages If the partial derivatives are known, linearization almost always gives a variance estimate for a statistic and can be applied in general sampling designs. Linearization methods have been used for a long time in statistics, and the theory is well developed. Software exists for calculating linearization variance estimates for many nonlinear functions of interest, such as ratios and regression coefficients; some software will be discussed in Section 9.6.

Disadvantages Calculations can be messy, and the method is difficult to apply for complex functions involving weights. You must either find analytical expressions for the partial derivatives of h or calculate the partial derivatives numerically. A separate variance formula is needed for each nonlinear statistic that is estimated, and that can require much special programming; a different method is needed for each statistic. In addition, not all statistics can be expressed as a smooth function of the population totals—the median and other quantiles, for example, do not fit into this framework. The accuracy of the linearization approximation depends on the sample size—the estimate of the variance is often biased downward if the sample is not large enough.

9.2
Random Group Methods

9.2.1 Replicating the Survey Design

Suppose the basic survey design is replicated independently R times. *Independently* here means that after each sample is drawn, the sampled units are replaced in the population so that they are available for later samples. Then, the R replicate samples produce R independent estimates of the quantity of interest; the variability among those estimates can be used to estimate the variance of $\hat{\theta}$. Mahalanobis (1946) describes early uses of the method, which he calls "replicated networks of sample units" and "interpenetrating sampling."

Let

$$\theta = \text{parameter of interest}$$

$$\hat{\theta}_r = \text{estimate of } \theta \text{ calculated from } r\text{th replicate}$$

$$\tilde{\theta} = \sum_{r=1}^{R} \frac{\hat{\theta}_r}{R}.$$

If $\hat{\theta}_r$ is an unbiased estimate of θ, so is $\tilde{\theta}$, and

$$\hat{V}_1(\tilde{\theta}) = \frac{1}{R} \frac{\sum_{r=1}^{R}(\hat{\theta}_r - \tilde{\theta})^2}{R-1} \tag{9.3}$$

is an unbiased estimate of $V(\tilde{\theta})$. Note that $\hat{V}_1(\tilde{\theta})$ is the sample variance of the R independent estimates of θ divided by R—the usual estimate of the variance of a sample mean.

EXAMPLE **9.3**   *The 1991 Information Please Almanac* listed enrollment, tuition, and room-and-board costs for every 4-year college in the United States. Suppose we want to estimate the ratio of nonresident tuition to resident tuition for public colleges and universities in the United States. In a typical implementation of the random group method, independent samples would be chosen using the same design and $\hat{\theta}$ found for each sample. Let's take four SRSs of size 10 each (Table 9.1). The four SRSs are without replacement, but the same college can appear in more than one of the four SRSs.

For this example,

$$\hat{\theta}_r = \frac{\text{average of nonresident tuitions for sample } r}{\text{average of resident tuitions for sample } r}.$$

Thus, $\hat{\theta}_1 = 2.3288$, $\hat{\theta}_2 = 2.5802$, $\hat{\theta}_3 = 2.4591$, and $\hat{\theta}_4 = 3.1110$. The sample average of the four independent estimates of θ is $\tilde{\theta} = 2.6198$. The sample standard deviation (SD) of the four estimates is 0.343, so the standard error (SE) of $\tilde{\theta}$ is $0.343/\sqrt{4} = 0.172$. The estimated variance is based on four independent observations, so a 95% confidence interval (CI) for the ratio is

$$2.6198 \pm 3.18(0.172)$$

where 3.18 is the appropriate t critical value with 3 degrees of freedom (df). Note that the small number of replicates causes the confidence interval to be wider than it would be if more replicate samples were taken, because the estimate of the variance with 3 df is not very stable.   ■

9.2.2   Dividing the Sample into Random Groups

In practice, subsamples are not usually drawn independently, but the complete sample is selected according to the survey design. The complete sample is then divided into R groups so that each group forms a miniature version of the survey, mirroring the sample design. The groups are then treated as though they are independent replicates of the basic survey design.

TABLE **9.1**

Four SRSs of Colleges, Used in Example 9.3

College	Enrollment	Resident Tuition	Nonresident Tuition
Columbus College	3,482	1,365	3,747
Southeastern Massachusetts University	5,354	1,677	4,983
U.S. Naval Academy	4,500	1,500	1,500
Athens State College	1,392	1,080	2,160
University of South Alabama	9,195	1,875	2,475
Virginia State University	3,308	3,071	5,135
SUNY College of Technology–Farmingdale	10,802	1,542	3,950
University of Houston	18,684	930	4,050
CUNY–Lehman College	7,841	1,340	4,140
Austin Peay State University	4,784	1,210	4,166
Average	6,934.2	1,559	3,630.6

College	Enrollment	Resident Tuition	Nonresident Tuition
SUNY–New Paltz	4,696	1,495	4,095
Indiana University–Southeast	4,931	1,350	3,342
University of Wisconsin–Platteville	5,080	1,658	4,740
University of California–Santa Barbara	16,853	1,578	5,799
Weber State College	12,783	1,308	3,513
Kennesaw College	8,404	1,296	3,678
South Dakota State University	6,366	1,835	3,363
Dickinson State University	1,402	1,659	4,731
Chadron State College	2,143	1,361	2,036
University of Alaska–Fairbanks	7,028	1,512	3,540
Average	6,968.6	1,505.2	3,883.7

College	Enrollment	Resident Tuition	Nonresident Tuition
University of Alaska–Anchorage	4,091	941	2,765
University of Maine–Fort Kent	594	1,710	4,140
Southern University–Baton Rouge	9,448	1,354	2,876
University of Oregon	13,786	1,782	5,043
Virginia State University	3,308	3,071	5,135
Glenville State College	2,185	1,150	2,900
Winston-Salem State University	2,532	896	4,268
Framingham State College	3,359	1,701	4,729
SUNY–Old Westbury	3,999	1,350	3,292
Northwest Missouri State University	4,600	1,320	2,415
Average	4,790.2	1,527.5	3,756.3

College	Enrollment	Resident Tuition	Nonresident Tuition
Central Washington University	6,398	1,674	5,712
Worcester State College	3,600	1,296	3,792
University of California–Davis	17,202	1,676	7,592
Sam Houston State University	12,359	1,060	4,180
University of Texas–Tyler	2,335	861	3,695
Southeastern Oklahoma State University	3,616	804	1,992
University of Southern Colorado	3,909	1,536	5,275
Pennsylvania State University	31,251	3,754	7,900
East Central University	3,606	1,200	4,140
Univ of Arkansas–Monticello	1,854	1,410	3,230
Average	8,613	1,527.1	4,750.8

If the sample is an SRS of size n, the groups are formed by randomly apportioning the n observations into R groups, each of size n/R. These pseudo–random groups are not quite independent replicates because an observation unit can only appear in one of the groups; if the population size is large relative to the sample size, however, the groups can be treated as though they are independent replicates. In a cluster sample, the psu's are randomly divided among the R groups. The psu takes all its observation units with it to the random group, so each random group is still a cluster sample. In a stratified multistage sample, a random group contains a sample of psu's from each stratum. Note that if k psu's are sampled in the smallest stratum, at most k random groups can be formed.

If θ is a nonlinear quantity, $\tilde{\theta}$ will not, in general, be the same as $\hat{\theta}$, the estimator calculated directly from the complete sample. For example, in ratio estimation, $\tilde{\theta} = (1/R)\sum_{r=1}^{R} \bar{y}_r/\bar{x}_r$, while $\hat{\theta} = \hat{\bar{y}}/\hat{\bar{x}}$. Usually, $\hat{\theta}$ is a more natural estimator than $\tilde{\theta}$. Sometimes $\hat{V}_1(\tilde{\theta})$ from (9.3) is used to estimate $V(\hat{\theta})$, although it is an overestimate. Another estimator of the variance is slightly larger but is often used:

$$\hat{V}_2(\hat{\theta}) = \frac{1}{R} \frac{\sum_{r=1}^{R}(\hat{\theta}_r - \hat{\theta})^2}{R - 1}. \tag{9.4}$$

EXAMPLE 9.4   The 1987 Survey of Youth in Custody, discussed in Example 7.4, was divided into seven random groups. The survey design had 16 strata. Strata 6–16 each consisted of one facility ($=$ psu), and these facilities were sampled with probability 1. In strata 1–5, facilities were selected with probability proportional to number of residents in the 1985 Children in Custody census.

It was desired that each random group be a miniature of the sampling design. For each self-representing facility in strata 6–16, random group numbers were assigned as follows: The first resident selected from the facility was assigned a number between 1 and 7. Let's say the first resident was assigned number 6. Then the second resident in that facility would be assigned number 7, the third resident 1, the fourth resident 2, and so on. In strata 1–5, all residents in a facility (psu) were assigned to the same random group. Thus, for the seven facilities sampled in stratum 2, all residents in facility 33 were assigned random group number 1, all residents in facility 9 were assigned random group number 2 (etc.). Seven random groups were formed because strata 2–5 each have seven psu's.

After all random group assignments were made, each random group had the same basic design as the original sample. Random group 1, for example, forms a stratified sample in which a (roughly) random sample of residents is taken from the self-representing facilities in strata 6–16, and a pps (probability proportional to size) sample of facilities is taken from each of strata 1–5.

To use the random group method to estimate a variance, $\hat{\theta}$ is calculated for each random group. The following table shows estimates of mean age of residents for each random group; each estimate was calculated using

$$\hat{\theta}_r = \frac{\sum w_i y_i}{\sum w_i},$$

where w_i is the final weight for resident i and the summations are over observations in random group r.

Random Group Number	Estimate of Mean Age, $\hat{\theta}_r$
1	16.55
2	16.66
3	16.83
4	16.06
5	16.32
6	17.03
7	17.27

The seven estimates, $\hat{\theta}_r$, are treated as independent observations, so

$$\tilde{\theta} = \frac{1}{7} \sum_{r=1}^{7} \hat{\theta}_r = 16.67$$

and

$$\hat{V}_1(\tilde{\theta}) = \frac{1}{7} \frac{\sum_{r=1}^{7}(\hat{\theta}_r - \tilde{\theta})^2}{6} = \frac{0.1704}{7} = 0.024.$$

Using the entire data set, we calculate $\hat{\theta} = 16.64$ with

$$\hat{V}_2(\hat{\theta}) = \frac{1}{7} \frac{\sum_{r=1}^{7}(\hat{\theta}_r - \hat{\theta})^2}{6} = \frac{0.1716}{7} = 0.025.$$

We can use either $\tilde{\theta}$ or $\hat{\theta}$ to calculate confidence intervals; using $\hat{\theta}$, a 95% CI for mean age is

$$16.64 \pm 2.45 \sqrt{0.025} = [16.3, 17.0]$$

(2.45 is the t critical value with 6 df). ∎

Advantages No special software is necessary to estimate the variance, and it is very easy to calculate the variance estimate. The method is well suited to multiparameter or nonparametric problems. It can be used to estimate variances for percentiles and nonsmooth functions, as well as variances of smooth functions of the population totals. Random group methods are easily used after weighting adjustments for nonresponse and undercoverage.

Disadvantages The number of random groups is often small—this gives imprecise estimates of the variances. Generally, you would like at least ten random groups to obtain a more stable estimate of the variance and to avoid inflating the confidence interval by using the t distribution rather than the normal distribution. Setting up the random groups can be difficult in complex designs, as each random group must have the same design structure as the complete survey. The survey design may limit the number of random groups that can be constructed; if two psu's are selected in each stratum, then only two random groups can be formed.

9.3
Resampling and Replication Methods

Random group methods are easy to compute and explain but are unstable if a complex sample can only be split into a small number of groups. Resampling methods treat the sample as if it were itself a population; we take different samples from this new "population" and use the subsamples to estimate a variance. All methods in this section calculate variance estimates for a sample in which psu's are sampled with replacement. If psu's are sampled without replacement, these methods may still be used but are expected to overestimate the variance and result in conservative confidence intervals.

9.3.1   Balanced Repeated Replication (BRR)

Some surveys are stratified to the point that only two psu's are selected from each stratum. This gives the highest degree of stratification possible while still allowing calculation of variance estimates in each stratum.

9.3.1.1   BRR in a Stratified Random Sample

We illustrate BRR for a problem we already know how to solve—calculating the variance for $\bar{y}_{\text{str}}$ from a stratified random sample. More complicated statistics from stratified multistage samples are discussed in Section 9.3.1.2.

Suppose an SRS of two observation units is chosen from each of seven strata. We arbitrarily label one of the sampled units in stratum h as y_{h1} and the other as y_{h2}. The sampled values are given in Table 9.2.

The stratified estimate of the population mean is

$$\bar{y}_{\text{str}} = \sum_{h=1}^{H} \frac{N_h}{N} \bar{y}_h = 4451.7.$$

Ignoring the fpc's (finite population corrections) in Equation (4.5) gives the variance

TABLE   **9.2**

A Small Stratified Random Sample, Used to Illustrate BRR

Stratum	$\dfrac{N_h}{N}$	y_{h1}	y_{h2}	$\bar{y}_h$	$y_{h1} - y_{h2}$
1	.30	2,000	1,792	1,896	208
2	.10	4,525	4,735	4,630	−210
3	.05	9,550	14,060	11,805	−4,510
4	.10	800	1,250	1,025	−450
5	.20	9,300	7,264	8,282	2,036
6	.05	13,286	12,840	13,063	446
7	.20	2,106	2,070	2,088	36

estimate

$$\hat{V}_{\text{str}}(\bar{y}_{\text{str}}) = \sum_{h=1}^{H} \left(\frac{N_h}{N}\right)^2 \frac{s_h^2}{n_h};$$

when $n_h = 2$, as here, $s_h^2 = (y_{h1} - y_{h2})^2/2$, so

$$\hat{V}_{\text{str}}(\bar{y}_{\text{str}}) = \sum_{h=1}^{H} \left(\frac{N_h}{N}\right)^2 \frac{(y_{h1} - y_{h2})^2}{4}.$$

Here, $\hat{V}_{\text{str}}(\bar{y}_{\text{str}}) = 55{,}892.75$. This may overestimate the variance if sampling is without replacement.

To use the random group method, we would randomly select one of the observations in each stratum for group 1 and assign the other to group 2. The groups in this situation are half-samples. For example, group 1 might consist of $\{y_{11}, y_{22}, y_{32}, y_{42}, y_{51}, y_{62}, y_{71}\}$ and group 2 of the other seven observations. Then,

$$\hat{\theta}_1 = (.3)(2000) + (.1)(4735) + \cdots + (.2)(2106) = 4824.7,$$

and

$$\hat{\theta}_2 = (.3)(1792) + (.1)(4525) + \cdots + (.2)(2070) = 4078.7.$$

The random group estimate of the variance—in this case, 139,129—has only 1 df for a two-psu-per-stratum design and is unstable in practice. If a different assignment of observations to groups had been made—had, for example, group 1 consisted of y_{h1} for strata 2, 3, and 5 and y_{h2} for strata 1, 4, 6, and 7—then $\hat{\theta}_1 = 4508.6$, $\hat{\theta}_2 = 4394.8$, and the random group estimate of the variance would have been 3238.

McCarthy (1966; 1969) notes that altogether 2^H possible half-samples could be formed and suggests using a balanced sample of the 2^H possible half-samples to estimate the variance. **Balanced repeated replication** uses the variability among R replicate half-samples that are selected in a balanced way to estimate the variance of $\hat{\theta}$.

To define balance, let's introduce the following notation. Half-sample r can be defined by a vector $\boldsymbol{\alpha}_r = (\alpha_{r1}, \ldots, \alpha_{rH})$: Let

$$y_h(\boldsymbol{\alpha}_r) = \begin{cases} y_{h1} & \text{if } \alpha_{rh} = 1. \\ y_{h2} & \text{if } \alpha_{rh} = -1. \end{cases}$$

Equivalently,

$$y_h(\boldsymbol{\alpha}_r) = \frac{\alpha_{rh} + 1}{2} y_{h1} - \frac{\alpha_{rh} - 1}{2} y_{h2}.$$

If group 1 contains observations $\{y_{11}, y_{22}, y_{32}, y_{42}, y_{51}, y_{62}, y_{71}\}$ as above, then $\boldsymbol{\alpha}_1 = (1, -1, -1, -1, 1, -1, 1)$. Similarly, $\boldsymbol{\alpha}_2 = (-1, 1, 1, 1, -1, 1, -1)$. The set of R replicate half-samples is **balanced** if

$$\sum_{r=1}^{R} \alpha_{rh}\alpha_{rl} = 0 \qquad \text{for all } l \neq h.$$

Let $\hat{\theta}(\boldsymbol{\alpha}_r)$ be the estimate of interest, calculated the same way as $\hat{\theta}$ but using only the observations in the half-sample selected by $\boldsymbol{\alpha}_r$. For estimating the mean of a stratified sample, $\hat{\theta}(\boldsymbol{\alpha}_r) = \sum_{h=1}^{H}(N_h/N)\bar{y}_h(\boldsymbol{\alpha}_r)$. Define the BRR variance estimator to be

$$\hat{V}_{\mathrm{BRR}}(\hat{\theta}) = \frac{1}{R}\sum_{r=1}^{R}[\hat{\theta}(\boldsymbol{\alpha}_r) - \hat{\theta}]^2.$$

If the set of half-samples is balanced, then $\hat{V}_{\mathrm{BRR}}(\bar{y}_{\mathrm{str}}) = \hat{V}_{\mathrm{str}}(\bar{y}_{\mathrm{str}})$. (The proof of this is left as Exercise 6.) If, in addition, $\sum_{r=1}^{R}\alpha_{rh} = 0$ for $h = 1,\ldots,H$, then $\frac{1}{R}\sum_{r=1}^{R}\bar{y}_{\mathrm{str}}(\boldsymbol{\alpha}_r) = \bar{y}_{\mathrm{str}}$.

For our example, the set of α's in the following table meets the balancing condition $\sum_{r=1}^{8}\alpha_{rh}\alpha_{rl} = 0$, for all $l \neq h$. The 8×7 matrix of -1's and 1's has orthogonal columns; in fact, it is the design matrix (excluding the column of 1's) for a fractional factorial design (Box et al. 1978). Designs described by Plackett and Burman (1946) give matrices with k orthogonal columns, for k a multiple of 4; Wolter (1985) explicitly lists some of these matrices.

		\multicolumn{7}{c}{Stratum (h)}
		1	2	3	4	5	6	7
	α_1	−1	−1	−1	1	1	1	−1
	α_2	1	−1	−1	−1	−1	1	1
	α_3	−1	1	−1	−1	1	−1	1
Half-Sample	α_4	1	1	−1	1	−1	−1	−1
(r)	α_5	−1	−1	1	1	−1	−1	1
	α_6	1	−1	1	−1	1	−1	−1
	α_7	−1	1	1	−1	−1	1	−1
	α_8	1	1	1	1	1	1	1

The estimate from each half-sample, $\hat{\theta}(\boldsymbol{\alpha}_r) = \bar{y}_{\mathrm{str}}(\boldsymbol{\alpha}_r)$ is calculated from the data in Table 9.2.

Half-Sample	$\hat{\theta}(\boldsymbol{\alpha}_r)$	$[\hat{\theta}(\boldsymbol{\alpha}_r) - \hat{\theta}]^2$
1	4732.4	78,792.5
2	4439.8	141.6
3	4741.3	83,868.2
4	4344.3	11,534.8
5	4084.6	134,762.4
6	4592.0	19,684.1
7	4123.7	107,584.0
8	4555.5	10,774.4
Average	4451.7	55,892.8

The average of $[\hat{\theta}(\boldsymbol{\alpha}_r) - \hat{\theta}]^2$ for the eight replicate half-samples is 55,892.75, which is the same as $\hat{V}_{\mathrm{str}}(\bar{y}_{\mathrm{str}})$ for sampling with replacement. Note that we can do the BRR estimation above by creating a new variable of weights for each replicate half-sample. The sampling weight for observation i in stratum h is $w_{hi} = N_h/n_h$,

and

$$\bar{y}_{\text{str}} = \frac{\displaystyle\sum_{h=1}^{H}\sum_{i=1}^{2} w_{hi} y_{hi}}{\displaystyle\sum_{h=1}^{H}\sum_{i=1}^{2} w_{hi}}.$$

In BRR with a stratified random sample, we eliminate one of the two observations in stratum h to calculate $y_h(\alpha_r)$. To compensate, we double the weight for the remaining observation. Define

$$w_{hi}(\alpha_r) = \begin{cases} 2w_{hi} & \text{if observation } i \text{ of stratum } h \text{ is in} \\ & \text{the half-sample selected by } \alpha_r. \\ 0 & \text{otherwise.} \end{cases}$$

Then,

$$\bar{y}_{\text{str}}(\alpha_r) = \frac{\displaystyle\sum_{h=1}^{H}\sum_{i=1}^{2} w_{hi}(\alpha_r) y_{hi}}{\displaystyle\sum_{h=1}^{H}\sum_{i=1}^{2} w_{hi}(\alpha_r)}.$$

Similarly, for any statistic $\hat{\theta}$ calculated using the weights w_{hi}, $\hat{\theta}(\alpha_r)$ is calculated exactly the same way, but using the new weights $w_{hi}(\alpha_r)$. Using the new weight variables instead of selecting the subset of observations simplifies calculations for surveys with many response variables—the same column $w(\alpha_r)$ can be used to find the rth half-sample estimate for all quantities of interest. The modified weights also make it easy to extend the method to stratified multistage samples.

9.3.1.2 BRR in a Stratified Multistage Survey

When $\bar{y}_U$ is the only quantity of interest in a stratified random sample, BRR is simply a fancy method of calculating the variance in Equation (4.5) and adds little extra to the procedure in Chapter 4. BRR's value in a complex survey comes from its ability to estimate the variance of a general population quantity θ, where θ may be a ratio of two variables, a correlation coefficient, a quantile, or another quantity of interest.

Suppose the population has H strata, and two psu's are selected from stratum h with unequal probabilities and with replacement. (In replication methods, we like sampling with replacement because the subsampling design does not affect the variance estimator, as we saw in Section 6.3.) The same method may be used when sampling is done without replacement in each stratum, but the estimated variance of $\hat{\theta}$, calculated under the assumption of with-replacement sampling, is expected to be larger than the without-replacement variance.

The data file for a complex survey with two psu's per stratum often resembles that shown in Table 9.3, after sorting by stratum and psu.

The vector α_r defines the half-sample r: If $\alpha_{rh} = 1$, then all observation units in psu 1 of stratum h are in half-sample r; if $\alpha_{rh} = -1$, then all observation units in psu 2 of stratum h are in half-sample r. The vectors α_r are selected in a balanced way, exactly as in stratified random sampling. Now, for half-sample r, create a new column of weights $w(\alpha_r)$:

$$w_i(\alpha_r) = \begin{cases} 2w_i & \text{if observation unit } i \text{ is in half-sample } r. \\ 0 & \text{otherwise.} \end{cases} \tag{9.5}$$

TABLE   9.3

Data Structure After Sorting

Observation Number	Stratum Number	psu Number	ssu Number	Weight, w_i	Response Variable 1	Response Variable 2	Response Variable 3
1	1	1	1	w_1	y_1	x_1	u_1
2	1	1	2	w_2	y_2	x_2	u_2
3	1	1	3	w_3	y_3	x_3	u_3
4	1	1	4	w_4	y_4	x_4	u_4
5	1	2	1	w_5	y_5	x_5	u_5
6	1	2	2	w_6	y_6	x_6	u_6
7	1	2	3	w_7	y_7	x_7	u_7
8	1	2	4	w_8	y_8	x_8	u_8
9	1	2	5	w_9	y_9	x_9	u_9
10	2	1	1	w_{10}	y_{10}	x_{10}	u_{10}
11	2	1	2	w_{11}	y_{11}	x_{11}	u_{11}
Etc.							

For the data structure in Table 9.3 and $\alpha_{r1} = -1$ and $\alpha_{r2} = 1$, the column $w(\alpha_r)$ will be

$$(0, 0, 0, 0, 2w_5, 2w_6, 2w_7, 2w_8, 2w_9, 2w_{10}, 2w_{11}, \ldots).$$

Now use the column $w(\alpha_r)$ instead of w to estimate quantities for half-sample r. The estimate of the population total of y for the full sample is $\sum w_i y_i$; the estimate of the population total of y for half-sample r is $\sum w_i(\alpha_r) y_i$. If $\theta = t_y/t_x$, then $\hat{\theta} = \sum w_i y_i / \sum w_i x_i$, and $\hat{\theta}(\alpha_r) = \sum w_i(\alpha_r) y_i / \sum w_i(\alpha_r) x_i$. We saw in Section 7.3 that the empirical distribution function is calculated using the weights

$$\hat{F}(y) = \frac{\text{sum of } w_i \text{ for all observations with } y_i \le y}{\text{sum of } w_i \text{ for all observations}}.$$

Then, the empirical distribution using half-sample r is

$$\hat{F}_r(y) = \frac{\text{sum of } w_i(\alpha_r) \text{ for all observations with } y_i \le y}{\text{sum of } w_i(\alpha_r) \text{ for all observations}}.$$

If θ is the population median, then $\hat{\theta}$ may be defined as the smallest value of y for which $\hat{F}(y) \ge 1/2$, and $\hat{\theta}(\alpha_r)$ is the smallest value of y for which $\hat{F}_r(y) \ge 1/2$.

For any quantity θ, we define

$$\hat{V}_{\text{BRR}}(\hat{\theta}) = \frac{1}{R} \sum_{r=1}^{R} [\hat{\theta}(\alpha_r) - \hat{\theta}]^2. \tag{9.6}$$

BRR can also be used to estimate covariances of statistics: If θ and η are two quantities of interest, then

$$\widehat{\text{Cov}}_{\text{BRR}}(\hat{\theta}, \hat{\eta}) = \frac{1}{R} \sum_{r=1}^{R} [\hat{\theta}(\alpha_r) - \hat{\theta}][\hat{\eta}(\alpha_r) - \hat{\eta}].$$

Other BRR variance estimators, variations of (9.6), are described in Exercise 7.

While the exact equivalence of $\hat{V}_{\text{BRR}}[\bar{y}_{\text{str}}(\alpha)]$ and $\hat{V}_{\text{str}}(\bar{y}_{\text{str}})$ does not extend to nonlinear statistics, Krewski and Rao (1981) and Rao and Wu (1985) show that if h is a smooth function of the population totals, the variance estimate from BRR is asymptotically equivalent to that from linearization. BRR also provides a consistent estimator of the variance for quantiles when a stratified random sample is taken (Shao and Wu 1992).

EXAMPLE 9.5   Bye and Gallicchio (1993) describe BRR estimates of variance in the U.S. Survey of Income and Program Participation (SIPP). SIPP, like the National Crime Victimization Survey (NCVS), has a stratified multistage cluster design. Self-representing (SR) strata consist of one psu that is sampled with probability 1, and one psu is selected with pps from each non-self-representing (NSR) stratum. Strictly speaking, BRR does not apply since only one psu is selected in each stratum, and BRR requires two psu's per stratum. To use BRR, "pseudostrata" and "pseudo–psu's" were formed. A typical pseudostratum was formed by combining an SR stratum with two similar NSR strata: The psu selected in each NSR stratum was randomly assigned to one of the two pseudo–psu's, and the segments in the SR psu were randomly split between the two pseudo–psu's. This procedure created 72 pseudostrata, each with two pseudo–psu's.

The 72 half-samples, each containing the observations from one pseudo–psu from each pseudostratum, were formed using a 71-factor Plackett–Burman (1946) design. This design is orthogonal, so the set of replicate half-samples is balanced.

About 8500 of the 54,000 persons in the 1990 sample said they received Social Security benefits; Bye and Gallicchio wanted to estimate the mean and median monthly benefit amount for persons receiving benefits, for a variety of subpopulations. The mean monthly benefit for married males was estimated as

$$\frac{\sum_{i \in S_M} w_i y_i}{\sum_{i \in S_M} w_i},$$

where y_i is the monthly benefit amount for person i in the sample, w_i is the weight assigned to person i, and S_M is the subset of the sample consisting of married males receiving Social Security benefits. The median benefit payment can be estimated from the empirical distribution function for the married men in the sample:

$$\hat{F}(y) = \frac{\text{sum of weights for married men with } 0 < y_i \leq y}{\text{sum of weights for all married men receiving benefits}}.$$

The estimate of the sample median, $\hat{\theta}$, satisfies $\hat{F}(\hat{\theta}) \geq 1/2$, but $\hat{F}(x) < 1/2$ for all $x < \hat{\theta}$.

Calculating $\hat{\theta}_r$ for a replicate is simple: Merely define a new weight variable $w(\alpha_r)$, as previously described, and use $w(\alpha_r)$ instead of w to estimate the mean and median. ∎

Advantages   BRR gives a variance estimate that is asymptotically equivalent to that from linearization methods for smooth functions of population totals and for quantiles. It requires relatively few computations when compared with the jackknife and the bootstrap.

Disadvantages   As defined earlier, BRR requires a two-psu-per-stratum design. In practice, though, it is often extended to other sampling designs by using more complicated balancing schemes. BRR, like the jackknife and bootstrap, estimates the with-replacement variance and may overestimate the variance if the N_h's, the number of psu's in stratum h in the population, are small.

9.3.2 The Jackknife

The **jackknife** method, like BRR, extends the random group method by allowing the replicate groups to overlap. The jackknife was introduced by Quenouille (1949; 1956) as a method of reducing bias; Tukey (1958) used it to estimate variances and calculate confidence intervals. In this section, we describe the *delete-1 jackknife;* Shao and Tu (1995) discuss other forms of the jackknife and give theoretical results.

For an SRS, let $\hat{\theta}_{(j)}$ be the estimator of the same form as $\hat{\theta}$, but not using observation j. Thus, if $\hat{\theta} = \bar{y}$, then $\hat{\theta}_{(j)} = \bar{y}_{(j)} = \sum_{i \neq j} y_i/(n-1)$. For an SRS, define the delete-1 jackknife estimator (so called because we delete one observation in each replicate) as

$$\hat{V}_{\text{JK}}(\hat{\theta}) = \frac{n-1}{n} \sum_{j=1}^{n} (\hat{\theta}_{(j)} - \hat{\theta})^2. \tag{9.7}$$

Why the multiplier $(n-1)/n$? Let's look at $\hat{V}_{\text{JK}}(\hat{\theta})$ when $\hat{\theta} = \bar{y}$. When $\hat{\theta} = \bar{y}$,

$$\bar{y}_{(j)} = \frac{1}{n-1} \sum_{i \neq j} y_i = \frac{1}{n-1} \left(\sum_{i=1}^{n} y_i - y_j \right) = \bar{y} - \frac{1}{n-1}(y_j - \bar{y}).$$

Then,

$$\sum_{j=1}^{n} (\bar{y}_{(j)} - \bar{y})^2 = \frac{1}{(n-1)^2} \sum_{j=1}^{n} (y_j - \bar{y})^2 = \frac{1}{n-1} s_y^2.$$

Thus, $\hat{V}_{\text{JK}}(\bar{y}) = s_y^2/n$, the with-replacement estimate of the variance of $\bar{y}$.

EXAMPLE 9.6   Let's use the jackknife to estimate the ratio of nonresident tuition to resident tuition for the first group of colleges in Table 9.1. Here, $\hat{\theta} = \bar{y}/\bar{x}$, $\hat{\theta}_{(j)} = \hat{B}_{(j)} = \bar{y}_{(j)}/\bar{x}_{(j)}$, and

$$\hat{V}_{\text{JK}}(\hat{B}) = \frac{n-1}{n} \sum (\hat{B}_{(j)} - \hat{B})^2.$$

For each jackknife group, omit one observation. Thus, $\bar{x}_{(1)}$ is the average of all x's except for x_1: $\bar{x}_{(1)} = (1/9) \sum_{i=2}^{9} x_i$ (Table 9.4).

Here, $\hat{B} = 2.3288$, $\sum (\hat{B}_{(j)} - \hat{B})^2 = 0.1043$, and $\hat{V}_{\text{JK}}(\hat{B}) = 0.0938$. ∎

TABLE 9.4

Jackknife Calculations for Example 9.6

j	x	y	$\bar{x}_{(j)}$	$\bar{y}_{(j)}$	$\hat{B}_{(j)}$
1	1365	3747	1580.6	3617.7	2.2889
2	1677	4983	1545.9	3480.3	2.2513
3	1500	1500	1565.6	3867.3	2.4703
4	1080	2160	1612.2	3794.0	2.3533
5	1875	2475	1523.9	3759.0	2.4667
6	3071	5135	1391.0	3463.4	2.4899
7	1542	3950	1560.9	3595.1	2.3032
8	930	4050	1628.9	3584.0	2.2003
9	1340	4140	1583.3	3574.0	2.2573
10	1210	4166	1597.8	3571.1	2.2350

How can we extend this to a cluster sample? One might think that you could just delete one observation unit at a time, but that will not work—deleting one observation unit at a time destroys the cluster structure and gives an estimate of the variance that is only correct if the intraclass correlation is zero. In any resampling method and in the random group method, keep observation units within a psu together while constructing the replicates—this preserves the dependence among observation units within the same psu. For a cluster sample, then, we would apply the jackknife variance estimator in (9.7) by letting n be the number of psu's and letting $\hat{\theta}_{(j)}$ be the estimate of θ that we would obtain by deleting all the observations in psu j.

In a stratified multistage cluster sample, the jackknife is applied separately in each stratum at the first stage of sampling, with one psu deleted at a time. Suppose there are H strata, and n_h psu's are chosen for the sample from stratum h. Assume these psu's are chosen with replacement.

To apply the jackknife, delete one psu at a time. Let $\hat{\theta}_{(hj)}$ be the estimator of the same form as $\hat{\theta}$ when psu j of stratum h is omitted. To calculate $\hat{\theta}_{(hj)}$, define a new weight variable: Let

$$w_{i(hj)} = \begin{cases} w_i & \text{if observation unit } i \text{ is not in stratum } h. \\ 0 & \text{if observation unit } i \text{ is in psu } j \text{ of stratum } h. \\ \dfrac{n_h}{n_h - 1} w_i & \text{if observation unit } i \text{ is in stratum } h \text{ but not in psu } j. \end{cases}$$

Then use the weights $w_{i(hj)}$ to calculate $\hat{\theta}_{(hj)}$, and

$$\hat{V}_{\text{JK}}(\hat{\theta}) = \sum_{h=1}^{H} \frac{n_h - 1}{n_h} \sum_{j=1}^{n_h} (\hat{\theta}_{(hj)} - \hat{\theta})^2. \tag{9.8}$$

EXAMPLE 9.7   Here we use the jackknife to calculate the variance of the mean egg volume from Example 5.6. We calculated $\hat{\theta} = \bar{y}_r = 4375.947/1757 = 2.49$. In that example, since we did not know the number of clutches in the population, we calculated the with-replacement variance.

First, find the weight vector for each of the 184 jackknife iterations. We have only one stratum, so $h = 1$ for all observations. For $\hat{\theta}_{(11)}$, delete the first psu. Thus, the new weights for the observations in the first psu are 0; the weights in all remaining psu's are the previous weights times $n_h/(n_h - 1) = 184/183$. Using the weights from Example 5.8, the new jackknife weight columns are shown in Table 9.5.

Note that the sums of the jackknife weights vary from column to column because the original sample is not self-weighting. We calculated $\hat{\theta}$ as $\left(\sum w_i y_i \right) / \sum w_i$; to find $\hat{\theta}_{(hj)}$, follow the same procedure but use $w_{i(hj)}$ in place of w_i. Thus, $\hat{\theta}_{(1,1)} = 4349.348/1753.53 = 2.48034$; $\hat{\theta}_{(1,2)} = 4345.036/1753.53 = 2.47788$; $\hat{\theta}_{(1,184)} = 4357.819/1754.54 = 2.48374$. Using (9.8) then, we calculate $\hat{V}_{\text{JK}}(\hat{\theta}) = 0.00373$. This results in a standard error of 0.061, the same as calculated in Example 5.6.   ∎

Advantages   This is an all-purpose method. The same procedure is used to estimate the variance for every statistic for which the jackknife can be used. The jackknife works in stratified multistage samples in which BRR does not apply because more than two psu's are sampled in each stratum. The jackknife provides a consistent

TABLE 9.5

Jackknife Weights, for Example 9.7

clutch	csize	relweight	w(1,1)	w(1,2)	. . .	w(1,184)
1	13	6.5	0	6.535519	. . .	6.535519
1	13	6.5	0	6.535519	. . .	6.535519
2	13	6.5	6.535519	0	. . .	6.535519
2	13	6.5	6.535519	0	. . .	6.535519
3	6	3	3.016393	3.016393	. . .	3.016393
3	6	3	3.016393	3.016393	. . .	3.016393
4	11	5.5	5.530055	5.530055	. . .	5.530055
4	11	5.5	5.530055	5.530055	. . .	5.530055
⋮	⋮	⋮	⋮	⋮		⋮
183	13	6.5	6.535519	6.535519	. . .	6.535519
183	13	6.5	6.535519	6.535519	. . .	6.535519
184	12	6	6.032787	6.032787	. . .	0
184	12	6	6.032787	6.032787	. . .	0
Sum	3514	1757	1753.53	1753.53	. . .	1754.54

estimator of the variance when θ is a smooth function of population totals (Krewski and Rao 1981).

Disadvantages The jackknife performs poorly for estimating the variances of some statistics. For example, the jackknife produces a poor estimate of the variance of quantiles in an SRS. Little is known about how the jackknife performs in unequal-probability, without-replacement sampling designs in general.

9.3.3 The Bootstrap

As with the jackknife, theoretical results for the **bootstrap** were developed for areas of statistics other than survey sampling; Shao and Tu (1995) summarize theoretical results for the bootstrap in complex survey samples. We first describe the bootstrap for an SRS with replacement, as developed by Efron (1979, 1982) and described in Efron and Tibshirani (1993). Suppose $\mathcal{S}$ is an SRS of size n. We hope, in drawing the sample, that it reproduces properties of the whole population. We then treat the sample $\mathcal{S}$ as if it were a population and take resamples from $\mathcal{S}$. If the sample really is similar to the population—if the empirical probability mass function (epmf) of the sample is similar to the probability mass function of the population—then samples generated from the epmf should behave like samples taken from the population.

EXAMPLE 9.8 Let's use the bootstrap to estimate the variance of the median height, θ, in the height population from Example 7.3, using the sample in the file ht.srs. The population median height is $\theta = 168$; the sample median from ht.srs is $\hat{\theta} = 169$. Figure 7.2, the probability mass function for the population, and Figure 7.3, the histogram of the sample, are similar in shape (largely because the sample size for the SRS is large), so we would expect that taking an SRS of size n with replacement from $\mathcal{S}$ would be like taking an SRS with replacement from the population. A resample from $\mathcal{S}$, though, will not be exactly the same as S because the resample is with replacement—some observations in S may occur twice or more in the resample, while other observations in S may not occur at all.

We take an SRS of size 200 with replacement from S to form the first resample. The first resample from S has an epmf similar to but not identical to that of S; the resample median is $\hat{\theta}_1^* = 170$. Repeating the process, the second resample from S has median $\hat{\theta}_2^* = 169$. We take a total of $R = 2000$ resamples from S and calculate the sample median from each sample, obtaining $\hat{\theta}_1^*, \hat{\theta}_2^*, \ldots, \hat{\theta}_R^*$. We obtain the following frequency table for the 2000 sample medians:

Median of resample	165.0	166.0	166.5	167.0	167.5	168.0	168.5	169.0	169.5	170.0	170.5	171.0	171.5	172.0
Frequency	1	5	2	40	15	268	87	739	111	491	44	188	5	4

The sample mean of these 2000 values is 169.3, and the sample variance of these 2000 values is 0.9148; this is the bootstrap estimator of the variance. The bootstrap distribution may be used to calculate a confidence interval directly: Since it estimates the sampling distribution of $\hat{\theta}$, a 95% CI is calculated by finding the 2.5 percentile and the 97.5 percentile of the bootstrap distribution. For this distribution, a 95% CI for the median is [167.5, 171]. ■

If the original SRS is without replacement, Gross (1980) proposes creating N/n copies of the sample to form a "pseudopopulation," then drawing R SRSs without replacement from the pseudopopulation. If n/N is small, the with-replacement and without-replacement bootstrap distributions should be similar.

Sitter (1992) describes and compares three bootstrap methods for complex surveys. In all these methods, bootstrapping is applied within each stratum. Here are steps for using one version of the rescaling bootstrap of Rao and Wu (1988) for a stratified random sample:

1 For each stratum, draw an SRS of size $n_h - 1$ with replacement from the sample in stratum h. Do this independently for each stratum.

2 For each resample $r (r = 1, 2, \ldots, R)$, create a new weight variable

$$w_i(r) = w_i \times \frac{n_h}{n_h - 1} m_i(r)$$

where $m_i(r)$ is the number of times that observation i is selected to be in the resample. Calculate $\hat{\theta}_r^*$, using the weights $w_i(r)$.

3 Repeat steps 1 and 2 R times, for R a large number.

4 Calculate

$$\hat{V}_B(\hat{\theta}) = \frac{1}{R-1} \sum_{r=1}^{R} (\hat{\theta}_r^* - \hat{\theta})^2.$$

Advantages The bootstrap will work for nonsmooth functions (such as quantiles) in general sampling designs. The bootstrap is well suited for finding confidence intervals directly: To get a 90% CI, merely take the 5th and 95th percentiles from $\hat{\theta}_1^*, \hat{\theta}_2^*, \ldots, \hat{\theta}_R^*$ or use a bootstrap-t method such as that described in Efron (1982).

Disadvantages The bootstrap requires more computations than BRR or jackknife, since R is typically a very large number. Compared with BRR and jackknife, less theoretical work has been done on properties of the bootstrap in complex sampling designs.

9.4
Generalized Variance Functions

In many large government surveys such as the U.S. Current Population Survey (CPS) or the Canadian Labour Force Survey, hundreds or thousands of estimates are calculated and published. The agencies analyzing the survey results could calculate standard errors for each published estimate and publish additional tables of the standard errors but that would add greatly to the labor involved in publishing timely estimates from the surveys. In addition, other analysts of the public-use tapes may wish to calculate additional estimates, and the public-use tapes may not provide enough information to allow calculation of standard errors.

Generalized variance functions (GVFs) are provided in a number of surveys to calculate standard errors. They have been used for the CPS since 1947. Here, we describe some GVFs in the 1990 NCVS.

Criminal Victimization in the United States, 1990 (U.S. Department of Justice 1992, 146) gives GVF formulas for calculating standard errors. If $\hat{t}$ is an estimated number of persons or households victimized by a particular type of crime or if $\hat{t}$ estimates a total number of victimization incidents,

$$\hat{V}(\hat{t}) = a\hat{t}^2 + b\hat{t}. \qquad (9.9)$$

If $\hat{p}$ is an estimated proportion,

$$\hat{V}(\hat{p}) = \left(\frac{b}{\hat{x}}\right)\hat{p}(1 - \hat{p}), \qquad (9.10)$$

where $\hat{x}$ is the estimated base population for the proportion. For the 1990 NCVS, the values of a and b were $a = -.00001833$ and $b = 3725$. For example, it was estimated that 1.23% of persons aged 20 to 24 were robbed in 1990 and that 18,017,100 persons were in that age group. Thus, the GVF estimate of $SE(\hat{p})$ is

$$\sqrt{\frac{3725}{18,017,100}(.0123)(1 - .0123)} = .0016.$$

Assuming that asymptotic results apply, this gives an approximate 95% CI of $.0123 \pm (1.96)(.0016)$, or $[.0091, .0153]$.

There were an estimated 800,510 completed robberies in 1990. Using (9.9), the standard error of this estimate is

$$\sqrt{(-.00001833)(800,510)^2 + 3725(800,510)} = 54,499.$$

Where do these formulas come from? Suppose t_i is the total number of observation units belonging to a class—say, the total number of persons in the United States who were victims of violent crime in 1990. Let $p_i = t_i/N$, the proportion of persons in the population belonging to that class. If d_i is the design effect (deff) in the survey for estimating p_i (see Section 7.5), then

$$V(\hat{p}_i) \approx d_i \frac{p_i(1-p_i)}{n} = \frac{b_i}{N} p_i(1-p_i), \tag{9.11}$$

where $b_i = d_i \times (N/n)$. Similarly,

$$V(\hat{t}_i) \approx d_i N^2 \frac{p_i(1-p_i)}{n} = a_i t_i^2 + b_i t_i,$$

where $a_i = -d_i/n$. If estimating a proportion in a domain—say, the proportion of persons in the 20–24 age group who were robbery victims—the denominator in (9.11) is changed to the estimated population size of the domain (see Section 3.3).

If the deff's are similar for different estimates so that $a_i \approx a$ and $b_i \approx b$, then constants a and b can be estimated that give (9.9) and (9.10) as approximations to the variance for a number of quantities. The general procedure for constructing a generalized variance function is as follows:

1   Using replication or some other method, estimate variances for k population totals of special interest, $\hat{t}_1, \hat{t}_2, \ldots, \hat{t}_k$. Let v_i be the relative variance for $\hat{t}_i$, $v_i = \hat{V}(\hat{t}_i)/\hat{t}_i^2 = \widehat{CV}(\hat{t}_i)^2$, for $i = 1, 2, \ldots, k$.

2   Postulate a model relating v_i to $\hat{t}_i$. Many surveys use the model

$$v_i = \alpha + \frac{\beta}{\hat{t}_i}.$$

This is a linear regression model with response variable v_i and explanatory variable $1/\hat{t}_i$. Valliant (1987) found that this model produces consistent estimates of the variances for the class of superpopulation models he studied.

3   Use regression techniques to estimate α and β. Valliant (1987) suggests using weighted least squares to estimate the parameters, giving higher weight to items with small v_i. The GVF estimate of variance, then, is the predicted value from the regression equation, $a + b/\hat{t}$.

The a_i and b_i for individual items are replaced by quantities a and b, which are calculated from all k items. For the 1990 NCVS, $b = 3725$. Most weights in the 1990 NCVS are between 1500 and 2500; b approximately equals the (average weight) $\times$ (deff), if the overall design effect is about 2.

Valliant (1987) found that if deff's for the k estimated totals are similar, the GVF variances were often more stable than the direct estimate, as they smooth out some of the fluctuations from item to item. If a quantity of interest does not follow the model in step 2, however, the GVF estimate of the variance is likely to be poor, and you can only know that it is poor by calculating the variance directly.

Advantages   The GVF may be used when insufficient information is provided on the public-use tapes to allow direct calculation of standard errors. The data collector can calculate the GVF, and the data collector often has more information for estimating variances than is released to the public. A generalized variance function saves a great deal of time and speeds production of annual reports. It is also useful for designing similar surveys in the future.

Disadvantages   The model relating v_i to $\hat{t}_i$ may not be appropriate for the quantity you are interested in, resulting in an unreliable estimate of the variance. You must be careful about using GVFs for estimates not included when calculating the regression parameters. If a subpopulation has an unusually high degree of clustering (and hence a high deff), the GVF estimate of the variance may be much too small.

9.5
Confidence Intervals

9.5.1 Confidence Intervals for Smooth Functions of Population Totals

Theoretical results exist for most of the variance estimation methods discussed in this chapter, stating that under certain assumptions $(\hat{\theta} - \theta)/\sqrt{\hat{V}(\hat{\theta})}$ asymptotically follows a standard normal distribution. These results and conditions are given in Binder (1983), for linearization estimates; in Krewski and Rao (1981) and Rao and Wu (1985), for jackknife and BRR; in Rao and Wu (1988) and Sitter (1992), for bootstrap. Consequently, when the assumptions are met, an approximate 95% confidence interval for θ may be constructed as

$$\hat{\theta} \pm 1.96 \sqrt{\hat{V}(\hat{\theta})}.$$

Alternatively, a t_{df} percentile may be substituted for 1.96, with df = (number of groups -1) for the random group method. Rust and Rao (1996) give guidelines for appropriate df's for other methods.

Roughly speaking, the assumptions for linearization, jackknife, BRR, and bootstrap are as follows:

1   The quantity of interest θ can be expressed as a smooth function of the population totals; more precisely, $\theta = h(t_1, t_2, \ldots, t_k)$, where the second-order partial derivatives of h are continuous.

2   The sample sizes are large: Either the number of psu's sampled in each stratum is large, or the survey contains a large number of strata. (See Rao and Wu 1985 for the precise technical conditions needed.) Also, to construct a confidence interval using the normal distribution, the sample sizes must be large enough so that the sampling distribution of $\hat{\theta}$ is approximately normal.

Furthermore, a number of simulation studies indicate that these confidence intervals behave well in practice. Wolter (1985) summarizes some of the simulation studies; others are found in Kovar et al. (1988) and Rao et al. (1992). These studies indicate that the jackknife and linearization methods tend to give similar estimates of the variance, while the bootstrap and BRR procedures give slightly larger estimates. Sometimes a transformation may be used so that the sampling distribution of a statistic is closer to a normal distribution: If estimating total income, for example, a log transformation may be used because the distribution of income is extremely skewed.

9.5.2 Confidence Intervals for Population Quantiles

The theoretical results described above for BRR, jackknife, bootstrap, and linearization do not apply to population quantiles, however, because they are not smooth functions of population totals. Special methods have been developed to construct confidence intervals for quantiles; McCarthy (1993) compares several confidence intervals for the median, and his discussion applies to other quantiles as well.

Let q be between 0 and 1. Then define the quantile θ_q as $\theta_q = F^{-1}(q)$, where $F^{-1}(q)$ is defined to be the smallest value y satisfying $F(y) \geq q$. Similarly, define $\hat{\theta}_q = \hat{F}^{-1}(q)$. Now F^{-1} and $\hat{F}^{-1}$ are *not* smooth functions, but we assume the population and sample are large enough so that they can be well approximated by continuous functions.

Some of the methods already discussed work quite well for constructing confidence intervals for quantiles. The random group method works well if the number of random groups, R, is moderate. Let $\hat{\theta}_q(r)$ be the estimated quantile from random group r. Then, a confidence interval for θ_q is

$$\hat{\theta}_q \pm t \sqrt{\frac{\sum\limits_r [\hat{\theta}_q(r) - \hat{\theta}_q]^2}{(R-1)R}}$$

where t is the appropriate percentile from a t distribution with $R-1$ df. Similarly, empirical studies by McCarthy (1993), Kovar et al. (1988), Sitter (1992), and Rao et al. (1992) indicate that in certain designs confidence intervals can be formed using

$$\hat{\theta}_q \pm 1.96 \sqrt{\hat{V}(\hat{\theta}_q)}$$

where the variance estimate is calculated using BRR or bootstrap.

An alternative interval can be constructed based on a method introduced by Woodruff (1952). For any y, $\hat{F}(y)$ is a function of population totals: $\hat{F}(y) = \sum w_i u_i / \sum w_i$, where $u_i = 1$ if $y_i \leq y$ and $u_i = 0$ if $y_i > y$. Thus, a method in this chapter can be used to estimate $V[\hat{F}(y)]$ for any value y, and an approximate 95% CI for $F(y)$ is given by

$$\hat{F}(y) \pm 1.96 \sqrt{\hat{V}[\hat{F}(y)]}.$$

Now let's use the confidence interval for $q = F(\theta_q)$ to obtain an approximate confidence interval for θ_q. Since we have a 95% CI,

$$0.95 \approx P\left\{ \hat{F}(\theta_q) - 1.96\sqrt{\hat{V}[\hat{F}(\theta_q)]} \leq q \leq \hat{F}(\theta_q) + 1.96\sqrt{\hat{V}[\hat{F}(\theta_q)]} \right\}$$

$$= P\left\{ q - 1.96\sqrt{\hat{V}[\hat{F}(\theta_q)]} \leq \hat{F}(\theta_q) \leq q + 1.96\sqrt{\hat{V}[\hat{F}(\theta_q)]} \right\}$$

$$\approx P\left(\hat{F}^{-1}\left\{ q - 1.96\sqrt{\hat{V}[\hat{F}(\theta_q)]} \right\} \leq \theta_q \leq \hat{F}^{-1}\left\{ q + 1.96\sqrt{\hat{V}[\hat{F}(\theta_q)]} \right\} \right).$$

FIGURE   9.2

Woodruff's confidence interval for the quantile θ_q if the empirical distribution function is continuous. Since $F(y)$ is a proportion, we can easily calculate a confidence interval (CI) for any value of y, shown on the vertical axis. We then look at the corresponding points on the horizontal axis to form a confidence interval for θ_q.

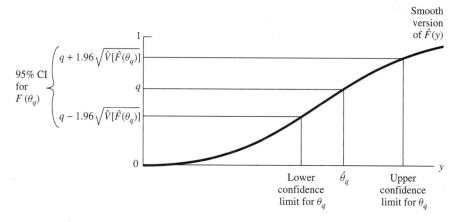

So an approximate 95% CI for the quantile θ_q is

$$\left[\hat{F}^{-1}\left\{ q - 1.96\sqrt{\hat{V}[\hat{F}(\hat{\theta}_q)]} \right\}, \ \hat{F}^{-1}\left\{ q + 1.96\sqrt{\hat{V}[\hat{F}(\hat{\theta}_q)]} \right\} \right].$$

The derivation of this confidence interval is illustrated in Figure 9.2.

Now we need several technical assumptions to use the Woodruff-method interval. These assumptions are stated by Rao and Wu (1987) and Francisco and Fuller (1991), who studied a similar confidence interval. Basically, the problem is that both F and $\hat{F}$ are step functions; they have jumps at the values of y in the population and sample. The technical conditions basically say that the jumps in F and in $\hat{F}$ should be small and that the sampling distribution of $\hat{F}(y)$ is approximately normal.

EXAMPLE 9.9   Let's use Woodruff's method to construct a 95% CI for the median height in the file ht.srs, discussed in Examples 7.3 and 9.8. Note that $\hat{F}(\theta_q)$ is the sample proportion of observations in the SRS that take on value at most θ_q; so, ignoring the fpc,

$$V[\hat{F}(\theta_q)] \approx \frac{1}{n}F(\theta_q)[1 - F(\theta_q)] = \frac{1}{n}q(1 - q).$$

Thus, for this sample,

$$1.96\sqrt{V[\hat{F}(\theta_{0.5})]} \approx 1.96\sqrt{\frac{(.5)(.5)}{200}} = 0.0693.$$

The lower confidence bound for the median is then $\hat{F}^{-1}(.5 - 0.0693)$, and the upper confidence bound for the median is $\hat{F}^{-1}(.5 + 0.0693)$. As heights were only measured to the nearest centimeter, we'll use linear interpolation to smooth the step function

$\hat{F}$. The following values were obtained for the empirical distribution function:

y	$\hat{F}(y)$
167	0.405
168	0.440
170	0.515
171	0.550
172	0.605

Then, interpolating,

$$\hat{F}^{-1}(0.4307) = 167 + \frac{.4307 - .405}{.44 - .405}(168 - 167) = 167.7,$$

and

$$\hat{F}^{-1}(0.5693) = 171 + \frac{.5693 - .55}{.605 - .55}(172 - 171) = 171.4.$$

Thus, an approximate 95% CI for the median is [167.7, 171.4]. ∎

9.5.3 Conditional Confidence Intervals

The confidence intervals presented so far in this chapter have been developed under the design-based approach. A 95% CI may be interpreted in the repeated-sampling sense that, if samples were repeatedly taken from the finite population, we would expect 95% of the resulting confidence intervals to include the true value of the quantity in the population.

Sometimes, especially in situations when ratio estimation or poststratification are used, you may want to consider constructing a conditional confidence interval instead. In poststratification as used for nonresponse (Section 8.5.2), the respondent sample sizes n_{hR} in the poststrata are unknown when the sample is selected; they are thus random variables, which may differ if a different sample is taken. In (8.3), the conditional variance, conditional on the values of n_{hR}, was presented. A 95% conditional confidence interval, constructed using the variance in (8.3), would have the interpretation that we would expect 95% of all samples having those specific values of n_{hR} to yield confidence intervals containing $\bar{y}_U$.

The theory of conditional confidence intervals is beyond the scope of this book; we refer the reader to Särndal et al. (1992, sec. 7.10), Casady and Valliant (1993), and Thompson (1997, sec. 5.12) for more discussion and bibliography.

9.6
Summary and Software

This chapter has briefly introduced you to some basic types of variance estimation methods that are used in practice: linearization, random groups, replication, and generalized variance functions. But this is just an introduction; you are encouraged to read some of the references mentioned in this chapter before applying these methods to your own complex survey. Much of the research done exploring properties and behavior of these methods has been done since 1980, and variance estimation methods are still a subject of research by statisticians.

Linearization methods are perhaps the most thoroughly researched in terms of theoretical properties and have been widely used to find variance estimates in complex surveys. The main drawback of linearization, though, is that the derivatives need to be calculated for each statistic of interest, and this complicates the programs for estimating variances. If the statistic you are interested in is not handled in the software, you must write your own code.

The random group method is an intuitively appealing method for estimating variances. Easy to explain and to compute, it can be used for almost any statistic of interest. Its main drawback is that we generally need enough random groups to have a stable estimate of the variance, and the number of random groups we can form is limited by the number of psu's sampled in a stratum.

Resampling methods for stratified multistage surveys avoid partial derivatives by computing estimates for subsamples of the complete sample. They must be constructed carefully, however, so that the correlation of observations in the same cluster is preserved in the resampling. Resampling methods require more computing time than linearization but less programming time: The same method is used on all statistics. They have been shown to be equivalent to linearization for large samples when the characteristic of interest is a smooth function of population totals.

The BRR method can be used with almost any statistic, but it is usually used only for two-psu-per-stratum designs or for designs that can be reformulated into two psu per strata. The jackknife and bootstrap can also be used for most estimators likely to be used in surveys (exception: the delete-1 jackknife may not work well for estimating the variance of quantiles) and may be used in stratified multistage samples in which more than two psu's are selected in each sample, but they require more computing than BRR.

Generalized variance functions are cheap and easy to use but have one major drawback: Unless you can calculate the variance using one of the other methods, you cannot be sure that your statistic follows the model used to develop the GVF.

All methods except GVFs assume that information on the clustering is available to the data analyst. In many surveys, such information is not released because it might lead to identification of the respondents. See Dippo et al. (1984) for a discussion of this problem.

Various software packages have been developed to assist in analyzing data from complex surveys. Cohen (1997), Lepkowski and Bowles (1996), and Carlson et al. (1993) evaluate PC-based packages for analysis of complex survey data.[1] SUDAAN (Shah et al. 1995), OSIRIS (Lepkowski 1982), Stata (StataCorp 1996), and PC-CARP (Fuller et al. 1989) all use linearization methods to estimate variances of nonlinear statistics. SUDAAN, for example, calculates variances of estimated population totals for various stratified multistage sampling designs that have H strata, unequal-probability cluster sampling with or without replacement at the first stage of sampling,

---

[1]Lepkowski and Bowles (1996) tell how to access the free (or almost-free) software packages CENVAR, CLUSTERS, Epi Info, VPLX, and WesVarPC through e-mail or from the Internet. Software for analysis of survey data is changing rapidly; the Survey Research Methods Section of the American Statistical Association (www.amstat.org) is a good resource for updated information.

and SRS with or without replacement at subsequent stages. The formula in (6.9) is used to estimate the variance for each stratum in with-replacement sampling, and the Sen–Yates–Grundy form in (6.15) is used for without-replacement variance. Then, the variances for the totals in the strata are added to estimate the variance for the estimated population total. SUDAAN then uses linearization to find variances for ratios, regression coefficients, and other nonlinear statistics. Recent versions of SUDAAN also implement BRR and jackknife.

OSIRIS also implements BRR and jackknife methods. The survey software packages WesVarPC (Brick et al. 1996; at press time, WesVarPC could be downloaded free from www.westat.com) and VPLX (Fay 1990) both use resampling methods to calculate variance estimates. A simple S-PLUS function for jackknife is given in Appendix D; this is not intended to substitute for well-tested commercial software but to give you an idea of how these calculations might be done. Then, after you understand the principles of the methods, you can use commercial software for your complex surveys.

9.7
Exercises

1   Which of the variance estimation methods in this chapter would be suitable for estimating the proportion of beds that have bed nets for the Gambia bed net survey in Example 7.1? Explain why each method is or is not appropriate.

2   As in Example 9.1, let $h(p) = p(1 - p)$.

   a   Find the remainder term in the Taylor expansion, $\int_a^x (x - t) h''(t)\, dt$, and use it to find an exact expression for $h(\hat{p})$.

   b   Is the remainder term likely to be smaller than the other terms? Explain.

   c   Find an exact expression for $V[h(\hat{p})]$ for an SRS with replacement. How does it compare with the approximation in Example 9.1?

3   The straight-line regression slope for the population is

$$B_1 = \frac{\displaystyle\sum_{i=1}^{N}(x_i - \bar{x}_U)(y_i - \bar{y}_U)}{\sqrt{\displaystyle\sum_{i=1}^{N}(x_i - \bar{x}_U)^2}}.$$

   a   Express B_1 as a function of population totals $t_1 = \sum_{i=1}^{N} x_i$, $t_2 = \sum_{i=1}^{N} y_i$, $t_3 = \sum_{i=1}^{N} x_i^2$, and $t_4 = \sum_{i=1}^{N} x_i y_i$ so that $B_1 = h(t_1, t_2, t_3, t_4)$.

   b   Let $\hat{B}_1 = h(\hat{t}_1, \hat{t}_2, \hat{t}_3, \hat{t}_4)$ and suppose that $E[\hat{t}_i] = t_i$, for $i = 1, 2, 3, 4$. Use the linearization method to find an approximation to the variance of $\hat{B}_1$. Express your answer in terms of $V(\hat{t}_i)$ and $\text{Cov}(\hat{t}_i, \hat{t}_j)$.

   c   What is the linearization approximation to the variance for an SRS of size n?

   d   Find a linearized variate q_i so that $\hat{V}(\hat{B}_1) = \hat{V}(\hat{t}_q)$.

4 The correlation coefficient for the population is

$$R = \frac{\displaystyle\sum_{i=1}^{N}(x_i - \bar{x}_U)(y_i - \bar{y}_U)}{\sqrt{\displaystyle\sum_{i=1}^{N}(x_i - \bar{x}_U)^2 \sum_{i=1}^{N}(y_i - \bar{y}_U)^2}}.$$

a Express R as a function of population totals $t_1 = \sum_{i=1}^{N} x_i$, $t_2 = \sum_{i=1}^{N} y_i$, $t_3 = \sum_{i=1}^{N} x_i^2$, $t_4 = \sum_{i=1}^{N} x_i y_i$, and $t_5 = \sum_{i=1}^{N} y_i^2$ so that $R = h(t_1, t_2, t_3, t_4, t_5)$.

b Let $\hat{R} = h(\hat{t}_1, \ldots, \hat{t}_5)$ and suppose that $E[\hat{t}_i] = t_i$, for $i = 1, \ldots, 5$. Use the linearization method to find an approximation to the variance of $\hat{R}$.

c What is the linearization approximation to the variance for an SRS of size n?

5 *Variance estimation with poststratification.* Suppose we poststratify the sample into L poststrata, with population counts $N_1, N_2, \ldots, N_L$. Then the poststratified estimator for the population total is

$$\hat{t}_{\text{post}} = \sum_{l=1}^{L} \frac{N_l}{\hat{N}_l} \hat{t}_l = h(\hat{t}_1, \ldots, \hat{t}_L, \hat{N}_1, \ldots, \hat{N}_L),$$

where

$$\hat{t}_l = \sum_{i \in \mathcal{S}_l} w_i y_i, \qquad \hat{N}_l = \sum_{i \in \mathcal{S}_l} w_i,$$

and $\mathcal{S}_l$ is the set of sample units that are in poststratum l. Show, using linearization, that

$$V(\hat{t}_{\text{post}}) \approx V\left(\hat{t} - \sum_{l=1}^{L} \frac{t_l}{N_l} \hat{N}_l\right).$$

6 Suppose a stratified random sample is taken with two observations per stratum. Show that if $\sum_{r=1}^{R} \alpha_{rh}\alpha_{rl} = 0$, for $l \neq h$, then

$$\hat{V}_{\text{BRR}}(\bar{y}_{\text{str}}) = \hat{V}_{\text{str}}(\bar{y}_{\text{str}}).$$

HINT: First note that

$$\bar{y}_{\text{str}}(\alpha_i) - \bar{y}_{\text{str}} = \sum_{h=1}^{H} \frac{N_h}{N} \alpha_{ih} \frac{y_{h1} - y_{h2}}{2}.$$

Then express $\hat{V}_{\text{BRR}}(\bar{y}_{\text{str}})$ directly using y_{h1} and y_{h2}.

7 Other BRR estimators of the variance are

$$\frac{1}{4R} \sum_{r=1}^{R} [\hat{\theta}(\alpha_r) - \hat{\theta}(-\alpha_r)]^2$$

and

$$\frac{1}{2R} \sum_{r=1}^{R} \{[\hat{\theta}(\alpha_r) - \hat{\theta}]^2 + [\hat{\theta}(-\alpha_r) - \hat{\theta}]^2\}.$$

For a stratified random sample with two observations per stratum, show that if $\sum_{r=1}^{R} \alpha_{rh} \, \alpha_{rl} = 0$ for $l \neq h$, then each of these variance estimators is equivalent to $\hat{V}_{str}(\bar{y}_{str})$.

8 Suppose the parameter of interest is $\theta = h(t)$, where $h(t) = at^2 + bt + c$ and t is the population total. Let $\hat{\theta} = h(\hat{t})$. Show, in a stratified random sample with two observations per stratum, that if $\sum_{r=1}^{R} \alpha_{rh} \, \alpha_{rl} = 0$ for $l \neq h$, then

$$\frac{1}{4R} \sum_{r=1}^{R} \left[\hat{\theta}(\alpha_r) - \hat{\theta}(-\alpha_r) \right]^2 = \hat{V}_L(\hat{\theta}),$$

which is the linearization estimate of variance (see: Rao and Wu 1985).

9 Use the random groups in the data file syc.dat to estimate the variances for the estimates of the proportion of youth who:

   a   Are age 14 or younger.

   b   Are held for a violent offense.

   c   Lived with both parents when growing up.

   d   Are male.

   e   Are Hispanic.

   f   Grew up primarily in a single-parent family.

   g   Have used illegal drugs.

10 The linearization method in Section 9.1 is the one historically used to find variances. Binder (1996) proposes proceeding directly to the estimate of the variance by evaluating the partial derivatives at the sample estimates rather than at the population quantities. What is Binder's estimate for the variance of the ratio estimator? Does it differ from that in Section 9.1?

11 Find a jackknife estimate of the population mean age of trees in a stand for the data in Exercise 4 of Chapter 3 and calculate the jackknife estimate of the variance. How do these estimates compare with those based on Taylor series methods? Be sure to include details about how you computed the jackknife estimates.

12 Use the jackknife to estimate the variances of your estimates in parts (a) and (b) of Exercise 17 of Chapter 5.

13 Use the jackknife to estimate the variance of the ratio estimator used in Example 3.2. How does it compare with the linearization estimator?

14 Use Woodruff's method to construct a confidence interval for the median weekday greens fee for nine holes, using the SRS in the file golfsrs.dat.

SURVEY Exercises

15 Draw an SRS of size 200 from Lockhart City (use the sample from Chapter 2 if you wish). We want to estimate $B = \bar{y}_U / \bar{x}_U$, the ratio of the price a household is willing to pay for cable TV (y) to the assessed value of the house (x). Use the linearization method to estimate the variance of $\hat{B} = \bar{y}/\bar{x}$.

16   Randomly divide your sample into 10 different subsamples, each of size 20. This can be done in SAS by creating a new variable *scramble,* which has 200 uniform random numbers between 0 and 1. Sort the data by the variable *scramble;* then assign the first 20 observations to group 1, the second 20 to group 2 (etc). The group means can be easily calculated by doing a one-way ANOVA on the data. Now find the ratio $\hat{B}_i = \bar{y}_i/\bar{x}_i$ for each group and use the random group method to estimate the variance of $\hat{B}$.

17   Calculate 200 different estimates $\hat{B}_{(j)}$ of B, each using all but one of the 200 data points. Calculate the jackknife estimate of the variance of $\hat{B} = \bar{y}/\bar{x}$.

18   How do your variance estimates from Exercises 15–17 compare?

10

Categorical Data Analysis in Complex Surveys*

But Statistics must be made otherwise than to prove a preconceived idea.

— Florence Nightingale, annotation in *Physique Sociale*, by A. Quetelet

Up to now we have mostly been looking at how to estimate summary quantities such as means, totals, and percentages in different sampling designs. Totals and percentages are important for many surveys, for they provide a description of the population: for instance, the percentage of the population victimized by crime or the total number of unemployed persons in the United States. Often, though, researchers are interested in multivariate questions: Is race associated with criminal victimization, or can we predict unemployment status from demographic variables? Such questions are typically answered in statistics using techniques in categorical data analysis or regression (which we will discuss in Chapter 11). The techniques you learned in an introductory statistics course, though, assumed that observations were all independent and identically distributed from some population distribution. These assumptions are no longer met in data from complex surveys; in this and the following chapter we examine the effects of the complex sampling design on commonly used statistical analyses.

Since much information from sample surveys is collected in the form of percentages, categorical data methods are extensively used in the analysis. In fact, many of the data sets used to illustrate the chi-square test in introductory statistics textbooks originate in complex surveys. Our greatest concern is with the effects of clustering on commonly used hypothesis tests and models for categorical data, as clustering usually decreases precision. We begin by reviewing various chi-square tests when a simple random sample (SRS) is taken from a large population.

10.1
Chi-Square Tests with Multinomial Sampling

EXAMPLE **10.1**   Each couple in an SRS of 500 married couples from a large population is asked whether (1) the household owns at least one personal computer and (2) the household subscribes to cable television. The following contingency table presents the outcomes:

		Computer?		
		Yes	No	
Cable?	Yes	119	188	307
	No	88	105	193
		207	293	500

Are households with a computer more likely to subscribe to cable? A chi-square test for independence is often used for such questions. Under the null hypothesis that owning a computer and subscribing to cable are independent, the expected counts for each cell in the contingency table are the following:

		Computer?		
		Yes	No	
Cable?	Yes	127.1	179.9	307
	No	79.9	113.1	193
		207	293	500

Pearson's chi-square test statistic is

$$X^2 = \sum_{\substack{\text{all} \\ \text{cells}}} \frac{(\text{observed count} - \text{expected count})^2}{\text{expected count}} = 2.281.$$

The **likelihood ratio chi-square test statistic** is

$$G^2 = 2 \sum_{\substack{\text{all} \\ \text{cells}}} (\text{observed count}) \ \ln\left(\frac{\text{observed count}}{\text{expected count}}\right) = 2.275.$$

The two test statistics are asymptotically equivalent; for large samples, each approximately follows a chi-square (χ^2) distribution with 1 degree of freedom (df) under the null hypothesis. The p-value for each statistic is 0.13, giving no reason to doubt the null hypothesis that owning a computer and subscribing to cable television are independent.

If owning a computer and subscribing to cable are independent events, the odds that a cable subscriber will own a computer should equal the odds that a non-cable-subscriber will own a computer. We estimate the odds of owning a computer if the household subscribes to cable as $119/188$ and estimate the odds of owning a computer if the household does not subscribe to cable as $88/105$. The **odds ratio** is therefore estimated as

$$\frac{\dfrac{119}{188}}{\dfrac{88}{105}} = 0.755.$$

If the null hypothesis of independence is true, we expect the odds ratio to be close to 1. Equivalently, we expect the logarithm of the odds ratio to be close to 0. The log odds is -0.28 with asymptotic standard error

$$\sqrt{\frac{1}{119} + \frac{1}{88} + \frac{1}{188} + \frac{1}{105}} = 0.186;$$

an approximate 95% confidence interval (CI) for the log odds is $-0.28 \pm 1.96(0.186)$ $= [-0.646, 0.084]$. This confidence interval includes zero, and confirms the result of the hypothesis test that there is no evidence against independence. ∎

Chi-square tests are commonly used in three situations; each assumes a form of random sampling. These tests are discussed in more detail in Lindgren (1993, chap. 10), Agresti (1990), and Christensen (1990).

10.1.1 Testing Independence of Factors

Each of n independent observations is cross-classified by two factors: row factor R with r levels and column factor C with c levels. Each observation has probability p_{ij} of falling into row category i and column category j, giving the following table of true probabilities. Here, $p_{i+} = \sum_{j=1}^{c} p_{ij}$ is the probability that a randomly selected unit will fall in row category i, and $p_{+j} = \sum_{i=1}^{r} p_{ij}$ is the probability that a randomly selected unit will fall in column category j:

		C				
		1	2	$\cdots$	c	
	1	p_{11}	p_{12}	$\cdots$	p_{1c}	p_{1+}
	2	p_{21}	p_{22}	$\cdots$	p_{2c}	p_{2+}
R	$\vdots$	$\vdots$	$\vdots$		$\vdots$	$\vdots$
	r	p_{r1}	p_{r2}	$\cdots$	p_{rc}	p_{r+}
		p_{+1}	p_{+2}	$\cdots$	p_{+c}	1

The observed count in cell (i, j) from the sample is x_{ij}. If all units in the sample are independent, the x_{ij}'s are from a multinomial distribution with rc categories; this sampling scheme is known as **multinomial sampling.** In surveys the assumptions for multinomial sampling are met in an SRS with replacement; they are approximately met in an SRS without replacement when the sample size is small compared with the population size. The latter situation occurred in Example 10.1: Independent multinomial sampling means we have a sample of 500 (approximately) independent households, and we observe to which of the four categories each household belongs.

The null hypothesis of independence is

$$H_0 : p_{ij} = p_{i+} p_{+j} \qquad \text{for } i = 1, \ldots, r \quad \text{and} \quad j = 1, \ldots, c. \qquad \textbf{(10.1)}$$

Let $m_{ij} = n p_{ij}$ represent the expected counts. If H_0 is true, $m_{ij} = n p_{i+} p_{+j}$, and m_{ij} can be estimated by

$$\hat{m}_{ij} = n \hat{p}_{i+} \hat{p}_{+j} = n \frac{x_{i+}}{n} \frac{x_{+j}}{n},$$

where $\hat{p}_{ij} = x_{ij}/n$ and $\hat{p}_{i+} = \sum_{j=1}^{c} \hat{p}_{ij}$. Pearson's chi-square test statistic is

$$X^2 = \sum_{i=1}^{r} \sum_{j=1}^{c} \frac{(x_{ij} - \hat{m}_{ij})^2}{\hat{m}_{ij}} = n \sum_{i=1}^{r} \sum_{j=1}^{c} \frac{(\hat{p}_{ij} - \hat{p}_{i+}\hat{p}_{+j})^2}{\hat{p}_{i+}\hat{p}_{+j}}. \qquad (10.2)$$

The likelihood ratio test statistic is

$$G^2 = 2 \sum_{i=1}^{r} \sum_{j=1}^{c} x_{ij} \ln\left(\frac{x_{ij}}{\hat{m}_{ij}}\right) = 2n \sum_{i=1}^{r} \sum_{j=1}^{c} \hat{p}_{ij} \ln\left(\frac{\hat{p}_{ij}}{\hat{p}_{i+}\hat{p}_{+j}}\right). \qquad (10.3)$$

If multinomial sampling is used with a sufficiently large sample size, X^2 and G^2 are approximately distributed as a chi-square random variable with $(r-1)(c-1)$ df under the null hypothesis. How large is "sufficiently large" depends on the number of cells and expected probabilities; Fienberg (1979) argues that p-values will be approximately correct if (a) the expected count in each cell is greater than 1 and (b) $n \geq 5 \times$ (number of cells).

An equivalent statement to (10.1) is that all odds ratios equal 1:

$$H_0 : \frac{p_{11}p_{ij}}{p_{1j}p_{i1}} = 1 \qquad \text{for all } i \geq 2 \text{ and } j \geq 2.$$

We may estimate any odds ratio $p_{ij}p_{kl}/p_{il}p_{kj}$ by substituting in estimated proportions: $\hat{p}_{ij}\hat{p}_{kl}/\hat{p}_{il}\hat{p}_{kj}$. If the sample is sufficiently large, the *logarithm* of the estimated odds ratio is approximately normally distributed with standard error

$$\sqrt{\frac{1}{x_{ij}} + \frac{1}{x_{kl}} + \frac{1}{x_{il}} + \frac{1}{x_{kj}}}.$$

10.1.2 Testing Homogeneity of Proportions

The Pearson and likelihood ratio test statistics in (10.2) and (10.3) may also be used when independent random samples from r populations are each classified into c categories. Multinomial sampling is done within each population, so the sampling scheme is called **product-multinomial sampling.** Product-multinomial sampling is equivalent to stratified random sampling when the sampling fraction for each stratum is small or when sampling is with replacement.

The difference between product-multinomial sampling and multinomial sampling is that the row totals p_{i+} and x_{i+} are fixed quantities in product-multinomial sampling—x_{i+} is the predetermined sample size for stratum i. The null hypothesis that the proportion of observations falling in class j is the same for all strata is

$$H_0 : \frac{p_{1j}}{p_{1+}} = \frac{p_{2j}}{p_{2+}} = \cdots = \frac{p_{rj}}{p_{r+}} = p_{+j} \qquad \text{for all } j = 1, \ldots, c. \qquad (10.4)$$

If the null hypothesis in (10.4) is true, again $m_{ij} = np_{i+}p_{+j}$ and the expected counts under H_0 are $\hat{m}_{ij} = n\hat{p}_{i+}\hat{p}_{+j}$, exactly as in the test for independence.

EXAMPLE **10.2**   The sample sizes used in Exercise 14 of Chapter 4, the stratified sample of nursing students and tutors, were the sample sizes for the respondents. Let's use a chi-square test for homogeneity of proportions to test the null hypothesis that the nonresponse rate is the same for each stratum. The four strata form the rows in the following contingency table:

	Nonrespondent	Respondent	
General student	46	222	268
General tutor	41	109	150
Psychiatric student	17	40	57
Psychiatric tutor	8	26	34
	112	397	509

The two chi-square test statistics are $X^2 = 8.218$, with p-value 0.042, and $G^2 = 8.165$, with p-value 0.043. There is thus evidence of different nonresponse rates among the four groups. However, the following table shows that the difference is not attributable to the main effect of either general/psychiatric or student/tutor:

	Nonresponse Rate	
	Student	Tutor
General	17%	27%
Psychiatric	30%	24%

Further investigation would be needed to explore the nonresponse pattern. ∎

10.1.3 Testing Goodness of Fit

Multinomial sampling is again assumed, with independent observations classified into k categories. The null hypothesis is

$$H_0 : p_i = p_i^{(0)} \qquad \text{for } i = 1, \ldots, k,$$

where $p_i^{(0)}$ is prespecified or is a function of parameters θ to be estimated from the data.

EXAMPLE 10.3 Webb (1955) examined the safety records for 17,952 U.S. Air Force pilots for an 8-year period around World War II and constructed the following frequency table.

Number of Accidents	Number of Pilots
0	12,475
1	4,117
2	1,016
3	269
4	53
5	14
6	6
7	2

If accidents occur randomly—if no pilots are more or less "accident prone" than others—a Poisson distribution should fit the data well. We estimate the mean of the Poisson distribution by the mean number of accidents per pilot in the sample, 0.40597. The observed and expected probabilities under the null hypothesis that the data follow a Poisson distribution are given in the following table. The expected probabilities are computed using the Poisson probabilities $e^{-\lambda}\lambda^x/x!$ with $\lambda = 0.40597$.

Number of Accidents	Observed Proportion, $\hat{p}_i$	Expected Probability Under H_0, $\hat{p}_i^{(0)}$
0	.6949	.6663
1	.2293	.2705
2	.0566	.0549
3	.0150	.0074
4	.0030	.0008
5+	.0012	.0001

The two chi-square test statistics are

$$X^2 = \sum_{\substack{\text{all} \\ \text{cells}}} \frac{(\text{observed count} - \text{expected count})^2}{\text{expected count}}$$

$$= \sum_{i=1}^{k} \frac{\left(n\hat{p}_i - n\hat{p}_i^{(0)}\right)^2}{n\hat{p}_i^{(0)}} \tag{10.5}$$

$$= n \sum_{i=1}^{k} \frac{\left(\hat{p}_i - \hat{p}_i^{(0)}\right)^2}{\hat{p}_i^{(0)}}$$

and

$$G^2 = 2n \sum_{i=1}^{k} \hat{p}_i \ln\left(\frac{\hat{p}_i}{\hat{p}_i^{(0)}}\right). \tag{10.6}$$

For the pilots, $X^2 = 756$ and $G^2 = 400$. If the null hypothesis is true, both statistics approximately follow a χ^2 distribution with 4 df (2 df are spent on n and $\hat{\lambda}$). Both p-values are less than 0.0001, providing evidence that a Poisson model does not fit the data. More pilots have no accidents, or more than two accidents, than would be expected under the Poisson model. Thus, evidence shows that some pilots are more accident-prone than would occur under the Poisson model. ∎

All the chi-square test statistics in (10.2), (10.3), (10.5), and (10.6) grow with n. If the null hypothesis is not exactly true in the population—if households with cable are even infinitesimally more likely to own a personal computer than households without cable—we can almost guarantee rejection of the null hypothesis by taking a large enough random sample. This property of the hypothesis test means that it will be sensitive to artificially inflating the sample size by ignoring clustering.

10.2
Effects of Survey Design on Chi-Square Tests

The survey design can affect both the estimated cell probabilities and the tests of association or goodness of fit. In complex survey designs, we no longer have the random sampling that gives both X^2 and G^2 an approximate χ^2 distribution. Thus, if we just run a standard statistical package to do our chi-square tests, the significance levels and p-values will be wrong. Clustering, especially, can have a strong effect on the p-values of chi-square tests. In a cluster sample with a positive intraclass correlation coefficient (ICC), the true p-value will often be much larger than the p-value reported by the statistical package under the assumption of independent multinomial sampling. Let's see what can happen to hypothesis tests if the survey design is ignored in a cluster sample.

EXAMPLE 10.4 Suppose both husband and wife are asked about the household's cable and computer status for the survey discussed in Example 10.1, and both give the same answer. While the assumptions of multinomial sampling were met for the SRS of couples, they are not met for the cluster sample of persons—far from being independent units, the husband and wife from the same household agree completely in their answers. The ICC for the cluster sample is 1.

What happens if we ignore the clustering? The contingency table for the observed frequencies is as follows:

		Computer?		
		Yes	No	
Cable?	Yes	238	376	614
	No	176	210	386
		414	586	1000

The estimated proportions and odds ratio are identical to those in Example 10.1: $\hat{p}_{11} = 238/1000 = 119/500$, and the odds ratio is

$$\frac{\dfrac{238}{376}}{\dfrac{176}{210}} = 0.755.$$

But $X^2 = 4.562$ and $G^2 = 4.550$ are twice the values of the test statistics in Example 10.1. If you ignored the clustering and compared these statistics to a χ^2 distribution with 1 df, you would report a "p-value" of 0.033 and conclude that the data provided evidence that having a computer and subscribing to cable are not independent. If playing this game, you could lower the "p-value" even more by interviewing both children in each household as well, thus multiplying the original test statistics by 4.

Can you attain an arbitrarily low p-value by observing more ssu's per psu? Absolutely not. The statistics X^2 and G^2 have a null χ_1^2 distribution *when multinomial sampling is used.* When a cluster sample is taken instead and when the ICC is positive, X^2 and G^2 do *not* follow a χ_1^2 distribution under the null hypothesis. For the 1000 husbands and wives, $X^2/2$ and $G^2/2$ follow a χ_1^2 distribution under H_0—this gives the same p-value found in Example 10.1. ∎

10.2.1 Contingency Tables for Data from Complex Surveys

The observed counts x_{ij} do not necessarily reflect the relative frequencies of the categories in the population unless the sample is self-weighting. Suppose an SRS of elementary school classrooms in Denver is taken, and each of ten randomly selected students in each classroom is evaluated for self-concept (high or low) and clinical depression (present or not). Students are selected for the sample with unequal probabilities—students in small classes are more likely to be in the sample than students from large classes. A table of observed counts from the sample, ignoring the probabilities of selection, would not give an accurate picture of the association between self-concept and depression in the population if the degree of association differs with class size. Even if the association between self-concept and depression is the same for different class sizes, the estimates of numbers of depressed students using the margins of the contingency table may be wrong.

Remember, though, that sampling weights can be used to estimate any population quantity. Here, they can be used to estimate the cell proportions. Estimate p_{ij} by

$$\hat{p}_{ij} = \frac{\sum_{k \in \mathcal{S}} w_k y_{kij}}{\sum_{k \in \mathcal{S}} w_k}, \tag{10.7}$$

where

$$y_{kij} = \begin{cases} 1 & \text{if observation unit } k \text{ is in cell } (i, j) \\ 0 & \text{otherwise} \end{cases}$$

and w_k is the weight for observation unit k. Thus,

$$\hat{p}_{ij} = \frac{\text{sum of weights for observation units in cell } (i, j)}{\text{sum of weights for all observation units in sample}}.$$

If the sample is self-weighting, $\hat{p}_{ij}$ will be the proportion of observation units falling in cell (i, j). Using the estimates $\hat{p}_{ij}$, construct the table

		C				
		1	2	⋯	c	
	1	$\hat{p}_{11}$	$\hat{p}_{12}$	⋯	$\hat{p}_{1c}$	$\hat{p}_{1+}$
	2	$\hat{p}_{21}$	$\hat{p}_{22}$	⋯	$\hat{p}_{2c}$	$\hat{p}_{2+}$
R	⋮	⋮				⋮
	r	$\hat{p}_{r1}$	$\hat{p}_{r2}$	⋯	$\hat{p}_{rc}$	$\hat{p}_{r+}$
		$\hat{p}_{+1}$	$\hat{p}_{+2}$	⋯	$\hat{p}_{+c}$	1

to examine associations, and estimate odds ratios by $\hat{p}_{ij}\hat{p}_{kl}/\hat{p}_{il}\hat{p}_{kj}$. A confidence interval for p_{ij} may be constructed by using any method of variance estimation discussed so far, or a design effect (deff) may be used to modify the SRS confidence interval, as in (7.7).

Do not throw the observed counts away, however. If the odds ratios calculated using the $\hat{p}_{ij}$ differ appreciably from the odds ratios calculated using the observed counts x_{ij}, you should explore why they differ. Perhaps the odds ratio for depression and self-concept differs for larger classes or depends on socioeconomic factors related to class size. If that is the case, you should include these other factors in a model for the data or perhaps test the association separately for large and small classes.

10.2.2 Effects on Hypothesis Tests and Confidence Intervals

We can estimate contingency table proportions and odds ratios using weights. The weights, however, provide no help in constructing hypothesis tests and confidence intervals—these depend on the clustering and (sometimes) stratification of the survey design.

Let's look at the effect of stratification first. If the strata are the row categories, the stratification poses no problem—we essentially have product-multinomial sampling, as described in Section 10.1.2, and can test for homogeneity of proportions the usual way.

In highly stratified surveys, though, the association between strata and other factors may not be of interest. In the National Crime Victimization Survey, for example, we may be interested in the association between gender and violent-crime victimization and want to include data from all strata in the examination. In general, stratification increases precision of the estimates over SRS. For an SRS, (10.2) gives

$$X^2 = n \sum_{i=1}^{r} \sum_{j=1}^{c} \frac{(\hat{p}_{ij} - \hat{p}_{i+}\hat{p}_{+j})^2}{\hat{p}_{i+}\hat{p}_{+j}}.$$

A stratified sample with n observation units provides the same precision for estimating p_{ij} as an SRS with n/d_{ij} observation units, where d_{ij} is the deff for estimating p_{ij}. If the stratification is worthwhile, the deff's will generally be less than 1. Consequently, if we use the SRS test statistics in (10.2) or (10.3) with the $\hat{p}_{ij}$ from the stratified sample, X^2 and G^2 will be smaller than they should be to follow a null $\chi^2_{(r-1)(c-1)}$ distribution; "p-values" calculated ignoring the stratification will be too large, and H_0 will not be rejected as often as it should be. Thus, while SAS PROC FREQ or another standard statistics package may give you a p-value of 0.04, the actual p-value may be 0.02. Ignoring the stratification results in a conservative test. Similarly, a confidence interval constructed for a log odds ratio is generally too large if the stratification is ignored. Your estimates are really more precise than the SRS confidence interval indicates.

Clustering usually has the opposite effect. Design effects for $\hat{p}_{ij}$ with a cluster sample are usually greater than 1—a cluster sample with n observation units gives the same precision as an SRS with fewer than n observations. If the clustering is ignored, X^2 and G^2 are expected to be larger than if the equivalently sized SRS were taken, and "p-values" calculated ignoring the clustering are likely to be too small. SAS may give you a p-value of 0.04, while the actual p-value may be 0.25. If you ignore the clustering, you may well declare an association to be statistically significant when it is really just due to random variation in the data. Confidence intervals for log odds ratios will be narrower than they should be—the estimates are not as precise as the confidence intervals from SAS would lead you to believe.

Ignoring clustering in chi-square tests is often more dangerous than ignoring stratification. An SRS-based chi-square test using stratified data will still indicate strong associations; it just will not uncover all weaker associations. Ignoring clustering, however, will lead to declaring associations statistically significant that really are not. Ignoring the clustering in goodness-of-fit tests may lead to adopting an unnecessarily complicated model to describe the data.

An investigator ignorant of sampling theory will often analyze a stratified sample correctly, using the strata as one of the classification variables. But the investigator may not even record the clustering, and too often simply runs the observed counts through SAS PROC FREQ or SPSS CROSSTABS and accepts the printed-out p-value as truth. To see how this could happen, consider an investigator wanting to replicate Basow and Silberg's (1987) study on whether male and female professors are evaluated differently by college students. (The original study was discussed in Example 5.1.) The investigator selects a stratified sample of male and female professors at the college and asks each student in those professors' classes to evaluate the professor's teaching. Over 2000 student responses are obtained, and the investigator cross-classifies those responses by professor gender and by whether the student gives the professor a high or low rating. The investigator, comparing Pearson's X^2 statistic on the observed counts to a χ_1^2 distribution, declares a statistically significant association between professor gender and student rating. The stratification variable *professor gender* is one of the classification variables, so no adjustments need be made for the stratification. But the reported p-value is almost certainly incorrect, for a number of reasons: (1) The clustering of students within a class is ignored—indeed, the investigator does not even record which professor is evaluated by a student but only records the professor's gender, so the investigation cannot account for the clustering. If student evaluations reflect teaching quality, students of a "good" professor would be expected to give higher ratings than students of a "bad" professor. The ICC for students is positive, and the equivalent sample size in an SRS is less than 2000. The p-value reported by the investigator is then much too small, and the investigator may be wrong in concluding faculty women receive a different mean level on student evaluations. (2) A number of students may give responses for more than one professor in the sample. It is unclear what effect these multiple responses would have on the test of independence. (3) Not all students attend class or turn in the evaluation. Some of the nonresponse may be missing completely at random (a student was ill the day of the study), but some may be related to perceived teaching quality (the student skips class because the professor is confusing).

The societal implications of reporting false positive results because clustering is ignored can be expensive. A university administrator may decide to give female faculty an unnecessary handicap when determining raises that are based in part on student evaluations. A medical researcher may conclude that a new medication with more side effects than the standard treatment is more effective for combating a disease, even though the statistical significance is due to the cluster inflation of the sample size. A government official may decide that a new social program is needed to remedy an "inequity" demonstrated in the hypothesis test. The same problem occurs outside of sample surveys as well, particularly in biostatistics. Clusters may correspond to pairs of eyes, to patients in the same hospital, or to repeated measures on the same person.

Is the clustering problem serious in surveys taken in practice? A number of studies have found that it can be. Holt et al. (1980) found that the actual significance levels for tests nominally conducted at the $\alpha = 0.05$ level ranged from 0.05 to 0.50. Fay (1985) references a number of studies demonstrating that the SRS-based test statistics "may give extremely erroneous results when applied to data arising from a complex sample design." The simulation study in Thomas et al. (1996) calculated actual significance levels attained for X^2 and G^2 when the nominal significance level was set at $\alpha = 0.05$—they found actual significance levels of about 0.30 to 0.40.

10.3
Corrections to Chi-Square Tests

A number of methods have been proposed to account for the survey design when testing for goodness of fit, homogeneity of populations, and independence of variables. Thomas et al. (1996) describe more than 25 methods that have been developed for testing independence in two-way tables and provide a useful bibliography. Some of these methods and variations are described in more detail in Rao and Thomas (1988; 1989). Fay (1985) describes an alternative method that involves jackknifing the test statistic itself.

In this section we outline some of the basic approaches for testing independence of variables. The theory for goodness-of-fit tests and tests for homogeneity of proportions is similar. In complex surveys, though, unlike in multinomial and product-multinomial sampling, the tests for independence and homogeneity of proportions are not necessarily the same. Holt et al. (1980) note that often (but not always) clustering has less effect on tests for independence than on tests for goodness of fit or homogeneity of proportions.

Recall from (10.1) that the null hypothesis of independence is

$$H_0 : p_{ij} = p_{i+}p_{+j} \quad \text{for } i = 1, \ldots, r \quad \text{and} \quad j = 1, \ldots, c.$$

For a 2×2 table, $p_{ij} = p_{i+}p_{+j}$ is equivalent to $p_{11}p_{22} - p_{12}p_{21} = 0$ for all i and j, so the null hypothesis reduces to a single equation. In general, the null hypothesis can be expressed as $(r-1)(c-1)$ distinct equations, which leads to $(r-1)(c-1)$ df for multinomial sampling. Let

$$\theta_{ij} = p_{ij} - p_{i+}p_{+j}.$$

Then, the null hypothesis of independence is

$$H_0 : \theta_{11} = 0, \; \theta_{12} = 0, \ldots, \theta_{r-1,c-1} = 0.$$

10.3.1 Wald Tests

The Wald (1943) test was the first to be used for testing independence in complex surveys (Koch et al. 1975). For the 2×2 table, the null hypothesis involves one quantity,

$$\theta = \theta_{11} = p_{11} - p_{1+}p_{+1} = p_{11}p_{22} - p_{12}p_{21},$$

and θ is estimated by

$$\hat{\theta} = \hat{p}_{11}\hat{p}_{22} - \hat{p}_{12}\hat{p}_{21}.$$

The quantity θ is a smooth function of population totals, so we can find an estimate of $V(\hat{\theta})$ by using one of the methods in Chapter 9. If the sample sizes are sufficiently large and $H_0 : \theta = 0$ is true, then

$$\frac{\hat{\theta}}{\sqrt{\hat{V}(\hat{\theta})}}$$

approximately follows a standard normal distribution. Equivalently, under H_0, the **Wald statistic**

$$X_W^2 = \frac{\hat{\theta}^2}{\hat{V}(\hat{\theta})} \tag{10.8}$$

approximately follows a χ^2 distribution with 1 df.

EXAMPLE 10.5   Let's look at the association between "Was anyone in your family ever incarcerated?" (variable *famtime*) and "Have you ever been put on probation or sent to a correctional institution for a violent offense?" (variable *everviol*) using data from the Survey of Youth in Custody. A total of $n = 2588$ youths in the survey had responses for both items. The following table gives the sum of the weights for each category. Note that this table can be calculated using SAS with the weight variable, but the chi-square test from SAS is completely wrong because it acts as though there are 24,699 observations. In this case, SAS, with the weights, gives $X^2 = G^2 = 11.6$, with incorrect "*p*-value" < 0.001.

		Ever Violent?		
		No	Yes	
Family Member	No	4,761	7,154	11,915
Incarcerated?	Yes	4,838	7,946	12,784
		9,599	15,100	24,699

This results in the following table of estimated proportions:

		Ever Violent?		
		No	Yes	
Family Member	No	.1928	.2896	.4824
Incarcerated?	Yes	.1959	.3217	.5176
		.3887	.6113	1.0000

Thus,

$$\hat{\theta} = \hat{p}_{11}\hat{p}_{22} - \hat{p}_{12}\hat{p}_{21} = \hat{p}_{11} - \hat{p}_{1+}\hat{p}_{+1} = 0.0053$$

One way to estimate the variance of $\hat{\theta}$ is to calculate $\hat{p}_{11}\hat{p}_{22} - \hat{p}_{12}\hat{p}_{21}$ for each of the seven random groups, as discussed in Example 9.4, and find the variance of the seven nearly independent estimates of θ. The seven estimates, with the average and standard deviation (SD), are

Random Group	$\hat{\theta}$
1	0.0132
2	0.0147
3	0.0252
4	−0.0224
5	0.0073
6	−0.0057
7	0.0135
Average	0.0065
SD	0.0158

Using the random group method, the standard error (SE) of $\hat{\theta}$ is $0.0158/\sqrt{7} = 0.0060$, so the test statistic is

$$\frac{\hat{\theta}}{\sqrt{\hat{V}(\hat{\theta})}} = 0.89.$$

Since our estimate of the variance from the random group method has only 6 df, we compare the test statistic to a t_6 distribution rather than to a standard normal distribution. This test gives no evidence of an association between the two factors, when we look at the population as a whole. But the hypothesis test does not say anything about possible associations between the two variables in subpopulations—it could occur, for example, that violence and incarceration of a family member are positively associated among older youth and negatively associated among younger youth—we would need to look at the subpopulations separately or fit a loglinear model to see if this was the case. ∎

For larger tables, let $\boldsymbol{\theta} = [\theta_{11}\, \theta_{12} \, \dots \, \theta_{r-1,c-1}]^T$ (the superscript T means "transpose") be the $(r-1)(c-1)$-vector of θ_{ij}, so that the null hypothesis is

$$H_0 : \boldsymbol{\theta} = \mathbf{0}.$$

The Wald statistic is then

$$X_W^2 = \hat{\boldsymbol{\theta}}^T \hat{V}(\hat{\boldsymbol{\theta}})^{-1} \hat{\boldsymbol{\theta}}$$

where $\hat{V}(\hat{\boldsymbol{\theta}})$ is the estimated covariance matrix of $\hat{\boldsymbol{\theta}}$. In very large samples under H_0, X_W^2 approximately follows a $\chi^2_{(r-1)(c-1)}$ distribution. But "large" in a complex survey refers to a large number of psu's, not necessarily to a large number of observation units. In a 4×4 contingency table, $\hat{V}(\hat{\boldsymbol{\theta}})$ is a 9×9 matrix and requires calculation of 45 different variances and covariances. If a cluster sample has only 50 psu's, the estimated covariance matrix will be very unstable. In practice, the Wald test for large contingency tables often performs poorly, and we do not recommend its use. Some modifications of the Wald test perform better; see Thomas et al. (1996) for details.

10.3.2 Bonferroni Tests

The null hypothesis of independence,

$$H_0 : \theta_{11} = 0,\ \theta_{12} = 0,\ \dots,\ \theta_{r-1,c-1} = 0,$$

has $m = (r-1)(c-1)$ components:

$$H_0(1) : \theta_{11} = 0$$
$$H_0(2) : \theta_{12} = 0$$
$$\vdots$$
$$H_0(m) : \theta_{(r-1)(c-1)} = 0.$$

Instead of using the estimated covariance of all $\hat{\theta}_{ij}$'s as in the Wald test, we can use the Bonferroni inequality to test each component $H_0(k)$ separately with significance level α/m (Thomas 1989). The Bonferroni procedure gives a conservative test. H_0 will be rejected at level α if any of the $H_0(k)$ is rejected at level α/m—that is, if

$$\frac{|\hat{\theta}_{ij}|}{\sqrt{\hat{V}(\hat{\theta}_{ij})}} > t_\kappa \left(\frac{\alpha}{2m}\right)$$

for any i and j. Each test statistic is compared to a t_κ distribution, where the estimator of the variance has κ df. If the random group method is used to estimate the variance, then κ equals (number of groups) -1; if another method is used, κ equals (number of psu's) $-$ (number of strata).

Even though this is a conservative test, it appears to work quite well in practice. In addition, it is easy to implement, particularly if a resampling method is used to estimate variances, as each test can be done separately.

EXAMPLE 10.6   In the Survey of Youth in Custody, let's look at the relationship between age and whether the youth was sent to the institution for a violent offense (using variable *crimtype, curviol* was defined to be 1 if *crimtype* = 1 and 0 otherwise). Using the weights, we estimate the proportion of the population falling in each cell:

		Age Class			
		≤ 15	16 or 17	≥ 18	
Violent Offense?	No	.1698	.2616	.1275	.5589
	Yes	.1107	.1851	.1453	.4411
		.2805	.4467	.2728	1.0000

The null hypothesis is

$$H_0 : \theta_{11} = p_{11} - p_{1+}p_{+1} = 0$$
$$\theta_{12} = p_{12} - p_{1+}p_{+2} = 0.$$

First, let's look at what happens if we ignore the clustering and pretend that the test statistic in (10.2) follows a χ^2 distribution with 2 df. With $n = 2621$ youths in the table, Pearson's X^2 statistic is

$$X^2 = n \sum_{i=1}^{2} \sum_{j=1}^{3} \frac{(\hat{p}_{ij} - \hat{p}_{i+}\hat{p}_{+j})^2}{\hat{p}_{i+}\hat{p}_{+j}} = 34.$$

Comparing this to a χ_2^2 distribution yields an incorrect "p-value" of 4×10^{-8}.

Now let's use the Bonferroni test. For these data, $\hat{\theta}_{11} = 0.0130$ and $\hat{\theta}_{12} = 0.0119$. Using the random group method to estimate the variances, as in Example 10.5, gives us the seven estimates:

Random Group	$\hat{\theta}_{11}$	$\hat{\theta}_{12}$
1	−0.0195	0.0140
2	0.0266	−0.0002
3	0.0052	0.0159
4	0.0340	0.0096
5	0.0197	0.0212
6	0.0025	0.0298
7	−0.0103	0.0143

Thus, $\text{SE}(\hat{\theta}_{11}) = 0.0074$, $\text{SE}(\hat{\theta}_{12}) = 0.0035$, $\hat{\theta}_{11}/\text{SE}(\hat{\theta}_{11}) = 1.8$, and $\hat{\theta}_{12}/\text{SE}(\hat{\theta}_{12}) = 3.4$. The 0.9875 percentile of a t distribution with 6 df is 2.97; because the test statistic for $H_0(2) : \theta_{12} = 0$ exceeds that critical value, we reject the null hypothesis at the 0.05 level. ∎

10.3.3 Matching the Moments to Chi-Square Distribution Moments

The test statistics X^2 and G^2 do not follow a $\chi^2_{(r-1)(c-1)}$ distribution in a complex survey under the null hypothesis of independence. But both statistics have a skewed distribution, and a multiple of X^2 or G^2 may approximately follow a χ^2 distribution.

We can obtain a **first-order correction** by matching the mean of the test statistic to the mean of the $\chi^2_{(r-1)(c-1)}$ distribution (Rao and Scott 1981; 1984). The mean of a $\chi^2_{(r-1)(c-1)}$ distribution is $(r-1)(c-1)$; we can calculate $E[X^2]$ or $E[G^2]$ under the complex sampling design when H_0 is true and compare the test statistic

$$X_F^2 = \frac{(r-1)(c-1)X^2}{E[X^2]}$$

or

$$G_F^2 = \frac{(r-1)(c-1)G^2}{E[G^2]}$$

to a $\chi^2_{(r-1)(c-1)}$ distribution. Bedrick (1983) and Rao and Scott (1984) show that under H_0,

$$E[X^2] \approx E[G^2]$$
$$\approx \sum_{i=1}^{r}\sum_{j=1}^{c}(1 - p_{ij})d_{ij} - \sum_{i=1}^{r}(1 - p_{i+})d_i^R - \sum_{j=1}^{c}(1 - p_{+j})d_j^C, \quad \text{(10.9)}$$

where d_{ij} is the deff for estimating p_{ij}, d_i^R is the deff for estimating p_{i+}, and d_j^C is the deff for estimating p_{+j}. In practice, if the estimator of the cell variances has κ df, it works slightly better to compare $X_F^2/(r-1)(c-1)$ or $G_F^2/(r-1)(c-1)$ to an F distribution with $(r-1)(c-1)$ and $(r-1)(c-1)\kappa$ df.

The first-order correction can often be used with published tables because you need to estimate only variances of the proportions in the contingency table—you need not estimate the full covariance matrix of the $\hat{p}_{ij}$, as is required for the Wald test. But we are only adjusting the test statistic so that its mean under H_0 is $(r-1)(c-1)$; p-values of interest come from the tail of the reference distribution, and it does not necessarily follow that the tail of the distribution of X_F^2 matches the tail of the

$\chi^2_{(r-1)(c-1)}$ distribution. Rao and Scott (1981) show that X_F^2 and G_F^2 have a null χ^2 distribution if and only if all the deff's for the variances and covariances of the $\hat{p}_{ij}$ are equal. Otherwise, the variance of X_F^2 is larger than the variance of a $\chi^2_{(r-1)(c-1)}$ distribution, and p-values from X_F^2 are often a bit smaller than they should be (but closer to the actual p-values than if no correction was done at all).

E X A M P L E   10.7   We can also conduct the hypothesis test in Example 10.6 using the first-order correction. The following design effects were estimated, using the random group method to estimate the cell variances:

			Age Class		
		≤ 15	16 or 17	≥ 18	
	No	20.2	1.9	2.8	5.7
Violent Offense?					
	Yes	5.3	8.4	2.4	5.7
		22.0	9.7	4.3	

Several of the deff's are very large, as might be expected because some facilities have mostly violent or mostly nonviolent offenders. All residents of facility 31, for example, are there for a violent offense. In addition, the facilities with primarily nonviolent offenders tend to be larger. We would expect the clustering, then, to have a substantial effect on the hypothesis test.

Using (10.9), we estimate $E[X^2]$ by 4.2 and use $X_F^2 = 2X^2/4.2 = 16.2$. Comparing $16.2/2$ to an $F_{2,12}$ distribution (the random group estimate of the variance has 6 df) gives an approximate p-value of 0.006. This p-value is probably still a bit too small, though, because of the wide disparity in the deff's.   ■

Rao and Scott (1981; 1984) also propose a **second-order correction**—matching the mean and variance of the test statistic to the mean and variance of a χ^2 distribution, as done for ANOVA model tests by Satterthwaite (1946). Satterthwaite compared a test statistic T with skewed distribution to a χ^2 reference distribution by choosing a constant k and degrees of freedom ν so that $E[kT] = \nu$ and $V[kT] = 2\nu$ (ν and 2ν are the mean and variance of a χ^2 distribution with ν df). Here, letting $m = (r-1)(c-1)$, we know that $E[kX_F^2] = km$ and

$$V[kX_F^2] = V\left(\frac{kmX^2}{EX^2}\right) = \frac{V[X^2]k^2m^2}{[E(X^2)]^2},$$

so matching the moments gives

$$\nu = 2\frac{[E(X^2)]^2}{V[X^2]} \quad \text{and} \quad k = \frac{\nu}{m}.$$

Then,

$$X_S^2 = \frac{\nu X_F^2}{(r-1)(c-1)} \tag{10.10}$$

is compared to a χ^2 distribution with ν df. The statistic G_S^2 is formed similarly. Again, if the estimator of the variances of the $\hat{p}_{ij}$ has κ df, it works slightly better to compare X_S^2/ν or G_S^2/ν to an F distribution with ν and $\nu\kappa$ df.

In general, estimating $V[X^2]$ is somewhat involved, and requires the complete covariance matrix of the $\hat{p}_{ij}$'s, so the second-order correction often cannot be used when the data are only available in published tables. If the design effects are all similar, the first- and second-order corrections will behave similarly. When the deff's vary appreciably, however, p-values using X_F^2 may be too small, and X_S^2 may perform better. Exercise 11 tells how the second-order correction can be calculated.

10.3.4 Model-Based Methods for Chi-Square Tests

All the methods discussed use the covariance estimates of the proportions to adjust the chi-square tests. A model-based approach may also be used. We describe a model due to Cohen (1976) for a cluster sample with two observation units per cluster. Extensions and other models that have been used for cluster sampling are described in Altham (1976), Brier (1980), Rao and Scott (1981), and Wilson and Koehler (1991). These models assume that the deff is the same for each cell and margin.

EXAMPLE 10.8   Cohen (1976) presents an example exploring the relationship between gender and diagnosis as a schizophrenic. The data consisted of 71 hospitalized pairs of siblings. Many mental illnesses tend to run in families, so we might expect that if one sibling is diagnosed as schizophrenic, the other sibling is more likely to be diagnosed as schizophrenic. Thus, any analysis that ignores the dependence among siblings is likely to give p-values that are much too small. If we just categorize the 142 patients by gender and diagnosis and ignore the correlation between siblings, we get the following table. Here, S means the patient was diagnosed as schizophrenic, and N means the patient was not diagnosed as schizophrenic.

	S	N	
Male	43	15	58
Female	32	52	84
	75	67	142

If analyzed in a standard statistics package (I used SAS), $X^2 = 17.89$ and $G^2 = 18.46$. Remember, though, that SAS assumes that all observations are independent, so the "p-value" of 0.00002 is incorrect.

We know the clustering structure for the 71 clusters, though. You can see in Table 10.1 that most of the pairs fall in the diagonal blocks: If one sibling has schizophrenia, the other is more likely to have it. In 52 of the sibling pairs, either both siblings are diagnosed as having schizophrenia, or both siblings are diagnosed as not having schizophrenia.

Let q_{ij} be the probability that a pair falls in the (i, j) cell in the classification of the pairs. Thus, q_{11} is the probability that both siblings are schizophrenic and male, q_{12} is the probability that the younger sibling is a schizophrenic female and the older sibling is a schizophrenic male (etc.). Then model the q_{ij}'s by

$$q_{ij} = \begin{cases} aq_i + (1-a)q_i^2 & \text{if } i = j \\ (1-a)q_i q_j & \text{if } i \neq j \end{cases} \tag{10.11}$$

where a is a clustering effect and q_i is the probability that an individual is in class i ($i = $ SM, SF, NM, NF). If $a = 0$, members of a pair are independent, and we can just do

TABLE   10.1

Cluster Information for the 71 Pairs of Siblings

| | | \multicolumn{4}{c}{Younger Sibling} | |
		SM	SF	NM	NF	
Older Sibling	SM	13	5	1	3	22
	SF	4	6	1	1	12
	NM	1	1	2	4	8
	NF	3	8	3	15	29
		21	20	7	23	71

the regular chi-square test using the individuals—the usual Pearson's X^2, calculated ignoring the clustering, would be compared to a $\chi^2_{(r-1)(c-1)}$ distribution. If $a = 1$, the two siblings are perfectly correlated, so we essentially have only one piece of information from each pair—$X^2/2$ would be compared to a $\chi^2_{(r-1)(c-1)}$ distribution. For a between 0 and 1, if the model holds, $X^2/(1 + a)$ approximately follows a $\chi^2_{(r-1)(c-1)}$ distribution if the null hypothesis is true.

The model may be fit by maximum likelihood (see Cohen 1976 for details). Then, $\hat{a} = .3006$, and the estimated probabilities for the four cells are the following:

	S	N	
Male	0.2923	0.1112	0.4035
Female	0.2330	0.3636	0.5966
	0.5253	0.4748	1.0000

We can check the model by using a goodness-of-fit test for the clustered data in Table 10.1. This model does not exhibit significant lack of fit, whereas the model assuming independence does. For testing whether gender and schizophrenia are independent in the 2×2 table, $X^2/1.3006 = 13.76$, which we compare to a χ^2_1 distribution. The resulting p-value is 0.0002, about ten times as large as the p-value from the analysis that pretended siblings were independent.   ∎

10.4
Loglinear Models

If there are more than two classification variables, we are often interested in seeing if there are more complex relationships in the data. Loglinear models are commonly used to study these relationships.

10.4.1   Loglinear Models with Multinomial Sampling

In a two-way table, if the row variable and the column variable are independent, then $p_{ij} = p_{i+}p_{+j}$. Equivalently,

$$\ln p_{ij} = \ln p_{i+} + \ln p_{+j}$$
$$= \mu + \alpha_i + \beta_j,$$

where

$$\sum_{i=1}^{r} \alpha_i = 0 \quad \text{and} \quad \sum_{j=1}^{c} \beta_j = 0.$$

This is called a **loglinear model** because the logarithms of the cell probabilities follow a linear model: The model for independence in a 2×2 table may be written as

$$y = X\beta,$$

where

$$y = \begin{bmatrix} \ln(p_{11}) \\ \ln(p_{12}) \\ \ln(p_{21}) \\ \ln(p_{22}) \end{bmatrix}$$

$$X = \begin{bmatrix} 1 & 1 & 1 \\ 1 & 1 & -1 \\ 1 & -1 & 1 \\ 1 & -1 & -1 \end{bmatrix}$$

$$\beta = \begin{bmatrix} \mu \\ \alpha_1 \\ \beta_1 \end{bmatrix}.$$

The parameters β are estimated using the estimated probabilities $\hat{p}_{ij}$. For the data in Example 10.1, the estimated probabilities are as follows:

		Computer? Yes	No	
Cable?	Yes	0.238	0.376	0.614
	No	0.176	0.210	0.386
		0.414	0.586	1.000

The parameter estimates are $\hat{\mu} = -1.428$, $\hat{\alpha}_1 = 0.232$, and $\hat{\beta}_1 = -0.174$. The fitted values of $\hat{p}_{ij}$ for the model of independence are then

$$\hat{p}_{ij} = \exp(\hat{\mu} + \hat{\alpha}_i + \hat{\beta}_j)$$

and are given in the following table:

		Computer? Yes	No	
Cable?	Yes	0.254	0.360	0.614
	No	0.160	0.226	0.386
		0.414	0.586	1.000

We would also like to see how well this model fits the data. We can do that in two ways:

1   Test the goodness of fit of the model using either X^2 in (10.5) or G^2 in (10.6): For a two-way contingency table, these statistics are equivalent to the statistics for testing independence. For the computer/cable example, the likelihood ratio statistic for goodness of fit is 2.27. In multinomial sampling, X^2 and G^2 approximately follow a $\chi^2_{(r-1)(c-1)}$ distribution if the model is correct.

2   A full, or saturated, model for the data can be written as

$$\ln p_{ij} = \mu + \alpha_i + \beta_j + (\alpha\beta)_{ij}$$

with $\sum_{i=1}^{r}(\alpha\beta)_{ij} = \sum_{j=1}^{c}(\alpha\beta)_{ij} = 0$. The last term is analogous to the interaction term in a two-way ANOVA model. This model will give a perfect fit to the observed cell probabilities because it has rc parameters. The null hypothesis of independence is equivalent to

$$H_0 : (\alpha\beta)_{ij} = 0 \qquad \text{for } i = 1, \ldots, r-1; \; j = 1, \ldots, c-1.$$

Standard statistical packages such as SAS give estimates of the $(\alpha\beta)_{ij}$'s and their asymptotic standard errors under multinomial sampling. For the saturated model in the computer/cable example, SAS PROC CATMOD gives the following:

Effect	Parameter	Estimate	Standard Error	Chi-Square	Prob
CABLE	1	0.2211	0.0465	22.59	0.0000
COMP	2	-0.1585	0.0465	11.61	0.0007
CABLE*COMP	3	-0.0702	0.0465	2.28	0.1313

The values in the column "Chi-Square" are the Wald test statistics for testing whether that parameter is zero. Thus, the p-value, under multinomial sampling, for testing whether the interaction term is zero is 0.1313—again, for this example, this is exactly the same as the p-value from the test for independence.

10.4.2   Loglinear Models in a Complex Survey

What happens in a complex survey? We obtain point estimates of the model parameters like we always do, by using weights. Thus, we estimate the p_{ij}'s by (10.7) and use the estimates $\hat{p}_{ij}$ in standard software for estimating the model parameters. But, as usual, the test statistics for goodness of fit and the asymptotic standard errors for the parameter estimates given by SAS are wrong. Scheuren (1973) discusses some of the challenges in fitting loglinear models to CPS data.

Many of the same corrections used for chi-square tests of independence can also be used for hypothesis tests in loglinear models. Rao and Thomas (1988; 1989) and Fay (1985) describe various tests of goodness of fit for contingency tables from complex surveys; these include Wald tests, jackknife, and first- and second-order corrections to X^2 and G^2.

The Bonferroni inequality may also be used to compare nested loglinear models. For testing independence in a two-way table, for example, we compare the saturated model with the reduced model of independence and test each of the $m = (r-1)(c-1)$ null hypotheses

$$H_0(1) : (\alpha\beta)_{11} = 0$$

$$\vdots$$

$$H_0(m) : (\alpha\beta)_{(r-1)(c-1)} = 0$$

separately at level α/m.

More generally, we can compare any two nested loglinear models using this method. For a three-dimensional $r \times c \times d$ table, let

$$\mathbf{y} = [\ln(p_{111}), \ln(p_{112}), \ldots, \ln(p_{rcd})]^T.$$

Suppose the smaller model is

$$\mathbf{y} = \mathbf{X}\boldsymbol{\beta},$$

and the larger model is

$$\mathbf{y} = \mathbf{X}\boldsymbol{\beta} + \mathbf{Z}\boldsymbol{\theta}$$

where $\boldsymbol{\theta}$ is a vector of length m. Then we can fit the larger model and perform m separate hypothesis tests of the null hypotheses

$$H_0 : \theta_i = 0,$$

each at level α/m, by comparing $\hat{\theta}_i/\mathrm{SE}(\hat{\theta}_i)$ to a t distribution.

EXAMPLE 10.9 Let's look at a three-dimensional table from the Survey of Youth in Custody, to examine relationships among the variables "Was anyone in your family ever incarcerated?" (*famtime*), "Have you ever been put on probation or sent to a correctional institution for a violent offense?" (*everviol*), and *age*, for observations with no missing data. The cell probabilities are p_{ijk}. The estimated probabilities $\hat{p}_{ijk}$, estimated using weights, are in the following table:

		Family Member Incarcerated?				
		No		Yes		
		Ever Violent?		Ever Violent?		
		No	Yes	No	Yes	
	≤15	0.0588	0.0698	0.0659	0.0856	0.2801
Age Class	16–17	0.0904	0.1237	0.0944	0.1375	0.4461
	≥18	0.0435	0.0962	0.0355	0.0986	0.2738
		0.1928	0.2896	0.1959	0.3217	1.0000

The saturated model for the three-way table is

$$\log p_{ijk} = \mu + \alpha_i + \beta_j + \gamma_k + (\alpha\beta)_{ij} + (\alpha\gamma)_{ik} + (\beta\gamma)_{jk} + (\alpha\beta\gamma)_{ijk}.$$

SAS PROC CATMOD, using the weights, gives the following parameter estimates for the saturated model:

```
                                     Standard    Chi-
Effect                    Parameter  Estimate    Error    Square    Prob
--------------------------------------------------------------------------
AGECLASS                      1      -0.1149    0.00980   137.45   0.0000
                              2       0.3441    0.00884  1515.52   0.0000
EVERVIOL                      3      -0.2446    0.00685  1275.26   0.0000
AGECLASS*EVERVIOL             4       0.1366    0.00980   194.27   0.0000
                              5       0.0724    0.00884    67.04   0.0000
FAMTIME                       6       0.0242    0.00685    12.51   0.0004
AGECLASS*FAMTIME              7       0.0555    0.00980    32.03   0.0000
                              8       0.0128    0.00884     2.10   0.1473
EVERVIOL*FAMTIME              9      -0.0317    0.00685    21.42   0.0000
AGECLAS*EVERVIOL*FAMTIME     10       0.0089    0.00980     0.82   0.3646
                             11       0.0161    0.00884     3.33   0.0680
```

Because this is a complex survey and because SAS acts as though the sample size is $\sum w_i$ when the weights are used, the standard errors and p-values given for the parameters are completely wrong. We can estimate the variance of each parameter, however, by refitting the loglinear model on each of the random groups and using the random group estimate of the variance to perform hypothesis tests on individual parameters. The random group standard errors for the 11 model parameters are given in Table 10.2. The null hypothesis of no interactions among variables is

$$H_0 : (\alpha\beta)_{ij} = (\alpha\gamma)_{ik} = (\beta\gamma)_{jk} = (\alpha\beta\gamma)_{ijk} = 0;$$

or, using the parameter numbering in SAS,

$$H_0 : \beta_4 = \beta_5 = \beta_7 = \beta_8 = \beta_9 = \beta_{10} = \beta_{11} = 0.$$

This null hypothesis has seven components; to use the Bonferroni test, we test each individual parameter at the $0.05/7$ level. The $(1 - .05/14)$ percentile of a t_6 distribution

T A B L E   10.2

Random Group Standard Errors for Example 10.9

Parameter	Estimate	Standard Error	Test Statistic
1	−0.1149	0.1709	−0.67
2	0.3441	0.0953	3.61
3	−0.2446	0.0589	−4.15
4	0.1366	0.0769	1.78
5	0.0724	0.0379	1.91
6	0.0242	0.0273	0.89
7	0.0555	0.0191	2.91
8	0.0128	0.0218	0.59
9	−0.0317	0.0233	−1.36
10	0.0089	0.0191	0.47
11	0.0161	0.0167	0.96

is 4.0; none of the test statistics $\hat{\beta}_i/\text{SE}(\hat{\beta}_i)$, for $i = 4, 5, 7, 8, 9, 10, 11$, exceed that critical value, so we would not reject the null hypothesis that all three variables are independent. We might want to explore the *ageclass*famtime* interaction further, however. ∎

The survey packages SUDAAN, PC CARP, and WesVarPC, among others, perform hypothesis tests using data from complex surveys. These packages were briefly discussed in Section 9.6.

10.5
Exercises

1 Find an example or exercise in an introductory statistics textbook that performs a chi-square test on data from a survey. What design do you think was used for the survey? Is a chi-square test for multinomial sampling appropriate for the data? Why, or why not?

2 Read one of the following articles or another research article in which a categorical data analysis is performed on data from a complex survey. Describe the sampling design and the method of analysis. Did the authors account for the design in their data analysis? Should they have?

Gold, M. R., R. Hurley, T. Lake, T. Ensor, and R. Berenson. 1995. A national survey of the arrangements managed-care plans make with physicians. *New England Journal of Medicine* 333: 1689–1683.

Koss, M. P., C. A. Gidycz, and N. Wisniewski. 1987. The scope of rape: Incidence and prevalence of sexual aggression and victimization in a national sample of higher education students. *Journal of Consulting and Clinical Psychology* 55: 162–170.

Lipton, R. B., W. F. Stewart, D. D. Celentano, and M. L. Reed. 1992. Undiagnosed migraine headaches: A comparison of symptom-based and reported physician diagnosis. *Archives of Internal Medicine* 152: 1273–1278.

Sarti, E., P. M. Schantz, A. Plancarte, M. Wilson, I. Gutierrez, A. Lopez, J. Roberts, and A. Flisser. 1992. Prevalence and risk factors for *Taenia Solium* taeniasis and cysticercosis in humans and pigs in a village in Morelos, Mexico. *American Journal of Tropical Medicine and Hygiene* 46: 677–685.

3 Schei and Bakketeig (1989) took an SRS of 150 women between 20 and 49 years of age from the city of Trondheim, Norway. Their goal was to investigate the relationship between sexual and physical abuse by a spouse and certain gynecological symptoms in the women. Of the 150 women selected to be in the sample, 15 had moved, 1 had died, 3 were excluded because they were not eligible for the study, and 13 refused to participate.

Of the 118 women who participated in the study, 20 reported some type of sexual or physical abuse from their spouse: Eight reported being hit, 2 being kicked or bitten, 7 being beaten up, and 3 being threatened or cut with a knife. Seventeen of the women in the study reported a gynecological symptom of irregular bleeding or pelvic pain. The numbers of women falling into the four categories of gynecological symptom and abuse by spouse are given in the following contingency table:

		Abuse		
		No	Yes	
	No	89	12	101
Gynecological Symptom Present?				
	Yes	9	8	17
		98	20	118

a If abuse and presence of gynecological symptoms are not associated, what are the expected probabilities in each of the four cells?

b Perform a chi-square test of association for the variables *abuse* and *presence of gynecological symptoms*.

c What is the response rate for this study? Which definition of response rate did you use? Do you think that the nonresponse might affect the conclusions of the study? Explain.

4 Samuels (1996) collected data to examine how well students do in follow-up courses if the prerequisite course is taught by a part-time or full-time instructor. The following table gives results for students in Math I and Math II.

Instructor for Math I	Instructor for Math II	Grade in Math II		
		A, B, or C	D, F, or Withdraw	
Full time	Full time	797	461	1258
Full time	Part time	311	181	492
Part time	Full time	570	480	1050
Part time	Part time	909	449	1358
		2587	1571	4158

a The null hypothesis here is that the proportion of students receiving an A, B, or C is the same for each of the four combinations of instructor type. Is this a test of independence, homogeneity, or goodness of fit?

b Perform a hypothesis test for the null hypothesis in part (a), assuming students are independent.

c Do you think the assumption that students are independent is valid? Explain.

5 Use the file winter.dat for this exercise. The data were first discussed in Exercise 20 of Chapter 4.

a Test the null hypothesis that *class* is not associated with *breakaga*. In the context of Section 10.1, what type of sampling was done?

b Now construct a 2×2 contingency table for the variables *breakaga* and *work*. Use the sampling weights to estimate the probabilities p_{ij} for each cell.

c Calculate the odds ratio using the $\hat{p}_{ij}$ from part (b). How does this compare with an odds ratio calculated using the observed counts (and ignoring the sampling weights)?

d Estimate $\theta = p_{11}p_{22} - p_{21}p_{12}$ using the $\hat{p}_{ij}$ you calculated in part (b).

e Test the null hypothesis $H_0 : \theta = 0$.

f How did the stratification affect the hypothesis test?

6 Use the file teachers.dat for this exercise. The data were first discussed in Exercise 16 of Chapter 5.

  a Construct a new variable *zassist,* which takes on the value 1 if a teacher's aide spends any time assisting the teacher, and 0 otherwise. Construct another new variable *zprep,* which takes on values low, medium, and high based on the amount of time the teacher spends in school on preparation.

  b Construct a 2×3 contingency table for the variables *zassist* and *zprep.* Use the sampling weights to estimate the probabilities p_{ij} for each cell.

  c Using the Bonferroni method, test the null hypothesis that *zassist* is not associated with *zprep.*

7 Some researchers have used the following method to perform tests of association in two-way tables. Instead of using the original observation weights w_k, define

$$w_k^* = \frac{n w_k}{\sum_{l \in \mathcal{S}} w_l},$$

where n is the number of observation units in the sample. The sum of the new weights w_k^*, then, is n. The "observed" count for cell (i, j) is

$$x_{ij} = \text{sum of the } w_k^* \text{ for observations in cell } (i, j),$$

and the "expected" count for cell (i, j) is

$$\hat{m}_{ij} = \frac{x_{i+}x_{+j}}{n}.$$

Then compare the test statistic

$$\sum_{i=1}^{r} \sum_{j=1}^{c} \frac{(x_{ij} - \hat{m}_{ij})^2}{\hat{m}_{ij}}$$

to a $\chi^2_{(r-1)(c-1)}$ distribution.
Does this test give correct *p*-values for data from a complex survey? Why, or why not? HINT: Try it out on the data in Examples 10.1 and 10.4.

*8 (Requires calculus.) Consider X_W^2 in (10.8).

  a Use the linearization method of Section 9.1 to approximate $V(\hat{\theta})$ in terms of $V(\hat{p}_{ij})$ and $\text{Cov}(\hat{p}_{ij}, \hat{p}_{kl})$.

  b What is the Wald statistic, using the linearization estimate of $V(\hat{\theta})$ in part (a), when multinomial sampling is used? (Under multinomial sampling, $V(\hat{p}_{ij}) = p_{ij}(1 - p_{ij})/n$ and $\text{Cov}(\hat{p}_{ij}, \hat{p}_{kl}) = -p_{ij}p_{kl}/n$.) Is this the same as Pearson's X^2 statistic?

9 (Requires calculus.) *Estimating the log odds ratio in a complex survey.* Let

$$\theta = \log\left(\frac{p_{11}p_{22}}{p_{12}p_{21}}\right) \quad \text{and} \quad \hat{\theta} = \log\left(\frac{\hat{p}_{11}\hat{p}_{22}}{\hat{p}_{12}\hat{p}_{21}}\right).$$

  a Use the linearization method of Section 9.1 to approximate $V(\hat{\theta})$ in terms of $V(\hat{p}_{ij})$ and $\text{Cov}(\hat{p}_{ij}, \hat{p}_{kl})$.

  b Using part (a), what is $\hat{V}(\hat{\theta})$ under multinomial sampling?

10 Show that for multinomial sampling, $X_F^2 = X^2$. HINT: What is $E[X^2]$ in (10.9) for a multinomial sample?

*11 (Requires mathematical statistics and linear algebra.) *Deriving the first- and second-order corrections to Pearson's X^2 (see Rao and Scott 1981).*

    a Suppose the random vector $\mathbf{Y}$ is normally distributed with mean $\mathbf{0}$ and covariance matrix $\mathbf{\Sigma}$. Then, if $\mathbf{C}$ is symmetric, show that $\mathbf{Y}^T \mathbf{C Y}$ has the same distribution as $\sum \lambda_i W_i$, where the W_i's are independent χ_1^2 random variables and the λ_i's are the eigenvalues of $\mathbf{C\Sigma}$.

    b Let $\hat{\boldsymbol{\theta}} = (\hat{\theta}_{11}, \ldots, \hat{\theta}_{1,(c-1)}, \ldots, \hat{\theta}_{(r-1),1}, \ldots, \hat{\theta}_{(r-1)(c-1)})^T$, where $\hat{\theta}_{ij} = \hat{p}_{ij} - \hat{p}_{i+}\hat{p}_{+j}$. Let $\mathbf{A}$ be the covariance matrix of $\hat{\boldsymbol{\theta}}$ if a multinomial sample of size n is taken and the null hypothesis is true. Using part (a), argue that asymptotically $\hat{\boldsymbol{\theta}}^T \mathbf{A}^{-1} \hat{\boldsymbol{\theta}}$ has the same distribution as $\sum \lambda_i W_i$, where the W_i's are independent χ_1^2 random variables and the λ_i's are the eigenvalues of $\mathbf{A}^{-1} V(\hat{\boldsymbol{\theta}})$.

    c What are $E\left[\hat{\boldsymbol{\theta}}^T \mathbf{A}^{-1} \hat{\boldsymbol{\theta}}\right]$ and $V\left[\hat{\boldsymbol{\theta}}^T \mathbf{A}^{-1} \hat{\boldsymbol{\theta}}\right]$ in terms of the λ_i's?

    d Find $E\left[\hat{\boldsymbol{\theta}}^T \mathbf{A}^{-1} \hat{\boldsymbol{\theta}}\right]$ and $V\left[\hat{\boldsymbol{\theta}}^T \mathbf{A}^{-1} \hat{\boldsymbol{\theta}}\right]$ for a 2×2 table. You may want to use your answer in Exercise *8.

12 We know the clustering structure for the data in Example 10.8. Use results from Chapter 5 (assume one-stage cluster sampling) to estimate the proportion for each cell and margin in the 2×2 table and find the variance for each estimated proportion. Now use estimated deff's to perform a hypothesis test of independence using X_F^2. How do the results compare with the model-based test?

13 The following data are from the Canada Health Survey and given in Rao and Thomas (1989, 107). They relate smoking status (current smoker, occasional smoker, never smoked) to fitness level for 2505 persons. Smokers who had quit were not included in the analysis. The estimated proportions in the table below were found by applying the sample weights to the sample. The deff's are in brackets. We would like to test whether smoking status and fitness level are independent.

		Fitness level:			
		Recommended	Minimum acceptable	Unacceptable	
Smoking status:	Current	.220 [3.50]	.150 [4.59]	.170 [1.50]	.540 [1.44]
	Occasional	.023 [3.45]	.010 [1.07]	.011 [1.09]	.044 [2.32]
	Never	.203 [3.49]	.099 [2.07]	.114 [1.51]	.416 [2.44]
		.446 [4.69]	.259 [5.96]	.295 [1.71]	1

    a What is the value of X^2 if you assume the 2505 observations were collected in a multinomial sample? Of G^2? What is the p-value for each statistic under multinomial sampling, and why are these p-values incorrect?

    b Using (10.9), find the approximate expected value of X^2 and G^2.

    c Calculate the corrected statistics X_F^2 and G_F^2 for these data and find p-values for the hypothesis tests. Does the clustering in the Canada Health Survey make a difference in the p-value you obtain?

14 The following data are from Rao and Thomas (1988) and were collected in the Canadian Class Structure Survey, a stratified multistage sample collected in 1982–1983 to study employment and social structure. Canada was divided into 35 strata by region and population size; two psu's were sampled in 34 of the strata, and one psu sampled in the 35th stratum. Variances were estimated using balanced repeated replication (BRR) using the 34 strata with two psu's. Estimated deff's are in brackets behind the estimated proportion for each cell.

	Males	Females	
Decision-making managers	0.103 [1.20]	0.038 [1.31]	0.141 [1.09]
Advisor-managers	0.018 [0.74]	0.016 [1.95]	0.034 [1.95]
Supervisors	0.075 [1.81]	0.043 [0.92]	0.118 [1.30]
Semi-autonomous workers	0.105 [0.71]	0.085 [1.85]	0.190 [1.44]
Workers	0.239 [1.42]	0.278 [1.15]	0.516 [1.86]
	0.540 [1.29]	0.460 [1.29]	

a What is the value of X^2 if you assume the 1463 persons were surveyed in an SRS? Of G^2? What is the p-value for each statistic under multinomial sampling, and why are these p-values incorrect?

b Using (10.9), find the approximate expected value of X^2 and G^2.

c How many degrees of freedom are associated with the BRR variance estimates?

d Calculate the first-order corrected statistics X_F^2 and G_F^2 for these data and find approximate p-values for the hypothesis tests. Does the clustering in the survey make a difference in the p-value you obtain?

e The second-order Rao–Scott correction gave test statistic $X_S^2 = 38.4$, with 3.07 df. How does the p-value obtained using the X_S^2 compare with the p-value from X_F^2?

SURVEY Exercises

15 Take an SRS of 400 households in Stephens County. Cross-classify the sample on two variables: whether the household has any children under age 12 and the number of televisions in the household (1, 2, or more than 2). Test the null hypothesis that the two variables are not associated.

16 Use your sample from the SURVEY Exercises in Chapter 5. Test the association between number of televisions (1, 2, 3 or more) and the price a household is willing to pay for cable TV (less than $10, $10 or more). What method did you use to account for the survey design?

Regression with Complex Survey Data*

Now he knew that he knew nothing fundamental and, like a lone monk stricken with a conviction of sin, he mourned, "If I only knew more! . . . Yes, and if I could only remember statistics!"

—Sinclair Lewis, *It Can't Happen Here*

EXAMPLE 11.1   How are maternal drug use and smoking related to birth weight and infant mortality? What variables are the best predictors of neonatal mortality? How is the birth weight of an infant related to that of older siblings?

In most of this book, we have emphasized estimating population means and totals—for example, how many low-birth-weight babies are born in the United States each year? Questions on the relation between variables, however, are often answered in statistics by using some form of a **regression analysis.** A response variable (for example, *birth weight*) is related to a number of explanatory variables (for example, *maternal smoking, family income,* and *maternal age*). We would like to use the resulting regression equation not only to identify the relationship among variables for our data but also to predict the value of the response for future infants or infants not included in the sample.

You know how to fit regression models if the "usual assumptions," reviewed in Section 11.1, are met. These assumptions are often not met for data from complex surveys, however. To answer the questions above, for example, you might want to use data from the 1988 Maternal and Infant Health Survey (MIHS) in the United States. The survey, collected by the Bureau of the Census for the National Center for Health Statistics, provides data on a number of factors related to pregnancy and infant health, including weight gain, smoking, and drug use during pregnancy; maternal exposure to toxic wastes and hazards; and complications during pregnancy and delivery (Sanderson et al. 1991). But, like most large-scale surveys, the MIHS is not a simple random sample (SRS). Stratified random samples were drawn from the 1988 vital records from the contiguous 48 states and the District of Columbia. The samples included 10,000 certificates of live birth from the 3,909,510 live births in 1988, 4000 reports of fetal death from the estimated 15,000 fetal deaths of 28 weeks' or more gestation, and 6000 certificates of death for infants under 1 year of age from the population of 38,910 such deaths. Because black infants have higher incidence of low birth weight and infant mortality than white infants, black infants had a higher sampling fraction than nonblack infants. Low-birth-weight infants were also over-sampled. Mothers in the sampled records were mailed a questionnaire asking about prenatal care; smoking, drinking, and drug use; family income; hospitalization; health of the baby; and a number of other related variables. After receiving permission from the mother, investigators also sent questionnaires to the prenatal-care providers and hospitals, asking about the mother's and baby's health before and after birth. ∎

As we found for analysis of contingency tables in Chapter 10, unequal probabilities of selection and the clustering and stratification of the sample complicate a statistical analysis. In the MIHS, the unequal-selection probabilities for infants in different strata may need to be considered when fitting regression models. If a survey involves clustering, as does the National Crime Victimization Survey (NCVS), then standard errors for the regression coefficients calculated under the assumption that observations are independent will be incorrect.

In this chapter, we explore how to do regression in complex sample surveys. We review the traditional model-based approach to regression analysis, as taught in intro-ductory statistics courses, in Section 11.1. In Section 11.2, we discuss a design-based approach to regression and give methods for calculating standard errors of regression coefficients. Section 11.3 contrasts design-based and model-based approaches, Section 11.4 presents a model-based approach, and Section 11.5 applies these ideas to logistic regression.

We already used regression estimation in Chapter 3. In Chapter 3, though, the emphasis was on using information in an auxiliary variable to increase the precision of the estimate of the population total, $t_y = \sum_{i=1}^{N} y_i$. In Sections 11.1 to 11.5, our primary interest is in exploring the relation among different variables, and thus in estimating the regression coefficients. In Section 11.6 we return to the use of regression for improving the precision of estimated totals.

11.1
Model-Based Regression in Simple Random Samples

As usually exposited in areas of statistics other than sampling, regression inference is based on a model that is assumed to describe the relationship between the explanatory variable, x, and the response variable, y. The straight-line model commonly used for a single explanatory variable is

$$Y_i \mid x_i = \beta_0 + \beta_1 x_i + \varepsilon_i, \tag{11.1}$$

where Y_i is a random variable for the response, x_i is an explanatory variable, and β_0 and β_1 are unknown parameters. The Y_i's are random variables; the data collected in the sample are one realization of those random variables, $y_i, i \in \mathcal{S}$. The ε_i's, the deviations of the response variable about the line described by the model, are assumed to satisfy conditions (A1) through (A3):

(A1) $E[\varepsilon_i] = 0$ for all i. In other words, $E[Y_i \mid x_i] = \beta_0 + \beta_1 x_i$.

(A2) $V[\varepsilon_i] = \sigma^2$ for all i. The variance about the regression line is the same for all values of x.

(A3) $\text{Cov}[\varepsilon_i, \varepsilon_j] = 0$ for $i \neq j$. Observations are uncorrelated.

Often, (A4) is also assumed: It implies (A1) through (A3) and adds the additional assumption of normally distributed ε_i's.

(A4) Conditionally on the x_i's, the ε_i's are independent and identically distributed from a normal distribution with mean 0 and variance σ^2.

The **ordinary least squares** (OLS) **estimates** of the parameters are the values of β_0 and β_1 that minimize the residual sum of squares $\sum [y_i - (\beta_0 + \beta_1 x_i)]^2$. Estimators of the slope β_1 and intercept β_0 are obtained by solving the **normal equations:** For the model in (11.1), these are

$$\beta_0 \, n \quad + \beta_1 \sum x_i = \sum y_i$$
$$\beta_0 \sum x_i + \beta_1 \sum x_i^2 = \sum x_i y_i.$$

Solving the normal equations gives the parameter estimates

$$\hat{\beta}_1 = \frac{\sum x_i y_i - \dfrac{\left(\sum x_i\right)\left(\sum y_i\right)}{n}}{\sum x_i^2 - \dfrac{\left(\sum x_i\right)^2}{n}} \tag{11.2}$$

$$\hat{\beta}_0 = \frac{\sum y_i - \hat{\beta}_1 \sum x_i}{n}.$$

Both $\hat{\beta}_1$ and $\hat{\beta}_0$ are linear estimates in y, as we can write each in the form $\sum a_i y_i$ for known constants a_i. Although not usually taught in this form, it is equivalent to (11.2) to write

$$\hat{\beta}_1 = \sum_i \left[\frac{x_i - \left(\sum x_j\right)/n}{\sum x_j^2 - \left(\sum x_j\right)^2/n} \right] y_i$$

and

$$\hat{\beta}_0 = \sum_i \frac{1}{n} \left[1 - \frac{x_i \sum x_j - \left(\sum x_j\right)^2/n}{\sum x_j^2 - \left(\sum x_j\right)^2/n} \right] y_i.$$

If assumptions (A1) to (A3) are satisfied, then $\hat{\beta}_0$ and $\hat{\beta}_1$ are the **best linear unbiased estimates**—that is, among all linear estimates that are unbiased under model (11.1), $\hat{\beta}_0$ and $\hat{\beta}_1$ have the smallest variance. If assumption (A4) is met, we can use the t distribution to construct confidence intervals and hypothesis tests for the slope and intercept of the "true" regression line. Under assumption (A4),

$$\frac{\hat{\beta}_1 - \beta_1}{\sqrt{\hat{V}_M(\hat{\beta}_1)}}$$

follows a t distribution with $n - 2$ degrees of freedom (df). The subscript M refers to the use of the model to estimate the variance; for model (11.1), a model-unbiased estimator of the variance is

$$\hat{V}_M(\hat{\beta}_1) = \frac{\sum(y_i - \hat{\beta}_0 - \hat{\beta}_1 x_i)^2/(n-2)}{\sum(x_i - \bar{x})^2}. \tag{11.3}$$

The coefficient of determination R^2 in straight-line regression is

$$R^2 = 1 - \frac{\displaystyle\sum_{i=1}^{n}(y_i - \hat{\beta}_0 - \hat{\beta}_1 x_i)^2}{\displaystyle\sum_{i=1}^{n}(y_i - \bar{y})^2}.$$

These are the results obtained from any good statistical software package.

EXAMPLE **11.2**   To illustrate regression in the setting just discussed, we use data from Macdonell (1901), giving the length (cm) of the left middle finger and height (inches) for 3000 criminals. At the end of the nineteenth century, it was widely thought that criminal tendencies might also be expressed in physical characteristics that were distinguishable from the physical characteristics of noncriminal classes. Macdonell compared means and correlations of anthropometric measurements of the criminals to those of Cambridge men (presumed to come from a different class in society). This is an important data set in the history of statistics—it is the one Student (1908) used to demonstrate the t distribution. The entire data set for the 3000 criminals is in the file anthrop.dat.

An SRS of 200 individuals (file anthsrs.dat) was taken from the 3000 observations. Fitting a straight-line model with $y = $ height and $x = $ (length of left middle finger) with S-PLUS results in the following output:

	Value	SE	t-value	$\Pr(>|t|)$
Intercept	30.3162	2.5668	11.8109	0.0000
x	3.0453	0.2217	13.7348	0.0000

The sample data are plotted along with the OLS regression line in Figure 11.1. The model appears to be a good fit to the data ($R^2 = 0.49$), and, using the model-based analysis, a 95% confidence interval for the slope of the line is

$$3.0453 \pm 1.972(0.2217) = [2.61, 3.48].$$

If we generated samples of size 200 from the model in (11.1) over and over again and constructed a confidence interval for the slope for each sample, we would expect 95% of the resulting confidence intervals to include the true value of β_1. ∎

FIGURE   **11.1**

A plot of height vs. finger length for an SRS of 200 observations. The area of each circle is proportional to the number of observations at that value of (x, y). The OLS regression line, drawn in, has equation $y = 30.32 + 3.05x$.

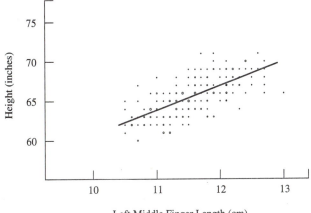

Left Middle Finger Length (cm)

Here are some remarks relevant to the application of regression to survey data:

1   No assumptions whatsoever are needed to calculate the estimates $\hat{\beta}_0$ and $\hat{\beta}_1$ from the data; these are simply formulas. The assumptions in (A1) to (A4) are needed to make *inferences* about the "true" but unknown parameters β_0 and β_1 and about predicted values of the response variable. So the assumptions are used only when we construct a confidence interval for β_1 or for a predicted value, or when we want to say, for example, that $\hat{\beta}_1$ is the best linear unbiased estimate of β_1.

The same holds true for other statistics we calculate. If we take a convenience sample of 100 persons, we may always calculate the average of those persons' incomes. But we cannot assess the accuracy of that statistic unless we make model assumptions about the population and sample. With a probability sample, however, we can use the sample design itself to make inferences and do not need to make assumptions about the model.

2   If the assumptions are not at least approximately satisfied, model-based inferences about parameters and predicted values will likely be wrong. For example, if observations are positively correlated rather than independent, the variance estimate from (11.3) is likely to be smaller than it should be. Consequently, regression coefficients are likely to be deemed statistically significant more often than they should be, as demonstrated in Kish and Frankel (1974).

3   We can partially check the assumptions of the model by plotting the residuals and using various diagnostic statistics as described in the regression books listed in the reference section. One commonly used plot is that of residuals versus predicted values, used to check (A1) and (A2). For the data in Example 11.2, this plot is shown in Figure 11.2 and gives no indication that the data in the sample violate assumptions (A1) or (A2). (This does not mean that the assumptions are true, just that we see nothing in the plot to indicate that they do not hold. Some of the assumptions,

FIGURE   11.2

A plot of residuals for model-based analysis of criminal height data, using the SRS plotted in Figure 11.1. No patterns are apparent, other than the diagonal lines caused by the integer-valued response variable.

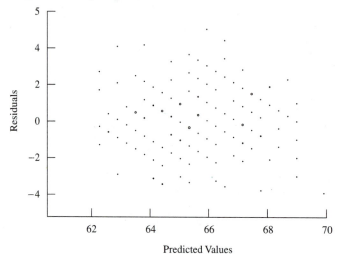

particularly independence, are quite difficult to check in practice.) However, we have no way of knowing whether observations not in the sample are fit by this model unless we actually see them.

4   Regression is not limited to variables related by a straight line. Let y be birth weight and let x take on the value 1 if the mother is black and 0 if the mother is not black. Then, the regression slope estimates the difference in mean birth weight for black and nonblack mothers, and the test statistic for $H_0 : \beta_1 = 0$ is the pooled t-test statistic for the null hypothesis that the mean birth weight for blacks is the same as the mean birth weight for nonblacks. Thus, comparison of means for subpopulations can be treated as a special case of regression analysis.

11.2
Regression in Complex Surveys

Many investigators performing regression analyses on complex survey data simply run the data through a standard statistical analysis program such as SAS or SPSS and report the model and standard errors given by the software. One may debate whether to take a model-based or design-based approach (and we shall, in Section 11.3), but the data structure needs to be taken into account in either approach.

What can happen in complex surveys?

1   Observations may have different probabilities of selection, π_i. If the probability of selection is related to the response variable y_i, then an analysis that does not account for the different probabilities of selection may lead to biases in the estimated regression parameters. This problem is discussed in detail by Nathan and Smith (1989), who give a bibliography of related literature.

For example, suppose an unequal-probability sample of 200 men is taken from the population described in Example 11.2 and that the selection probabilities are higher for the shorter men. (For illustration purposes, I used the y_i's to set the selection probabilities, with π_i proportional to 24 for $y < 65$, 12 for $y = 65$, 2 for $y = 66$ or 67, and 1 for $y > 67$, with data in the file anthuneq.dat.) Figure 11.3 shows a scatterplot of the data from this sample, along with the OLS regression line described in Section 11.1. The OLS regression equation is $y = 43.41 + 1.79x$, compared with the equation $y = 30.32 + 3.05x$ for the SRS in Example 11.2. Ignoring the selection probabilities in this example leads to a very different estimate of the regression line.

Nonrespondents, who may be thought of as having zero probability of selection, can distort the relationship for much the same reason. If the nonrespondents in the MIHS are more likely to have low-birth-weight infants, then a regression model predicting birth weight from explanatory variables may not fit the nonrespondents. Item nonresponse may have similar effects.

The stratification of the MIHS would also need to be taken into account. The survey was stratified because the investigators wanted to ensure an adequate sample size for blacks and low-birth-weight infants. It is certainly plausible that each stratum may have its own regression line, and postulating a single straight line to fit all the data may hide some of the information in the data.

2 Even if the estimators of the regression parameters are approximately design unbiased, the standard errors given by SAS or SPSS will likely be wrong if the survey design involves clustering. Usually, with clustering, the design effect (deff) for regression coefficients will be greater than 1.

FIGURE 11.3

A plot of y vs. x for an unequal-probability sample of 200 criminals. The area of each circle is proportional to the number of observations at that data point. The OLS line is $y = 43.41 + 1.79x$. The smaller slope of this line, when compared with the slope 3.05 for the SRS in Figure 11.1, reflects the undersampling of tall men.

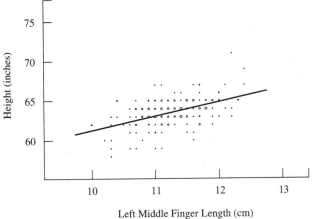

Left Middle Finger Length (cm)

11.2.1 Point Estimation

Traditionally, design-based sampling theory has been concerned with estimating quantities from a finite population, quantities such as $t_y = \sum_{i=1}^{N} y_i$ or $\bar{y}_U = t_y/N$. In that descriptive spirit, then, the finite population quantities of interest for regression are the least squares coefficients for the population, B_0 and B_1, that minimize

$$\sum_{i=1}^{N} (y_i - B_0 - B_1 x_i)^2$$

over the entire finite population. It would be nice if the equation $y = B_0 + B_1 x$ summarizes useful information about the population (otherwise, why are you really interested in B_0 and B_1?), but no assumptions are necessary to say that these could be the quantities of interest. As in Section 11.1, the normal equations are

$$B_0 N \qquad + B_1 \sum_{i=1}^{N} x_i = \sum_{i=1}^{N} y_i$$

$$B_0 \sum_{i=1}^{N} x_i + B_1 \sum_{i=1}^{N} x_i^2 = \sum_{i=1}^{N} x_i y_i,$$

and B_0 and B_1 can be expressed as functions of the population totals:

$$B_1 = \frac{\sum_{i=1}^{N} x_i y_i - \left(\sum_{i=1}^{N} x_i\right)\left(\sum_{i=1}^{N} y_i\right)\Big/ N}{\sum_{i=1}^{N} x_i^2 - \left(\sum_{i=1}^{N} x_i\right)^2 \Big/ N} = \frac{t_{xy} - \dfrac{t_x t_y}{N}}{t_{x^2} - \dfrac{(t_x)^2}{N}} \qquad (11.4)$$

$$B_0 = \frac{\sum_{i=1}^{N} y_i - B_1 \sum_{i=1}^{N} x_i}{N} = \frac{t_y - B_1 t_x}{N}. \qquad (11.5)$$

We know the values for the entire population for the sample drawn in Example 11.2. These population values are plotted in Figure 11.4, along with the population least squares line $y = 30.179 + 3.056x$.

As both B_0 and B_1 are functions of population totals, we can use methods derived in earlier chapters to estimate each total separately and then substitute estimates into (11.4) and (11.5). We estimate each population total in (11.4) and (11.5) using weights, to obtain

$$\hat{B}_1 = \frac{\displaystyle\sum_{i\in\mathcal{S}} w_i x_i y_i - \frac{\left(\displaystyle\sum_{i\in\mathcal{S}} w_i x_i\right)\left(\displaystyle\sum_{i\in\mathcal{S}} w_i y_i\right)}{\displaystyle\sum_{i\in\mathcal{S}} w_i}}{\displaystyle\sum_{i\in\mathcal{S}} w_i x_i^2 - \frac{\left(\displaystyle\sum_{i\in\mathcal{S}} w_i x_i\right)^2}{\displaystyle\sum_{i\in\mathcal{S}} w_i}} \qquad (11.6)$$

FIGURE   11.4

A plot of population for 3000 criminals. The area of each circle is proportional to the number of population observations at those coordinates. The population OLS regression line is $y = 30.18 + 3.06x$.

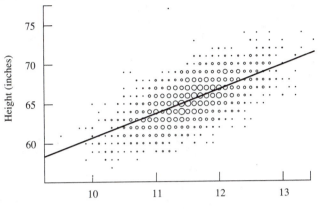

Left Middle Finger Length (cm)

$$\hat{B}_0 = \frac{\displaystyle\sum_{i \in S} w_i y_i - \hat{B}_1 \sum_{i \in S} w_i x_i}{\displaystyle\sum_{i \in S} w_i}. \tag{11.7}$$

Computational Note   Although (11.6) and (11.7) are correct expressions for the estimators, they are subject to roundoff error and are not as good for computation as other algorithms that have been developed. In practice, use professional software designed for estimating regression parameters in complex surveys. If you do not have access to such software, use any statistical regression package that calculates weighted least squares estimates. If you use weights w_i in the weighted least squares estimation, you will obtain the same point estimates as in (11.6) and (11.7); however, in complex surveys, the standard errors and hypothesis tests the software provides will be incorrect and should be ignored.

Plotting the Data   In any regression analysis, you *must* plot the data. Plotting multivariate data is challenging even for data from an SRS (Cook and Weisberg 1994 discuss regression graphics in depth). Data from a complex survey design—with stratification, unequal weights, and clustering—have even more features to incorporate into plots. In Figure 11.5, we indicate the weighting by circle area. The unequal-probability sample used on page 353 and in Example 11.3 has no clustering or stratification, though. If a survey has relatively few clusters or strata, you can plot the data separately for each, or indicate cluster membership using color. Graphics for survey data is an area currently being researched. Korn and Graubard (1998) independently develop some of the plots shown here and discuss other possible plots.

EXAMPLE   11.3   Let's estimate the finite population quantities B_0 and B_1 for the unequal-probability sample plotted in Figure 11.3. The point estimates, using the weights, are $\hat{B}_0 = 30.19$ and $\hat{B}_1 = 3.05$. If we ignored the weights and simply ran the observed data through a

FIGURE   11.5

A plot of data from an unequal-probability sample. The area of each circle is proportional to the sum of the weights for observations with that value of x and y. Note that the taller men in the sample also have larger weights, so the slope of the regression line using weights is drawn upward.

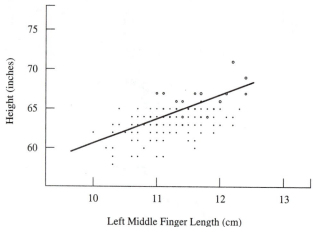

Left Middle Finger Length (cm)

standard regression program such as SAS PROC REG, we get very different estimates: $\hat{B}_0 = 43.41$ and $\hat{B}_1 = 1.79$.

Figure 11.5 shows why the weights, which were related to y, made a difference here. Taller men had lower selection probabilities and thus not as many of them appeared in the unequal-probability sample. However, the taller men that were selected had higher sampling weights; a 69-inch man in the sample represented 24 times as many population units as a 60-inch man in the sample. When the weights are incorporated, estimates of the parameters are computed as though there were actually w_i data points with values (x_i, y_i). ∎

11.2.2   Standard Errors

Let's now examine the effect of the complex sampling design on the standard errors. As $\hat{B}_0$ and $\hat{B}_1$ are functions of estimated population totals, methods from Chapter 9 may be used to calculate variance estimates.

For any method of estimating the variance, under certain regularity conditions an approximate $100(1 - \alpha)\%$ CI for B_1 is given by

$$\hat{B}_1 \pm t_{\alpha/2}\sqrt{\hat{V}(\hat{B}_1)},$$

where $t_{\alpha/2}$ is the upper $\alpha/2$ point of a t distribution with degrees of freedom associated with the variance estimate. For linearization, jackknife, or BRR (balanced repeated replication) in a stratified multistage sample, we would use (number of sampled psu's) − (number of strata) as the degrees of freedom. For the random group method of estimating the variance, the appropriate degrees of freedom would be (number of groups) − 1.

11.2.2.1 Standard Errors Using Linearization

The linearization variance estimator for the slope may be used because B_1 is a function of four population totals t_{xy}, t_x, t_y, and t_{x^2}. Using linearization, then, as you showed in Exercise 3 from Chapter 9,

$$V_L(\hat{B}_1) \approx V\left[\frac{\partial B_1}{\partial t_{xy}}(\hat{t}_{xy} - t_{xy}) + \frac{\partial B_1}{\partial t_x}(\hat{t}_x - t_x) + \frac{\partial B_1}{\partial t_y}(\hat{t}_y - t_y) + \frac{\partial B_1}{\partial t_{x^2}}(\hat{t}_{x^2} - t_{x^2}) \right]$$

$$= V\left[\left(t_{x^2} - \frac{(t_x)^2}{N}\right)^{-1} \sum_{i \in S} w_i(y_i - B_0 - B_1 x_i)\left(x_i - \frac{t_x}{N}\right) \right].$$

Define

$$q_i = (y_i - \hat{B}_0 - \hat{B}_1 x_i)(x_i - \hat{\bar{x}}),$$

where $\hat{\bar{x}} = \hat{t}_x/\hat{N}$. Then, we may use

$$\hat{V}_L(\hat{B}_1) = \frac{\hat{V}\left(\sum_{i \in S} w_i q_i\right)}{\left[\sum_{i \in S} w_i x_i^2 - \frac{\left(\sum_{i \in S} w_i x_i\right)^2}{\sum_{i \in S} w_i}\right]^2} \tag{11.8}$$

to estimate the variance of $\hat{B}_1$.

Note that the design-based variance estimator in (11.8) differs from the model-based variance estimator in (11.3), even if an SRS is taken. In an SRS of size n, ignoring the fpc (finite population correction),

$$\hat{V}\left(\sum_{i \in S} w_i q_i\right) = \hat{V}(\hat{t}_q) = \frac{N^2 s_q^2}{n},$$

with

$$s_q^2 = \frac{\sum_{i \in S}(x_i - \bar{x}_S)^2(y_i - \hat{B}_0 - \hat{B}_1 x_i)^2}{n - 1}.$$

Thus, if we ignore the fpc, (11.8) gives

$$\hat{V}_L(\hat{B}_1) = \frac{n \sum_{i \in S}(x_i - \bar{x}_S)^2(y_i - \hat{B}_0 - \hat{B}_1 x_i)^2}{(n - 1)\left[\sum_{i \in S}(x_i - \bar{x}_S)^2\right]^2}.$$

However, from (11.3),

$$\hat{V}_M(\hat{\beta}_1) = \frac{\sum_{i \in S}(y_i - \hat{\beta}_0 - \hat{\beta}_1 x_i)^2}{(n - 2)\sum_{i \in S}(x_i - \bar{x})^2}.$$

Why the difference? The design-based estimator of the variance $\hat{V}_L$ comes from the selection probabilities of the design, while $\hat{V}_M$ comes from the average squared deviation over all possible realizations of the model. Confidence intervals constructed from the two variance estimates have different interpretations. With the design-based confidence interval

$$\hat{B}_1 \pm t_{\alpha/2}\sqrt{\hat{V}_L(\hat{B}_1)},$$

the confidence level is $\sum u(\mathcal{S})P(\mathcal{S})$, where the sum is over all possible samples $\mathcal{S}$ that can be selected using the sampling design, $P(\mathcal{S})$ is the probability that sample $\mathcal{S}$ is selected, and $u(\mathcal{S}) = 1$ if the confidence interval constructed from sample $\mathcal{S}$ contains the population characteristic B_1 and $u(\mathcal{S}) = 0$ otherwise. In an SRS, the design-based confidence level is the proportion of possible samples that result in a confidence interval that includes B_1, from the set of all SRSs of size n from the finite population of fixed values $\{(x_1, y_1), (x_2, y_2), \ldots, (x_N, y_N)\}$.

For the model-based confidence interval

$$\hat{\beta}_1 \pm t_{\alpha/2}\sqrt{\hat{V}_M(\hat{\beta}_1)},$$

the confidence level is the expected proportion of confidence intervals that will include β_1, from the set of all samples that could be generated from the model in (A1) to (A3). Thus, the model-based estimator assumes that (A1) to (A3) hold for the infinite population mechanism that generates the data. The SRS design of the sample makes assumption (A3) (uncorrelated observations) reasonable. If a straight-line model describes the relation between x and y, then (A1) is also plausible. A violation of assumption (A2) (equal variances), however, can have a large effect on inferences. The linearization estimator of the variance is more robust to assumption (A2), as explored in Exercise 16.

EXAMPLE 11.4   For the SRS in Example 11.2, the model-based and design-based estimates of the variance are quite similar, as the model assumptions appear to be met for the sample and population. For these data, $\hat{B}_1 = \hat{\beta}_1$ because $w_i = 3000/200$ for all i; $\hat{V}_L(\hat{B}_1) = 0.048$; and $\hat{V}_M(\hat{\beta}_1) = (0.2217)^2 = 0.049$. In other situations, however, the estimates of the variance can be quite different; usually, if there is a difference, the linearization estimate of the variance is larger than the model-based estimate of the variance.

For the unequal-probability sample of 200 criminals, we define the new variable

$$q_i = (y_i - \hat{B}_0 - \hat{B}_1 x_i)(x_i - \hat{\bar{x}}) = (y_i - 30.1859 - 3.0541 x_i)(x_i - 11.51359).$$

(Note that $\hat{\bar{x}}$ is the estimate of $\bar{x}_U$ calculated using the unequal probabilities; the sample average of the 200 x_i's in the sample is 11.2475, which is quite a bit smaller.) Then, $\hat{V}\left(\sum_{i \in \mathcal{S}} w_i q_i\right) = 238{,}161$, and

$$\left[\sum_{i \in \mathcal{S}} w_i x_i^2 - \frac{\left(\sum_{i \in \mathcal{S}} w_i x_i\right)^2}{\sum_{i \in \mathcal{S}} w_i}\right]^2 = 688{,}508,$$

so $\hat{V}_L(\hat{B}_1) = 0.346$. If the weights are ignored, then the OLS analysis gives $\hat{\beta}_1 = 1.79$ and $\hat{V}_M(\hat{\beta}_1) = 0.05121169$. The estimated variance is much smaller using the model, but $\hat{\beta}_1$ is biased as an estimator of B_1. ∎

11.2.2.2   Standard Errors Using Jackknife

Suppose we have a stratified multistage sample, with weights w_i and H strata. A total of n_h psu's are sampled in stratum h. Recall (see Section 9.3.2) that for jackknife iteration j in stratum h, we omit all observation units in psu j and recalculate the estimate using the remaining units. Define

$$
w_{i(hj)} = \begin{cases} w_i & \text{if observation unit } i \text{ is not in stratum } h. \\ 0 & \text{if observation unit } i \text{ is in psu } j \text{ of stratum } h. \\ \dfrac{n_h}{n_h - 1} w_i & \text{if observation unit } i \text{ is in stratum } h \text{ but not in psu } j. \end{cases}
$$

Then, the jackknife estimator of the with-replacement variance of $\hat{B}_1$ is

$$
\hat{V}_{\text{JK}}(\hat{B}_1) = \sum_{h=1}^{H} \frac{n_h - 1}{n_h} \sum_{j=1}^{n_h} (\hat{B}_{1(hj)} - \hat{B}_1)^2, \tag{11.9}
$$

where $\hat{B}_1$ is defined in (11.6) and $\hat{B}_{1(hj)}$ is of the same form but with $w_{i(hj)}$ substituted for every occurrence of w_i in (11.6).

EXAMPLE 11.5   For our two samples of size 200 from the 3000 criminals,

$$
\hat{V}_{\text{JK}}(\hat{B}_1) = \frac{199}{200} \sum_{j=1}^{200} (\hat{B}_{1(j)} - \hat{B}_1)^2,
$$

where $\hat{B}_{1(j)}$ is the estimated slope when observation j is deleted and the other observations reweighted accordingly. The difference between the SRS and the unequal-probability sample is in the weights. For the SRS, the original weights are $w_i = 3000/200$; consequently, $w_{i(j)} = 200 w_i / 199 = 3000/199$ for $i \neq j$. Thus, for the SRS, $\hat{B}_{1(j)}$ is the OLS estimate of the slope when observation j is omitted. For the SRS, we calculate $\hat{V}_{\text{JK}}(\hat{B}_1) = 0.050$.

For the unequal-probability sample, the original weights are $w_i = 1/\pi_i$ and $w_{i(j)} = 200 w_i / 199$ for $i \neq j$. The new weights $w_{i(j)}$ are used to calculate $\hat{B}_{1(j)}$ for each jackknife iteration, giving $\hat{V}_{\text{JK}}(\hat{B}_1) = 0.461$. The jackknife estimated variance is larger than the linearization variance, as often occurs in practice. ∎

11.2.3   Multiple Regression Using Matrices

Now let's give results for multiple regression in general. We rely heavily on matrix results found in the linear models and regression books listed in the references at the end of the book. If you are not well versed in regression theory, you should learn that material before reading this section.

Suppose we wish to find a relation between y_i and a p-dimensional vector of explanatory variables $\mathbf{x}_i$, where $\mathbf{x}_i = [x_{i1}, x_{i2}, \ldots, x_{ip}]^T$. We wish to estimate the

p-dimensional vector of population parameters, $\mathbf{B}$, in the model $y = \mathbf{x}^T \mathbf{B}$. Define

$$
\mathbf{y}_U = \begin{bmatrix} y_1 \\ y_2 \\ \vdots \\ y_N \end{bmatrix} \quad \text{and} \quad \mathbf{X}_U = \begin{bmatrix} \mathbf{x}_1^T \\ \mathbf{x}_2^T \\ \vdots \\ \mathbf{x}_N^T \end{bmatrix}.
$$

The normal equations for the entire population are

$$
\mathbf{X}_U^T \mathbf{X}_U \mathbf{B} = \mathbf{X}_U^T \mathbf{y}_U ,
$$

and the finite population quantities of interest are, assuming that $(\mathbf{X}_U^T \mathbf{X}_U)^{-1}$ exists,

$$
\mathbf{B} = \left(\mathbf{X}_U^T \mathbf{X}_U \right)^{-1} \mathbf{X}_U^T \mathbf{y}_U ,
$$

which are the least squares estimates for the entire population.

Both $\mathbf{X}_U^T \mathbf{X}_U$ and $\mathbf{X}_U^T \mathbf{y}_U$ are matrices of population totals. The (j, k)th element of the $p \times p$ matrix $\mathbf{X}_U^T \mathbf{X}_U$ is $\sum_{i=1}^N x_{ij} x_{ik}$, and the kth element of the p-vector $\mathbf{X}_U^T \mathbf{y}_U$ is $\sum_{i=1}^N x_{ik} y_i$.

Thus, we can estimate the matrices $\mathbf{X}_U^T \mathbf{X}_U$ and $\mathbf{X}_U^T \mathbf{y}_U$ using weights. Let $\mathbf{X}_S$ be the matrix of explanatory values for the sample, $\mathbf{y}_S$ be the response vector of sample observations, and $\mathbf{W}_S$ be a diagonal matrix of the sample weights w_i. Then, the (j, k)th element of the $p \times p$ matrix $\mathbf{X}_S^T \mathbf{W}_S \mathbf{X}_S$ is $\sum_{i \in S} w_i x_{ij} x_{ik}$, which estimates the population total $\sum_{i=1}^N x_{ij} x_{ik}$; the kth element of the p-vector $\mathbf{X}_S^T \mathbf{W}_S \mathbf{y}_S$ is $\sum_{i \in S} w_i x_{ik} y_i$, which estimates the population total $\sum_{i=1}^N x_{ik} y_i$. Then, analogously to (11.6) and (11.7), define the estimator of $\mathbf{B}$ to be

$$
\hat{\mathbf{B}} = \left(\mathbf{X}_S^T \mathbf{W}_S \mathbf{X}_S \right)^{-1} \mathbf{X}_S^T \mathbf{W}_S \mathbf{y}_S . \tag{11.10}
$$

Let

$$
\mathbf{q}_i = \mathbf{x}_i^T \left(y_i - \mathbf{x}_i^T \hat{\mathbf{B}} \right).
$$

Then, using linearization, as shown in Shah et al. (1977),

$$
\hat{V} \left(\hat{\mathbf{B}} \right) = \left(\mathbf{X}_S^T \mathbf{W}_S \mathbf{X}_S \right)^{-1} \hat{V} \left(\sum_{i \in S} w_i \mathbf{q}_i \right) \left(\mathbf{X}_S^T \mathbf{W}_S \mathbf{X}_S \right)^{-1} . \tag{11.11}
$$

Confidence intervals for individual parameters may be constructed as

$$
\hat{B}_k \pm t \sqrt{\hat{V}(\hat{B}_k)},
$$

where t is the appropriate percentile from the t distribution. Korn and Graubard (1990) suggest using the Bonferroni method for simultaneous inference about m regression parameters, constructing a $100(1 - \alpha/m)\%$ CI for each of the parameters.

11.2.4 Regression Using Weights versus Weighted Least Squares

Many regression textbooks discuss regression estimation using weighted least squares as a remedy for unequal variances. If the model generating the data is

$$
Y_i = \mathbf{x}_i^T \beta + \varepsilon_i
$$

with ε_i independent and normally distributed with mean 0 and variance σ_i^2, then ε_i / σ_i follows a normal distribution with mean 0 and variance 1. The weighted least squares estimate is

$$\hat{\beta}_{\text{WLS}} = \left(\mathbf{X}^T \mathbf{\Sigma}^{-1} \mathbf{X}\right)^{-1} \mathbf{X}^T \mathbf{\Sigma}^{-1} \mathbf{y}$$

with $\mathbf{\Sigma} = \text{diag}(\sigma_1^2, \sigma_2^2, \dots, \sigma_n^2)$. The weighted least squares estimate minimizes $\sum(y_i - \mathbf{x}_i^T \beta)^2 / \sigma_i^2$ and gives observations with smaller variance more weight in determining the regression equation. If the model holds, then under weighted least squares theory,

$$V(\hat{\beta}) = \left(\mathbf{X}^T \mathbf{\Sigma}^{-1} \mathbf{X}\right)^{-1}.$$

We are *not* using weighted least squares in this sense, even though our point estimator is the same. Our weights come from the sampling design, not from an assumed covariance structure. Our estimated variance of the coefficients is not $(\mathbf{X}^T \hat{\mathbf{\Sigma}}^{-1} \mathbf{X})^{-1}$, the estimated variance under weighted least squares theory, but is

$$\left(\mathbf{X}_S^T \mathbf{W}_S \mathbf{X}_S\right)^{-1} \hat{V} \left[\sum_{i \in S} w_i \mathbf{x}_i^T \left(y_i - \mathbf{x}_i^T \hat{\mathbf{B}}\right) \right] \left(\mathbf{X}_S^T \mathbf{W}_S \mathbf{X}_S\right)^{-1}.$$

One may, of course, combine the weighted least squares approach as taught in regression courses with the finite population approach by defining the population quantities of interest to be

$$\mathbf{B} = \left(\mathbf{X}_U^T \mathbf{\Sigma}_U^{-1} \mathbf{X}_U\right)^{-1} \mathbf{X}_U^T \mathbf{\Sigma}_U^{-1} \mathbf{y}_U,$$

thus generalizing the regression model. This is essentially what is done in ratio estimation, using $\mathbf{\Sigma}_U = \text{diag}(x_1, x_2, \dots, x_N)$, as will be shown in Example 11.9.

11.2.5 Software for Regression in Complex Surveys

Several software packages among those discussed in Section 9.6 will calculate regression coefficients and their standard errors for complex survey data. SUDAAN and PC CARP both use linearization to calculate the estimated variances of parameter estimates. OSIRIS and WesVarPC use replication methods to estimate variances.

Before you use software written by someone else to perform a regression analysis on sample survey data, investigate how it deals with missing data. For example, if an observation is missing one of the x-values, SUDAAN, like SAS, excludes the observation from the analysis. If your survey has a large amount of item nonresponse on different variables, it is possible that you may end up performing your regression analysis using only 20 of the observations in your sample. You may want to consider amount of item nonresponse as well as scientific issues when choosing covariates for your model.

Many surveys conducted by government organizations do not release enough information on the public-use tapes to allow you to calculate estimated variances for regression coefficients. The 1990 NCVS public-use data set, for example, contains weights for each household and person in the sample but does not provide clustering information. Such surveys, however, often provide information on deff's for estimating population totals. In this situation, estimate the regression parameters using the provided weights. Then estimate the variance for the regression coefficients as though an SRS were taken and multiply each estimated variance by an overall deff for population totals. In general, deff's for regression coefficients tend to be (but do not have to be) smaller than deff's for estimating population means and totals, so multiplying estimated variances of regression coefficients by the deff often results in a conservative estimate of the variance (see Skinner 1989). Intuitively, this can be explained because a good regression model may control for some of the cluster-to-cluster variability in the response variable. For example, if part of the reason households in the same cluster tend to have more similar crime-victimization experiences is the average income level of the neighborhood, then we would expect that adjusting for income in the regression might account for some of the cluster-to-cluster variability. The residuals from the model would then show less effect from the clustering.

11.3
Should Weights Be Used in Regression?

In most areas of statistics, a regression analysis generally has one of three purposes:

1   It describes the relationship between two or more variables. Of interest may be the relationship between family income and the infant's birth weight or the relationship between education level, income, and likelihood of being a victim of violent crime. The interest is simply in a summary statistic that describes the association between the explanatory and response variables.

2   It predicts the value of y for a future observation. If we know the values for a number of demographic and health variables for an expectant mother, can we predict the birth weight of the infant or the probability of the infant's survival?

3   It allows us to control future values of y by changing the values of the explanatory variables. For this purpose, we would like the regression equation to give us a cause-and-effect relationship between x and y.

Survey data can be used for the first and second purposes, but they generally cannot be used to establish definitive causal relationships among variables.[1] Sample surveys generally provide observational, not experimental, data. We observe a subset of possible explanatory variables, and these do not necessarily include the variables that are the root causes of changes in y. In a health survey intended to study the relationship between nutrition, exercise, and cancer incidence, survey participants may be asked about their diet and exercise habits (or the researcher may observe them) and be followed up later to see whether they have contracted cancer. Suppose a regression analysis indicates a significant negative association between vitamin E intake and cancer incidence, after adjusting for other variables such as age. The analysis only establishes association, not causation; you cannot conclude that cancer incidence will decrease if you start feeding people vitamin E. Although vitamin E could be the cause of the decreased cancer incidence, the cause could also be one of the unmeasured variables that is associated with both vitamin E intake and cancer incidence. To conclude that vitamin E affects cancer incidence, you need to perform

---

[1] Many statisticians would say that survey data cannot be used to make causal statements in any shape or form. Experimental units must be randomly assigned to treatments in order to infer causation. Some surveys, however, such as the study in Example 8.2, include experimentation, and for these we can often conclude that a change in the treatment caused a change in the response.

an experiment: Randomly assign study participants to vitamin E and no-vitamin-E groups and observe the cancer incidence at a later time.

The purpose of a regression analysis often differs from that of an analysis to estimate population means and totals. When estimating the total number of unemployed persons from a survey, we are interested in the finite population quantity t_y; we want to estimate how many persons in the population in August 1994 were unemployed. But in a regression analysis, are you interested in B_0 and B_1, the summary statistics for the finite population? Or are you interested in uncovering a "universal truth"—to be able to say, for example, that not only do you find a positive association between amount of fat in diet and systolic blood pressure for the population studied, but also that you would expect a similar association in other populations? Cochran notes this point for comparison of domain means: "It is seldom of scientific interest to ask whether [the finite population domain means are equal], because these means would not be exactly equal in a finite population, except by rare chance. Instead, we test the null hypothesis that the two domains were drawn from *infinite* populations having the same mean" (1977, 39). Comparing domain means is a special case of linear regression (see Exercise 13), and Cochran's comments apply equally well to linear regression in general.

Many survey statisticians have debated whether the sampling weights are relevant for inference in regression; some of the papers involved in the debate are in the references for this chapter. Brewer and Mellor (1973) present an entertaining and insightful dialogue between a model-based and a design-based statistician who eventually reach a compromise; this dialogue is an excellent starting point for further study. These references provide a much deeper discussion of the issues involved than we present in this section; we try to summarize the various approaches and present the contributions of each to a good analysis of survey data.

Two basic approaches have been advocated:

1 *Design-based.* The design-based position was presented in the previous section. The quantities of interest are the finite population characteristics **B,** regardless of how well the model fits the population. Inferences are based on repeated sampling from the finite population, and the probability structure used for inference is that defined by the random variables indicating inclusion in the sample. A model that generates the data may exist, but we do not necessarily know what it is, so the analysis does not rely on any theoretical model. Weights are needed for estimating population means and totals and by analogy should be used in linear regression as well.

2 *Model-based.* A stochastic model describes the relation between y_i and x_i that holds for every observation in the population. One possible model is $Y_i \mid \mathbf{x}_i = \mathbf{x}_i^T \beta + \varepsilon_i$, with the ε_i's independent and normally distributed with constant variance. If the observations in the population really follow the model, then the sample design should have no effect as long as the probabilities of selection depend on y only through the x's. The value **B** is merely the least squares estimate of β if values for the whole population were known; since only a sample is known, use the OLS estimates

$$\hat{\beta}_{\text{OLS}} = \left(\mathbf{X}_S^T \mathbf{X}_S\right)^{-1} \mathbf{X}_S^T \mathbf{y}_S.$$

Search for a model that can be thought to generate the population and then estimate the parameters for that model.

Särndal et al. (1992) adopt a *model-assisted* approach; for that approach, a model is used to specify the parameters of interest, but all inference is based on the survey design. Thus, you fit a particular model because you believe it a plausible candidate for generating the population but use the sampling weights to estimate the parameters and the sample design to estimate variances of the estimate. As inference is made using the sample design, we consider the model-assisted approach to be part of the design-based approach in this section.

The distinction between the approaches is important for the survey analyst because most software packages use either a design-based or a model-based approach. Standard statistical software such as SAS, S-PLUS, BMDP, or SPSS assumes a model-based approach to regression, as exposited in Section 11.1. Survey packages such as SUDAAN, PC CARP, and WesVarPC are based on estimating the finite population parameters using the approach in Section 11.2. Thus, knowing which approach you wish to take is important. Blindly running your data through software, without understanding what the software is estimating, can lead to misinterpreted results.

Most statisticians agree that it is a good thing if a regression model describes the true state of nature. Thus, if it were known that a model would describe every possible observation involving x and y, then that model should be adopted. In the physical sciences, many models such as force = mass $\times$ acceleration can be theoretically derived. As long as you stay away from near-light velocity, any observation for which force, mass, and acceleration are accurately measured should be fit by the model. The design for how observations are sampled should then make little difference for finding the point estimates of regression coefficients, as every possible observation is described by the model.[2]

Unfortunately, theoretically derived models known to hold for all observations do not often exist for survey situations. An economist may conjecture a relationship between number of children, income, and amount spent on food, but there is no guarantee that this model will be appropriate for every subgroup in the population. Other variables may be related to the amount spent on food (such as educational level or amount of time away from home), but not measured in the survey. In addition, the true relation among the variables might not be exactly linear. Thus, the main challenge to model-based inference is specifying the model.

If taking a model-based approach, then, examine the model assumptions carefully and do everything you can to check the adequacy of the model for your data. This includes plotting the data and residuals, performing diagnostic tests, and using sampling designs that allow estimation of alternative models that may provide a better description of the relationship between variables. (Of course, you should also plot the data if adopting a design-based approach.) Inference about observations not in the sample is based solely on the assumption that the model you have adopted applies to them, and you need to be very careful about generalizing outside the sampled data. You must assume that the nonsampled population units can also be described by the model, and this is a very strong assumption.

Much is attractive about the model-based approach for regression: It links with sociological theories of the investigator, is consistent with other areas of statistics, and provides a mechanism for accounting for nonresponse. The model-based approach provides a framework for comparing theories about structural relationships. In addition, model-based estimates can be used with relatively small samples and with nonprobability samples. Although design-based inference does not depend on model

---

[2]The sampling design, however, can affect the variances of the point estimates.

assumptions, it does require large sample sizes in practice to be able to construct confidence intervals. The standard errors of the model-based parameter estimates are generally lower than those of design-based estimates incorporating weights.

But model misspecification and omitted covariates are of concern for a model-based analysis, and missing covariates may not show up in standard residual analyses. Moreover, in a complex survey design, the needed missing predictors may be related to the design and the survey weights. For example, for our unequal-probability sample in Figure 11.3, the selection probabilities we used depend on the value of y. Now, you can think of height as being determined by many, many variables $x_1, x_2, \ldots,$ but the data set has only one of those possible explanatory variables. If all the other variables were included in the model, then the unequal-selection probabilities would be irrelevant; because they are not, however, the probabilities of selection have useful information for estimating the regression slope.

Pfeffermann and Holmes (1985), DuMouchel and Duncan (1983), and Kott (1991) argue that using sampling weights in regression can provide robustness to model mis-specification: The weighted estimates are relatively unaffected if some independent variables are left out of the model.[3] Kott (1991) argues that sampling weights are needed in linear regression because the choice of covariates in survey data is limited to variables collected in the survey: If necessary covariates are omitted, $\hat{B}$ and $\hat{\beta}_{OLS}$ are both biased estimators of β, but the bias of $\hat{B}$ is a decreasing function of the sample size, while $\hat{\beta}_{OLS}$ is only asymptotically unbiased if the probabilities of selection are not related to the missing covariates. Rubin (1985), Smith (1988), and Little (1991) adopt a model-based perspective but argue that sampling weights are useful in model-based inference as summaries of covariates describing the mechanism by which units are included in the sample.

One point is clear: If the model you are using really does describe the mechanism generating the data, then the finite population quantity $\mathbf{B}$ should be close to the theoretical parameter β. Thus, if the model is a good one, we would expect that the point estimate of β using the model should be similar to the point estimate $\hat{\mathbf{B}}$ calculated using sampling weights. We suggest fitting a model both with and without weights. If the parameter estimates differ, then you should explore alternatives to the model you have adopted. A difference in the weighted and unweighted estimates can tell you that the proposed model does not fit well for part of the population. Lohr and Liu (1994) explore this issue for the NCVS.

EXAMPLE **11.6**   Korn and Graubard (1995b) illustrate the difference that including weights can make in a regression analysis, using data from the live-birth component of the 1988 MIHS. As mentioned in Example 11.1, black infants and low-birth-weight infants are over-sampled, so their sampling weights are lower than the weights for white, normal-birth-weight infants. Figure 11.6 shows a plot of the data and estimated regression line when weights are used in calculating the regression parameters; Figure 11.7 ignores the weights. The weighted regression pulls the regression line to where the population is estimated to be; in the unweighted regression, the line provides the best least squares fit to the sample data but does not describe the population as well. It

---

[3]But this robustness comes at a price; as mentioned earlier, the design-based variance, using the weights, is generally larger than the model-based variance. Kish (1992) gives a good overview of the variance inflation due to using weighted estimates rather than estimates without weights.

FIGURE   11.6

Plot of weighted mean gestational age versus weighted mean birthweight for successive groups of approximately 500 observations. Areas of bubbles are proportional to the estimated population sizes of the groups. The straight line is the weighted linear regression fit to the original (ungrouped) data.

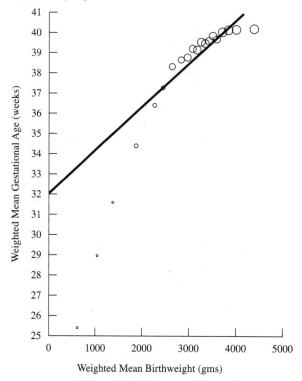

Weighted Mean Birthweight (gms)

SOURCE: From "Examples of Differing Weighted and Unweighted Estimates from a Sample Survey," by E. L. Korn and B. I. Graubard, 1995, *The American Statistician,* vol. 49, pp. 291–295. Copyright © 1995 American Statistical Association. Reprinted by permission.

is clear from examining the plots that the regression lines differ to such an extent because a straight-line model is not appropriate for the data; if a quadratic regression were fit instead, then the models from the weighted and unweighted regressions would show greater agreement. In this example, then, the differences between the parameter estimates with weights and without weights arise because the straight-line model adopted is inappropriate.   ∎

Each of the approaches to inference about regression parameters in complex surveys can be appropriate, depending on the desired use of the regression model. You may want to consider the following questions when deciding on your approach:

1   Are you performing a regression to generate official statistics that will be used to determine public policy? If so, you may want to use the weights to estimate parameters and the design to make inferences about the parameters. If you are using weights to estimate population and domain means, you may also want to use them to estimate regression parameters so that the results from different analyses

FIGURE **11.7**

Plot of mean gestational age versus mean birthweight for successive groups of approximately 500 observations. Areas of bubbles are proportional to the sample sizes of the groups. The straight line is the unweighted linear regression fit to the original (ungrouped) data.

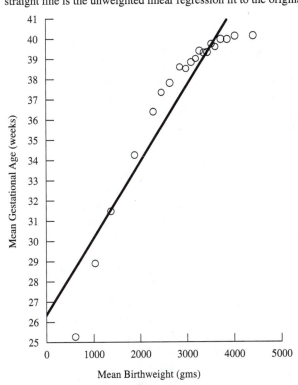

SOURCE: From "Examples of Differing Weighted and Unweighted Estimates from a Sample Survey," by E. L. Korn and B. I. Graubard, 1995, *The American Statistician,* vol. 49, pp. 291–295. Copyright © 1995 American Statistical Association. Reprinted by permission.

are consistent (see Alexander 1991). As noted above, **B** should be close to β for a good model and large finite population, so a design-based estimate of **B** should also estimate β.

2 Was a probability sample taken? If not, then you must use a model-based approach.

3 How large is the sample size? The design-based theory relies on large sample sizes to make inferences about the parameters. If you have a small sample, you should probably use a model-based approach.

4 How extensively has the subject been studied before? If scientific theory and previous empirical investigations support the model you are proposing, you may trust your model more and have more confidence in a model-based approach.

However, a mistake is often made by investigators who have heard the message that sampling weights are irrelevant in regression analysis but have ignored the rest of the discussion: They ignore the weights *and* the clustering in the data by simply running the survey data through standard regression software. This is incorrect under any approach: Whether or not weights are used to construct an estimator, the dependence in the data reflected in the clustering *must* be considered when calculating standard errors. A model-based approach that incorporates the positive correlation among observations in the same cluster is discussed in the next section.

11.4
Mixed Models for Cluster Samples

In Chapters 5 and 6 we discussed using a random-effect model as a superpopulation model for cluster sampling. We can use this approach for regression analyses as well, by allowing different clusters to have their own regression equations but relating the different regression equations for the clusters through a model.

EXAMPLE 11.7 The National Assessment of Educational Progress (NAEP) collects data on student background and achievement in the United States. It is sometimes referred to as "The Nation's Report Card" because it provides a scale for measuring student progress and comparing student achievement among different states and over time. A wealth of information is collected for each student, teacher, and participating school. Besides proficiency scores for various subjects, the student-level data include information on the students' gender, race, ethnicity, courses taken, and variables related to socioeconomic status. School-level information includes fiscal resources, instructional methods, student-body characteristics, and expectations of academic achievement.

The NAEP data can be used to identify school- and student-level variables that are associated with mathematics achievement among eighth-grade students. For simplicity, let's consider one student-level characteristic, gender; and one school-level characteristic, average amount of time spent in class on math tests. In practice, of course, you would probably include more variables in the model, as you would expect a number of characteristics to be associated with the tested mathematics achievement. Let Y_{ij} be the mathematics proficiency score of student j at school i in the sample and let $x_{ij} = 1$ if student j at school i is female and 0 if student j at school i is male.

We expect a clustering effect in these data—measuring all variables that might be associated with student achievement scores in mathematics is impossible, and the characteristics of the schools, teachers, and neighborhoods that are not included in the model induce a positive correlation in the test scores within a school. For example, the seventh- and eighth-grade mathematics teacher in one school might be superb at inspiring students to learn mathematics, but that excellence would not be recorded in the survey. The students from that class might then all perform better than average on the proficiency test, so their scores are more similar, even after adjusting for known covariates, than scores of a random sample of students from the population. When unmeasured characteristics such as these are considered over all schools, the result is a positive intraclass correlation coefficient.

Thus, a model $Y_{ij} = \beta_0 + x_{ij}\beta_1 + \varepsilon_{ij}$, with the ε_{ij}'s independent random variables with mean 0 and variance σ^2, is likely to be inappropriate for these data. If this erroneous model is adopted and the data for all students run through SAS PROC REG, then the p-values for parameter estimates will be far too small. In addition, the model does not allow for different relations between gender and test score in different

schools—which may certainly occur, as some schools may encourage students of one
gender more than students of the other gender.

A model that incorporates cluster effects and allows schools to have different
slopes for gender is

$$Y_{ij} = \beta_{0i} + (x_{ij} - \bar{x}_i)\beta_{1i} + \varepsilon_{ij}.$$

Here, the ε_{ij}'s are assumed to be independent $N(0, \sigma^2)$ random variables; the mean
of x_{ij} for school i, $\bar{x}_i$, is subtracted from each x_{ij} so that β_{0i} can be interpreted as
the average test score in school i. School i has its own straight-line regression model
with intercept β_{0i} and slope β_{1i}. But the slopes and intercepts from different schools
are also related through a model. A simple model for the slopes and intercepts allows
them to essentially be randomly distributed about a mean:

$$\beta_{0i} = \beta_0 + \delta_{0i}; \quad \beta_{1i} = \beta_1 + \delta_{1i},$$

with δ_{0i} and δ_{1i} following a bivariate normal distribution with $E_M[\delta_{0i}] = E_M[\delta_{1i}] = 0$,
$V_M[\delta_{0i}] = \tau_{00}$, $V_M[\delta_{1i}] = \tau_{11}$, and $\text{Cov}_M(\delta_{0i}, \delta_{1i}) = \tau_{01}$. Under this situation, the
model may then be written as

$$Y_{ij} = \beta_0 + (x_{ij} - \bar{x}_i)\beta_1 + \delta_{0i} + (x_{ij} - \bar{x}_i)\delta_{1i} + \varepsilon_{ij}. \tag{11.12}$$

The parameter β_0 represents the mean test score for schools; β_1 represents the mean
slope for gender for schools. The random effects δ_{0i} and δ_{1i} represent the difference
in the intercept and slope between school i and the average values for intercept and
slope for all schools; they measure the school effect. Finally, ε_{ij} refers to additional
deviation from the mean due to the individual student, after the effect of gender and
school have been accounted for.

Note that if $\tau_{00} = \tau_{11} = 0$, there is no school effect on test score, and the model
then reduces to a regular straight-line regression model. In most applications, however,
the slopes and intercepts will vary from school to school. ■

In statistics, the model in (11.12) is an example of a **mixed linear model;** it
has both fixed (β_0 and β_1) and random (δ_{0i}, δ_{1i}, and ε_{ij}) effects. In econometrics,
(11.12) is often referred to as a **random-coefficient regression model;** in the social
sciences, it is called a **multilevel** or **hierarchical linear model.** The summer 1995
issue of the *Journal of Educational and Behavioral Statistics* was devoted to multilevel
models; these articles contain a useful bibliography and are a good starting point
for further reading. Other references providing a good introduction to the subject
include deLeeuw and Kreft (1986), Goldstein (1987), Goldstein and Silver (1989),
and Bryk and Raudenbush (1992). These models may be fit in SAS PROC MIXED or
in specialized packages such as HLM (Bryk et al. 1988) or ML3 (Prosser et al. 1992).

The mixed model in (11.12) is a superpopulation model and is assumed to hold
for all schools and students in the population. One advantage of using such a model
is that it does not require that the schools be randomly selected, as long as the model
describes the population. A mixed model approach is also congenial to testing different
theories about mathematics education.

The model in (11.12) may also be used as a starting point for further investigation.
The random effects δ_{0i} and δ_{1i} may be estimated for each school; the investigator

may want to examine schools with unusually high or low values to try to conjecture why those schools might be different. The investigator may also want to include other predictor variables when estimating the intercepts and slopes for the different schools. For example, it might be conjectured that having more math tests at a school might lead to better mathematics proficiency scores and might also lead to a smaller gender difference in the school. This extra predictor can easily be included in the mixed model. Let z_i be the average amount of time spent on math tests at school i. Then, the intercept and slope at school i can be modeled as

$$\beta_{0i} = \beta_0 + \gamma_0 z_i + \delta_{0i} \quad \text{and} \quad \beta_{1i} = \beta_1 + \gamma_1 z_i + \delta_{1i},$$

where γ_0 represents the effect of time spent on math tests on the intercept and δ_{0i} represents the remaining school effect after adjusting for z_i.

11.5
Logistic Regression

In linear regression, the response variable is usually considered to be approximately continuous—for example, birth weight, income, or leaf area. In surveys, however, many variables of interest are dichotomous, with y_i taking only values of 1 (yes) or 0 (no). **Logistic regression** (see Hosmer and Lemeshow 1989 for a general reference) is often used to predict probabilities of having response 1 for dichotomous variables.

Let $\mathbf{x}$ be a vector of independent variables and $\boldsymbol{\beta}$ be the vector of unknown parameters. Then the standard logistic regression model takes the form

$$p(\mathbf{x}) = \frac{\exp\left(\mathbf{x}^T \boldsymbol{\beta}\right)}{1 + \exp\left(\mathbf{x}^T \boldsymbol{\beta}\right)}, \tag{11.13}$$

where $p(\mathbf{x})$ represents the probability that a unit with covariates $\mathbf{x}$ will have a response of 1. Alternatively, the model may be expressed in logit scale, where $\text{logit}(p) = \ln[p/(1-p)]$:

$$\text{logit}[p(\mathbf{x})] = \mathbf{x}^T \boldsymbol{\beta}. \tag{11.14}$$

EXAMPLE 11.8   For the data in Example 10.1, let $y_i = 1$ if household i has a computer and $y_i = 0$ if household i does not have a computer. Let $x_i = 1$ if household i subscribes to cable and $x_i = 0$ if household i does not subscribe to cable. The fitted logistic regression model is

$$\widehat{\text{logit}}[p_i] = -0.177 - 0.281 x_i.$$

Note that the slope, -0.28, is the log odds ratio from Example 10.1. It is easy to transform back to predicted conditional probabilities: When $x = 1$, then $\ln[\hat{p}/(1-\hat{p})] = -0.4573184$ so that

$$\hat{p} = \frac{\exp(-0.4573184)}{1 + \exp(-0.4573184)} = 0.388 = \frac{119}{307}. \quad \blacksquare$$

Much of the previous discussion in this chapter on linear regression also applies to logistic regression—a complex sample design will affect standard errors of the logistic regression coefficients, just as it affects standard errors of the linear regression coefficients. Logistic regression with one dichotomous independent variable is essentially equivalent to finding the odds ratio in a 2×2 contingency table, so the discussion in Chapter 10 about how the sampling design affects standard goodness-of-fit tests also applies to testing the significance of logistic regression coefficients.

Binder (1983), Chambless and Boyle (1985), and Roberts et al. (1987) give design-based theory for estimating logistic regression parameters. Just as the design-based theory for linear regression started with defining the population quantities of interest using the normal equations, here the quantities of interest are defined in terms of the likelihood function that would be adopted if the entire population were available for study. If there are N units in the population, this likelihood (assuming independence) is

$$\mathcal{L}(\beta) = \prod_{i=1}^{N} p_i^{y_i} (1 - p_i)^{1-y_i} \tag{11.15}$$

where $p_i = \exp(\mathbf{x}_i^T \beta)/[1 + \exp(\mathbf{x}_i^T \beta)]$ represents the probability that a unit with covariates $\mathbf{x}_i$ has a response of 1. The finite population parameter $\mathbf{B}$ is then defined to be the maximum likelihood estimate of β using (11.15). The parameter $\mathbf{B}$ is the solution to the system of equations

$$\sum_{i=1}^{N} x_{ij} \left[y_i - \frac{\exp\left(\mathbf{x}_i^T \mathbf{B}\right)}{1 + \exp\left(\mathbf{x}_i^T \mathbf{B}\right)} \right] = 0 \qquad \text{for } j = 1, \dots, p \tag{11.16}$$

if all elements in the population could be observed.

Now that $\mathbf{B}$ is defined, estimate it by substituting estimates for the population totals. A design-based estimate of $\mathbf{B}$ is given by the solution $\hat{\mathbf{B}}$ to

$$\sum_{i \in \mathcal{S}} w_i x_{ij} \left[y_i - \frac{\exp\left(\mathbf{x}_i^T \hat{\mathbf{B}}\right)}{1 + \exp\left(\mathbf{x}_i^T \hat{\mathbf{B}}\right)} \right] = 0 \qquad \text{for } j = 1, \dots, p, \tag{11.17}$$

where $\mathcal{S}$ denotes the units included in the sample. The ith observation in the sample represents w_i observations in the population.

For a model-based estimate of β, the weights are simply omitted: $\hat{\beta}$ is the solution to

$$\sum_{i \in \mathcal{S}} x_{ij} \left[y_i - \frac{\exp\left(\mathbf{x}_i^T \hat{\beta}\right)}{1 + \exp\left(\mathbf{x}_i^T \hat{\beta}\right)} \right] = 0 \qquad \text{for } j = 1, \dots, p. \tag{11.18}$$

Variance estimation for logistic regression is discussed in the references cited above. Rao et al. (1998) present a modified version of score tests for testing the significance of logistic regression coefficients.

Logistic regression has one important difference from linear regression. In Section 11.2 we noted the bias that can occur in estimating linear regression parameters if the probabilities of selection are related to the response variable, but the unequal probabilities of selection are not accounted for in the analysis. In a health survey, for example, blood pressure might be used as a stratification variable, and a higher sampling fraction used in the high-blood-pressure stratum than in the low-blood-pressure stratum. If we ignore the selection probabilities and fit a linear regression model predicting the continuous variable *blood pressure* from covariates such as *age, diet,* and *smoking history,* the regression coefficients may be severely biased for estimating **B**.

Prentice and Pyke (1979), however, show that if a logistic regression model is valid and contains an intercept term, then the intercept is the only parameter estimate affected by a sample design that depends on the y's. Such sample designs are particularly common in epidemiology and economics, where they are referred to as *case-control studies* and *choice-based sampling.* In an epidemiology application, the population may be divided into two strata: persons with lung cancer, and persons without lung cancer. A sample is selected from each stratum; as lung cancer is rare, the stratified sample has a far greater sampling fraction (and lower sampling weights) in the cancer stratum than in the noncancer stratum. But if the primary interest is in estimating the coefficients of age, diet, and smoking history in a logistic regression, the disproportionate sampling makes no difference in a model-based analysis. We would expect that if the model is good, the only difference between a weighted and unweighted analysis would appear in the intercept terms. Of course, if a cluster sample is used, the dependence of the data induced by clustering will need to be considered in the logistic regression model for variance estimation, as discussed by Scott and Wild (1989).

11.6
Generalized Regression Estimation for Population Totals

In Chapter 3 we introduced ratio and regression estimation in the setting of SRSs, with estimators

$$\hat{t}_{yr} = \frac{\hat{t}_y}{\hat{t}_x} t_x$$

$$\hat{t}_{yreg} = \hat{t}_y + \hat{B}_1 (t_x - \hat{t}_x).$$

Now let's extend these estimates to complex survey samples. We want to improve on the estimator $\hat{t}_y = \sum_{i \in \mathcal{S}} w_i y_i$ by including auxiliary information through the model

$$Y_i \mid \mathbf{x}_i = \mathbf{x}_i^T \boldsymbol{\beta} + \varepsilon_i, \tag{11.19}$$

with $\mathbf{x}_i^T = (x_{i1}, x_{i2}, \ldots, x_{ip})$ and $V_M(\varepsilon_i) = \sigma_i^2$. We assume that the true population totals $\mathbf{t}_x$ are known and thus can be used to adjust the estimate $\hat{t}_y$. We allow the variances to differ so that ratio estimation and poststratification also fit into this general framework. We are using the model-assisted approach further described in Särndal et al. (1992, chap. 6 and 7).

Define

$$\mathbf{B} = \left(\mathbf{X}_U^T \boldsymbol{\Sigma}_U^{-1} \mathbf{X}_U \right)^{-1} \mathbf{X}_U^T \boldsymbol{\Sigma}_U^{-1} \mathbf{y}_U,$$

where $\boldsymbol{\Sigma}_U$ is a diagonal matrix with ith diagonal element σ_i^2. The finite population parameter **B** is the weighted least squares estimate of $\boldsymbol{\beta}$ for observations in the population, using the model in (11.19). Thus, the form of **B** is inspired by (11.19), but we then treat **B** as a finite population quantity. The (jk)th entry of $(\mathbf{X}_U^T \boldsymbol{\Sigma}_U^{-1} \mathbf{X}_U)$ is

$\sum_{i=1}^{N} x_{ij} x_{ik} / \sigma_i^2$. Now estimate $\mathbf{B}$ by

$$\hat{\mathbf{B}} = \left(\mathbf{X}_\mathcal{S}^T \mathbf{W}_\mathcal{S} \boldsymbol{\Sigma}_\mathcal{S}^{-1} \mathbf{X}_\mathcal{S} \right)^{-1} \mathbf{X}_\mathcal{S}^T \mathbf{W}_\mathcal{S} \boldsymbol{\Sigma}_\mathcal{S}^{-1} \mathbf{y}_\mathcal{S}. \tag{11.20}$$

The **generalized regression estimator** of the population total is

$$\hat{t}_{y\text{greg}} = \hat{t}_y + (\mathbf{t}_x - \hat{\mathbf{t}}_x)^T \hat{\mathbf{B}}. \tag{11.21}$$

Using linearization,

$$V(\hat{t}_{y\text{greg}}) = V \left[\hat{t}_y + (\mathbf{t}_x - \hat{\mathbf{t}}_x)^T \hat{\mathbf{B}} \right] \approx V \left(\hat{t}_y - \hat{\mathbf{t}}_x^T \mathbf{B} \right).$$

Let $e_i = y_i - \mathbf{x}_i^T \hat{\mathbf{B}}$ be the ith residual. The variance may then be estimated by

$$\hat{V}(\hat{t}_{y\text{greg}}) = \hat{V} \left(\sum_{i \in \mathcal{S}} w_i e_i \right).$$

If the model is a good one, we expect the variability in the residuals to be smaller than the variability in the original observations, so the generalized regression estimator will be more efficient than $\hat{t}_y$. In an SRS, for example,

$$\hat{V}_{\text{SRS}}(\hat{t}_y) = \frac{N^2}{n} \left(1 - \frac{n}{N} \right) \frac{\sum_{i \in \mathcal{S}} (y_i - \bar{y})^2}{n - 1},$$

but

$$\hat{V}_{\text{SRS}}(\hat{t}_{y\text{greg}}) = \frac{N^2}{n} \left(1 - \frac{n}{N} \right) \frac{\sum_{i \in \mathcal{S}} e_i^2}{n - 1};$$

if the residuals tend to be smaller than the deviations of y_i about the mean, then the estimated variance is smaller for the generalized regression estimator.

EXAMPLE 11.9   *Ratio Estimation*

Adopt the model

$$y_i = \beta x_i + \varepsilon_i, \quad V_M(\varepsilon_i) = \sigma^2 x_i.$$

Then,

$$\hat{B} = \left(\sum_{i \in \mathcal{S}} \frac{w_i x_i^2}{x_i} \right)^{-1} \sum_{i \in \mathcal{S}} \frac{w_i x_i y_i}{x_i} = \frac{\sum_{i \in \mathcal{S}} w_i y_i}{\sum_{i \in \mathcal{S}} w_i x_i} = \frac{\hat{t}_y}{\hat{t}_x}.$$

The generalized regression estimator of the population total is

$$\hat{t}_{y\text{greg}} = \hat{t}_y + (t_x - \hat{t}_x) \frac{\hat{t}_y}{\hat{t}_x} = \frac{t_x \hat{t}_y}{\hat{t}_x},$$

which is the standard ratio estimator.   ∎

EXAMPLE 11.10   *Poststratification*

Suppose we know the population counts N_c for C poststrata, $c = 1, \ldots, C$. Define the variables $x_{ic} = 1$ if observation unit i is in poststratum c and 0 otherwise. Consider the model

$$y_i = \beta_1 x_{i1} + \beta_2 x_{i2} + \cdots + \beta_C x_{iC} + \varepsilon_i,$$

with $V_M(\varepsilon_i) = \sigma^2$. Then,

$$\sigma^2 \mathbf{X}_U^T \boldsymbol{\Sigma}_U^{-1} \mathbf{X}_U = \mathbf{X}_U^T \mathbf{X}_U = \text{diag}(N_1, \ldots, N_C),$$

and

$$\sigma^2 \mathbf{X}_{\mathcal{S}}^T \mathbf{W}_{\mathcal{S}} \boldsymbol{\Sigma}_{\mathcal{S}}^{-1} \mathbf{X}_{\mathcal{S}} = \mathbf{X}_{\mathcal{S}}^T \mathbf{W}_{\mathcal{S}} \mathbf{X}_{\mathcal{S}} = \text{diag}(\hat{N}_1, \ldots, \hat{N}_C).$$

As a result, $\hat{B}_c = \hat{t}_{yc}/\hat{N}_c$, where $\hat{t}_{yc} = \sum_{i \in \mathcal{S}} w_i x_{ic} y_i$ is the estimated population total in poststratum c and $\hat{N}_c = \sum_{i \in \mathcal{S}} w_i x_{ic}$ is the estimated population count in poststratum c. The generalized regression estimator is

$$\hat{t}_{y\text{greg}} = \hat{t}_y + \sum_{c=1}^{C} (N_c - \hat{N}_c) \frac{\hat{t}_{yc}}{\hat{N}_c} = \sum_{c=1}^{C} \frac{N_c \hat{t}_{yc}}{\hat{N}_c}. \quad \blacksquare$$

Often, the auxiliary variables are useful for many of the response variables of interest. You may want to poststratify by age, race, and gender groups when estimating every population total for your survey. This is easily implemented because the generalized regression estimator is a linear estimator in y. To see this, define

$$g_i = 1 + (\mathbf{t}_x - \hat{\mathbf{t}}_x)^T \left(\mathbf{X}_{\mathcal{S}}^T \mathbf{W}_{\mathcal{S}} \boldsymbol{\Sigma}_{\mathcal{S}}^{-1} \mathbf{X}_{\mathcal{S}} \right)^{-1} \frac{\mathbf{x}_i}{\sigma_i^2}.$$

Then,

$$\hat{t}_{y\text{greg}} = \sum_{i \in \mathcal{S}} w_i g_i y_i,$$

where the g_i's do not depend on values of the response variable. To estimate totals with the generalized regression estimator, form a new column in the data with values $a_i = w_i g_i$. Then use the vector of a_i as the weight vector for estimating the population total of any variable.

11.7
Exercises

1   Read one of the following articles or another article in which regression or logistic regression is used on data from a complex survey.

Stevens, R. G., D. Y. Jones, M. S. Micozzi, and P. R. Taylor. 1988. Body iron stores and the risk of cancer. *New England Journal of Medicine* 319: 1047–1052.

Martorell, R., F. Mendoza, and R. O. Castillo. 1989. Genetic and environmental determinants of growth in Mexican-Americans. *Pediatrics* 84: 864–871.

Patterson, C. J., J. S. Kupersmidt, and N. A. Vaden. 1990. Income level, gender, ethnicity, and household composition as predictors of children's school-based competence. *Child Development* 61: 485–494.

Tymms, P. B., and C. T. Fitz-Gibbon. 1992. The relationship between part-time employment and A-level results. *Educational Research* 34: 193–199.

Breen, N., and L. Kessler. 1994. Changes in the use of screening mammography: Evidence from the 1987 and 1990 National Health Interview Surveys. *American Journal of Public Health* 84: 62–67.

Subar, A. F., R. G. Ziegler, B. H. Patterson, G. Ursin, and B. Graubard. 1994. US dietary patterns associated with fat intake: The 1987 National Health Interview Survey. *American Journal of Public Health* 84: 359–366.

Bachman, R., and A. L. Coker. 1995. Police involvement in domestic violence: The interactive effects of victim injury, offender's history of violence, and race. *Violence and Victims* 10: 91–106.

Flegal, K. M., R. P. Troiano, E. R. Pamuk, R. J. Kuczmarksi, and S. M. Campbell. 1995. The influence of smoking cessation on the prevalence of overweight in the United States. *New England Journal of Medicine* 333: 1165–1170.

Sashi, C. M., and L. W. Stern. 1995. Product differentiation and market performance in producer goods industries. *Journal of Business Research* 33: 115–127.

Singhapakdi, A., K. L. Kraft, S. J. Vitell, and K. C. Rallapalli. 1995. The perceived importance of ethics and social responsibility on organizational effectiveness: A survey of marketers. *Journal of the Academy of Marketing Science* 23: 49–56.

Wang, X., B. Zuckerman, G. A. Coffman, and M. J. Corwin. 1995. Familial aggregation of low birth weight among whites and blacks in the United States. *New England Journal of Medicine* 333: 1744–1749.

Write a critique of the article. What is the purpose and design of the survey? What is the goal of the analysis? How do the authors use information from the survey design in the analysis? Do you think that the data analysis is done well? If so, why? If not, how could it have been improved? Are the conclusions drawn in the article justified?

2 An investigator wants to study the relationship between a child's age, number of siblings, and the dollar amount of the child's Christmas list presented to Santa Claus. She also wants to estimate the total number of children that visit Santa Claus and the total dollar amount of all children's requests. It would be very difficult to construct a sampling frame of children who will visit Santa Claus between December 1 and 24, but the investigator has a list of shopping malls and stores in which Santa will appear in the city, as well as the times that Santa will be at each location. The Santa sites are divided into four categories: 23 department stores, 19 discount stores, 15 toy stores, and 5 shopping malls. The investigator wants you to help design the sample of children.

a What questions would you ask the investigator to clarify the problem?

b Assuming any answers you like to the questions you asked, suggest a design for the survey.

c How will your survey design affect the regression analysis of the data? How do you propose to analyze the data? Are there other explanatory variables that you would suggest to the investigator?

3 Use the data in the file anthrop.dat for this problem.

  a Construct a population from the 3000 observations in anthrop.dat in which the 1000 individuals with the highest value of y have been removed. Now take an SRS of size 200 from the remaining 2000 individuals and plot the data along with the OLS regression line. How does this line compare to the population regression line?

  b Repeat part (a) but use as the population the 2000 individuals with the lowest value of x.

  c Is there a difference in the regression equations in parts (a) and (b)? Explain and relate your findings to the model in (11.1).

4 Use the data in the file nybight.dat (see Exercise 19 of Chapter 4) for this problem. Using the 1974 data, estimate the coefficients in a straight-line regression model predicting weight of the catch from the number of fish caught. Give standard errors for your estimates. (Be sure to plot the data!)

5 Perform a model-based analysis for the setting in Exercise 4. Examine the residuals and postulate an appropriate variance structure for the model.

6 Repeat Exercise 4 for predicting the number of species caught from the surface temperature.

7 Repeat Exercise 5 for predicting the number of species caught from the surface temperature.

8 Use the data in the file teachers.dat (described in Exercise 16 of Chapter 5) for this problem.

  a Estimate the coefficients in a straight-line regression model predicting *preprmin* from *size*. Give standard errors for your estimates. Is there evidence that the two variables are related? (Be sure to plot the data!)

  b Perform a model-based analysis of the same data. Examine the residuals and postulate an appropriate variance structure for the model.

9 Use the data in the file books.dat (described in Exercise 6 of Chapter 5) for this problem.

  a Plot *replace* vs. *purchase* for the raw data.

  b Plot *replace* vs. *purchase* using the sampling weights.

  c Using a design-based approach, estimate the regression equation for predicting *replace* from *purchase,* along with standard errors. How many degrees of freedom would you use in constructing a confidence interval for the slope?

10 For the situation in Exercise 9, postulate a model for the variance structure. Using your model, estimate the slope of the regression line predicting *replace* from *purchase*. How do your estimate and its standard error compare with your answers in Exercise 9?

11 Use your data set from Exercise 13 of Chapter 4 for this problem. Using the weights, fit a regression model predicting *acres92* from *largef92*. Give a standard error for the estimated slope. Now ignore the sampling design and calculate the OLS estimate of the slope. Do your point estimates differ? Explain why or why not by examining plots of the data.

12 Lush (1945, 95) discusses different estimates of heritability for milk-fat percentage in dairy cattle herds. *Heritability* is defined to be the percentage of variability in fat percentage that is attributable to differences in the heredity of different individuals; the remainder of the variability is attributed to differences in environment. He notes that when the herd was treated as an SRS, the estimate of heritability was about 0.8; when fat percentage for daughters was regressed on fat percentage for dams and where each dam was represented by only one record, the estimate of heritability decreased to below 0.3.

From a sampling perspective, why are these estimates so different? Discuss how you would analyze the full-herd data from both a design-based and a model-based perspective.

13 *Comparison of domain means.* Suppose the population may be divided into two groups, with respective sizes N_1 and N_2 and population means $\bar{y}_{1U}$ and $\bar{y}_{2U}$. The overall population mean is $\bar{y}_U = (N_1\bar{y}_{1U} + N_2\bar{y}_{2U})/N$, with $N = N_1 + N_2$. Let $x_i = 1$ if observation unit i is in group 1 and $x_i = 0$ if it is in group 2. The weight for observation unit i is w_i.

Show that $B_1 = \bar{y}_{1U} - \bar{y}_{2U}$ and $B_0 = \bar{y}_{2U}$. Also show that

$$\hat{B}_1 = \frac{\displaystyle\sum_{i\in S} w_i x_i y_i}{\displaystyle\sum_{i\in S} w_i x_i} - \frac{\displaystyle\sum_{i\in S} w_i(1-x_i)y_i}{\displaystyle\sum_{i\in S} w_i(1-x_i)} = \hat{\bar{y}}_1 - \hat{\bar{y}}_2$$

and $\hat{B}_0 = \hat{\bar{y}}_2$.

14 Consider the SRS data in the file uneqvar.dat.

a Plot y vs. x.

b Find the fitted regression line under the assumption of equal variances.

c Calculate $\hat{V}_M(\hat{B}_1)$ and $\hat{V}_L(\hat{B}_1)$. How do they compare?

15 Show that (11.10) is equivalent to (11.6) and (11.7) for straight-line regression.

*16 (Requires theory of linear models.) Suppose the "true" model describing the relation between x and y is

$$Y_i \mid x_i = \beta_0 + \beta_1 x_i + \varepsilon_i,$$

where the ε_i's are independently generated from a $N(0, \sigma_i^2)$ distribution. Let Σ be a matrix with diagonal entries $\sigma_1^2, \sigma_2^2, \ldots, \sigma_n^2$. What is the covariance matrix for the OLS parameter estimates? How does this relate to the discussion of different estimators of the variance on pages 357–358?

17 The coefficient of determination R^2 is often reported for regression analyses. For a straight-line regression, the finite population quantity R^2 is defined to be

$$R^2 = \frac{B_1 \displaystyle\sum_{i=1}^{N} (x_i - \bar{x}_U)(y_i - \bar{y}_U)}{\displaystyle\sum_{i=1}^{N} (y_i - \bar{y}_U)^2}.$$

a Show that R^2 is the square of the population correlation coefficient R defined in (3.1).

b Write R^2 as a function of population totals.

c Give an estimator $\hat{R}^2$ of R^2 for data from a complex sample, using weights.

18 Fienberg (1980) says, "We know of no justification whatsoever for applying standard multivariate methods to weighted data . . . the automatic insertion of a matrix of sample-based weights into a weighted least-squares analysis is more often than not misleading, and possibly even incorrect." Which approach to regression inference does Fienberg advocate? What is your reaction?

19 Assuming a model

$$y_i = \beta_0 + \beta_1 x_i + \varepsilon_i,$$

with $V_M(\varepsilon_i) = \sigma^2$, what is the generalized regression estimator of t_y? Show that $\hat{t}_{x\text{greg}} = t_x$.

SURVEY Exercises

20 Use your stratified sample with optimal allocation from Exercise 28 of Chapter 4 and fit a regression model predicting the amount a household is willing to spend for cable TV from the assessed value of the house. As part of your analysis, plot the data. Give standard errors for your parameter estimates. Does it make a difference for the parameter estimates whether you include the weights or not? Should you consider different regression models for the different strata?

21 Repeat Exercise 20, using the cluster sample from Exercise 30 of Chapter 6. What effect does the clustering have on the regression coefficients and their standard errors?

12

Other Topics in Sampling*

Nearly the whole of the states have now returned their census. I send you the result, which as far as founded on actual returns is written in black ink, and the numbers not actually returned, yet pretty well known, are written in red ink. Making a very small allowance for omissions, we are upwards of four millions; and we know in fact that the omissions have been very great.

—Thomas Jefferson, letter to David Humphreys, August 23, 1791

12.1
Two-Phase Sampling

Sometimes, you would like to use stratification, unequal-probability sampling, or ratio estimation to increase the precision of your estimator, but the sampling frame has no information on useful auxiliary variables. For example, suppose you want to sample businesses with probability proportional to income but do not have income information in the sampling frame. Or you want to estimate the total timber volume that has been cut in the forest by measuring the total volume in a sample of truckloads of logs. Timber volume in a truck is related to the weight of the truckload, so you would expect to gain precision by using ratio estimation with y_i = timber volume in truck i and x_i = weight of truck i. But the ratio estimate $\hat{t}_{yr} = t_x \hat{t}_y / \hat{t}_x$ requires that the total weight for all truckloads be known, and weighing every truck in the population is impractical.

Two-phase sampling, also called **double sampling,** provides a solution. Two-phase sampling, as introduced by Neyman (1938), is useful when the variable of interest y is relatively expensive to measure, but a correlated variable x can be measured fairly easily and used to improve the precision of the estimator of t_y.

Suppose the population has N observation units. The sample is taken in two phases:

1 *Phase I sample.* Take a probability sample of $n^{(1)}$ units, called the phase I sample. Measure the auxiliary variables $\mathbf{x}$ for every unit in the phase I sample. In the survey of businesses, you could take a random sample of tax records and record the reported income for each business in the sample. For measuring timber volume, you could weigh a sample of trucks selected either randomly or with probability proportional to estimated timber volume. The phase I sample is generally relatively large (and can be large because the auxiliary information is inexpensive to obtain) and should provide accurate information about the distribution of the x's.

2 *Phase II sample.* Now act as though the phase I sample is a population and select a probability sample from the phase I sample. Measure the variables of interest for each unit in the subsample, called the phase II sample. Since you are treating the phase I sample as the population from which the phase II sample is drawn, you may use the auxiliary information gathered in phase I when designing the phase II sample. You might select the businesses to be contacted with probability proportional to the income measured in the phase I sample. Alternatively, you might use the income information to stratify the businesses in the phase I sample and then contact a randomly selected subset of the businesses in each income stratum to obtain the desired information on variables such as total expenses. You could select the truckloads on which timber volume is to be measured with probability proportional to weight, or you could use the information in the phase I sample to obtain a better estimate of total weight and use ratio estimation. In each case, the y variables are relatively expensive to measure, but y is correlated with x.

Two-phase sampling can save time and money if the auxiliary information is relatively inexpensive to obtain and if having that auxiliary information can increase the precision of the estimates for quantities of interest.

E X A M P L E   12.1 Stockford and Page (1984) used two-phase sampling to estimate the percentage of Vietnam-era veterans in U.S. Veterans Administration (VA) hospitals who actually served in Vietnam.

The 1982 VA Annual Patient Census (APC) included a random sample of 20% of the patients in VA hospitals. The following question was included: "If period of service is 'Vietnam era,' was service in Vietnam?" with answer categories "yes," "no," and "not available." The answers to the question were obtained from patients' medical records. The response from medical records could be inaccurate, however, for several reasons: (1) The medical record classification was largely self-reported, and the patient may not have been able to recall the location of service due to medical problems or may have been confused about the definition of Vietnam service (some pilots whose duty station was officially recorded as Thailand flew missions over Vietnam); (2) a patient might misstate Vietnam service because he or she thought the answer might affect VA benefits; or (3) errors might be made in recording the response in the medical record. In addition, a large number of patients had "not available" for the answer. Thus, the answer to the question on Vietnam service in the APC survey was unsatisfactory for estimating the percentage of Vietnam-era veterans in VA hospitals who served in Vietnam.

Stockford and Page checked the military records for a stratified subsample of the hospitalized veterans to determine the true classification of Vietnam service. The information in the original survey was used for the stratification, as different percentages with Vietnam service were expected in the "yes," "no," and "not available" groups in the APC survey. Military records for all patients in the "not available" stratum were checked. It was expected that the within-stratum variances would be relatively low in the "yes" and "no" strata—even though the APC survey data are inaccurate, you would expect a higher percentage of "yes" respondents to have served in Vietnam than "no" respondents—and military records for a 10% subsample were checked for each of those two strata.

The results for the question "Was service in Vietnam?" were as follows:

APC Group	APC Survey Classification	Subsample Size	Vietnam Service in Subsample
Yes	755	67	49
No	804	72	11
Not available	505	505	211
Total	2064	644	271

As expected, the percentage of veterans with Vietnam service differed for the three groups: Of the veterans with a "yes" response to the APC survey question, 73% actually served in Vietnam, compared with 15% for the "no" group and 42% for the veterans for whom the information was not available. ■

EXAMPLE 12.2 Two-phase sampling is often used in forestry surveys. Aerial photographs are available for the region of interest, and points are systematically distributed across the photographs. Areas around the points are inspected on the photographs and classified by land class: forest land, unproductive forest land, nonforest land, and water. A phase I sample of points is then drawn from the grid, with a higher sampling fraction for grid points classified as forest land than those classified as nonforest land. Areas in the phase I sample are examined more closely to classify them by stand size and density. Then, a subsample is taken of the points in the phase I sample, and ground measurements such as land use, volume, and mortality taken; the percentage of area that is forest from the phase II ground sample may differ somewhat from the photo estimate in phase I, and ratio estimation can be used in the phase II sample to increase the precision of the estimate. ■

EXAMPLE 12.3 We have already seen two-phase sampling used in nonresponse adjustment, in Section 8.3. A probability sample is taken from the population; the sampled units are then divided into the two strata of respondents and nonrespondents. Then, a subsample is taken of the nonrespondents. The phase I sample is the original probability sample. The variable

$$x_i = \begin{cases} 1 & \text{if observation } i \text{ responds} \\ 0 & \text{if observation } i \text{ is a nonrespondent} \end{cases}$$

is observed for everyone in the phase I sample. The information about x_i is then used in the phase II sample. The value of interest y_i is observed for all observations with $x_i = 1$; a subsample is taken for observations with $x_i = 0$. ■

12.1.1 Two-Phase Sampling Theory

We first state the results in general and then for the case when both phase I and phase II samples are simple random samples (SRSs). A general framework for two-phase sampling is given in Särndal and Swensson (1987).

Let $\mathcal{S}^{(1)}$ denote the phase I sample; the units selected for the sample are determined by the random variables

$$Z_i = \begin{cases} 1 & \text{if unit } i \text{ is in the phase I sample.} \\ 0 & \text{if unit } i \text{ is not in the phase I sample.} \end{cases}$$

Let $w_i^{(1)}$ be the sampling weight for the phase I sample: $w_i^{(1)} = 1/[P(Z_i = 1)]$. We observe a vector of auxiliary characteristics $\mathbf{x}_i = [x_{i1}, x_{i2}, \ldots, x_{ik}]^T$ for each observation unit in the phase I sample. Using the theory developed in earlier chapters, we can estimate the population total for auxiliary variable j as

$$\hat{t}_{x_j}^{(1)} = \sum_{i \in \mathcal{S}^{(1)}} w_i^{(1)} x_{ij} = \sum_{i=1}^{N} Z_i w_i^{(1)} x_{ij}.$$

Now, indicate membership in the phase II sample $\mathcal{S}^{(2)}$ by the random variable

$$D_i = \begin{cases} 1 & \text{if unit } i \text{ is in the phase II sample.} \\ 0 & \text{if unit } i \text{ is not in the phase II sample.} \end{cases}$$

The probability that a unit is in the phase II sample depends on whether it is in the phase I sample and also may depend on auxiliary information collected in the phase I sample; we denote this dependence by writing $P(D_i = 1 \mid \mathbf{Z})$, where $\mathbf{Z}$ is the vector $(Z_1, Z_2, \ldots, Z_N)^T$. Thus, when we find an expectation conditional on $\mathbf{Z}$, we are treating the information from the phase I sample as known. The subsampling weights for the final, phase II sample also depend on which units were selected to be in the phase I sample:

$$w_i^{(2)} = w_i^{(2)}(\mathbf{Z}) = \begin{cases} \dfrac{1}{P(D_i = 1 \mid \mathbf{Z})} & \text{if } Z_i = 1. \\ 0 & \text{if } Z_i = 0. \end{cases}$$

An analog of the Horvitz–Thompson estimator for two-phase sampling is

$$\hat{t}_y^{(2)} = \sum_{i \in \mathcal{S}^{(2)}} w_i^{(1)} w_i^{(2)} y_i = \sum_{i=1}^{N} Z_i D_i w_i^{(1)} w_i^{(2)} y_i. \tag{12.1}$$

We use the following device to find properties of two-phase estimates. Define

$$\hat{t}_y^{(1)} = \sum_{i \in \mathcal{S}^{(1)}} w_i^{(1)} y_i = \sum_{i=1}^{N} Z_i w_i^{(1)} y_i.$$

Now, we do not know what $\hat{t}_y^{(1)}$ is, because we only observe the y_i's in the phase II sample. But $\hat{t}_y^{(1)}$ serves as the "population total" estimated in phase II—if we knew y_i for all units in the phase I sample, we would estimate t_y by $\hat{t}_y^{(1)}$. Treating the phase I sample as known, we have

$$E\left[\hat{t}_y^{(2)} \mid \mathbf{Z}\right] = \sum_{i=1}^{N} Z_i w_i^{(1)} w_i^{(2)} y_i E[D_i \mid \mathbf{Z}] = \sum_{i=1}^{N} Z_i w_i^{(1)} y_i = \hat{t}_y^{(1)}.$$

Then, using successive conditioning (see p. 434, Section B.4),

$$E\left[\hat{t}_y^{(2)}\right] = E\left\{E\left[\hat{t}_y^{(2)} \mid \mathbf{Z}\right]\right\} = E\left[\sum_{i=1}^{N} Z_i w_i^{(1)} y_i\right] = t_y.$$

Also, from property 5 in Section B.4,

$$V\left(\hat{t}_y^{(2)}\right) = V\left(E\left[\hat{t}_y^{(2)} \mid \mathbf{Z}\right]\right) + E\left(V\left[\hat{t}_y^{(2)} \mid \mathbf{Z}\right]\right) = V\left(\hat{t}_y^{(1)}\right) + E\left(V\left[\hat{t}_y^{(2)} \mid \mathbf{Z}\right]\right).$$

The first term is the variance that would be obtained if y_i had been observed for every observation in $S^{(1)}$; the second term is the additional variance from subsampling in phase II.

12.1.2 Two-Phase Sampling with Ratio Estimation

Define $S^{(1)}$, $S^{(2)}$, Z_i, and D_i as above. The auxiliary variable x_i is measured for each observation in the phase I sample; from that sample, we may estimate the population total $t_x = \sum_{i=1}^{N} x_i$ by

$$\hat{t}_x^{(1)} = \sum_{i \in S^{(1)}} w_i^{(1)} x_i = \sum_{i=1}^{N} Z_i w_i^{(1)} x_i.$$

Now select the phase II subsample and measure y_i on units in the subsample. From the phase II sample $S^{(2)}$, we can calculate $\hat{t}_y^{(2)}$ using (12.1) and

$$\hat{t}_x^{(2)} = \sum_{i \in S^{(2)}} w_i^{(1)} w_i^{(2)} x_i = \sum_{i=1}^{N} Z_i D_i w_i^{(1)} w_i^{(2)} x_i.$$

Then,

$$\hat{t}_{yr}^{(2)} = \frac{\hat{t}_x^{(1)} \hat{t}_y^{(2)}}{\hat{t}_x^{(2)}}.$$

Note that this estimator is very similar to the ratio estimator in (3.2); we use $\hat{t}_x^{(1)}$ from the phase I sample instead of the unknown quantity t_x.

Using linearization,

$$\hat{t}_{yr}^{(2)} \approx t_y + \frac{t_x}{t_x}\left(\hat{t}_y^{(2)} - t_y\right) + \frac{t_y}{t_x}\left(\hat{t}_x^{(1)} - t_x\right) - \frac{t_y t_x}{t_x^2}\left(\hat{t}_x^{(2)} - t_x\right).$$

Then,

$$V\left(\hat{t}_{yr}^{(2)}\right) \approx V\left[\hat{t}_y^{(2)} + \frac{t_y}{t_x}\left(\hat{t}_x^{(1)} - \hat{t}_x^{(2)}\right)\right]$$

$$= V\left\{E\left[\hat{t}_y^{(2)} + \frac{t_y}{t_x}\left(\hat{t}_x^{(1)} - \hat{t}_x^{(2)}\right)\,\middle|\, \mathbf{Z}\right]\right\} + E\left\{V\left[\hat{t}_y^{(2)} + \frac{t_y}{t_x}\left(\hat{t}_x^{(1)} - \hat{t}_x^{(2)}\right)\,\middle|\, \mathbf{Z}\right]\right\}$$

$$= V\left[\hat{t}_y^{(1)}\right] + E\left[V\left(\hat{t}_y^{(2)} - \frac{t_y}{t_x}\hat{t}_x^{(2)}\,\middle|\, \mathbf{Z}\right)\right]$$

$$= V\left[\hat{t}_y^{(1)}\right] + E\left[V(\hat{t}_d^{(2)} \mid \mathbf{Z})\right],$$

where $d_i = y_i - (t_y/t_x)x_i$. Thus, the variance of the two-phase ratio estimator is the variance that would be calculated for $\hat{t}_y^{(1)}$ if we observed y_i for every unit in the phase

I sample, plus an extra term involving the variance of the residuals from the ratio model. Exercise 2 gives the variance and an estimator of the variance if the sample design for both phases is an SRS. Rao and Sitter (1995) and Sitter (1997) derived other variance estimators for ratio and regression estimators in two-phase sampling.

12.1.3 Two-Phase Sampling for Stratification

For simplicity, assume that an SRS is taken in phase I and that simple random sampling is used for the subsamples in phase II. (Särndal et al. 1992 give a more general treatment, allowing unequal-probability sampling for either phase.) Define $\mathcal{S}^{(1)}$, $\mathcal{S}^{(2)}$, Z_i, and D_i as above. If an SRS of size n is taken in phase I,

$$P(Z_i = 1) = \frac{n}{N}.$$

The observation units are divided among H strata, but we do not know which stratum a unit belongs to until it is selected in phase I. In the population, however, stratum h has N_h units (N_h is unknown), and $N = \sum_{h=1}^{H} N_h$ (assume N is known). Let

$$x_{ih} = \begin{cases} 1 & \text{if unit } i \text{ is in stratum } h. \\ 0 & \text{if unit } i \text{ is not in stratum } h. \end{cases}$$

Observe x_{ih}, $h = 1, \ldots, H$ for each unit in the phase I sample; assume that at least two units from each stratum are sampled. The number of units in the phase I sample that belong to stratum h is a random variable:

$$n_h = \sum_{i=1}^{N} Z_i x_{ih}.$$

Now take a simple random subsample of size m_h in stratum h; m_h may depend on the first phase of the sampling. The subsamples in different strata are selected independently, given the information in the phase I sample. With random subsampling,

$$P(D_i = 1 \mid \mathbf{Z}) = Z_i \sum_{h=1}^{H} x_{ih} \frac{m_h}{n_h}.$$

Although $P(D_i = 1 \mid \mathbf{Z})$ is written as a sum, all but one of the x_{ih} for $h = 1, \ldots, H$ will equal zero because each unit belongs to exactly one stratum. The sampling weight for a phase II unit in stratum h is $w_i^{(2)} = n_h/m_h$; in general, $w_i^{(2)} = Z_i \sum_{h=1}^{H} x_{ih} n_h/m_h$.

The two-phase-sampling stratified estimator of the population total is

$$\hat{t}_{\text{str}}^{(2)} = \sum_{i=1}^{N} Z_i D_i w_i^{(1)} w_i^{(2)} y_i$$

$$= \sum_{i=1}^{N} \sum_{h=1}^{H} Z_i D_i \frac{N}{n} \frac{n_h}{m_h} x_{ih} y_i \qquad \textbf{(12.2)}$$

$$= N \sum_{h=1}^{H} \frac{n_h}{n} \bar{y}_h^{(2)},$$

where $\bar{y}_h^{(2)} = \sum_{i \in \mathcal{S}^{(2)}} x_{ih} y_i / m_h$ is the average of the phase II units in stratum h. The corresponding estimator of the population mean is

$$\hat{\bar{y}}_{str}^{(2)} = \sum_{h=1}^{H} \frac{n_h}{n} \bar{y}_h^{(2)}. \tag{12.3}$$

Recall that a stratified random sampling estimator of the population total from (4.1) is

$$\hat{t}_{str} = N \sum_{h=1}^{H} \frac{N_h}{N} \bar{y}_h;$$

the two-phase sampling estimator simply substitutes n_h/n for N_h/N. As was shown for the estimator in (12.1), $E[\hat{t}_{str}^{(2)} \mid \mathbf{Z}] = \hat{t}_y^{(1)}$, so $E[\hat{t}_{str}^{(2)}] = t_y$.

The variance is again computed conditionally:

$$V\left(\hat{t}_{str}^{(2)}\right) = V\left(E\left[\hat{t}_{str}^{(2)} \mid \mathbf{Z}\right]\right) + E\left(V\left[\hat{t}_{str}^{(2)} \mid \mathbf{Z}\right]\right)$$

$$= V\left(\hat{t}_y^{(1)}\right) + N^2 E\left(V\left[\sum_{h=1}^{H} \frac{n_h}{n} \bar{y}_h^{(2)} \mid \mathbf{Z}\right]\right)$$

$$= N^2 \left(1 - \frac{n}{N}\right) \frac{S_y^2}{n} + N^2 E\left[\sum_{h=1}^{H} \left(\frac{n_h}{n}\right)^2 \left(1 - \frac{m_h}{n_h}\right) \frac{s_h^{2(1)}}{m_h}\right].$$

The first term is the variance from the phase I SRS; the second term is the additional variance resulting from the subsampling in phase II. Here, $S_y^2 = \sum_{i=1}^{N} (y_i - \bar{y}_U)^2 / (N - 1)$ is the population variance of the y's;

$$s_h^{2(1)} = \frac{\sum\limits_{i \in \mathcal{S}^{(1)}} x_{ih}\left(y_i - \bar{y}_h^{(1)}\right)^2}{n_h - 1}$$

would be the sample variance of the y_i's in stratum h in the phase I sample if we observed all of them. The variance of $\hat{t}_{str}^{(2)}$ is left as an expectation because n_h and m_h are random variables.

Rao (1973) gives the estimated variance in two-phase sampling as

$$\hat{V}\left(\hat{t}_{str}^{(2)}\right) = N(N-1) \sum_{h=1}^{H} \left(\frac{n_h - 1}{n - 1} - \frac{m_h - 1}{N - 1}\right) \frac{n_h}{n} \frac{s_h^{2(2)}}{m_h}$$

$$+ \frac{N^2}{n-1}\left(1 - \frac{n}{N}\right) \sum_{h=1}^{H} \frac{n_h}{n} \left(\bar{y}_h^{(2)} - \hat{\bar{y}}_{str}^{(2)}\right)^2 \tag{12.4}$$

where $s_h^{2(2)}$ is the sample variance of the y_i's in stratum h. If we can ignore the fpc's (finite population corrections),

$$\hat{V}\left(\hat{\bar{y}}_{str}^{(2)}\right) \approx \sum_{h=1}^{H} \frac{n_h - 1}{n - 1} \frac{n_h}{n} \frac{s_h^{2(2)}}{m_h} + \frac{1}{n-1} \sum_{h=1}^{H} \frac{n_h}{n} \left(\bar{y}_h^{(2)} - \hat{\bar{y}}_{str}^{(2)}\right)^2. \tag{12.5}$$

EXAMPLE 12.4   Let's apply these results to the data in Example 12.1. Because $\bar{y}_h^{(2)} = \hat{p}_h$ is a proportion, $s_h^{2(2)} = m_h \hat{p}_h (1 - \hat{p}_h)/(m_h - 1)$. The statistics from the phase II sample are as follows:

Stratum	n_h	m_h	$\hat{p}_h$	$s_h^{2(2)}$
Yes	755	67	0.7313	0.1995
No	804	72	0.1528	0.1313
Not available	505	505	0.4178	0.2437
Total	2064	644		

The estimated percentage of Vietnam-era VA hospital patients who served in Vietnam is, from (12.3),

$$\hat{\bar{y}}_{str}^{(2)} = \left(\frac{755}{2064}\right)(0.7313) + \left(\frac{804}{2064}\right)(0.1528) + \left(\frac{505}{2064}\right)(0.4178) = 0.4293.$$

The phase I sample is an SRS with $n/N = 0.2$, so the fpc should be included in the variance estimate. Calculating the terms in (12.4),

$$\sum_{h=1}^{H} \left(\frac{n_h - 1}{n - 1} - \frac{m_h - 1}{N - 1}\right) \frac{n_h}{n} \frac{s_h^{2(2)}}{m_h} = 0.000391 + 0.000271 + 0.0000231 = 0.000686,$$

and

$$\frac{1}{n-1}\left(1 - \frac{n}{N}\right) \sum_{h=1}^{H} \frac{n_h}{n} \left(\bar{y}_h^{(2)} - \hat{\bar{y}}_{str}^{(2)}\right)^2$$

$$= (1.29 \times 10^{-5}) + (1.16 \times 10^{-5}) + (1.24 \times 10^{-8}) = 0.0000245.$$

Thus, $\hat{V}(\hat{\bar{y}}_{str}^{(2)}) = 0.000686 + 0.0000245 = 0.00071$, and $\text{SE}(\hat{\bar{y}}_{str}^{(2)}) = 0.027$.

Was two-phase sampling more efficient here? Had an SRS of size 644 been taken directly from the records and had $\hat{p} = 0.429$ been observed, the standard error would have been $\text{SE}(\hat{p}) = 0.019$, which is actually smaller than the standard error from the two-phase sampling design. If you look at the individual terms in the variance estimates, you can see why two-phase sampling did not increase efficiency in this example. All the phase I units in the "not available" stratum were subsampled, giving a very low value of $s_h^{2(2)}/m_h$ for that stratum. But the sample sizes in the other two strata were too small, leading to relatively large contributions to the overall variance from those two strata.

Suppose proportional allocation had been used in the phase II sample instead and that the same sample proportions had been observed. Then, you would subsample 236 records in the "yes" stratum, 251 records in the "no" stratum, and 157 records in the "not available" stratum. In that case, if the sample proportions remained the same, the standard error from the two-phase sample would have been 0.017, a modest decrease from the standard error of an SRS of size 644. More savings could possibly have been achieved if some sort of optimal allocation had been used (see Exercise 5).   ∎

12.2
Capture-Recapture Estimation

EXAMPLE 12.5   Suppose we want to estimate N, the number of fish in a lake. One method is as follows: Catch and mark 200 fish in the lake, then release them. Allow the marked and released fish to mix with the other fish in the lake. Then, take a second, independent sample of 100 fish. Suppose that 20 of the fish in the second sample are marked. Then, assuming that the population of fish has not changed between the two samples and that each catch gives an SRS of fish in the lake, estimate that 20% of the fish in the lake are marked and therefore the 200 fish tagged in the original sample represent approximately 20% of the population of fish. The population size N is then estimated to be approximately 1000.   ■

This method for estimating the size of a population is called two-sample **capture-recapture estimation.** Other names sometimes used are tag- or mark-recapture, the Petersen (1896) method, or the Lincoln (1930) index. The method relies on the following assumptions:

1   The population is *closed*—no fish enter or leave the lake between the samples. This means that N is the same for each sample.

2   Each sample of fish is an SRS from the population. This means that each fish is equally likely to be chosen in a sample—it is not the case, for example, that smaller or less healthy fish are more likely to be caught. Also, there are no "hidden fish" in the population that are impossible to catch.

3   The two samples are independent. The marked fish from the first sample become re-mixed in the population so that the marking status of a fish is unrelated to the probability that the fish is selected in the second sample. Also, fish included in the first sample do not become "trap-shy" or "trap-happy"—the probability that a fish will be caught in the second sample does not depend on its capture history.

4   Fish do not lose their markings, and marked fish can be identified as such. Water-soluble paint, for example, would not be a good choice for marking material.

In this simple form, capture-recapture is a special case of ratio estimation of a population total, and results from Chapter 3 may be used when the samples and population are large. Let n_1 be the size of the first sample, n_2 the size of the second sample, and m the number of marked fish caught in the second sample. In Example 12.5, $n_1 = 200$, $n_2 = 100$, $m = 20$, and we used the estimate $\hat{N} = n_1 n_2 / m$. To see how this estimate fits into the framework of Chapter 3, let

$$y_i = 1 \text{ for every fish in the lake.}$$

$$x_i = \begin{cases} 1 & \text{if fish } i \text{ is marked.} \\ 0 & \text{if fish } i \text{ is not marked.} \end{cases}$$

Then estimate $N = t_y = \sum_{i=1}^{N} y_i$ by $\hat{t}_{yr} = t_x \hat{B}$, where $t_x = \sum_{i=1}^{N} x_i = n_1$ and

$\hat{B} = \bar{y}/\bar{x} = n_2/m$. This ratio estimate,

$$\hat{N} = \hat{t}_{yr} = \frac{n_1 n_2}{m}, \tag{12.6}$$

is also the maximum likelihood estimate (see Exercises 8 and 9). Applying (3.7) to the second SRS and ignoring the fpc,

$$\hat{V}(\hat{N}) = t_x^2 \hat{V}(\hat{B}) = \left(\frac{n_1 n_2}{m}\right)^2 \frac{n_2 - m}{m(n_2 - 1)} \approx \frac{n_1^2 n_2 (n_2 - m)}{m^3}.$$

For the data in Example 12.5, $\hat{V}(\hat{N}) = 40{,}000$.

Being a ratio estimator, though, $\hat{N}$ is biased, and the bias can be large in wildlife applications with small sample sizes. Indeed, it is possible for the second sample to consist entirely of unmarked animals, making the estimate in (12.6) infinite. Chapman (1951) proposes the less biased estimate

$$\tilde{N} = \frac{(n_1 + 1)(n_2 + 1)}{m + 1} - 1. \tag{12.7}$$

A variance estimate for $\tilde{N}$ (Seber 1970) is

$$\hat{V}(\tilde{N}) = \frac{(n_1 + 1)(n_2 + 1)(n_1 - m)(n_2 - m)}{(m + 1)^2 (m + 2)}. \tag{12.8}$$

The estimates in (12.7) and (12.8) are often used in wildlife applications. For the fish data, $\tilde{N} = (201)(101)/21 - 1 = 966$, and $\hat{V}(\tilde{N}) = 30{,}131$.

Many researchers have constructed confidence intervals for the population size using either

$$\hat{N} \pm 1.96\sqrt{\hat{V}(\hat{N})} \quad \text{or} \quad \tilde{N} \pm 1.96\sqrt{\hat{V}(\tilde{N})}.$$

These are not entirely satisfactory, however, because both require that $\hat{N}$ or $\tilde{N}$ be approximately normally distributed, and the normal distribution may not be a good approximation to the distribution of $\hat{N}$ or $\tilde{N}$ for small populations and samples. We'll discuss confidence intervals in Section 12.2.2; first, however, let's look at another approach for these data that will be useful in developing confidence intervals.

12.2.1 Contingency Tables for Capture-Recapture Experiments

Fienberg (1972) suggests viewing capture-recapture data in an incomplete contingency table. For the data in Example 12.5, the table is as follows:

		In Sample 2?		
		Yes	No	
In Sample 1?	Yes	20	180	200
	No	80	?	?
		100	?	N

In general, if x_{ij} is the observed count in cell (i, j), the contingency table looks as follows. An asterisk indicates that we do not observe that cell.

		In Sample 2?		
		Yes	No	
In Sample 1?	Yes	$x_{11}(=m)$	x_{12}	$x_{1+}(=n_1)$
	No	x_{21}	x_{22}^*	x_{2+}^*
		$x_{+1}(=n_2)$	x_{+2}^*	x_{++}^*

The expected counts are the following:

		In Sample 2?		
		Yes	No	
In Sample 1?	Yes	m_{11}	m_{12}	m_{1+}
	No	m_{21}	m_{22}^*	m_{2+}^*
		m_{+1}	m_{+2}^*	$m_{++}^* = N$

To estimate the expected counts then, we would use $\hat{m}_{11} = x_{11}$, $\hat{m}_{12} = x_{12}$, and $\hat{m}_{21} = x_{21}$. If presence in sample 1 is independent of presence in sample 2, then the odds of being in sample 2 are the same for marked fish as for unmarked fish: $m_{11}/m_{21} = m_{12}/m_{22}$. Consequently, under independence, the estimated count in the cell of fish not included in either sample is

$$\hat{m}_{22} = \frac{\hat{m}_{12}\hat{m}_{21}}{\hat{m}_{11}} = \frac{x_{12}x_{21}}{x_{11}},$$

and

$$\hat{N} = \hat{m}_{11} + \hat{m}_{12} + \hat{m}_{21} + \hat{m}_{22} = \frac{x_{+1}x_{1+}}{x_{11}}.$$

The estimate $\hat{N}$ is calculated based on the assumption that the two samples are independent; unfortunately, this assumption cannot be tested because only three of the four cells of the contingency table are observed.

12.2.2 Confidence Intervals for N

In many applications of capture-recapture, confidence intervals (CIs) have been constructed using

$$\hat{N} \pm 1.96\sqrt{\hat{V}(\hat{N})} \quad \text{or} \quad \tilde{N} \pm 1.96\sqrt{\hat{V}(\tilde{N})}.$$

If we use the first interval for the data in Example 12.5, $\hat{V}(\hat{N}) = 40{,}000$, and an asymptotic 95% CI would be $1000 \pm 1.96(200) = [608, 1392]$. The confidence interval using the normal distribution and $\tilde{N}$ is $[626, 1306]$. Unfortunately, confidence intervals based on the assumption that $\hat{N}$ or $\tilde{N}$ follow a normal distribution often have poor coverage probability in small samples because the distribution of $\hat{N}$ and $\tilde{N}$ is actually quite skewed, as you will see in Exercise 13. In general, we do not recommend using these confidence intervals.

An additional shortcoming of confidence intervals based on the normal distribution can occur in small samples. For example, suppose $n_1 = 30, n_2 = 20$, and $m = 15$.

Then $\hat{N} = (30)(20)/15 = 40$, and $\hat{V}(\hat{N}) = 26.7$. Using a normal approximation to the distribution of $\hat{N}$ results in the confidence interval [30, 50]. The lower bound of 30 is silly, however; a total of 35 distinct animals were observed in the two samples, so we know that N must be at least 35.

Cormack (1992) discusses using the Pearson or likelihood ratio chi-square test for independence to construct a confidence interval. Using this method, we fill in the missing observation x_{22} by some value u and perform a chi-square test for independence on the artificially completed data set. The 95% CI for m_{22} is then all values of u for which the null hypothesis of independence for the two samples would not be rejected at the 0.05 level. For the data in Example 12.5, let's try the value $u = 600$. With this value, the "completed" contingency table is

		In Sample 2? Yes	No	
In Sample 1?	Yes	20	180	200
	No	80	600	680
		100	780	880

We can easily perform Pearson's chi-square test for independence on this table, obtaining a p-value of 0.49. As $0.49 > 0.05$, the value 600 would be inside the 95% CI for u, and the value 880 would be inside the 95% CI for N. Setting u equal to 1500, though, gives p-value 0.0043, so 1500 is outside the 95% CI for u, and 1780 is thus outside the 95% CI for N. Continuing in this manner, we find that values of u between 430 and 1198 are the only ones that result in p-value > 0.05, so [430, 1198] is a 95% CI for m_{22}. The corresponding confidence interval for N is obtained by adding the number of observed animals in the other cells, 280, to the endpoints of the confidence interval for m_{22}, resulting in the interval [710, 1478].

The likelihood ratio test may be used in a similar manner, by including in the confidence interval all values of u for which the p-value from the likelihood ratio test exceeds 0.05. Using the S-PLUS code given in Appendix D, we find that values of u between 437 and 1233 give a likelihood ratio p-value exceeding 0.05. The confidence interval for N, using the likelihood ratio test, is then [717, 1513].

Another alternative for confidence intervals is to use the bootstrap (Buckland 1984). To apply bootstrap here, resample from the observed individuals in the second sample. Take R samples of size 100 with replacement from the 20 tagged and 80 untagged fish we observed. Calculate $\hat{N}^*$ for each of the R resamples and find the 2.5 and 97.5 percentage points of the R values. With $R = 999$, the 95% CI was the 25th and 975th values from the ordered list of the $\hat{N}^*$, [714, 1538].

Note that all three of these confidence intervals resulting from Pearson's chi-square test, the likelihood ratio chi-square test, and the bootstrap are similar, but all differ from the confidence intervals based on the asymptotic normality of $\hat{N}$ or $\tilde{N}$.

12.2.3 Using Capture-Recapture on Lists

Capture-recapture estimation is not limited to estimating wildlife populations. It can also be used when the two samples are lists of individuals, provided that the assumptions for the method are met. Suppose you want to estimate the number of statisticians in the United States, and you obtain membership lists from the American Statistical Association (ASA) and the Institute for Mathematical Statistics (IMS). Every statistician either is or is not a member of the ASA, and either is or is not a member of the IMS. (There are other worthy statistical organizations, but for simplicity let's limit the discussion here to these two.) Then, n_1 is the number of ASA members, n_2 the number of IMS members, and m is the number of persons on both lists. We can estimate the number of statisticians using $\hat{N} = n_1 n_2 / m$, exactly as if statisticians were fish. The assumptions for this estimate are as above, but with slightly different implications than in wildlife settings:

1 The population is closed. In wildlife surveys, this assumption may not be met because animals often die or migrate between samples. When treating lists as the samples, though, we can usually act as though the population is closed if the lists are from the same time period.

2 Each list provides an SRS from the population of statisticians. This assumption is more of a problem; it implies that the probability of belonging to ASA is the same for all statisticians and the probability of belonging to IMS is the same for all statisticians. It does not allow for the possibility that a group of statisticians may refuse to belong to either organization or for the possibility that subgroups of statisticians may have different probabilities of belonging to an organization.

3 The two lists are independent. Here, this means that the probability that a statistician is in ASA does not depend on his or her membership in IMS. This assumption is also often not met—it may be that statisticians tend to belong to only one organization and therefore that ASA members are less likely to belong to IMS than non-ASA members.

4 Individuals can be matched on the lists. This sounds easy but often proves surprisingly difficult. Is J. Smith on list 1 the same person as Jonquil Smith on list 2?

EXAMPLE 12.6 The Bureau of the Census tries to enumerate as many persons as possible in the decennial census. Inevitably, however, persons are missed, leading population estimates from the census to underestimate the true population count. Moreover, it is thought that the undercount rate is not uniform; the undercount is thought to be greater for inner-city areas and minority groups and varies among different regions of the United States. Because congressional representatives, billions of dollars of federal funding, and other resources are apportioned based on census results, many state and local governments are concerned that the population counts be accurate. Capture-recapture estimation, called **dual-system estimation** in this context, has been used since 1950 to evaluate the coverage of the decennial census. In recent years there has been considerable controversy, culminating in lawsuits, over whether these methods should also be used to adjust the population estimates from the census. Fienberg (1992) gives a bibliography for dual-system estimation; articles in the November 1994 issue of *Statistical Science* discuss the controversy.

Hogan (1993) describes the 1990 Post-Enumeration Survey (PES) used by the Census Bureau. A similar procedure, called the Reverse Record Check, is used in Canada. Two samples are taken. The P sample is taken directly from the population, independently of the census, and is used to estimate number of persons missed by the census. The E sample is taken from the census enumeration itself and is used to estimate errors in the census, such as nonexistent persons or duplicates.

Separate population estimates are derived for each of the 1392 poststrata, where the population is poststratified by region, race, ownership of dwelling unit, age, and other variables. Poststrata are used because it is hoped that assumption 2 of equal recapture probabilities is approximately satisfied within each poststratum; we know it is not satisfied for the population as a whole because of the differential undercount rates in the census. The population table for a poststratum is as follows:

		In Census Enumeration?		
		Yes	No	
In PES?	Yes	N_{11}	N_{12}	N_{1+}
	No	N_{21}	N_{22}^*	N_{2+}^*
		N_{+1}	N_{+2}^*	N

The census enumeration, the P sample, and the E sample are all used to fill in the cells of the table. Then,

$$\hat{N} = \frac{\hat{N}_{+1}\hat{N}_{1+}}{\hat{N}_{11}}.$$

The quantities $\hat{N}_{1+}$ and $\hat{N}_{11}$ are estimates from the P sample: $\hat{N}_{1+}$ is the estimate of the poststratum total, using weights, from the P sample, and $\hat{N}_{11}$ is a weighted estimate of matches between the P sample and the census enumeration. Here, $\hat{N}_{+1}$ is not the actual count from the census but is the census count adjusted using the E sample to remove duplicates and fictitious persons. Many sample sizes in poststrata were small, leading to large variances for the estimates of population count, so the estimates were smoothed and adjusted using regression models.

The preceding assumptions need to be met for dual-system estimation to give a better estimate of the population than the original census data. It is hoped that assumption 2 holds within the poststrata. Assumption 3 is also of some concern, though, as the P sample also has nonresponse. Freedman and Navidi (1992) and Breiman (1994) discuss this problem, as well as concerns about the regression adjustment of the estimates. Another concern is the ability to match persons in the P sample to persons in the census. Because P-sample persons not matched are assumed to have been missed by the census, errors in matching persons in the two samples can lead to biases in the population estimates. Ding and Fienberg (1994; 1996) derive models for matching errors in dual-system estimation.

The debate over the use of sampling to improve the accuracy of census counts continues. For the year 2000 census, a panel of the National Academy of Sciences has recommended enumerating the population in each county until a 90% response rate for housing units has been attained, then sampling the remaining 10%. One bill before Congress as of this writing, however, would prohibit use of any funds "to plan or otherwise prepare for the use of sampling in taking the 2000 decennial census." ∎

12.2.4 Multiple-Recapture Estimation

The assumptions for the two-sample capture-recapture estimate described above are strong: The population must be closed and the two random samples independent. Moreover, these assumptions cannot be tested because we observe only three of the four cells in the contingency table—we need all four cells to test for the independence of samples.

More complicated models may be fit if $K > 2$ random samples are taken and especially if different markings are used for individuals caught in the different samples. With fish, for example, the left pectoral fin might be marked for fish caught in the first sample, the right pectoral fin marked for fish caught in the second sample, and a dorsal fin marked for fish caught in the third sample. A fish caught in sample 4 that had markings on the left pectoral fin and dorsal fin would then be known to have been caught in sample 1 and sample 3, but not sample 2.

Schnabel (1938) first discussed how to estimate N when K samples are taken. She found the maximum likelihood estimate of N to be the solution to

$$\sum_{i=1}^{K} \frac{(n_i - r_i)M_i}{N - M_i} = \sum_{i=1}^{K} r_i,$$

where n_i is the size of sample i, r_i is the number of recaptured fish in sample i, and M_i is the number of tagged fish in the lake when sample i is drawn.

If individual markings are used, we can also explore issues of immigration or emigration from the population and test some of the assumptions of independence.

EXAMPLE 12.7 Domingo-Salvany et al. (1995) used capture-recapture to estimate the prevalence of opiate addiction in Barcelona, Spain. One of their data sets consisted of three samples from 1989: (1) a list of opiate addicts from emergency rooms (E list); (2) a list of persons who started treatment for opiate addiction during 1989, reported to the Catalonia Information System on Drug Abuse (T list); (3) a list of heroin-overdose deaths registered by the forensic institute in 1989 (D list). A total of 2864 distinct persons were on the three lists. Persons on the three lists were matched, with the following results:

		In D List?			
		Yes		No	
		In T List?		In T List?	
		Yes	No	Yes	No
In E List?	Yes	6	27	314	1728
	No	8	69	712	?

It is unclear whether these data will fulfill the assumptions for the two-sample capture-recapture method. The assumption of independence among the samples may not be met—if treatment is useful, treated persons are less likely to appear in one of the other samples. In addition, persons on the death list are much less likely to subsequently appear on one of the other lists; the closed population assumption is also not met because one of the samples is a death list. Nevertheless, an analysis using the imperfectly met assumptions can provide some information on the number of opiate addicts. Because there are more than two samples, we can assess the assumptions of independence among different samples by using loglinear models. There is one assumption, though, that we can *never* test: The missing cell follows the same model as the rest of the data. ∎

If three samples are taken, the expected counts are:

		In Sample 3?			
		Yes		No	
		In Sample 2?		In Sample 2?	
		Yes	No	Yes	No
In Sample 1?	Yes	m_{111}	m_{121}	m_{112}	m_{122}
	No	m_{211}	m_{221}	m_{212}	m^*_{222}

Loglinear models were discussed in Section 10.4. The saturated model for three samples is:

$$\ln m_{ijk} = \mu + \alpha_i + \beta_j + \gamma_k + (\alpha\beta)_{ij} + (\alpha\gamma)_{ik} + (\beta\gamma)_{jk} + (\alpha\beta\gamma)_{ijk}.$$

This model cannot be fit, however, as it requires 8 degrees of freedom (df) and we only have seven cells. The following models may be fit, with α referring to the E list, β referring to the T list, and γ referring to the D list.

1   *Complete independence.*

$$\ln m_{ijk} = \mu + \alpha_i + \beta_j + \gamma_k.$$

This model implies that presence on any of the lists is independent of presence on any of the other lists. The independence model must always be adopted in two-sample capture-recapture.

2   *One list is independent of the other two.*

$$\ln m_{ijk} = \mu + \alpha_i + \beta_j + \gamma_k + (\alpha\beta)_{ij}.$$

Presence on the E list is related to the probability that an individual is on the T list, but presence on the D list is independent of presence on the other lists. There are three versions of this model; the other two substitute $(\alpha\gamma)_{ik}$ or $(\beta\gamma)_{jk}$ for $(\alpha\beta)_{ij}$.

3   *Two samples are independent given the third.*

$$\ln m_{ijk} = \mu + \alpha_i + \beta_j + \gamma_k + (\alpha\beta)_{ij} + (\alpha\gamma)_{ik}.$$

Three models of this type exist; the other two substitute either $(\alpha\beta)_{ij} + (\beta\gamma)_{ik}$ or $(\alpha\gamma)_{ij} + (\beta\gamma)_{ik}$ for $(\alpha\beta)_{ij} + (\alpha\gamma)_{ik}$. Presence on the death and treatment lists are conditionally independent given the E-list status—once we know that a person is on the emergency room list, knowing that he or she is on the death list gives us no additional information about the probability that he or she will be on the treatment list.

4   *All two-way interactions.*

$$\ln m_{ijk} = \mu + \alpha_i + \beta_j + \gamma_k + (\alpha\beta)_{ij} + (\alpha\gamma)_{ik} + (\beta\gamma)_{jk}.$$

This model will always fit the data perfectly: It has the same number of parameters as there are cells in the contingency table.

Unfortunately, in none of these models can we test the hypothesis that the missing cell follows the model. But at least we can examine hypotheses of pairwise independence among the samples. For the addiction data, the following loglinear models were fit from the data, using the function glim in S-PLUS (any loglinear model program that finds estimates using maximum likelihood will work):

Model	G^2	df	p-Value	$\hat{m}_{222}$	$\hat{N}$	95% CI
1 Independence	1.80	3	0.62	3,967	6,831	[6,322, 7,407]
2a E*T	1.09	2	0.58	4,634	7,499	[5,992, 9,706]
2b E*D	1.79	2	0.41	3,959	6,823	[6,296, 7,425]
2c T*D	1.21	2	0.55	3,929	6,793	[6,283, 7,373]
3a E*T, E*D	0.19	1	0.67	6,141	9,005	[5,921, 16,445]
3b E*T, T*D	0.92	1	0.34	4,416	7,280	[5,687, 9,820]
3c E*D, T*D	1.20	1	0.27	3,918	6,782	[6,253, 7,388]
4 E*T, E*D, T*D	—	0	—	7,510	10,374	[4,941, 25,964]

Here, G^2 is the likelihood ratio test statistic for that model. Somewhat surprisingly, the model of independence fits the data well. The predicted cell counts under model 1, complete independence, are as follows:

		In D List?			
		Yes		No	
		In T List?		In T List?	
		Yes	No	Yes	No
In E list?	Yes	5.1	28.3	310.8	1730.7
	No	11.7	64.9	712.4	3966.7

These predicted cell counts lead to the estimate

$$\hat{N} = 2864 + 3967 = 6831$$

if the model of independence is adopted. The values of $\hat{N}$ for the other models are calculated similarly, by estimating the value in the missing cell from the model and adding that estimate to the known total for the other cells, 2864.

We can use an inverted likelihood ratio test (Cormack 1992) to construct a confidence interval for N, using any of the models. A 95% CI for the missing cell consists of those values u for which a 0.05-level hypothesis test of $H_0 : m_{222} = u$ would not be rejected for the loglinear model adopted. Let $G^2(u)$ be the likelihood ratio test statistic (deviance) for the completed table with u substituted for the missing cell, let t be the total of the seven observed cells, and let $\hat{u}$ be the estimate of the missing cell using that loglinear model. Cormack shows that the set

$$\left\{ u : G^2(u) - G^2(\hat{u}) + \log\left(\frac{u}{t+u}\right) - \log\left(\frac{\hat{u}}{t+\hat{u}}\right) < q_1(\alpha) \right\}$$

—where $q_1(\alpha)$ is the percentile of the χ_1^2 distribution with right-tail area α—is an approximate $100(1-\alpha)\%$ CI for m_{222}. We give an S-PLUS function for calculating Cormack's confidence interval in Appendix D. This confidence interval is conditional on the model selected and does not include uncertainty associated with the choice of model. Cormack also discusses extending the inverted Pearson chi-square test for goodness of fit, which produces a similar interval. Buckland and Garthwaite (1991) discuss using the bootstrap to find confidence intervals for multiple recapture using loglinear models; they incorporate the model-selection procedure into each bootstrap iteration.

For these data, the point estimate and confidence interval appear to rely heavily on the particular model fit, even though all seem to fit the observed cells. Note that the estimate $\hat{N}$ is larger and the confidence intervals much wider for models including the E*T interaction, even though that interaction is not statistically significant. The good fit of the independence model is somewhat surprising because you would not expect the assumptions for independence to be satisfied. In addition, the population is not closed, but we have little information on migration in and out of the population.

In this section we have presented only an introduction to estimating population size, under the assumption that the population is closed. Much other research has been done in capture-recapture estimation, including models for populations with births, deaths, and migrations; good sources for further reading are Seber (1982), Pollock (1991), and the review paper by the International Working Group for Disease Monitoring and Forecasting (1995).

12.3
Estimation in Domains, Revisited

12.3.1   Domain Means in Complex Surveys

In most surveys, estimates are desired not only for the population as a whole but also for subpopulations, called **domains** in survey sampling. We discussed estimation in subpopulations in Section 3.3 for SRSs and showed that estimating domain means was a special case of ratio estimation because the sample size in the domain varies from sample to sample. But it was noted that if the sample size for the domain in an SRS was large enough, we could essentially act as though the sample size was fixed for inference about the domain mean.

In complex surveys with many domains, it's not quite that simple. One worry is that the sample size for a given domain will be too small to provide a useful estimate. An investigator using the National Crime Victimization Survey (NCVS) to estimate victimization rates for race × gender groups separately in each state will find some empty cells even with a sample of 90,000 persons. In addition, even if the domain is not completely empty, it is possible in a complex survey that some psu's and even some strata contain no one in the domain, so variance estimates must be calculated with care.

Let y_i be the variable of interest and let

$$x_{id} = \begin{cases} 1 & \text{if observation unit } i \text{ is in domain } d. \\ 0 & \text{if observation unit } i \text{ is not in domain } d. \end{cases}$$

Then, using the theory we have developed throughout this book, estimate the population total for domain d by

$$\hat{t}_d = \sum_{i \in S} w_i x_{id} y_i$$

and the population mean for domain d, assuming the sample has some observations in domain d, by

$$\hat{\bar{y}}_d = \frac{\hat{t}_d}{\sum_{i \in S} w_i x_{id}}.$$

Because $\hat{\bar{y}}_d$ is a ratio, the variance is estimated using linearization (see Example 9.2) as

$$\hat{V}(\hat{\bar{y}}_d) = \frac{1}{\hat{N}_d^2} \hat{V}\left[\sum_{i \in S} w_i x_{id}(y_i - \hat{\bar{y}}_d)\right]. \tag{12.9}$$

The sample size in domain d must be large if the linearization variance is to be accurate.

As discussed in Chapter 3, if we ignore the fpc and an SRS is taken, (12.9) gives

$$\hat{V}(\hat{\bar{y}}_d) \approx \frac{s_d^2}{n_d},$$

where n_d is the number of sample observations in domain d and s_d^2 is the sample variance for the sample observations in domain d.

Warning   In an SRS, if you create a new data set that consists solely of sampled observations in domain d and then apply the standard variance formula, your variance estimate is approximately unbiased. Do *not* adopt this approach for estimating the variance of domain means in complex samples. It is quite common for a sampled psu to contain no observations in domain d; if you eliminate such psu's and then apply the standard variance formula, you will likely underestimate the variance.

Sometimes, when using published tables or public-use data files, you cannot calculate standard errors for each domain because you are not provided with enough information about the sample design. One possible solution is to multiply the standard error under simple random sampling by $\sqrt{\text{deff}}$ (design effect) for the overall mean. As noted by Kish and Frankel (1974), this approach may often overestimate the standard error, as the cluster effect may be reduced within the domain. For small domains, and especially for differences, the deff's tend toward 1.

12.3.2   Small Area Estimation

In the preceding discussion, we used linearization to approximate the variance of the ratio $\hat{t}_d / \hat{N}_d$. The validity of this approximation depends on having a sufficiently large sample size in the domain. In practice, the sample size in domain d may be so small that the variance of $\hat{\bar{y}}_d$ is extremely large. Some domains of interest may have no observations at all.

Many large government surveys provide very accurate estimates at the national level. The NCVS, for example, gives reliable information on the incidence of different types of criminal victimizations in the United States. However, if you are interested in estimates of violent-crime rates at the state level, to be used in allocating federal funds for additional police officers, the sample sizes for some states are so small that direct estimates of the violent-crime rate for those states are of very little use. You might conjecture, though, that crime rates are similar in neighboring states with similar characteristics and use information from other states to improve the estimate of violent-crime rate for the state with a small sample size. You could also incorporate information on crime rate from other sources, such as police statistics, to improve your estimate.

Similarly, the National Assessment of Educational Progress (NAEP; see Example 11.7) data collected on students in New York may be sufficient for estimating eighth-grade mathematics achievement for students in the state, but not for a direct assessment of mathematics achievement in individual cities such as Rochester. The survey data from Rochester, though, can be combined with estimates from other cities and with school administrative data (scores on other standardized tests, for example, or information about mathematics instruction in the schools) to produce an estimate of eighth-grade mathematics achievement for Rochester that we hope has smaller mean squared error (MSE).

Small area estimation techniques, in which estimates are obtained for domains with small sample sizes, have in recent years been the focus of intense research in statistics. A number of techniques have been proposed; a detailed description of the techniques and a bibliography for further reading is given in Ghosh and Rao (1994). Here, we summarize some of the proposed approaches. In the following, the quantities of interest are the domain totals t_d, for $d = 1, \dots, D$; the indicator variables for membership in domain d are x_{id}, as defined earlier.

1 Direct estimators. A direct estimator of t_d depends only on the sampled observations in domain d; as exposited above,

$$\hat{t}_d(\text{dir}) = \sum_{i \in S} w_i x_{id} y_i.$$

This direct estimator is unbiased, but the small sample size can lead to an unacceptably large variance (especially if domain d has no sampled observations!).

2 Synthetic estimators. Assume that we have some quantity associated with t_d for each domain d. For estimating violent-crime-victimization rates, we might use $u_d =$ total amount of violent crime in domain d obtained from police reports. Then, if the ratios t_d/u_d are similar in different domains and if each ratio is similar to the ratio of population totals t_y/t_u, then a simple form of synthetic estimator

$$\hat{t}_d(\text{syn}) = \left(\frac{\hat{t}_y}{\hat{t}_u} \right) u_d$$

may be more accurate than $\hat{t}_d(\text{dir})$. Certainly, the variance of $\hat{t}_d(\text{syn})$ will be relatively small, as $(\hat{t}_y/\hat{t}_u)$ is estimated from the entire sample and is expected to be precise. If the ratios are not homogeneous, however—if, for example, the proportion of violent-crime victimizations reported to the police varies greatly from domain to domain—then the synthetic estimator may have large bias.

You can also use synthetic estimation in subsets of the population and then combine the synthetic estimators for each subset. For estimating violent-crime victimization in small areas, you could divide the population into different age-race-gender classes. Then find a synthetic estimate of the total violent-crime victimization in domain d for each age-race-gender class and sum the estimates for the age-race-gender classes to estimate the total violent-crime victimizations in small area d. It is hoped that the ratios (violent-crime victimizations in domain d for age-race-gender class c from NCVS)/(violent-crime victimizations in domain d for age-race-gender class c from police reports) are more homogeneous than the ratios t_d/u_d.

3   *Composite estimators.* The direct estimator is unbiased but has large variance; the synthetic estimator has smaller variance but may have large bias. They may be combined to form a composite estimator:

$$\hat{t}_d(\text{comp}) = \alpha_d \hat{t}_d(\text{dir}) + (1 - \alpha_d)\hat{t}_d(\text{syn})$$

for $0 \le \alpha_d \le 1$. The relative optimal weights α_d are difficult to estimate, but one possible solution has α_d related to the sample size in domain d. Then, if too few units are observed in domain d, α_d will be close to zero and more reliance will be placed on the synthetic estimator.

4   *Model-based estimators.* In a model-based approach, a superpopulation model is used to predict values in domain d. The model often "borrows strength" from the data in closely related domains or incorporates auxiliary information from administrative data or other surveys.

Mixed models, described in Section 11.4, are often used in small area estimation. In the NAEP, if Y_{jd} is the mathematics achievement of student j in domain d in the population, you might postulate a model such as

$$Y_{jd} = \beta_{0d} + (u_{jd} - \bar{u}_d)\beta_1 + \varepsilon_{jd},$$

where $\beta_{0d} = \beta_0 + z_d\gamma_0 + \delta_{0d}$, the ε_{jd}'s are independent random variables with mean 0 and variance σ^2, the δ_{0d}'s are independent random variables with mean 0 and variance σ_δ^2, and ε_{jd} and δ_{0d} are independent of each other. The student-level covariate u_{jd} (we just used one covariate for simplicity, but several covariates could of course be included) could come from administrative records—for example, the student's score on an achievement test given to all students in the state or the student's grades in mathematics classes. A domain-level covariate z_d could be, for example, an assessment of the socioeconomic status of the domain or a variable related to methods of mathematics teaching in the domain. The mixed model approach allows the estimate for domain d to borrow strength from other domains through the model for β_{0d}; a common regression equation is assumed for predicting the mean achievement in domain d, and all domains in the area of interest contribute to estimating the parameters in that regression equation. Similarly, in this example, all students sampled in the area of interest contribute to the estimation of β_1.

Indirect estimation—whether synthetic, composite, or model-based—is essentially an exercise in predicting missing data. Indirect estimators are thus highly dependent on the model used to predict the missing data—the synthetic estimator, for example, assumes that the ratios are homogeneous across domains. When possible, the model assumptions should be checked empirically; one method for exploring validity of the model assumptions is to pretend that some of the data you have are actually not available and to compare the indirect estimator with the direct estimator computed with all the data.

12.4
Sampling for Rare Events

Sometimes you would like to investigate characteristics of a population that is difficult to find or that is dispersed widely in the target population. For example, relatively few people are victims of violent crime in a given year, but you may want to obtain information about the population of violent-crime victims. In an epidemiology survey, you may want to estimate the incidence of a rare disease and to make sure you have enough cases of the disease in your sample to analyze how the persons with the disease differ from persons without the disease.

One possibility, of course, is to take a very large sample. That is done in the NCVS, which is used to estimate victimization rates. As it was intended to estimate victimization rates for many different types of victimizations and to investigate households' victimization experiences over time, the NCVS was designed to be approximately self-weighting. If you are interested in domestic-violence victims, however, the sample size is very small. The NCVS would need to be prohibitively expensive to remain a self-weighting survey and still give sufficient sample sizes for all different types of crime victims.

A number of methods have been proposed to allow estimation of the prevalence of the rare characteristic and to estimate quantities of interest for the rare populations. Many of these ideas are discussed in Kalton and Anderson (1986), and several are based on concepts we have already discussed in this book. We briefly describe some of these methods, so you have a general idea of what is available and where to look to learn more.

12.4.1   Stratified Sampling with Disproportional Allocation

Sometimes strata can be constructed so that the rare characteristic is much more prevalent in one of the strata (say, in stratum 1). Then, a stratified sample in which the sampling fraction is higher in stratum 1 can give a more accurate estimate of the prevalence of the rare characteristic in the general population. The higher sampling fraction in stratum 1 also increases the domain sample size for population members with the rare characteristic. The National Maternal and Infant Health Survey (MIHS), discussed in Example 11.1, sampled a higher fraction of records from low-birth-weight infants to ensure an adequate sample size of such infants.

Disproportional stratified sampling may work well when the allocation is efficient for all items of interest. For example, in the MIHS, a major concern was low-birth-weight infants, who have many more health problems. But disproportional stratification may not be helpful for all items of interest in other surveys. A design in which New York City and San Francisco are oversampled is sensible for estimating prevalence of AIDS and obtaining information about persons with AIDS, as New York City and San Francisco are thought to have the highest AIDS prevalence in the

United States; the design would not be as efficient for estimating the prevalence of Alzheimer's disease, which is rare but not concentrated in New York City and San Francisco.

12.4.2 Two-Phase Sampling

Screen the phase I sample units to determine whether they have the rare characteristic or not. Then subsample all (or a high sampling fraction) of the units with the rare characteristic for the phase II sample. If the screening technique is completely accurate, use the phase I sample to estimate prevalence of the rare characteristic and the phase II sample to estimate other quantities for the rare population.

What if the screening technique is not completely accurate? If sampling Arctic regions for presence of walruses, it is possible that you will not see walruses in some of the sectors from the air because the walruses are under the ice. Asking persons whether they have diabetes will not always produce an accurate response, because persons do not always know whether they have the disease. As Deming (1977) points out, placing a person with diabetes in the "no-diabetes" stratum is more serious than placing a person without diabetes in the "diabetes" stratum: If only the "diabetes" stratum is subsampled, it is likely that the persons without diabetes who have been erroneously placed in that stratum will be discovered, while the error for the diabetic misclassified into the "no-diabetes" stratum will not be found. One possible solution is to broaden the screening criterion so that it encompasses all units that might have the rare characteristic. Another is to subsample both strata in phase II but to use a much higher sampling fraction in the "likely to have diabetes" stratum.

12.4.3 Multiple Frame Surveys

Even though you may not have a list of all members of the rare population, you may have some incomplete sampling frames that contain a high percentage of units with the rare characteristic. You can sometimes combine these incomplete frames, omitting duplicates, to construct a complete sampling frame for the population. Alternatively, you can select samples independently from the frames, then combine sample estimates from the incomplete frames (and, possibly, a complete frame) to obtain general population estimates. This idea was first explored by Hartley (1962).

For example, suppose you want to estimate the prevalence of Alzheimer's disease in the noninstitutionalized population. Because many users of adult day-care centers have Alzheimer's disease, you would expect that a sample of adult day-care centers would yield a higher percentage of persons with Alzheimer's disease than a general population survey. But not all persons with Alzheimer's attend an adult day-care center. Thus, you might have two sampling frames: frame A, which is the sampling frame for the general population survey, and frame B, which is the sampling frame of adult day-care centers. As all persons covered in frame B are presumed to also be in the frame for the general population survey, there are two domains: ab, which consists of persons in frame A and in frame B, and a, which consists of persons in frame A but not in frame B.

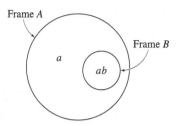

When taking the survey, determine whether each person sampled from frame A is also in frame B. Then estimate the population total by $\hat{t}_a + \hat{t}_{ab}$, where $\hat{t}_a$ is an estimate of the total in domain a and $\hat{t}_{ab}$ is an estimate of the total in domain ab. A variety of estimates can be used to estimate the two domain totals; Skinner and Rao (1996) describe some of these.

Iachan and Dennis (1993) describe the use of multiple frames to sample the homeless population in Washington, D.C. Four frames were used: (1) homeless shelters, (2) soup kitchens, (3) encampments such as vacant buildings and locations under bridges, and (4) streets, sampled by census blocks. Theoretically, the four frames together should capture much of the homeless population; homeless persons are mobile, however, and some may be actively hiding.

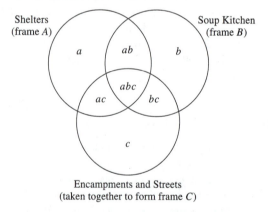

Membership in more than one frame was estimated by asking survey respondents whether they had been or expected to be in soup kitchens, shelters, or on the street in the 24-hour period of sampling.

12.4.4 Network Sampling

In a household survey such as the NCVS, each household provides information only on victimizations that have occurred to members of that household. In a network sample to study crime victimization (Czaja and Blair 1990; see Sudman et al. 1988 for the general method), each household in the population is linked to other units in the population; the sampled household can also provide information on units linked to it (called the *network* for that household). For example, the network of a household might be defined to be the siblings of adult household members.

Suppose an equal-probability sample of households is taken. Each adult member of a household selected to be in the sample is asked to provide information about crime incidents that occurred to him or her and to his or her siblings. Information about Rob Victim could be obtained, then, because his household is selected for the sample or because one of his sibling's households is selected. The probability that Rob is included in the sample depends on the number of separate households in which he has siblings; if he has many siblings in different households, the weight assigned to him will be smaller than the weight of a person with no siblings.

12.4.5   Snowball Sampling

Snowball sampling is based on the premise that members of the rare population know one another. To take a snowball sample of homeless persons, you would find a few homeless persons. Ask each of those persons to identify other homeless persons for your sample, then ask the new persons in your sample to identify additional homeless persons, etc., until a desired sample size is attained. Snowball sampling can create a fairly large sample of a rare population, but it is not a probability sample; strong modeling assumptions (that are usually not met!) need to be made to generalize results from a snowball sample to the population. However, snowball sampling can be useful in early stages of an investigation, to learn something about the rare population.

12.4.6   Sequential Sampling

In sequential sampling, observations or psu's are sampled one or a few at a time, and information from previously drawn psu's can be used to modify the sampling design for subsequently selected psu's. In one method dating back to Stein (1945) and Cox (1952), an initial sample is taken, and results from that sample are used to estimate the additional sample size necessary to achieve a desired precision. If it is desired that the sample contain a certain number of members from the rare population, the initial sample could be used to obtain a preliminary estimate of prevalence, and that estimate of prevalence is used to estimate the necessary size of the second sample. After the second sample is collected, it is combined with the initial sample to obtain estimates for the population. A sequential sampling scheme generally needs to be accounted for in the estimation; in Cox's method, for example, the sample variance obtained after combining the data from the initial and second samples is biased downward (Lohr 1990). The book by Wetherill and Glazebrook (1986) is a good starting point for further reading about sequential methods.

Adaptive cluster sampling assumes that the rare population is aggregated—caribou are in herds, an infectious disease is concentrated in regions of the country, or artifacts are clustered at specific sites of an archaeological dig. An initial probability sample of psu's (often quadrats, in wildlife applications) is selected. For each psu in the initial sample, a response is measured, such as the number of caribou in the psu. If the number of caribou in psu i exceeds a predetermined figure, then neighboring units are added to the sample. Again, the adaptive nature of the sampling scheme needs to be accounted for when estimating population quantities—if you estimate caribou density by (number of caribou observed)/(number of psu's sampled), your estimate of caribou density will be far too high. Thompson and Seber (1996) describe various approaches of adaptive cluster sampling and give a bibliography for the subject.

12.4.7   Nonresponse When Sampling Rare Populations

We never like nonresponse, but it can be an especial hazard for surveys of rare populations. If population members with the rare characteristic are more likely to be nonrespondents than members without the rare characteristic, estimates of prevalence will be biased. In some health surveys, the characteristic itself can lead to nonresponse—a survey of cancer patients may have nonresponse because the illness prevents persons from responding.

12.5
Randomized Response

Sometimes you want to conduct a survey asking very sensitive questions, such as "Do you use cocaine?" or "Have you ever shoplifted?" or "Did you understate your income on your tax return?"

These are all questions that "yes" respondents could be expected to lie about. A question form that encourages truthful answers but makes people comfortable is desired. Horvitz et al. (1967), in a variation of Warner's (1965) original idea, suggest using two questions—the sensitive question and an innocuous question—and using a randomizing device (such as a coin flip) to determine which question the respondent should answer. If a coin flip is used as the randomizing device, the respondent might be instructed to answer the question "Did you use cocaine in the past week?" if the coin is heads, and "Is the second hand on your watch between 0 and 30?" if the coin is tails. The interviewer does not know whether the coin was heads or tails and hence does not know which question is being answered. It is hoped that the randomization and the knowledge that the interviewer does not know which question is being answered will encourage respondents to tell the truth if they have used cocaine in the past week.

The randomizing device can be anything, but it must have known probability P that the person is asked the sensitive question and probability $1 - P$ that the person is asked the innocuous question. Other forms of randomized response are described in Fox and Tracy (1986).

The key to randomized response is that the probability that the person responds yes to the innocuous question, p_I, is known. We want to estimate p_S, the proportion responding yes to the sensitive question.

E X A M P L E   12.8   In one implementation of randomized response (Duffy and Waterton 1988), the respondent was given a deck of 50 cards. Ten cards had the instruction "Say 'Yes,'" 10 had the instruction "Say 'No,'" and the other 30 contained the sensitive question "Have you ever drunk more than the legal limit immediately before driving a car?" The respondent was asked to examine the deck (so he or she would know that there were indeed some cards that did not ask the sensitive question), to shuffle the cards, and then select one. The respondent did not show the card to the interviewer but was asked to answer the sensitive question truthfully if it was on the card, and otherwise to say yes or no as the card directed. In this setting,

$$P = P(\text{asked sensitive question}) = 0.6,$$

and

$$p_I = P(\text{say yes} \mid \text{asked innocuous question}) = 0.5. \quad \blacksquare$$

If everyone answers the questions truthfully, then

$$\phi = P(\text{respondent replies yes})$$

$$= P(\text{yes} \mid \text{asked sensitive question})P(\text{asked sensitive question})$$
$$\quad + P(\text{yes} \mid \text{asked innocuous question})P(\text{asked innocuous question})$$

$$= p_S P + p_I(1 - P).$$

Let $\hat{\phi}$ be the estimated proportion of "yesses" from the sample. Because P is known and p_I is known, p_S can be estimated by

$$\hat{p}_S = \frac{\hat{\phi} - (1 - P)p_I}{P}. \tag{12.10}$$

Then, the estimated variance of $\hat{p}_S$ is

$$\hat{V}(\hat{p}_S) = \frac{\hat{V}(\hat{\phi})}{P^2}.$$

The "penalty" for randomized response appears in the factor $1/P^2$ in the estimated variance. If $P = 1/3$, for example, the variance is nine times as great as it would have been had everyone in the sample been asked the sensitive question and responded truthfully.

You need to think before choosing P: The larger P is, the smaller the variance of $\hat{p}_S$. But if P is too large, respondents may think that the interviewer will know which question is being answered. Some respondents may think that only a $P = 0.5$ is "fair" and that no other probabilities exist when choosing between two items.

EXAMPLE 12.9 An SRS of high school seniors is selected. Each senior in the sample is presented with a card containing the following two questions:

Question 1:   Have you ever cheated on an exam?

Question 2:   Were you born in July?

We know from birth records that $p_I = 0.085$. Suppose the randomizing device is a spinner, with $P = 1/5$. Of the 800 people surveyed, 175 say yes to whichever question the spinner indicated they should answer. Then, $\hat{\phi} = 175/800$. Because this is an SRS,

$$\hat{V}(\hat{\phi}) = \frac{\hat{\phi}(1 - \hat{\phi})}{n - 1} = 0.0002139.$$

Thus,

$$\hat{p}_S = \frac{\dfrac{175}{800} - \left(\dfrac{4}{5}\right).085}{\dfrac{1}{5}} = .75375,$$

and

$$\hat{V}[\hat{p}_S] = \frac{0.0002139}{\left(\dfrac{1}{5}\right)^2} = 0.0053. \quad \blacksquare$$

Before using randomized response methods in your survey, though, test the method with persons in your population to see if the extra complication does indeed increase compliance and appear to reduce bias. Brown and Harding (1973), comparing randomized response with an anonymous questionnaire asking the questions directly, found that estimates of drug use among army officers were higher for the randomized response method than for the questionnaire. It is presumed that a higher estimate in this situation has less bias. Not all field tests, however, show that randomized response is an improvement.

EXAMPLE 12.10   Duffy and Waterton (1988) used a two-stage cluster sample to select respondents in their survey to estimate incidence of various alcohol-related problems in Edinburgh, Scotland. The 20 psu's (polling districts) were selected with probability proportional to the number of registered voters. Then 75 persons were randomly selected from each selected district, and persons in hospitals and other institutions were eliminated from the sample. One-fifth of the respondents were randomly assigned to be asked direct questions; the others participated in randomized response. Because this was a cluster sample, formulas from Chapter 6 should be used to estimate $\hat{\phi}$ and $V(\hat{\phi})$, with $\hat{V}(\hat{p}_S) = \hat{V}(\hat{\phi})/P^2$. For this study, the response rate was 81.1% for the direct question group and 76.5% for the randomized response group. The estimates of p_S, the proportion who had drunk more than the legal limit immediately before driving a car, were 0.469 for the direct question group and 0.382 for the randomized response group (the difference in these proportions was not statistically significant). In this study then, the investigators found that randomized response did not increase the response rate, nor did it increase the estimated incidence of the sensitive characteristic.

Randomized response did, however, increase the complexity of the interviews. Interviewers reported that few persons in the randomized response group examined the cards before choosing one. A number of respondents, particularly older and less well-educated respondents, had difficulty understanding the method. In addition, many respondents answered "Say yes" or "Say no" rather than "Yes" or "No" when they drew one of the innocuous question cards, so the interviewer knew which card had been selected. Duffy and Waterton suggest that the skills of the interviewer may be more important than the survey technique in obtaining truthful answers and high response rates. ▪

12.6
Exercises

*1 (Requires probability.) Suppose the phase I sample is an SRS of size $n^{(1)}$ and the phase II subsample is an SRS of size $n^{(2)}$, with $n^{(2)} < n^{(1)}$. Show that

$$V(\hat{t}_y^{(2)}) = N^2\left(1 - \frac{n^{(2)}}{N}\right)\frac{S_y^2}{n^{(2)}}$$

is the same variance that would result if an SRS of size $n^{(2)}$ were taken directly.

*2 (Requires probability.) For two-phase sampling with ratio estimation (page 383), suppose the phase I sample is an SRS of size $n^{(1)}$ and the phase II sample is an SRS of fixed size $n^{(2)}$.

a Show that $P(Z_i = 1) = n^{(1)}/N$, and $P(D_i = 1 \mid \mathbf{Z}) = Z_i n^{(2)}/n^{(1)}$.

b Show that the variance of the estimator is

$$V(\hat{t}_{yr}^{(2)}) \approx N^2\left(1 - \frac{n^{(1)}}{N}\right)\frac{S_y^2}{n^{(1)}} + N^2\left(1 - \frac{n^{(2)}}{n^{(1)}}\right)\frac{S_d^2}{n^{(2)}}$$

where S_d^2 is the population variance of the d_i's and $d_i = y_i - (t_y/t_x)x_i$.

c Let $e_i = y_i - (\hat{t}_y^{(2)}/\hat{t}_x^{(2)})x_i$ and let s_y^2 and s_e^2 be the sample variances of the y_i's and the e_i's from the phase II sample. Show that

$$\hat{V}(\hat{t}_{yr}^{(2)}) = N^2\left(1 - \frac{n^{(1)}}{N}\right)\frac{s_y^2}{n^{(1)}} + N^2\left(1 - \frac{n^{(2)}}{n^{(1)}}\right)\frac{s_e^2}{n^{(2)}}$$

estimates $V(\hat{t}_{yr}^{(2)})$.

*3 *Estimating the variance in two-phase sampling for stratification.* Show that (12.4) is an unbiased estimator of $V(\hat{t}_{str}^{(2)})$ in Section 12.1.3. HINT: Use the result from Chapter 4 (page 105) that $S^2 = \left[\sum_{h=1}^{H}(N_h - 1)S_h^2 + \sum_{h=1}^{H} N_h(\bar{y}_{hU} - \bar{y}_U)^2\right]/(N - 1)$.

*4 (Requires calculus.) *Optimal allocation for two-phase sampling with stratification.* Efficiency gains for two-phase sampling arise when more observations are subsampled in strata with large variance, large values of N_h, or low cost. Rao (1973) proposes letting $m_h = v_h n_h$ for stratum h, with v_h, $h = 1, \ldots, H$, being constants to be determined before sampling.

a Let c be the cost to sample a unit in the phase I sample and to determine its stratum membership. Let c_h be the cost of measuring y for a unit in stratum h in phase II. Assume the total cost will be a linear function:

$$C = cn + \sum_{h=1}^{H} c_h m_h.$$

The total cost C varies from sample to sample, because the m_h are only determined after the phase I sample is taken. Show that the expected cost is

$$E[C] = cn + n\sum_{h=1}^{H} c_h v_h W_h, \tag{12.11}$$

where $W_h = N_h/N$.

b   With v_h fixed,

$$V\left(\hat{\bar{y}}_{\text{str}}^{(2)}\right) = S^2\left(\frac{1}{n} - \frac{1}{N}\right) + \frac{1}{n}\sum_{h=1}^{H} W_h S_h^2\left(\frac{1}{v_h} - 1\right).$$

Show that $V(\hat{\bar{y}}_{\text{str}}^{(2)})$ is minimized, subject to the constraint in (12.11), when

$$v_h = \sqrt{\frac{cS_h^2}{c_h\left(S^2 - \sum_{j=1}^{H} W_j S_j^2\right)}}.$$

HINT: Use Lagrange multipliers. Alternatively, use (12.11) to express n as a function of expected cost and the other values, substitute this expression for n in $V(\hat{\bar{y}}_{\text{str}}^{(2)})$, and then take partial derivatives.

c   For a given expected cost C^*, determine the value of n.

Other forms of optimal allocation have been proposed; see Treder and Sedransk (1993) for other methods and algorithms.

5   Use the results of Exercise 4 to determine an optimal allocation for a follow-up survey similar to that in Example 12.1. Assume that the relative costs are $c = 1$ and $c_h = 20$, for $h = 1, 2, 3$. Use the data in Example 12.1 to estimate quantities such as W_h and S_h^2. How does your allocation differ from the one used? From proportional allocation?

6   Note that in (12.6), $\hat{N} = n_1/\hat{p}$, where $\hat{p}$ is the sample proportion of individuals in the second sample that are tagged. Use linearization to find an estimate of $V(\hat{N})$.

7   The distribution of $\hat{N}$ in (12.6) is often not approximately normal. The distribution of $\hat{p} = m/n_2$, however, is often close to normality, and confidence intervals for p are easily constructed. For the data in Example 12.5, find a 95% CI for $\hat{p}$. How can you use that interval to obtain a confidence interval for N? How does the resulting confidence interval compare with others we calculated? Is the interval symmetric about $\hat{N}$?

*8   (Requires probability.) In a lake with N fish, n_1 of them tagged, the probability of obtaining m recaptured and $n_2 - m$ previously uncaught fish in an SRS of size n_2 is

$$\mathcal{L}(N \mid n_1, n_2) = \frac{\binom{n_1}{m}\binom{N - n_1}{n_2 - m}}{\binom{N}{n_2}}.$$

The maximum likelihood estimate $\hat{N}$ of N is the value that maximizes $\mathcal{L}(N)$—it is the value that makes the observed value of m appear most probable if we know n_1 and n_2. Find the maximum likelihood estimate of N. HINT: When is $\mathcal{L}(N) \geq \mathcal{L}(N - 1)$?

*9   (Requires mathematical statistics.) *Maximum likelihood estimation of N in large samples.* Suppose that n_1 of the N fish in a lake are marked. An SRS of n_2 fish is then taken, and m of those fish are found to be marked. Assume that N, n_1, and n_2 are all "large." Then, the probability that m of the fish in the sample are marked is approximately

$$\mathcal{L}(N) = \binom{n_2}{m}\left(\frac{n_1}{N}\right)^m\left(1 - \frac{n_1}{N}\right)^{n_2-m}.$$

a Show that $\hat{N} = n_1 n_2/m$ is the maximum likelihood estimate of N.

b Using maximum likelihood theory, show that the asymptotic variance of $\hat{N}$ is approximately $N^2(N - n_1)/(n_1 n_2)$.

*10 (Requires calculus.) Suppose the cost of catching a fish is the same for each fish in the first and second samples and you have enough resources to catch a total of $n_1 + n_2 = C$ fish altogether. If N and C are known and $C < N$, what should n_1 and n_2 be to minimize the variance in Exercise 9(b)?

*11 (Requires probability.)

a For Chapman's estimate $\tilde{N}$ in (12.7), let X be the random variable denoting the number of marked individuals in the second sample. What is the probability distribution of X?

b Show that $E[\tilde{N}] = N$ if $n_2 \geq N - n_1$.

12 Investigators in the Wisconsin Department of Natural Resources (1993) used capture-recapture to estimate the number of fishers in the Monico Study Area in Wisconsin.

a In the first study, 7 fishers were captured between August 11, 1981, and January 31, 1982. Twelve fishers were captured between February 1 and February 19, 1982; of those 12, 4 had also been captured in the first sample. Give an estimate of the total number of fishers in the area, along with a 95% CI for your estimate.

b In the second study, 16 fishers were captured between September 28 and October 31, 1982, and 19 fishers were captured between November 1 and November 17, 1982. Eleven of the 19 fishers in the second sample had also been caught in the first sample. Give an estimate of the total number of fishers in the area, along with a 95% CI for your estimate.

c What assumptions are you making to calculate these estimates? What do these assumptions mean in terms of fisher behavior and "catchability"?

13 Suppose the lake has N fish, and n_1 of them are marked. A sample of size n_2 is then drawn from the lake. Choose three values of N, n_1, and n_2. Approximate the distribution of $\hat{N}$ by drawing 1000 different samples of size n_2 from the population of N units and drawing a histogram of the $\hat{N}$'s that result from the different samples. Repeat this for other values of N, n_1, and n_2. When does the histogram appear approximately normally distributed?

[An alternative version of this problem is to calculate the probability distribution of $\hat{N}$ for different values of N, n_1, and n_2 using the hypergeometric distribution given in Exercise 8. You may want to use Stirling's formula (see Durrett 1994, 156) to approximate the factorials.]

14 Try out the two-sample capture-recapture method to estimate the total number of popcorn kernels or dried beans in a package or to estimate the total number of coins in a jar. Describe fully what you did and give the estimate of the population size along with a 95% CI for N. How did you select the sizes of the two samples?

15 Repeat Exercise 14, using three samples and loglinear models. Would you expect the model of complete independence to fit well? Does it?

16 Domingo-Salvany et al. (1995) also used capture-recapture on the emergency room survey by dividing the list into four samples according to trimester (TR). The following data are from table 1 of their paper:

	TR1 yes TR2 yes	TR1 yes TR2 no	TR1 no TR2 yes	TR1 no TR2 no
TR3 yes, TR4 yes	29	35	35	96
TR3 yes, TR4 no	48	58	80	400
TR3 no, TR4 yes	25	77	50	376
TR3 no, TR4 no	97	357	312	?

Fit loglinear models to these data. Which model do you think is best? Use your model to estimate the number of persons in the missing cell and construct a 95% CI for your estimate.

17 Cochi et al. (1989) recorded data on congenital rubella syndrome from two sources. The National Congenital Rubella Syndrome Registry (NCRSR) obtained data through voluntary reports from state and local health departments. The Birth Defects Monitoring Program (BDMP) obtained data from hospital discharge records from a subset of hospitals. Below are data from 1970 to 1985, from the two systems:

Year	NCRSR	BDMP	Both	Year	NCRSR	BDMP	Both
1970	45	15	2	1978	18	9	2
1971	23	3	0	1979	39	11	2
1972	20	6	2	1980	12	4	1
1973	22	13	3	1981	4	0	0
1974	12	6	1	1982	11	2	0
1975	22	9	1	1983	3	0	0
1976	15	7	2	1984	3	0	0
1977	13	8	3	1985	1	0	0

a The authors state that the NCRSR and the BDMP are independent sources of information. Do you think that is plausible? What about the other assumptions for capture-recapture?

b Use Chapman's estimate (12.7) to find $\tilde{N}$ for each year.

c Now aggregate the data for all the years and estimate the total number of cases of congenital rubella syndrome between 1970 and 1985. How does your estimate from the aggregated data compare with the sum of the estimates from part (b)? Which do you think is more reliable?

d Is there evidence of a decline in congenital rubella syndrome? Provide a statistical analysis to justify your answer.

18 Frank (1978) reports on the following experiment to estimate the number of minnows in a tank. The first two samples used a minnow trap to catch fish, while the third used a net to catch the minnows. Minnows trapped in the first sample were marked by clipping their caudal fin, and minnows trapped in the second sample were marked by clipping the left pectoral fin.

Sample 1?	Sample 2?	Sample 3?	Number of Fish
Y	Y	Y	17
Y	N	Y	28
N	Y	Y	52
N	N	Y	234
Y	Y	N	80
Y	N	N	223
N	Y	N	400

Which loglinear model provides the best fit to these data? Using that model, estimate the total number of fish and provide a 95% CI for your estimate.

19 In the experiment in Exercise 18, what does it mean in terms of fish behavior if there is an interaction between presence in sample 1 and presence in sample 2? Between presence in sample 1 and presence in sample 3?

20 Egeland et al. (1995) used capture-recapture to estimate the total number of fetal alcohol syndrome cases among Alaska Natives born between 1982 and 1989. Two sources of cases were used: thirteen cases identified by private physicians and 45 cases identified by the Indian Health Service (IHS). Eight cases were on both lists.

  a Estimate the total number of fetal alcohol syndrome cases. Give a 95% CI for your estimate, using either the inverted chi-square test or the bootstrap method.

  b The capture-recapture estimate relies on the assumption that the two sources of data are independent—that is, a child on the IHS list has the same probability of appearing on the private physicians' list as a child not on the IHS list. Do you think this assumption will hold here? Why, or why not? What advice would you give the investigators if they were concerned about independence?

  c Suppose that children who are seen by private physicians are less likely to be seen by the IHS. Is $\hat{N}$ then likely to underestimate or to overestimate the number of children with fetal alcohol syndrome? Explain.

21 A university wishes to estimate the proportion of its students who have used cocaine. Students were classified into one of three groups—undergraduate, graduate, or professional school (that is, medical or law school)—and were sampled randomly within the groups. Since there was some concern that students might be unwilling to disclose their use of cocaine to a university official, the following method was used. Thirty red balls, sixteen blue balls, and four white balls were placed in a box and mixed well. The student was then asked to draw one ball from the box. If the ball drawn was red, the person answered question 1. Otherwise question 2 was answered.

Question 1: Have you ever used cocaine?

Question 2: Is the ball you drew white?

The results are as follows:

Group	Undergraduates	Graduates	Professional
Total number of students in group	8972	1548	860
Number of students sampled	900	150	80
Number answering yes	123	27	27

Assuming that all responses were truthful, estimate the proportion of students who have used cocaine and report the standard error of your estimate. Compare this standard error with the standard error you would expect to have if you asked the sample students question 1 directly and if all answered truthfully.

Now suppose that all respondents answer truthfully with the randomized response method, but 25% of those who have used cocaine deny the fact when asked directly. Which method gives an estimate of the overall proportion of students who have used cocaine with the smallest MSE?

22 Kuk (1990) proposes the following randomized response method. Ask the respondent to generate two independent binary variables X_1 and X_2 with $P(X_1 = 1) = \theta_1$ and $P(X_2 = 1) = \theta_2$. The probabilities θ_1 and θ_2 are known. Now ask the respondent to tell you the value of X_1 if she is in the sensitive class and X_2 if she is not in the sensitive class. Suppose the true proportion of persons in the sensitive class is p_S.

a   What is the probability that the respondent reports 1?

b   Using your answer to part (a), give an estimate $\hat{p}_S$ of p_S. What conditions must θ_1 and θ_2 satisfy?

c   What is $V(\hat{p}_S)$ if an SRS is taken?

SURVEY Exercises

23 Draw an SRS of size 500 from Lockhart City. Pretend that you do not see the price a household is willing to pay for cable TV; you only see the assessed value of the house. Use the assessed value to divide the phase I sample into five strata of approximately equal size.

24 Draw a stratified phase II sample, with proportional allocation, of 100 observations. Estimate the average amount that households are willing to spend on cable, along with the standard error. How does the precision of this estimate compare with that of an SRS with the same overall cost?

25 Repeat Exercise 24, after determining the optimal allocation using results in Exercise 4. Use information from the samples drawn in Chapter 4 to estimate the S_h^2 or postulate a model for them.

A

The SURVEY Program

Two thirds of Americans tell researchers they get "most of their information" about the world from television, and the other statistics are so familiar we hardly notice them—more American homes have TVs than plumbing and they're on an average of seven hours a day; children spend more time watching TV than doing anything else save sleeping; on weekday evenings in the winter half the American population is sitting in front of the television; as many as 12 percent of adults (that is, one in eight) feel they are physically addicted to the set, watching an average of fifty-six hours a week; and so on.

—Bill McKibben, *The Age of Missing Information*

The computer program SURVEY,[1] developed by Theodore Chang, simulates the results and costs that might be experienced in actual sample surveys. The exercises using SURVEY are designed to provide a practical illustration of the theoretical aspects of survey design and to allow comparisons between the different designs discussed in the course. FORTRAN code for the program is available on the diskette in the programs survey.for and addgen.for, and updated versions may be obtained from the publications server at www.stat.virginia.edu.

Stephens County is a fictitious county in the midwestern part of the United States with a population of approximately 103,000. It has two main cities: Lockhart City, population 57,500, and Eavesville, population 11,700. Both cities are commercial and transportation centers and boast a variety of light industries. Among the county's industrial products are farm chemicals, pet foods, cable and wire, aircraft radios, greeting cards, corrugated paper boxes, industrial gases, and pipe organs. The county has three smaller municipalities: Villegas, Weldon, and Routledge with populations between 1000 and 2000. These cities are local commercial centers. The surrounding areas are agricultural, although a sizable number of persons commute to the larger cities. The county's main agricultural products are beef cattle, wheat, sorghum, and soybeans.

Stephens County has been organized into 75 districts, with the houses within a district numbered consecutively starting with 1. For the purposes of these exercises,

---

[1] Part of the material in this appendix previously appeared in Chang et al. (1992), which introduced the SURVEY program. The computer programs SURVEY and ADDGEN, and many of the exercises using the SURVEY program, are included, in either the data disk or in the text or in both, with the permission of Dr. Theodore C. Chang of the University of Virginia.

FIGURE   **A.1**

A district map of Stephens County

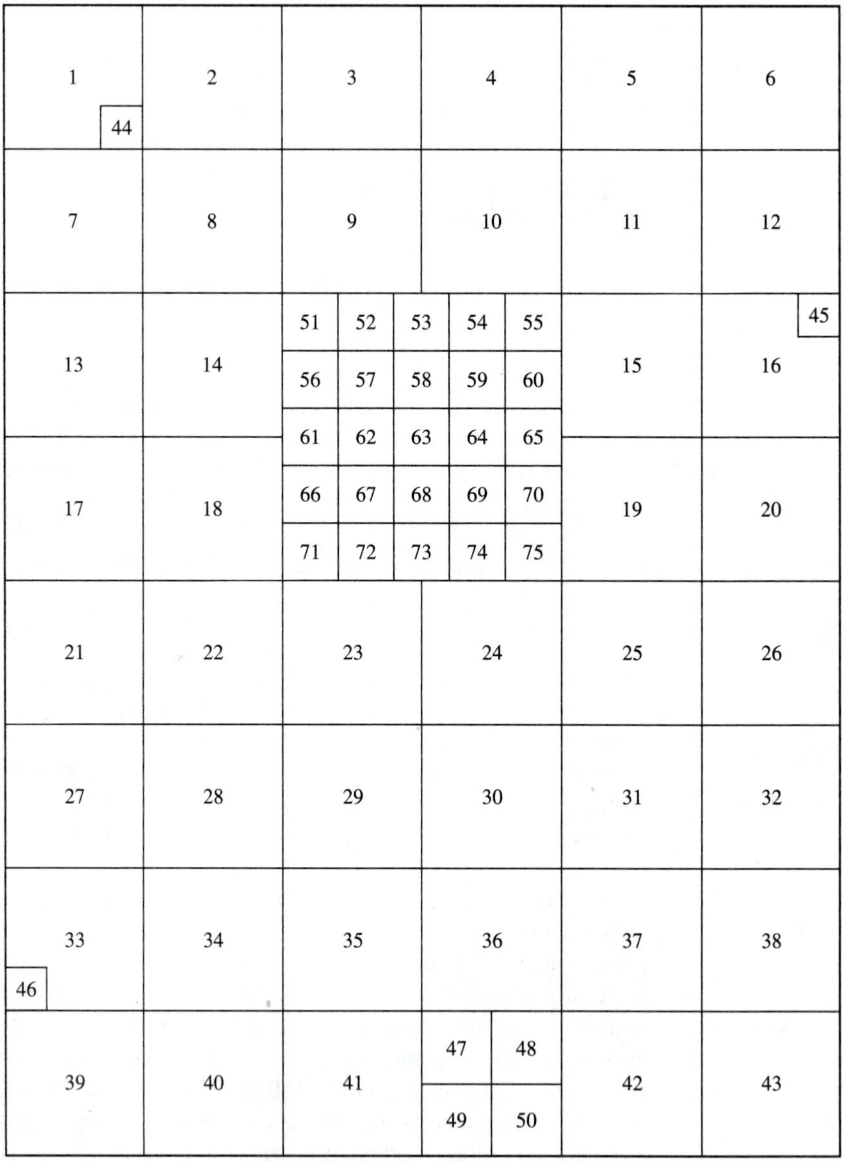

Area	Districts	Number of houses
Rural areas	1–43	7,932
Lockhart City	51–75	19,664
Eavesville	47–50	3,236
Villegas	44	283
Weldon	45	562
Routledge	46	312

you may assume that houses in the same district with close numbers are physically close. The district map of Stephens County is provided in Figure A.1, and information about each district is in Figure A.2.

The Stephens County Cablevision Company (SCCC) has been formed to provide cable TV service to Stephens County. It has commissioned this survey to help it with its pricing and programming decisions.

The Interview Questionnaire   SCCC has supplied an interview questionnaire for your use, shown in Figure A.3.

In addition, for each surveyed household, the SCCC has obtained from the county tax assessor the assessed valuation of that household's living quarters. This information is meant to provide a measure of family income (without having to ask about it). Note that Stephens County is somewhat behind the rest of the United States in terms of cable TV and satellite dishes; this may be because the original SURVEY program was written in 1982.

Survey Program Assumptions   To make as realistic a simulation as possible, certain assumptions have been programmed into SURVEY (Figure A.4). These assumptions should be used in efficient design. Assumptions 1 and 2 are obvious; the others seem reasonable.

Costs of Sampling in Stephens County   Of course, one does not obtain information from survey respondents for free. SURVEY has built-in costs for sampling various units:

```
SAMPLING COSTS IN STEPHENS COUNTY
$60 per rural district visited (1-46)
$20 per urban district visited (47-50)
$6 per rural household visited (whether home or not)
$3 per urban household visited (whether home or not)
$10 processing cost per completed interview
```

As an example of the preceding costs, if the addresses visited and interviewed were 3–47, 3–25, 5–16, 51–25, and 51–36, the sampling cost printed at the end of the output from the program SURVEY would be 2*60 + 1*20 + 3*6 + 2*3 + 5*10 = $214.

Running the SURVEY Program   The FORTRAN source code is provided on the data disk, and is also available from the publications server at www.stat.virginia.edu. You need to compile SURVEY, using a FORTRAN compiler available for your system. After compilation, type survey.exe to run the program on an IBM PC.

SURVEY first asks you to enter the desired nonresponse rates. For now, we're assuming that everyone in Stephens County is always at home and cooperative, so type

```
0 0 0
```

and press the Enter key. Then, when asked, type the address of each household

(text continues on page 418)

Stephens County district information

Column 1: District number
Column 2: Number of houses
Column 3: Cumulative house count

Column 4: Population
Column 5: Mean assessed
house valuation

(1)	(2)	(3)	(4)	(5)	(1)	(2)	(3)	(4)	(5)
1	142	142	526	65248.	39	95	7390	312	57174.
2	153	295	624	58759.	40	130	7520	446	55702.
3	135	430	508	62319.	41	152	7672	533	53285.
4	128	558	560	59416.	42	169	7841	672	56866.
5	110	668	455	57202.	43	91	7932	371	50710.
6	103	771	404	59290.	44	283	8215	1029	60057.
7	105	876	421	71122.	45	562	8777	2079	57233.
8	385	1261	1488	79265.	46	312	9089	1149	52719.
9	296	1557	1112	75921.	47	897	9986	3263	62034.
10	287	1844	994	68254.	48	734	10720	2623	60764.
11	253	2097	929	60660.	49	963	11683	3490	60010.
12	172	2269	628	53569.	50	642	12325	2318	54498.
13	198	2467	768	65182.	51	525	12850	1825	95123.
14	432	2899	1595	77907.	52	726	13576	2497	68406.
15	248	3147	864	65739.	53	674	14250	1948	53634.
16	251	3398	915	53771.	54	585	14835	1219	48643.
17	221	3619	864	68257.	55	553	15388	1090	43493.
18	297	3916	1099	78449.	56	583	15971	1977	95110.
19	235	4151	812	70772.	57	911	16882	2691	84394.
20	171	4322	687	52711.	58	1051	17933	2663	57657.
21	135	4457	525	66739.	59	918	18851	1824	36706.
22	254	4711	923	66249.	60	799	19650	1636	44308.
23	203	4914	708	74757.	61	545	20195	1853	101906.
24	244	5158	825	75766.	62	895	21090	2588	74815.
25	202	5360	799	68989.	63	1313	22403	2642	55560.
26	103	5463	388	56994.	64	968	23371	2457	62813.
27	102	5565	398	58940.	65	717	24088	2203	69846.
28	115	5680	448	60448.	66	651	24739	2197	93771.
29	180	5860	693	69111.	67	886	25625	2711	82902.
30	190	6050	766	69685.	68	912	26537	2750	76832.
31	152	6202	633	70276.	69	898	27435	2671	72062.
32	141	6343	572	63819.	70	759	28194	2650	79887.
33	143	6486	610	58636.	71	722	28916	2568	87383.
34	135	6621	491	55554.	72	753	29669	2652	80341.
35	178	6799	699	62361.	73	793	30462	2763	79833.
36	221	7020	811	60052.	74	725	31187	2560	83354.
37	174	7194	719	55699.	75	802	31989	2870	80522.
38	101	7295	390	53322.					

Districts	Districts	Number of Houses	Population	Mean Assessed House Valuation
Rural	1-43	7932	29985	65511
Villegas, Weldon, Routledge	44-46	1157	4257	56706
Eavesville	47-50	3236	11694	59649
Lockhart City	51-75	19664	57505	71117
Stephens County	1-75	31989	103441	68045

The Interview Questionnaire

I am doing a survey for Stephens County Cablevision. As you may know, Stephens County will soon have cable service; you can help us make sure that the programming we offer meets your needs by answering the following questions.

1   How many persons aged 12 or older live at this address? Please include any persons you consider to be part of your family; do not include persons renting rooms from you.

2   How many persons aged 11 or younger live at this address?

3   How many television sets are in this household?

4   If cable television service cost $5 per month, would your household subscribe? If it cost $10 per month? $15? $20? $25? [The interviewer records the highest price the respondent would be willing to pay for cable.]

5   How many hours did you, personally, spend watching TV last week, in the period from _____ to _____? Your spouse? Each child? Other persons living in the household? [The interviewer sums these amounts and records the sum. If other persons are available, they are asked directly.]

For the following types of programming, the total number of hours spent watching the type of programming is recorded.

1   How many hours did you watch news and "public affairs" programming last week? What about other members of the household?

2   Sports?

3   Children's programming?

4   Movies?

Assumptions for Survey

1   Each occupied address has at least one adult.

2   Only households with televisions will be willing to subscribe to cable service.

3   All other factors being equal, a household with a higher income will tend to have a more expensive house.

4   Assessed valuation is a reasonably accurate estimate of house price.

5   All other factors being equal, a household with a higher income will tend to be willing to pay more for cable service.

6   All other factors being equal, a household with a higher income will tend to own more television sets. This tendency is much weaker than that of assumption 5 because of the low cost and longevity of most TV sets.

7   Larger families tend to be more willing to subscribe to cable TV.

8   All other factors being equal, a family's willingness to subscribe to cable TV decreases as the other entertainment options available to it increase. These options decrease the further one moves from the population concentrations in Stephens County.

9   Due to zoning and development practices, urban neighborhoods tend to be more homogeneous than rural neighborhoods.

to be questioned in the form

```
district number, house number
```

and press Enter. SURVEY responds DONE to each correctly entered address. When you have finished your list of houses, enter 0 for the district number, followed by any house number, and press Enter. The following shows a sample run.

```
DEMONSTRATION EDUCATIONAL SAMPLE SURVEY PROGRAM
COPYRIGHT (C) 1992, TED CHANG AND SHARON LOHR
ENTER FILENAME CONTAINING ADDRESSES—8 OR FEWER LETTERS
IF ENTERING FROM TERMINAL, TYPE T
t
ENTER FILENAME FOR OUTPUT—8 OR FEWER LETTERS
myoutput
ENTER DESIRED THREE NONRESPONSE RATES:
NOT-AT-HOMES, REFUSALS, RANDOM ANSWERS
0 0 0
ENTER DISTRICT NUMBER, HOUSE NUMBER
23,45
DONE
ENTER DISTRICT NUMBER, HOUSE NUMBER
22,96
DONE
ENTER DISTRICT NUMBER, HOUSE NUMBER
53,47
DONE
ENTER DISTRICT NUMBER, HOUSE NUMBER
583,22
DISTRICT NUMBERS MUST BE BETWEEN -75 AND 75.
RE-ENTER DISTRICT NUMBER, HOUSE NUMBER
SET DISTRICT NUMBER = 0 TO STOP PROGRAM.
ENTER DISTRICT NUMBER, HOUSE NUMBER
55,9999
IN DISTRICT 55 HOUSE NUMBERS MUST BE BETWEEN 1 AND 553
RE-ENTER DISTRICT NUMBER, HOUSE NUMBER
SET DISTRICT NUMBER = 0 TO STOP PROGRAM.
ENTER DISTRICT NUMBER, HOUSE NUMBER
0,0

THE COST OF THIS SESSION IS 185 DOLLARS.
```

The file myoutput is displayed below.

ADDRESS		VALUE	1	2	3	4	5	6	7	8	9
23	45	59722	1	0	0	0	0	0	0	0	0
22	96	101571	6	0	1	25	125	12	54	0	35
53	47	50366	1	0	0	0	0	0	0	0	0

```
THE COST OF THIS SESSION IS 185 DOLLARS.
```

In the file myoutput, VALUE = house value, and the numbers in columns labeled 1 through 9 are the household's responses to questions 1 through 9. SURVEY places the answers that each house gives in the file you have specified. You may then edit or print the file using a word processor or spreadsheet.

To analyze the data, use a computer package or program that has subset selection capabilities and allows you to write your own programs. Most programs that only have menus but no programming language are not flexible enough to be useful in survey sampling.

Analyzing SURVEY Data Using a Spreadsheet   Spreadsheets are ideal for learning the basic concepts of sampling and for analyzing the data from small stratified or cluster samples. Guidelines are given throughout the text for using a basic spreadsheet; all widely used spreadsheets have a number of features that may simplify your data analysis, but you don't need the latest version to be able to analyze samples.

To read in the data, use the instructions in your spreadsheet for importing a text (ASCII) file. In some spreadsheets, all 12 columns of data from SURVEY will be in column A of the worksheet. For you to analyze the data, the answers to the 12 questions need to be in separate columns. In your spreadsheet, you need to "parse" the columns. Column G then contains the information about the price of cable TV. The functions you will be using the most will be average, standard deviation, and sum.

■   **Excel**   To find the average and standard deviation of cable price for a sample of size 200, use the functions AVERAGE(G1:G200) and STDEV(G1:G200). When asked to find the proportion of households willing to spend at least $10, it may be easiest to create a new column using the command =IF(G1>9,1,0) in the first row and then to copy the command to the other 199 rows of the data set. You can then find the mean of that column of 0s and 1s.

■   **Quattro Pro**   The functions @SUM(G1..G200), @AVG(G1..G200), and @STDS(G1..G200) find the sum, mean, and standard deviation for the 200 observations of cable price in column G. The subset selection command is @IF(G1>9,1,0).

Analyzing SURVEY Data in a Statistical Package   To use any statistical package, you must first use a text editor to remove the last line of the output from SURVEY. For some packages, you must remove the first line, with the variable names, as well.

■   **S-PLUS**   In S-PLUS for Windows, you can import the output file into a data frame using menu commands. To access any variable—say, cable price—use `survout$cable`; the average cable price is `mean(survout$cable)`. Subset selection is simple in S-PLUS: To find the proportion of households willing to spend at least $10, type and enter

```
survout$cable[survout$cable >= 10]
```

■   **SAS**   If the output is in the file myoutput in the directory from which you started SAS, you can read in the data and calculate summary statistics for the variables by typing (and entering)

```
data tv;
  infile 'myoutput';
```

```
      input dist house value over12 under12 numtv cable hourstv
                            news sports child movies;
proc means data=tv;
```

One way to select a subset of the data is to define a new variable; for example, to find the proportion of households willing to spend at least $10, revise the above code as:

```
data tv;
   infile 'myoutput';
   input dist house value over12 under12 numtv cable hourstv
                         news sports child movies;
     if cable >= 10 then highcab = 1;
     else highcab = 0;
proc means data=tv;
```

Computer Generation of Random Addresses   For any sampling scheme to work effectively, the units must be selected randomly. This is a laborious process, and many sample surveys are ruined by attempts to shortcut it.

C. G. McLaren wrote the program ADDGEN to randomly select addresses from any specified set of districts. ADDGEN asks the user for a random start. This is any integer between 1 and 1,000,000 that the program uses as a start point in a long table of random numbers. Given the same start, districts, sample size, and type of computer, ADDGEN always produces the same sample of addresses. It is extremely important that you record the start in order to repeat a particular sample for further analysis in future assignments. The random start is written on the last line of the output file from ADDGEN.

The program then asks for the districts that you wish to sample. Any subset of the districts 1–75 can be specified. Simply enter the desired district numbers along a line separated by commas. If you want consecutive districts, type only the first and last district numbers separated by a – (dash symbol). If you need to continue your list onto a new line, simply end the previous line with a $ (dollar symbol), press Enter, and continue on the next line. Finally, the program asks for the number of addresses to be selected from the specified districts.

The program ADDGEN generates an output file named by you in a format suitable for input into the SURVEY program. When running SURVEY, merely type in the name of the file you created using ADDGEN.

Sample Run of ADDGEN   The following is a journal of a sample run, which was made using the above procedure. ADDGEN was used to create a random sample of size 5 from districts 1–49,60,70. The output file address from ADDGEN can be fed to SURVEY.

```
ENTER FILENAME FOR ADDRESS SET—8 OR FEWER LETTERS
address
ENTER RANDOM START—ANY INTEGER BETWEEN 1 AND 1000000
219654
ENTER DISTRICTS FROM WHICH YOU WISH TO SAMPLE
1-49,60,70
```

```
     51 DISTRICTS WITH 13241 HOUSEHOLDS HAVE BEEN SPECIFIED
ENTER NUMBER OF ADDRESSES TO BE GENERATED (MAX 1000)
5
DO YOU WANT TO SPECIFY A NEW DISTRICT SET
ANSWER YES OR NO
no
     5 RANDOM ADDRESSES GENERATED WITH RANDOM START 219654
```

Below are the contents of the output file address:

```
 4   67
 8  246
18   94
18  191
24  244
 0    0              219654
```

B

Probability Concepts Used in Sampling

I recollect nothing that passed that day, except Johnson's quickness, who, when Dr. Beattie observed, as something remarkable which had happened to him, that he had chanced to see both No. 1, and No. 1000, of the hackney-coaches, the first and the last; "Why, Sir, (said Johnson,) there is an equal chance for one's seeing those two numbers as any other two." He was clearly right; yet the seeing of the two extremes, each of which is in some degree more conspicuous than the rest, could not but strike one in a stronger manner than the sight of any other two numbers."

—James Boswell, *The Life of Samuel Johnson*

The essence of probability sampling is that we can calculate the probability with which any subset of observations in the population will be selected as the sample. Most of the randomization theory results used in this book depend on probability concepts for their proof. In this appendix we present a brief review of some of the basic ideas used. The reader should consult a more comprehensive reference on probability, such as Ross (1998) or Durrett (1994), for more detail and for derivations and proofs.

Because all work in randomization theory concerns discrete random variables, only results for discrete random variables are given in this section. We use the results in Sections B.1–B.3 in Chapters 2–4, and the results in Section B.4 in Chapters 5 and 6.

B.1
Probability

Consider performing an experiment in which you can write out all outcomes that could possibly happen, but you do not know exactly which one of those outcomes will occur. You might flip a coin, draw a card from a deck, or pick three names out of a hat containing 20 names. Probabilities are assigned to the different outcomes and to sets composed of outcomes (called **events**), in accordance with the likelihood that the events will occur. Let Ω be the **sample space,** the list of all possible outcomes. For flipping a coin, $\Omega = \{\text{heads, tails}\}$. Probabilities in finite sample spaces have three basic properties:

1 $P(\Omega) = 1$.

2 For any event $A, 0 \le P(A) \le 1$.

3 If the events $A_1, \ldots, A_k$ are disjoint, then $P(\cup_{i=1}^{k} A_i) = \sum_{i=1}^{k} P(A_i)$.

In sampling, we have a population of N units and use a probability sampling scheme to select n of those units. We can think of those N units as balls labeled 1 through N in a box, and we draw n balls from the box. For illustration, suppose $N = 5$ and $n = 2$. Then we draw two labeled balls out of the box:

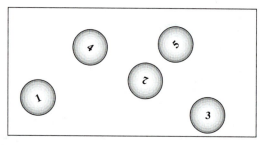

If we take a simple random sample (SRS) of one ball, each ball has an equal probability $1/N$ of being chosen as the sample.

B.1.1   Simple Random Sampling with Replacement

In sampling with replacement, we put a ball back after it is chosen, so the same population is used on successive draws from the population. For the box with $N = 5$, there are 25 possible samples (a, b) in Ω, where a represents the first ball chosen and b represents the second ball chosen:

(1, 1)	(2, 1)	(3, 1)	(4, 1)	(5, 1)
(1, 2)	(2, 2)	(3, 2)	(4, 2)	(5, 2)
(1, 3)	(2, 3)	(3, 3)	(4, 3)	(5, 3)
(1, 4)	(2, 4)	(3, 4)	(4, 4)	(5, 4)
(1, 5)	(2, 5)	(3, 5)	(4, 5)	(5, 5)

Since we are taking a random sample, each of the possible samples has the same probability, $1/25$, of being the one chosen. When we take a sample, though, we usually do not care whether we chose unit 4 first and unit 5 second, or the other way around. Instead, we are interested in the probability that our sample consists of 4 and 5 in either order, which we write as $S = \{4, 5\}$. By the third property in the definition of a probability,

$$P(\{4, 5\}) = P[(4, 5) \cup (5, 4)] = P[(4, 5)] + P[(5, 4)] = \frac{2}{25}.$$

Suppose we want to find $P(\text{unit 2 is in the sample})$. We can either count that nine of the outcomes above contain 2, so the probability is $9/25$, or we can use the **addition formula:**

$$P(A \cup B) = P(A) + P(B) - P(A \cap B). \tag{B.1}$$

Here, let $A = \{$unit 2 is chosen on the first draw$\}$ and let $B = \{$unit 2 is chosen on the second draw$\}$.

$$P(\text{unit 2 is in the sample}) = P(A) + P(B) - P(A \cap B)$$
$$= \frac{1}{5} + \frac{1}{5} - \frac{1}{25} = \frac{9}{25}.$$

Note that, for this example,

$$P(A \cap B) = P(A) \times P(B).$$

That occurs in this situation because events A and B are **independent**—that is, whatever happens on the first draw has no effect on the probabilities of what will happen on the second draw. Independence of the draws occurs in finite population sampling only when we sample with replacement.

B.1.2 Simple Random Sampling Without Replacement

Most of the time, we sample without replacement because it is more efficient—if Heather is already in the sample, why should we use resources by sampling her again? If we plan to take an SRS without replacement of our population with N balls, the ten possible samples (ignoring the ordering) are

$$\{1, 2\} \quad \{1, 3\} \quad \{1, 4\} \quad \{1, 5\} \quad \{2, 3\}$$
$$\{2, 4\} \quad \{2, 5\} \quad \{3, 4\} \quad \{3, 5\} \quad \{4, 5\}$$

Since there are ten possible samples and we are sampling with equal probabilities, the probability that a given sample will be chosen is $1/10$.

In general, there are

$$\binom{N}{n} = \frac{N!}{n!(N-n)!} \tag{B.2}$$

possible samples of size n that can be drawn without replacement and with equal probabilities from a population of size N, where

$$k! = k(k-1)(k-2) \cdots 1 \quad \text{and} \quad 0! = 1.$$

For our example, there are

$$\binom{5}{2} = \frac{5!}{2!(5-2)!} = \frac{5 \cdot 4 \cdot 3 \cdot 2 \cdot 1}{(2 \cdot 1)(3 \cdot 2 \cdot 1)} = 10$$

possible samples of size 2, as we found when we listed them.

Note that in sampling without replacement, successive draws are *not* independent. For this example,

$$P(\text{2 chosen on first draw, 4 chosen on second draw}) = \frac{1}{20}.$$

But $P(\text{2 chosen on first draw}) = 1/5$, and $P(\text{4 chosen on second draw}) = 1/5$, so the product of the probabilities of the two events is not the probability of the intersection.

EXAMPLE **B.1**   Players of the Arizona State Lottery game "Fantasy 5" choose 5 numbers without replacement from the numbers 1 through 35. If the 5 numbers you choose match the 5 official winning numbers, you win $50,000. What is the probability you will win $50,000? You could select a total of

$$\binom{35}{5} = \frac{35!}{5!30!} = 324{,}632$$

possible sets of 5 numbers. But only

$$\binom{5}{5} = 1$$

of those sets will match the official winning numbers, so your probability of winning $50,000 is 1/324,632.

Cash prizes are also given if you match 3 or 4 of the numbers. To match 4, you must select 4 numbers out of the set of 5 winning numbers and the remaining number out of the set of 30 nonwinning numbers, so the probability is

$$P(\text{match exactly 4 numbers}) = \frac{\binom{5}{4}\binom{30}{1}}{\binom{35}{5}} = \frac{150}{324{,}632}. \quad \blacksquare$$

EXERCISE **B1**   What is the probability you match exactly 3 of the numbers? Match at least 1 of the numbers? ∎

EXERCISE **B2**   *Calculating the Sampling Distribution in Example 2.3*

A box has eight balls; three of the balls contain the number 7. You select an SRS without replacement of size 4. What is the probability that your sample contains no 7s? Exactly one 7? Exactly two 7s? ∎

B.2
Random Variables and Expected Value

A **random variable** is a function that assigns a number to each outcome in the sample space. Which number the random variable will actually assume is determined only after we conduct the experiment and depends on a random process: Before we conduct the experiment, we only know probabilities with which the different outcomes can occur. The set of possible values of a random variable, along with the probability with which each value occurs, is called the **probability distribution** of the random variable. Random variables are denoted by capital letters in this book to distinguish them from the fixed values y_i. If X is a random variable, then $P(X = x)$ is the probability that the random variable takes on the value x. The quantity x is sometimes called a **realization** of the random variable X; x is one of the values that could occur if we performed the experiment.

EXAMPLE B.2   In the lottery game "Fantasy 5," let X be the amount of money you will win from your selection of numbers. You win \$50,000 if you match all 5 winning numbers, \$500 if you match 4, \$5 if you match 3, and nothing if you match fewer than 3. Then the probability distribution of X is given in the following table:

x	0	5	500	50,000
$P(X = x)$	$\dfrac{320{,}131}{324{,}632}$	$\dfrac{4350}{324{,}632}$	$\dfrac{150}{324{,}632}$	$\dfrac{1}{324{,}632}$

∎

If you played "Fantasy 5" many, many times, what would you expect your average winnings per game to be? The answer is the **expected value** of X, defined by

$$E(X) = EX = \sum_x x P(X = x). \tag{B.3}$$

For "Fantasy 5,"

$$EX = \left(0 \times \frac{320{,}131}{324{,}632}\right) + \left(5 \times \frac{4350}{324{,}632}\right) + \left(500 \times \frac{150}{324{,}632}\right)$$
$$+ \left(50{,}000 \times \frac{1}{324{,}632}\right) = \frac{146{,}750}{324{,}632} = 0.45.$$

Think of a box containing 324,632 balls, in which 1 ball has the number 50,000 inside it, 150 balls have the number 500, 4350 balls have the number 5, and the remaining balls have the number 0. The expected value is simply the average of the numbers written inside all the balls in the box. One way to think about expected value is to imagine repeating the experiment over and over again and calculating the long-run average of the results. If you play "Fantasy 5" many, many times, you would expect to win about 45¢ per game, even though 45¢ is not one of the possible realizations of X.

Variance, covariance, correlation, and the coefficient of variation are defined directly in terms of the expected value:

$$V(X) = E[(X - EX)^2] = \mathrm{Cov}(X, X). \tag{B.4}$$
$$\mathrm{Cov}(X, Y) = E[(X - EX)(Y - EY)]. \tag{B.5}$$
$$\mathrm{Corr}(X, Y) = \frac{\mathrm{Cov}(X, Y)}{\sqrt{V(X)V(Y)}}. \tag{B.6}$$
$$\mathrm{CV}(X) = \frac{\sqrt{V(X)}}{EX} \qquad \text{for } EX \neq 0. \tag{B.7}$$

Expected value and variance have a number of properties that follow directly from the definitions above.

Properties of Expected Value

1   If g is a function, then $E[g(X)] = \sum_x g(x)P(X = x)$.

2   If a and b are constants, then $E(aX + b) = aE(X) + b$.

3   If X and Y are independent, then $E(XY) = (EX)(EY)$.

4   $\mathrm{Cov}(X, Y) = E(XY) - (EX)(EY)$.

5 $\text{Cov}\left(\sum_{i=1}^{n} a_i X_i + b_i, \sum_{j=1}^{m} c_j Y_j + d_j\right) = \sum_{i=1}^{n} \sum_{j=1}^{m} a_i c_j \, \text{Cov}(X_i, Y_j).$

6 $V(X) = E(X^2) - (EX)^2.$

7 $V(X + Y) = V(X) + V(Y) + 2\,\text{Cov}(X, Y).$

8 $-1 \leq \text{Corr}(X, Y) \leq 1.$

EXERCISE B3 Prove properties 1 through 8 using the definitions in (B.3) through (B.7). ∎

In sampling, we often use estimators that are ratios of two random variables. But $E(Y/X)$ usually does not equal EY/EX. To illustrate this, consider the following probability distribution for X and Y:

x	y	$\dfrac{y}{x}$	$P(X = x, Y = y)$
1	2	2	$\dfrac{1}{4}$
2	8	4	$\dfrac{1}{4}$
3	6	2	$\dfrac{1}{4}$
4	8	2	$\dfrac{1}{4}$

Then, $EY/EX = 6/2.5 = 2.4$, but $E(Y/X) = 2.5$. In this example, the values are close but are not equal.

The random variable we use most frequently in this book is

$$Z_i = \begin{cases} 1 & \text{if unit } i \text{ is in the sample.} \\ 0 & \text{if unit } i \text{ is not in the sample.} \end{cases} \tag{B.8}$$

This indicator variable tells us whether the ith unit is in the sample or not. In an SRS without replacement, n of the random variables $Z_1, Z_2, \ldots, Z_N$ will take on the value 1, and the remaining $N - n$ will be 0. For Z_i to equal 1, one of the units in the sample must be unit i, and the other $n - 1$ units must come from the remaining $N - 1$ units in the population, so

$$P(Z_i = 1) = P(i\text{th unit is in the sample}) \tag{B.9}$$

$$= \frac{\binom{1}{1}\binom{N-1}{n-1}}{\binom{N}{n}}$$

$$= \frac{n}{N}.$$

Thus,

$$E(Z_i) = 0 \times P(Z_i = 0) + 1 \times P(Z_i = 1)$$
$$= P(Z_i = 1) = \frac{n}{N}.$$

Similarly, for $i \neq j$,

$$P(Z_i Z_j = 1) = P(Z_i = 1, Z_j = 1)$$
$$= P(i\text{th unit is in the sample, and } j\text{th unit is in the sample})$$
$$= \frac{\binom{2}{2}\binom{N-2}{n-2}}{\binom{N}{n}}$$
$$= \frac{n(n-1)}{N(N-1)}.$$

Thus, for $i \neq j$,

$$E(Z_i Z_j) = 0 \times P(Z_i Z_j = 0) + 1 \times P(Z_i Z_j = 1)$$
$$= P(Z_i Z_j = 1) = \frac{n(n-1)}{N(N-1)}.$$

EXERCISE B4 Show that

$$V(Z_i) = \text{Cov}(Z_i, Z_i) = \frac{n(N-n)}{N^2}$$

and that, for $i \neq j$,

$$\text{Cov}(Z_i, Z_j) = -\frac{n(N-n)}{N^2(N-1)}. \quad \blacksquare$$

The properties of expectation and covariance may be used to prove many results in finite population sampling. One result, used in Chapters 2 and 3, is given below.

Covariance of $\bar{x}$ and $\bar{y}$ from an SRS Let

$$\bar{x}_U = \frac{1}{N}\sum_{i=1}^{N} x_i, \qquad \bar{y}_U = \frac{1}{N}\sum_{j=1}^{N} y_j,$$

$$\bar{x} = \frac{1}{n}\sum_{i=1}^{N} Z_i x_i, \qquad \bar{y} = \frac{1}{n}\sum_{j=1}^{N} Z_j y_j,$$

$$R = \frac{\sum_{i=1}^{N}(x_i - \bar{x}_U)(y_i - \bar{y}_U)}{(N-1)S_x S_y}.$$

Then,

$$\text{Cov}(\bar{x}, \bar{y}) = \left(1 - \frac{n}{N}\right)\frac{R S_x S_y}{n}. \tag{B.10}$$

We use properties 5 and 6 of expected value, along with some algebra, to show (B.10):

$$\text{Cov}(\bar{x}, \bar{y}) = \frac{1}{n^2}\text{Cov}\left(\sum_{i=1}^{N} Z_i x_i, \sum_{j=1}^{N} Z_j y_j\right)$$

$$= \frac{1}{n^2} \sum_{i=1}^{N} \sum_{j=1}^{N} x_i y_j \, \mathrm{Cov}(Z_i, Z_j)$$

$$= \frac{1}{n^2} \sum_{i=1}^{N} x_i y_i V(Z_i) + \frac{1}{n^2} \sum_{i=1}^{N} \sum_{j \neq i}^{N} x_i y_j \, \mathrm{Cov}(Z_i, Z_j)$$

$$= \frac{1}{n} \frac{N-n}{N^2} \sum_{i=1}^{N} x_i y_i - \frac{1}{n} \frac{N-n}{N^2(N-1)} \sum_{i=1}^{N} \sum_{j \neq i}^{N} x_i y_j$$

$$= \frac{1}{n} \left[\frac{N-n}{N^2} + \frac{N-n}{N^2(N-1)} \right] \sum_{i=1}^{N} x_i y_i - \frac{1}{n} \frac{N-n}{N^2(N-1)} \sum_{i=1}^{N} \sum_{j=1}^{N} x_i y_j$$

$$= \frac{1}{n} \frac{N-n}{N(N-1)} \sum_{i=1}^{N} x_i y_i - \frac{1}{n} \frac{N-n}{N-1} \bar{x}_U \bar{y}_U$$

$$= \frac{1}{n} \frac{N-n}{N(N-1)} \sum_{i=1}^{N} (x_i - \bar{x}_U)(y_i - \bar{y}_U)$$

$$= \frac{1}{n} \left(1 - \frac{n}{N} \right) R S_x S_y.$$

EXERCISE B5   Show that

$$\mathrm{Corr}(\bar{x}, \bar{y}) = R. \quad \blacksquare \tag{B.11}$$

B.3
Conditional Probability

In an SRS without replacement, successive draws from the population are **dependent:** The unit we choose on the first draw changes the probabilities of selecting the other units on subsequent draws. For our box of five balls, each ball has probability $1/5$ of being chosen on the first draw. If we choose ball 2 on the first draw and sample without replacement, then

$$P(\text{ball 3 on second draw} \mid \text{ball 2 on first draw}) = \frac{1}{4}.$$

(Read as "the conditional probability that ball 3 is selected on the second draw *given that* ball 2 is selected on the first draw equals $1/4$.") Conditional probability allows us to adjust the probability of an event if we know that a related event occurred.

The **conditional probability** of A given B is defined to be

$$P(A \mid B) = \frac{P(A \cap B)}{P(B)}. \tag{B.12}$$

In sampling we usually use this definition the other way around:

$$P(A \cap B) = P(A \mid B) P(B). \tag{B.13}$$

If events A and B are independent—that is, knowing whether A occurred gives us absolutely no information about whether B occurred—then $P(A \mid B) = P(A)$ and $P(B \mid A) = P(B)$.

Suppose we have a population with eight households (HHs) and 15 persons living in the households, as follows:

Household	Persons
1	1, 2, 3
2	4
3	5
4	6, 7
5	8
6	9, 10
7	11, 12, 13, 14
8	15

In a one-stage cluster sample, as discussed in Chapter 5, we might take an SRS of two households, then interview each person in the selected households. Then,

$$P(\text{interview person 10}) = P(\text{select HH 6}) \, P(\text{interview person 10} \mid \text{select HH 6})$$
$$= \left(\frac{2}{8}\right)\left(\frac{2}{2}\right) = \frac{2}{8}.$$

In fact, the probability that any individual in the population is interviewed is the same value, 2/8, because the probability a person is selected is the same as the probability that the household is selected.

If we interview only one randomly selected person in each selected household, though, we are more likely to interview persons living alone than those living with others:

$$P(\text{interview person 4}) = P(\text{select HH 2}) \, P(\text{interview person 4} \mid \text{select HH 2})$$
$$= \left(\frac{2}{8}\right)\left(\frac{1}{1}\right) = \frac{2}{8},$$

but

$$P(\text{interview person 12}) = P(\text{select HH 7}) \, P(\text{interview person 12} \mid \text{select HH 7})$$
$$= \left(\frac{2}{8}\right)\left(\frac{1}{4}\right) = \frac{2}{32}.$$

These calculations extend to multistage cluster sampling because of the general result

$$P(A_1 \cap A_2 \cap \cdots \cap A_k)$$
$$= P(A_1 \mid A_2, \ldots, A_k)P(A_2 \mid A_3, \ldots, A_k)\cdots P(A_k). \qquad \textbf{(B.14)}$$

Suppose we take a three-stage cluster sample of grade school students. First, we take an SRS of schools, then sample classes within schools, then sample students within classes. Then, the event {Joe will be selected in the sample} is the same as {Joe's school is selected $\cap$ Joe's class is selected $\cap$ Joe is selected}, and we can find

Joe's probability of inclusion by

$P(\text{Joe in sample}) = P(\text{Joe's school is selected})$

$\times P(\text{Joe's class is selected} \mid \text{Joe's school is selected})$

$\times P(\text{Joe is selected} \mid \text{Joe's school and class are selected}).$

If we sample 10% of the schools, 20% of classes within selected schools, and 50% of students within selected classes, then

$$P(\text{Joe in sample}) = (0.10)(0.20)(0.50) = 0.01.$$

B.4
Conditional Expectation

Conditional expectation is used extensively in the theory of cluster sampling. Let X and Y be random variables. Then, using the definition of conditional probability,

$$P(Y = y \mid X = x) = \frac{P(Y = y \cap X = x)}{P(X = x)}. \qquad \text{(B.15)}$$

This gives the **conditional distribution** of Y given that $X = x$. The **conditional expectation** of Y given that $X = x$ simply follows the definition of expectation using the conditional distribution:

$$E(Y \mid X = x) = \sum_y y P(Y = y \mid X = x). \qquad \text{(B.16)}$$

The **conditional variance** of Y given that $X = x$ is defined similarly:

$$V(Y \mid X = x) = \sum_y [y - E(Y \mid X = x)]^2 P(Y = y \mid X = x). \qquad \text{(B.17)}$$

EXAMPLE B.3 Consider a box with two balls:

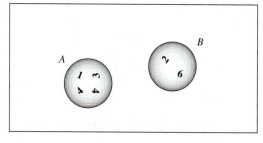

Choose one of the balls at random, then choose one of the numbers inside that ball. Let Y = the number that we choose and let

$$Z = \begin{cases} 1 & \text{if we choose ball A.} \\ 0 & \text{if we choose ball B.} \end{cases}$$

Then,

$$P(Y = 1 \mid Z = 1) = \frac{1}{4},$$

$$P(Y = 3 \mid Z = 1) = \frac{1}{4},$$

$$P(Y = 4 \mid Z = 1) = \frac{1}{2},$$

and

$$E(Y \mid Z = 1) = \left(1 \times \frac{1}{4}\right) + \left(3 \times \frac{1}{4}\right) + \left(4 \times \frac{1}{2}\right) = 3.$$

Similarly,

$$P(Y = 2 \mid Z = 0) = \frac{1}{2},$$

$$P(Y = 6 \mid Z = 0) = \frac{1}{2},$$

so

$$E(Y \mid Z = 0) = \left(2 \times \frac{1}{2}\right) + \left(6 \times \frac{1}{2}\right) = 4.$$

In short, if we know that ball A is picked, then the conditional expectation of Y is the average of the numbers in ball A since an SRS is taken; the conditional expectation of Y given that ball B is picked is the average of the numbers in ball B. ■

Note that $E(Y \mid X = x)$ is a function of x; call it $g(x)$. Define the conditional expectation of Y given X, $E(Y \mid X)$, to be $g(X)$, the same function but of the random variable instead. $E(Y \mid X)$ is a random variable and gives us the conditional expected value of Y for the general random variable X: For each possible value of x, the value $E(Y \mid X = x)$ occurs with probability $P(X = x)$.

EXAMPLE B.4   In Example B.3, we know the probability distribution of Z and can thus use the conditional expectations calculated to write the probability distribution of $E(Y \mid Z)$:

z	$E(Y \mid Z = z)$	Probability
0	4	$\frac{1}{2}$
1	3	$\frac{1}{2}$

■

In sampling, we need this general concept of conditional expectation largely so we can use the following properties of conditional expectation to find expected values and variances in cluster samples.

Properties of Conditional Expectation

1   $E(X \mid X) = X$.
2   $E[f(X)Y \mid X] = f(X)E(Y \mid X)$.
3   If X and Y are independent, then $E(Y \mid X) = E(Y)$.
4   $E(Y) = E[E(Y \mid X)]$.
5   $V(Y) = V[E(Y \mid X)] + E[V(Y \mid X)]$.

Conditional expectation can be confusing, so let's talk about what these properties mean. The interested reader should see Ross (1998) or Durrett (1994) for proofs of these properties.

1 $E(X \mid X) = X$. If we know what X is already, then we expect X to be X. The probability distribution of $E(X \mid X)$ is the same as the probability distribution of X.

2 $E[f(X)Y \mid X] = f(X)E(Y \mid X)$. If we know what X is, then we know X^2, or $\log X$, or any function $f(X)$ of X.

3 If X and Y are independent, then $E(Y \mid X) = E(Y)$. If X and Y are independent, then knowing X gives us no information about Y. Thus, the expected value of Y, the average of all the possible outcomes of Y in the experiment, is the same no matter what X is.

4 $E(Y) = E[E(Y \mid X)]$. This property, called **successive conditioning,** and property 5 are the ones we use the most in sampling to show that certain estimates in cluster sampling are unbiased and to calculate their variances. Successive conditioning simply says that if we average the conditional averages the result is the average of the response of interest. You use successive conditioning every time you take a weighted average of a quantity over subpopulations: If a population has 60 women and 40 men, and if the average height of the women is 64 inches and the average height of the men is 69 inches, then the average height for the class is

$$64 \times 0.6 + 69 \times 0.4 = 66 \text{ inches.}$$

In this example, 64 is the conditional expected value of height given that the person is a woman, and 66 is the expected value of height for all persons in the population.

5 $V(Y) = V[E(Y \mid X)] + E[V(Y \mid X)]$. This property gives an easy way of calculating variances in two-stage cluster samples. It says that the total variability has two parts: the variability that arises because (a) $E(Y \mid X = x)$ varies with different values of x and (b) not all y's associated with the same value of x have the same value.

E X A M P L E B.5 Here's how conditional expectation properties work in Example B.3. Successive conditioning implies that

$$EY = E(Y \mid Z = 1)P(Z = 1) + E(Y \mid Z = 0)P(Z = 0)$$

$$= \left(3 \times \frac{1}{2}\right) + \left(4 \times \frac{1}{2}\right) = 3.5.$$

We can also find the distribution of $V(Y \mid Z)$, using property 6 of expected value:

$$V(Y \mid Z = 0) = E(Y^2 \mid Z = 0) - [E(Y \mid Z = 0)]^2$$

$$= \left(4 \times \frac{1}{2}\right) + \left(36 \times \frac{1}{2}\right) - (4)^2 = 4.$$

$$V(Y \mid Z = 1) = E(Y^2 \mid Z = 1) - [E(Y \mid Z = 1)]^2$$

$$= \left(1 \times \frac{1}{4}\right) + \left(9 \times \frac{1}{4}\right) + \left(16 \times \frac{1}{2}\right) - (3)^2 = 1.5.$$

z	$V(Y \mid Z = z)$	Probability
0	4	$\dfrac{1}{2}$
1	1.5	$\dfrac{1}{2}$

Thus,

$$V[E(Y \mid Z)] = \left(16 \times \frac{1}{2}\right) + \left(9 \times \frac{1}{2}\right) - \{E[E(Y \mid Z)]\}^2$$

$$= \left(16 \times \frac{1}{2}\right) + \left(9 \times \frac{1}{2}\right) - (3.5)^2 = 0.25,$$

$$E[V(Y \mid Z)] = \left(4 \times \frac{1}{2}\right) + \left(1.5 \times \frac{1}{2}\right) = 2.75;$$

so

$$V(Y) = 0.25 + 2.75 = 3.00. \quad \blacksquare$$

If we did not have the properties of conditional expectation, we would need to find the unconditional probability distribution of Y to calculate its expectation and variance—a relatively easy task for the small number of options in Example B.3 but cumbersome to do for general multistage cluster sampling.

C

Data Sets

In some cases, the data sets used in this book are a subset of the original data; in others, the information has been modified to protect the confidentiality of the respondents. They are included for instructional purposes only. Anyone wishing to investigate the subject matter further should obtain the original data from the source. Neither the data collectors nor the distributors bear any responsibility for analyses presented in this book.

All data sets use commas as a separator between fields.

agpop.dat Data from the U.S. 1992 Census of Agriculture. In columns 3–14, the value −99 denotes missing data.

Column	Name	Value
1	county	county name
2	state	state abbreviation
3	acres92	number of acres devoted to farms, 1992
4	acres87	number of acres devoted to farms, 1987
5	acres82	number of acres devoted to farms, 1982
6	farms92	number of farms, 1992
7	farms87	number of farms, 1987
8	farms82	number of farms, 1982
9	largef92	number of farms with 1000 acres or more, 1992
10	largef87	number of farms with 1000 acres or more, 1987
11	largef82	number of farms with 1000 acres or more, 1982
12	smallf92	number of farms with 9 acres or fewer, 1992
13	smallf87	number of farms with 9 acres or fewer, 1987
14	smallf82	number of farms with 9 acres or fewer, 1982
15	region	S = south, W = west, NC = north central, NE = northeast

agsrs.dat Data from an SRS of size 300 from the U.S. 1992 Census of Agriculture. The variables are as in columns 1–14 of the file agpop.dat. In columns 3–14, the value −99 denotes missing data.

agstrat.dat   Data from a stratified random sample of size 300 from the U.S. 1992 Census of Agriculture. In columns 3–14, the value −99 denotes missing data.

Column	Name	Value
1	county	county name
2	state	state abbreviation
3	acres92	number of acres devoted to farms, 1992
4	acres87	number of acres devoted to farms, 1987
5	acres82	number of acres devoted to farms, 1982
6	farms92	number of farms, 1992
7	farms87	number of farms, 1987
8	farms82	number of farms, 1982
9	largef92	number of farms with 1000 acres or more, 1992
10	largef87	number of farms with 1000 acres or more, 1987
11	largef82	number of farms with 1000 acres or more, 1982
12	smallf92	number of farms with 9 acres or fewer, 1992
13	smallf87	number of farms with 9 acres or fewer, 1987
14	smallf82	number of farms with 9 acres or fewer, 1982
15	region	S = south, W = west, NC = north central, NE = northeast
16	rn	random numbers used to select sample in each stratum
17	weight	sampling weight for each county in sample

anthrop.dat   Length of left middle finger and height for 3000 criminals (see Macdonell 1901). This data set contains information for the entire population.

Column	Name	Value
1	finger	length of left middle finger (cm)
2	height	height (inches)

anthsrs.dat   Length of left middle finger and height for an SRS of size 200 from the file anthrop.dat. The variables are the same as for anthrop.dat.

anthuneq.dat   Length of left middle finger and height for a with-replacement unequal-probability sample of size 200 from the file anthrop.dat. The probability of selection, ψ_i, was proportional to 24 for $y < 65$, 12 for $y = 65$, 2 for $y = 66$ or 67, and 1 for $y > 67$.

Column	Name	Value
1	finger	length of left middle finger (cm)
2	height	height (inches)
3	prob	probability of selection

audit.dat   Selection of accounts for audit in Example 6.11.

Column	Name	Value
1	account	audit unit
2	bookval	book value of account
3	cumbv	cumulative book value
4	rn1	random number 1 selecting account
5	rn2	random number 2 selecting account
6	rn3	random number 3 selecting account

books.dat   Data from home owner's survey to estimate total number of books, used in Exercise 6 of Chapter 5.

Column	Name	Value
1	shelf	shelf number
2	number	number of the book selected
3	purchase	purchase cost of book
4	replace	replacement cost of book

certify.dat   Data from the 1994 Survey of ASA Membership on Certification. The full data set is on Statlib (Web address: lib.stat.cmu.edu/asacert/certsurvey). For questions 1 through 5, the responses are coded: 0 = no response, 1 = yes, 2 = possibly, 3 = no opinion, 4 = unlikely, and 5 = no. Missing values for other questions are coded as blanks.

Column	Name	Value
1	certify	Should the ASA develop some form of certification?
2	approve	Would you approve of a certification program similar to that described in the July 1993 issue of *Amstat News*?
3	speccert	Should there be specific certification programs for statistics subdisciplines?
4	wouldyou	If the ASA developed a certification program, would you attempt to become certified?
5	recert	If the ASA offered certification, should recertification be required every several years?
6	subdisc	Major subdiscipline: BA = Bayesian, BE = business & economic, BI = biometrics, BP = biopharmaceutical, CM = computing, EN = environment, EP = epidemiology, GV = government, MR = marketing, PE = physical & engineering sciences, QP = quality & productivity, SE = statistical education, SG = statistical graphics, SP = sports, SR = survey research, SS = social statistics, TH = teaching statistics in health sciences, O = other
7	college	Highest collegiate degree: B = BS or BA, M = MS, N = none, P = Ph.D., O = other
8	employ	Employment status: E = employed, I = in school, R = retired, S = self-employed, U = unemployed, O = other
9	workenv	Primary work environment: A = academia, G = government, I = industry, O = other
10	workact	Primary work activity: C = consultant, E = educator, P = practitioner, R = researcher, S = student, O = other
11	yearsmem	For how many years have you been a member of the ASA?

coots.dat   Selected information on egg size, from a larger study by Arnold (1991). (Data provided courtesy of Todd Arnold.) Not all observations are used for this data set, so results may not agree with those in Arnold (1991).

Column	Name	Value
1	clutch	clutch number from which eggs were subsampled
2	csize	number of eggs in clutch (M_i)
3	length	length of egg (mm)
4	breadth	maximum breadth of egg (mm)
5	volume	calculated as $0.000507 \times$ length $\times$ breadth^2
6	tmt	$= 1$ if received supplemental feeding, 0 otherwise

counties.dat   Data from an SRS of 100 of the 3141 counties in the United States (U.S. Bureau of the Census 1994). Missing values are coded as -99.

Column	Name	Value
1	RN	random number used to select the county
2	State	
3	County	
4	landarea	land area, 1990 (square miles)
5	totpop	total persons, 1992
6	physician	active nonfederal physicians on Jan. 1, 1990
7	enroll	school enrollment in elementary or high school, 1990
8	percpub	percent of school enrollment in public schools
9	civlabor	civilian labor force, 1991
10	unemp	number unemployed, 1991
11	farmpop	farm population, 1990
12	numfarm	number of farms, 1987
13	farmacre	acreage in farms, 1987
14	fedgrant	total expenditures in federal funds and grants, 1992 (millions of dollars)
15	fedciv	civilians employed by federal government, 1990
16	milit	military personnel, 1990
17	veterans	number of veterans, 1990
18	percviet	percent of veterans from Vietnam era, 1990

divorce.dat   Data from a sample of divorce records for states in the Divorce Registration Area (National Center for Health Statistics 1987).

Column	Name	Value
1	state	state name
2	abbrev	state abbreviation
3	samprate	sampling rate for state
4	numrecs	number of records sampled in state
5	hsblt20	number of records in sample with husband's age < 20

(continued)

Column	Name	Value
6	hsb20-24	number of records with $20 \leq$ husband's age ≤ 24
7	hsb25-29	number of records with $25 \leq$ husband's age ≤ 29
8	hsb30-34	number of records with $30 \leq$ husband's age ≤ 34
9	hsb35-39	number of records with $35 \leq$ husband's age ≤ 39
10	hsb40-44	number of records with $40 \leq$ husband's age ≤ 44
11	hsb45-49	number of records with $45 \leq$ husband's age ≤ 49
12	hsbge50	number of records with husband's age ≥ 50
13	wflt20	number of records with wife's age < 20
14	wf20-24	number of records with $20 \leq$ wife's age ≤ 24
15	wf25-29	number of records with $25 \leq$ wife's age ≤ 29
16	wf30-34	number of records with $30 \leq$ wife's age ≤ 34
17	wf35-39	number of records with $35 \leq$ wife's age ≤ 39
18	wf40-44	number of records with $40 \leq$ wife's age ≤ 44
19	wf45-49	number of records with $45 \leq$ wife's age ≤ 49
20	wfge50	number of records with wife's age ≥ 50

golfsrs.dat   An SRS of 120 golf courses, taken from the population on the Web site www.golfcourse.com.

Column	Name	Value
1	RN	random number used to select golf course for sample
2	state	state name
3	holes	number of holes
4	type	type of course: priv(ate), semi(-private), pub(lic), mili(tary), res(ort)
5	yearblt	year course was built
6	wkday18	greens fee for 18 holes during week
7	wkday9	greens fee for 9 holes during week
8	wkend18	greens fee for 18 holes on weekend
9	wkend9	greens fee for 9 holes on weekend
10	backtee	back-tee yardage
11	rating	course rating
12	par	par for course
13	cart18	golf cart rental fee for 18 holes
14	cart9	golf cart rental fee for 9 holes
15	caddy	Are caddies available? (y or n)
16	pro	Is a golf pro available? (y or n)

htpop.dat   Height and gender of 2000 persons in an artificial population.

Column	Name	Value
1	height	height of person, cm
2	gender	M = male, F = female

htsrs.dat   Height and gender for an SRS of 200 persons, taken from the file htpop.dat.

Column	Name	Value
1	rn	random number used to select unit
2	height	height of person, cm
3	gender	M = male, F = female

htstrat.dat   Height and gender for a stratified random sample of 160 women and 40 men, taken from the file htpop.dat. The columns and names are as in htsrs.dat.

journal.dat   Types of sampling used for articles in a sample of journals (Jacoby and Handlin 1991). Note that columns 2 and 3 do not always sum to column 1; for some articles, the investigators could not determine which type of sampling was used. When working with these data, you may wish to create a fourth column, "Indeterminate," which equals column 1 − (column 2 + column 3).

Column	Name	Value
1	numemp	number of articles in 1988 that used sampling
2	prob	number of articles that used probability sampling
3	nonprob	number of articles that used nonprobability sampling

measles.dat   Roberts et al. (1995) report on the results of a survey of parents whose children had not been immunized against measles during a recent campaign to immunize all children in the first five years of secondary school. The original data were unavailable; univariate and multivariate summary statistics from these artificial data, however, are consistent with those in the paper. All variables are coded as 1 for "yes," 0 for "no," and 9 for "no answer." A parent who refused consent (variable 4) was asked why, with responses in variables 5–10. If a response in variables 5–10 was checked, it was assigned value 1; otherwise, it was assigned value 0. A parent could give more than one reason for not having the child immunized.

Column	Name	Value
1	school	school attended by child
2	form	parent received consent form
3	returnf	parent returned consent form
4	consent	parent gave consent for measles immunization
5	hadmeas	child had already had measles
6	previmm	child had been immunized against measles
7	sideeff	parent concerned about side effects
8	gp	parent wanted GP (general practitioner) to give vaccine
9	noshot	child did not want injection
10	notser	parent thought measles not a serious illness
11	gpadv	GP advised that vaccine was not needed

ncvs.dat   Selected variables for victimization incidents in the July–December 1989 NCVS. (SOURCE: Incident-level concatenated file, NCS8864I, in NCJ-130915, U.S. Department of Justice 1991.) NOTE: Some variables are recoded from the original data file. Missing data are indicated by −9.

Column	Name	Value
1	wt	incident weight
2	sex	= 1 if victim male, 2 if victim female
3	violent	= 1 if violent crime, 0 if not violent crime
4	injury	= 1 if victim had injuries, 0 if no injuries
5	medcare	= 1 if received medical care for injuries, 0 otherwise
6	reppol	= 1 if incident reported to the police, 0 otherwise
7	numoff	number of offenders involved in crime: 1 = only one, 2 = more than one, 3 = don't know

nybight.dat   Data collected in the New York Bight for June 1974 and June 1975 (Wilk et al. 1977). Two of the original strata were combined because of insufficient sample sizes. For variable *catchwt,* weights less than 0.5 were recorded as 0.5 kg.

Column	Name	Value
1	year	
2	stratum	stratum membership, based on depth
3	catchnum	number of fish caught during trawl
4	catchwt	total weight (kg) of fish caught during trawl
5	numspp	number of species of fish caught during trawl
6	depth	depth of station (m)
7	temp	surface temperature (degrees C)

otters.dat   Data on number of holts (dens) in Shetland, United Kingdom, used in Kruuk et al. (1989). (Data courtesy of Hans Kruuk.)

Column	Name	Value
1	section	coastline section
2	habitat	type of habitat (stratum)
3	holts	number of holts

ozone.dat   Hourly ozone readings in parts per billion (ppb) from Eskdalemuir, Scotland, for 1994 and 1995. Missing values are coded as blanks. (SOURCE: Air Quality Information Centre: www.aeat.co.uk.)

Column	Value
1	date (day/month/year)
2	ozone reading at 1:00 GMT
3	ozone reading at 2:00 GMT
⋮	⋮
25	ozone reading at 24:00 GMT

samples.dat   All possible SRSs that can be generated from the population in Example 2.1.

Column	Value
1	sample number
2–5	sampled units in $\mathcal{S}$
6–9	$\{y_i, i \in \mathcal{S}\}$
10	$\hat{t}_{\mathcal{S}}$

seals.dat   Data on number of breathing holes found in sampled areas of Svalbard fjords, reconstructed from summary statistics given in Lydersen and Ryg (1991). These data are used in Exercise 11 in Chapter 4.

Column	Name	Value
1	zone	zone number for sampled area
2	holes	number of breathing holes Imjak found in area

selectrs.dat   Steps used in selecting the SRS in Example 2.4.

Column	Value
1	random number generated between 0 and 1
2	ceiling(3078*RN)
3	distinct values in column 2
4	new values generated to replace duplicates
5	set of 300 distinct values to be used in sample

statepop.dat   Unequal-probability sample of counties in the United States; counties selected with probability proportional to 1992 population.

Column	Name	Value
1	state	state abbreviation
2	county	county
3	landarea	land area of county, 1990 (square miles)
4	popn	population of county, 1992
5	phys	number of physicians, 1990
6	farmpop	farm population, 1990
7	numfarm	number of farms, 1987
8	farmacre	number of acres devoted to farming, 1987
9	veterans	number of veterans, 1990
10	percviet	percent of veterans from Vietnam era, 1990

statepps.dat   Number of counties, land area, and population for the 50 states plus the District of Columbia.

Column	Name	Value
1	state	state name
2	counties	number of counties in state
3	cumcount	cumulative number of counties
4	landarea	land area of state, 1990 (square miles)
5	cumland	cumulative land area
6	popn	population of state, 1992
7	cumpopn	cumulative population

syc.dat   Selected variables from the Survey of Youth in Custody. (SOURCE: Inter-University Consortium on Political and Social Research, NCJ-130915, U.S. Department of Justice 1989.)

Column	Name	Value
1	stratum	stratum number
2	psu	psu (facility) number
3	psusize	number of eligible residents in psu
4	initwt	initial weight
5	finalwt	final weight
6	randgrp	random group number
7	age	age of resident (99 = missing)
8	race	race of resident: 1 = white; 2 = black; 3 = Asian/Pacific Islander; 4 = American Indian, Aleut, Eskimo; 5 = other; 9 = missing
9	ethnicty	1 = Hispanic, 2 = not Hispanic, 9 = missing
10	educ	highest grade attended before sent to correctional institution: 00 = never attended school, 01–12 = highest grade attended, 13 = GED, 14 = other, 99 = missing
11	sex	1 = male, 2 = female, 9 = missing
12	livewith	Who did you live with most of the time you were growing up? 1 = mother only, 2 = father only, 3 = both mother and father, 4 = grandparents, 5 = other relatives, 6 = friends, 7 = foster home, 8 = agency or institution, 9 = someone else, 99 = missing
13	famtime	Has anyone in your family, such as your mother, father, brother, sister, ever served time in jail or prison? 1 = yes, 2 = no, 7 = don't know, 9 = missing
14	crimtype	most serious crime in current offense: 1 = violent (e.g., murder, rape, robbery, assault) 2 = property (e.g., burglary, larceny, arson, fraud, motor vehicle theft) 3 = drug (drug possession or trafficking) 4 = public order (weapons violation, perjury, failure to appear in court) 5 = juvenile-status offense (truancy, running away, incorrigible behavior) 9 = missing
15	everviol	Ever put on probation or sent to correctional institution for violent offense? 1 = yes, 0 = no
16	numarr	number of times arrested (99 = missing)

(continued)

(continued)

Column	Name	Value
17	probtn	number of times on probation (99 = missing)
18	corrinst	number of times previously committed to correctional institution (99 = missing)
19	evertime	Prior to being sent here, did you ever serve time in a correctional institution? 1 = yes, 2 = no, 9 = missing
20	prviol	= 1 if previously arrested for violent offense
21	prprop	= 1 if previously arrested for property offense
22	prdrug	= 1 if previously arrested for drug offense
23	prpub	= 1 if previously arrested for public-order offense
24	prjuv	= 1 if previously arrested for juvenile-status offense
25	agefirst	age first arrested (99 = missing)
26	usewepn	Did you use a weapon … for this incident? 1 = yes, 2 = no, 9 = missing
27	alcuse	Did you drink alcohol at all during the year before being sent here this time? 1 = yes; 2 = no, didn't drink during year before; 3 = no, don't drink at all; 9 = missing
28	everdrug	Ever used illegal drugs? 0 = no, 1 = yes, 9 = missing

teachers.dat   Selected variables from a study on elementary school teacher workload in Maricopa County, Arizona. (Data courtesy of Rita Gnap 1995.) The psu sizes are given in file teachmi.dat. The large stratum had 245 schools; the small/medium stratum had 66 schools. Missing values are coded as −9. The study is described in Exercise 16 of Chapter 5.

Column	Name	Value
1	dist	school district size: large or medium/small
2	school	school identifier
3	hrwork	number of hours required to work at school per week
4	size	class size
5	preprmin	minutes spent per week in school on preparation
6	assist	minutes per week that a teacher's aide works with the teacher in the classroom

teachmi.dat   Cluster sizes for data in the file teachers.dat.

Column	Name	Value
1	dist	school district size: large or medium/small
2	school	school identifier
3	popteach	number of teachers in that school
4	ssteach	number of surveys returned from that school

teachnr.dat Data from a follow-up study of nonrespondents from Gnap (1995). See teachers.dat for a description.

Column	Name	Value
1	hrwork	number of hours required to work at school per week
2	size	class size
3	preprmin	minutes spent per week in school on preparation
4	assist	minutes per week that a teacher's aide works with the teacher in the classroom

uneqvar.dat Artificial data used in Exercise 14 of Chapter 11.

Column	Name
1	x
2	y

winter.dat Selected variables from the Arizona State University Winter Closure Survey, taken in January 1995 (provided courtesy of the ASU Office of University Evaluation). This survey was taken to investigate the attitudes and opinions of university employees toward the closing of the university between December 25 and January 1. Missing values are coded as 9. For the yes/no questions, the responses are coded as $1 = $ no, $2 = $ yes. The variables *treatsta* and *treatme* were coded as $1 = $ strongly agree, $2 = $ agree, $3 = $ undecided, $4 = $ disagree, $5 = $ strongly disagree. The variables *process* and *satbreak* were coded as $1 = $ very satisfied, $2 = $ satisfied, $3 = $ undecided, $4 = $ dissatisfied, $5 = $ very dissatisfied. Variables *ownsupp* through *offclose* were coded as 1 if the person checked that the statement applied to him/her, and as 2 if the statement was not checked.

Column	Name	Value
1	class	stratum number: $1 = $ faculty, $2 = $ classified staff, $3 = $ administrative staff, $4 = $ academic professional
2	yearasu	number of years worked at ASU: $1 = $ 1–2 years, $2 = $ 3–4 years, $3 = $ 5–9 years, $4 = $ 10–14 years, $5 = $ 15 or more years
3	vacation	In the past, have you *usually* taken vacation days the entire period between December 25 and January 1?
4	work	Did you work on campus during Winter Break Closure?
5	havediff	Did the Winter Break Closure cause you any difficulty/concerns?
6	negaeffe	Did the Winter Break Closure *negatively* affect your work productivity?
7	ownsupp	I was unable to obtain staff support in my department/office.
8	othersup	I was unable to obtain staff support in other departments/offices.
9	utility	I was unable to access computers, copy machine, etc. in my department/office.
10	environ	I was unable to endure environmental conditions—e.g., not properly climatized.

(continued)

(continued)

Column	Name	Value
11	uniserve	I was unable to access university services necessary to my work.
12	workelse	I was unable to work on my assignments because I work in another department/office.
13	offclose	I was unable to work on my assignments because my office was closed.
14	treatsta	Compared to other departments/offices, I feel staff in my department/office were treated fairly.
15	treatme	Compared to other people working in my department/office, I feel I was treated fairly.
16	process	How satisfied are you with the process used to inform staff about Winter Break Closure?
17	satbreak	How satisfied are you with the fact that ASU had a Winter Break Closure this year?
18	breakaga	Would you want to have Winter Break Closure again?

Computer Code Used for Examples

EXAMPLE 2.4 S-PLUS provides the easiest way to select a simple random sample (SRS) from a set of *N* units. For the SRS from the Census of Agriculture in Example 2.4, the command

```
sample(1:3078, 300, replace=F)
```

gives an SRS of size 300 from the integers 1 to 3078.

Minitab also has a command called sample. To select an SRS of size 300 from $N = 3078$ units, type

```
MTB > set c1
DATA > 1:3078
DATA > end
MTB > sample 300 c1 c2
```

Column c2 will then contain an SRS of 300 of the numbers between 1 and 3078. ∎

In SAS, in another statistical package, or in a spreadsheet, selecting an SRS can be more work. If the population is small, one way to select an SRS is to generate a random number between 0 and 1 for each unit in the population, then select the units with the *n* smallest random numbers as the sample.

EXAMPLE 2.5 Analyzing the data from an SRS is easy: Use descriptive statistics from any spreadsheet or statistical package. Here is the SAS code used to obtain summary statistics for the SRS from the Census of Agriculture, also in the file agsrs.sas on the data disk.

```
data agsrs;
    infile 'agsrs.csv' delimiter= ',' ;
    input randnum county $ state $ acres92 acres87 acres82 farms92
                    farms87 farms82 largef92 largef87
                    largef82 smallf92 smallf87 smallf82;
    if acres92 = -99 then acres92 = . ; /* check for missing values*/
proc univariate data = agsrs plot;
    var acres92;  ∎
```

CHAPTER **2**

EXERCISE **14**   In Exercise 14 a statistical technique known as the bootstrap (see Efron and Tibshirani 1993) was used to repeatedly take samples of size 300 with replacement from the data and then to display the estimated sampling distribution of $\bar{y}$ in Figure 2.5. S-PLUS code used to construct this figure is given below:

```
nboot_1000
sampsize_300
boot _ matrix(sample(agsrs[,3],size=sampsize*nboot, replace=T),
          nrow=nboot)
ybarstar <- apply(boot, 1, mean)
hist(ybarstar, nclass=14, xlab="Estimated sampling
          distribution of ybar")
```

Version 4.5 of S-PLUS has a function bootstrap that will now do the calculations in the above code. The Web site www.sas.com/techsup/download/stat/jackboot.sas provides an SAS macro for using the bootstrap with an SRS. ■

EXAMPLE **3.2**   In SAS, add two lines to the bottom of the data set to obtain agsrs1.csv. These lines tell SAS to calculate the predicted value and standard error for the model-based analysis. The lines are

```
.,MEAN,.,.,313343.283,.,.,.,.,.,.,.,.,.,.,.,.,.
.,TOTAL,.,.,964470625,.,.,.,.,.,.,.,.,.,.,.,.,.
```

The SAS code is in the file agratio.sas on the data disk.

```
data agsrs;
   infile 'agsrs1.csv' delimiter= ',';
   input randnum county $ state $ acres92 acres87 acres82 farms92
                        farms87 farms82 largef92 largef87
                        largef82 smallf92 smallf87 smallf82;
   if acres92 = -99 then acres92 = . ; /* check for missing values */
   if acres87 = -99 then acres87 = . ;
   if acres82 = -99 then acres82 = . ;
   if farms92 = -99 then farms92 = . ;
   if farms87 = -99 then farms87 = . ;
   if farms82 = -99 then farms82 = . ;
   if largef92 = -99 then largef92 = . ;
   if largef87 = -99 then largef87 = . ;
   if largef82 = -99 then largef82 = . ;
   if smallf92 = -99 then smallf92 = . ;
   if smallf87 = -99 then smallf87 = . ;
   if smallf82 = -99 then smallf82 = . ;

   if acres87 > 0 then recacr87 = 1.0/acres87;
```

```
/* Obtain summary statistics for x and y */

   proc univariate data = agsrs plot;
       var acres92 acres87;
   proc corr data=agsrs;
       var acres92 acres87;

   /* ALWAYS plot the data ! */

   proc plot data = agsrs;
       plot acres92*acres87;

/* Use weighted least squares to estimate parameters */

   proc reg data=agsrs;
       model acres92=acres87 / noint r p clm;
       weight recacr87;
       output out = resids residual = residual;

/* Examine residuals (used in model-based analysis) */

   data resids;
       set resids;
       if acres87 <= 0 then delete;
       wtresid = residual/sqrt(acres87);
   proc plot data = resids;
       plot wtresid*acres87;  ■
```

EXAMPLE 4.1   Here is the SAS code for obtaining summary statistics for each stratum, in the file agstrat.sas.

```
infile 'agstrat.csv' delimiter= ',';
   input county $ state $ acres92 acres87 acres82 farms92 farms87 farms82
           largef92 largef87 largef82 smallf92 smallf87 smallf82
           region $ rn;
   if acres92 = -99 then acres92 = . ;      /* check for missing values */
   if acres87 = -99 then acres87 = . ;
   if acres82 = -99 then acres82 = . ;
   if farms92 = -99 then farms92 = . ;
   if farms87 = -99 then farms87 = . ;
   if farms82 = -99 then farms82 = . ;
   if largef92 = -99 then largef92 = . ;
   if largef87 = -99 then largef87 = . ;
   if largef82 = -99 then largef82 = . ;
   if smallf92 = -99 then smallf92 = . ;
```

```
if smallf87 = -99 then smallf87 = . ;
if smallf82 = -99 then smallf82 = . ;

/* Obtain summary statistics for each stratum */

proc sort data=agstrat;
   by region;
proc univariate data = agstrat plot;
   var acres92;
   by region;

/* Construct ANOVA table (optional) */

proc glm data=agstrat;
   class region;
   model acres92=region;
   means region;   ■
```

EXAMPLE 5.6   The following is SAS code (in the file coots.sas) for obtaining summary statistics for each cluster and for fitting the model in Example 5.14.

```
data coots;
   infile "coots.csv" delimiter=",";
   input clutch csize length breadth volume trt;
   wt = csize/2;
proc univariate data=coots; /*use weights to estimate total*/
   var volume;
   weight wt;
proc glm data=coots;
   class clutch;
   model volume=clutch;
   means clutch;
proc mixed data=coots method=reml;
   class clutch;
   model volume=;
   random clutch;   ■
```

CHAPTER 6   *An S-PLUS Function for Using Lahiri's Method to Draw Samples*

This is found in the file lahiri.spl on the data disk.

```
lahiri.design <- function(relsize, n, clnames = seq(1:length(relsize)))
# Arguments:
#  relsize   vector of relative sizes of population psu's
#  n         desired sample size
#  clnames   vector of psu names for population
#
#  Return
```

```
#  clusters vector of n psu's selected with replacement and
#           with probability proportional to relsize.
#  ********************************************************
{
maxrel <- max(relsize)
sizeratio <- maxrel/mean(relsize)
numpsu <- length(relsize)
size <- 0
clusters <- NULL
while(size < n) {
   ss <- ceiling((n - size) * sizeratio)
   temp <- sample(seq(1:numpsu), ss, replace = T)
   temp1 <- clnames[temp[relsize[temp] > runif(ss, min = 0, max = maxrel)]]
   clusters <- append(clusters, temp1[!is.na(temp1)])
   size <- length(clusters)
   }
clusters[1:n]
}
```

> S-PLUS code for drawing the graphs in Chapter 6 is in the file chap6.spl. Professor Ted Chang has written S-PLUS programs to calculate the Horvitz–Thompson estimate; these programs are available from the publications site at www.stat.virginia. edu.

CHAPTER 7 To calculate the empirical probability mass function, we used the following function in S-PLUS:

```
emppmf _ function(y, weight)
{ tapply(weight, y, sum, na.rm=T)/sum(weight[!is.na(y)],na.rm=T) }
```

> This function, and S-PLUS functions I used to construct the plots in Chapter 7, are in the file chap7.spl. ∎

CHAPTER 9 *A Jackknife Function for Stratified Multistage Samples (File jack.spl)*

> This program requires that observations be sorted by stratum and psu.

```
jkvar <- function(fcn, ymat, wt, strata, psu) {
# *******************************************
# Arguments:
#  fcn       function calculated from estimated totals
#  ymat      matrix of y's
#  wt        vector of weights
#  strata    vector indicating stratum membership
#  psu       vector indicating psu membership
#
# Return:
#  list with four components
#
#  thetahat estimate of theta from full data
```

```
#  jkvar     jackknife variance estimate
#  jkmean    average of jackknife iterations
#  jktheta   estimate of theta from jackknife iterations
#
# Form of user-supplied function 'fcn':
#
#  Arguments to fcn:
#      totalmat    (R x k) matrix of estimated totals; each row
#                  is from a jackknife iteration
#
#  Return from fcn:
#
#      fvalue      vector of function of the k estimated totals,
# *******************************************
# identify the distinct psu's
  temp <- paste(as.character(strata), as.character(psu), sep = "**")
  consecpsu <- match(temp, unique(temp))
  numpsu <- length(unique(temp))
  psucol <- 1:numpsu
  strcol <- strata[!duplicated(consecpsu)]
  temp <- rle(strcol)$lengths
  nsubh <- rep(temp, temp)
  nhdnhm1 <- nsubh/(nsubh-1) # Construct matrix of replicate weights
  wtmat <- matrix(wt, ncol = numpsu, nrow = length(wt))
  samestr <- strata[row(wtmat)] == strcol[col(wtmat)]

 # Replace weights in stratum h by nh/ (nh-1) wt
  wtmat[samestr] <- wtmat[samestr] * matrix(nhdnhm1, ncol = numpsu,
        nrow = length(wt), byrow = T) [samestr]

 # Replace weight in psu (h,j) by 0
  wtmat[consecpsu[row(wtmat)] == psucol[col(wtmat)]] <- 0
  thaty <- crossprod(wt, ymat)    #calculate est. total for each variable
  jktots <- crossprod(wtmat, yat) #calculate est. totals for jk iterations
  thetahat <- fcn(matrix(thaty, nrow = 1))
  thetajk <- fcn(jktots)
  jkvar <- sum((thetajk - thetahat)^2/nhdnhm1)
  list(thetahat = thetahat, jkvar = jkvar, jkmean = mean(thetajk),
      jktheta = thetajk)
}
```

To use the jackknife to estimate a ratio, I used the function

```
ratiojk <- function(totalmat)   {
  totalmat[, 2]/totalmat[, 1]
} ∎
```

E X A M P L E 9.7 *Applying the Function jkvar to the Coots Data*

```
cootsjk <- jkvar(ratiojk,cbind(rep(1,length(coots$volume)),coots$volume),
    coots$csize/2,rep(1,length(coots$volume)),coots$clutch)  ■
```

E X A M P L E 12.7 *SAS Code (Partial) (File opium.sas)*

```
data opium;
  input er treat death count;
cards;
1 1 1 6
1 1 2 314
1 2 1 27
1 2 2 1728
2 1 1 8
2 1 2 712
2 2 1 69
;
proc catmod data=opium;
  weight count;
  model er*treat*death = _response_ /ml pred=freq freq prob;
  loglin er treat death;
proc catmod data=opium;
  weight count;
  model er*treat*death = _response_ /ml pred=freq freq prob;
  loglin er treat death er*treat;  ■
```

E X A M P L E 12.7 The S-PLUS function recapci.spl, used to find confidence intervals for N, is on the data disk. ■

E

Statistical Table

TABLE E.1

Random Numbers

74970	06996	11136	26428	23607	97462
74077	63454	45058	20708	42772	61311
13557	72942	59693	42635	69187	17870
66824	77092	51315	11910	91362	85877
36135	62333	37762	06766	52006	48746
06176	37697	40726	66014	78540	03503
17371	29089	26149	86755	36502	45455
21223	60124	07325	61085	61663	93814
31842	75317	58670	07821	75722	75152
20516	27594	21126	21262	14847	85513
99277	64548	70107	01059	34794	89863
01991	83000	27894	43577	82087	71504
54377	90482	39785	75722	20978	72511
20121	24555	25752	35312	85403	46189
11571	25668	34005	60874	72564	27470
93725	16472	21779	22432	71132	58118
65299	19900	21083	77915	20234	57314
36671	66533	86361	01327	80226	67405
49870	72912	20126	71728	86130	22113
50647	27134	56117	08650	91732	56189
17834	90311	00470	25024	20604	55526
27421	59467	69163	36665	26139	59445
26586	93561	52994	91112	74191	53986
51769	19891	46105	60143	63230	43817
41635	22882	85301	06875	58116	90778
04382	75863	37867	86246	58449	47432
48736	95362	21908	86094	43262	82826
49226	85080	33783	98388	62526	04014

(continued)

TABLE   E.1

(continued)

20854	80874	15061	24566	72654	83590
50093	79411	58243	12538	16000	81354
32746	91894	87531	03933	08670	35011
45655	67247	49062	80256	21828	70217
96268	69668	23518	85192	81640	19832
43792	70776	17047	10233	44527	40725
66726	38354	88229	52784	48167	43464
00305	60732	03985	83552	83744	33572
47203	23522	41528	72453	88184	97289
94417	00980	76255	09103	55746	57149
28492	27329	28987	08292	22457	27594
15068	78906	13085	52751	42272	10144
86628	62686	03694	38080	35208	10638
70099	52095	34944	74139	92323	24202
59642	03751	88891	73720	90197	48857
21373	68891	89516	31394	29618	13531
62249	55787	68112	51338	09111	84084
15068	28465	20985	64222	79260	22767
35078	08613	30709	07408	99171	30553
19643	91937	12828	53404	07541	10589
75025	72481	37200	27222	92688	11164
71553	58597	83573	12991	32797	24758

References

The goal of this book has been to introduce you to the subject of sampling. Now it's up to you to continue learning about the subject. The references have been organized into two sections:

■ *Books and articles for further exploration of the chapter topics.* These have been chosen for their importance in the historical development of sampling and for their clarity in presenting material in more detail. Often, recent papers on a subject are more difficult to read than the early papers, in which the researchers were struggling with the initial concepts. Much insight can be gained from reading original papers in which a concept was introduced or exposited in a new way; many of the original papers are given as chapter references.

■ *All references cited in the book.* Many of these are more mathematically involved than those listed as further reading for each chapter. The reference section of the book does not list all of the important contributions made to sampling, and I apologize to researchers whose work is not cited. Were I to include every substantial contribution to survey sampling, however, a better title for the book would be *A Bibliography of Survey Sampling,* as all pages would be filled entirely with references. Instead, I have tried to mention one or two references for each topic that will start you on your exploration. Bon voyage!

For Further Exploration
Sampling Theory and Techniques

Books

Cochran, W. G. 1977. *Sampling techniques.* 3d ed. New York: Wiley. Cochran's book has been an indispensable reference for persons doing sampling since the first edition was published in 1953. The book belongs in the library of any serious student of the subject.

Deming, W. E. 1950. *Some theory of sampling.* New York: Dover. One of the classic books on sampling.

Hansen, M. H., W. N. Hurwitz, and W. G. Madow. 1953. *Sample survey methods and theory.* New York: Wiley. Vols. I and II. Two of the earliest books on sampling, full of ideas and insights.

Jessen, R. J. 1978. *Statistical survey techniques.* New York: Wiley.

Kish, L. 1965. *Survey sampling.* New York: Wiley. A useful, classic reference for the survey practitioner, with many examples in social applications.

Lehtonen, R., and E. J. Pahkinen. 1995. *Practical methods for design and analysis of complex surveys.* New York: Wiley. Discusses software and additional applications for methods discussed in Chapters 9–11 of this book.

Madow, W. G., I. Olkin, and D. B. Rubin, eds. 1983. *Incomplete data in sample surveys.* New York: Academic Press. Many of the articles in this three-volume set provide descriptions of various causes of and remedies for nonresponse.

Raj, D. 1968. *Sampling theory.* New York: McGraw-Hill. A concise and insightful treatment of the theory of sampling methods, as known in 1968.

———. 1971. *The design of sample surveys.* New York: McGraw-Hill.

Särndal, C. E., B. Swensson, and J. Wretman. 1992. *Model assisted survey sampling.* New York: Springer-Verlag. This book has something to say on almost any topic you would want to know about, from unequal-probability sampling to optimal design to models for nonresponse. The authors take a "model-assisted" approach; they use models to drive designs and methods of point estimation but use randomization theory results for estimating standard errors. Highly recommended.

Skinner, C. J., D. Holt, and T. M. F. Smith, eds. 1989. *Analysis of complex surveys.* New York: Wiley. This is the most complete book to date on doing secondary analyses on complex survey data.

Stuart, A. 1984. *The ideas of sampling.* New York: Oxford University Press. Stuart's book is a short, nontechnical introduction to the concepts of sampling.

Sudman, S. 1976. *Applied sampling.* San Diego: Academic Press. A wonderful guide to how to use survey samples in practice, with many examples from real surveys.

Sukhatme, P. V., B. V. Sukhatme, S. Sukhatme, and C. Asok. 1984. *Sampling theory of surveys with applications.* 3rd ed. Ames: Iowa State University Press.

Thompson, M. E. 1997. *Theory of sample surveys.* London: Chapman & Hall. Thompson's book gives a concise treatment of classical and recent theoretical work in sample surveys. If you are interested in foundational issues or a more theoretical approach to survey sampling, you should buy this book. Highly recommended.

Thompson, S. K. 1992. *Sampling.* New York: Wiley.

Yates, F. 1981. *Sampling methods for censuses and surveys.* 4th ed. New York: Macmillan. Stuart says of this book: "This is a comprehensive work on the principles and techniques of survey design and analysis which cannot be too highly recommended to the practitioner, for whom it is compulsory bedside reading" (1984, 90).

Journals

Many statistical journals, such as *Journal of the American Statistical Association, Biometrika,* and *Journal of the Royal Statistical Society*, publish technical papers on sampling theory and practice. *The American Statistician, Statistical Science,* and *International Statistical Review* often carry review articles and articles that will help the survey practitioner. Two excellent journals are devoted exclusively to issues in survey research: *Journal of Official Statistics* and *Survey Methodology.* The International Association of Survey Statisticians publishes *The Survey Statistician,* a newsletter devoted to activities and issues of interest to survey statisticians.

Mathematical Statistics and Probability

Bain, L., and M. Engelhardt. 1992. *Introduction to probability and mathematical statistics.* 2d ed. Boston: PWS-Kent.

Durrett, R. 1994. *The essentials of probability.* Belmont, Calif.: Duxbury Press.
Lindgren, B. 1993. *Statistical theory.* 4th ed. New York: Chapman & Hall.
Ross, S. 1998. *A first course in probability.* 5th ed. Upper Saddle River, N.J.: Prentice-Hall.

History of Sampling

Converse, J. M. 1987. *Survey research in the United States: Roots and emergence 1890–1960.* Berkeley: University of California Press. Focuses on the history of social survey research in the United States, with emphasis on the history of polling and the persons involved in developing methods of public opinion research.
Duncan, J. W., and W. C. Shelton. 1978. *Revolution in United States government statistics, 1926–1976.* Washington, D.C.: Department of Commerce.
Fienberg, S. E., and J. M. Tanur. 1996. Reconsidering the fundamental contributions of Fisher and Neyman on experimentation and sampling. *International Statistical Review* 64: 237–253.
Hansen, M. H., and W. G. Madow. 1976. Some important events in the historical development of sample surveys. In *On the history of statistics and probability.* Edited by D. B. Owen, 75–102. New York: Marcel Dekker.
Kruskal, W., and F. Mosteller. 1980. Representative sampling, IV: The history of the concept in statistics, 1895–1939. *International Statistical Review* 48: 169–195.
Rao, J. N. K., and D. R. Bellhouse. 1990. History and development of the theoretical foundations of survey based estimation and analysis. *Survey Methodology* 16: 3–29.
Seng, Y. P. 1951. Historical survey of the development of sampling theories and practice. *Journal of the Royal Statistical Society,* ser. A, 114: 214–231.
Snedecor, G. W. 1939. Design of sampling experiments in the social sciences. *Journal of Farm Economics* 21: 846–855.
Yates, F. 1946. A review of recent statistical developments in sampling and sampling surveys. *Journal of the Royal Statistical Society* 109: 12–30.

Expanding Chapter-Content Knowledge

Chapter 1

The American Statistical Association series on "What Is a Survey?" provides an introduction to survey sampling, with examples of many of the concepts discussed in Chapter 1. In particular, see the pamphlet on "Judging the Quality of a Survey." These pamphlets are available by contacting

> Section on Survey Research Methods
> American Statistical Association
> 1429 Duke Street
> Alexandria, VA 22314–3402 USA

or from the Survey Research Methods Section's home page at www.amstat.org.

Three other good sources for further reading on nonsampling errors are the following:

Biemer, P. P., R. M. Groves, L. E. Lyberg, N. A. Mathiowetz, and S. Sudman. 1991. *Measurement errors in surveys.* New York: Wiley.
Groves, R. M. 1989. *Survey errors and survey costs.* New York: Wiley.
Lessler, J. T., and W. D. Kalsbeek. 1992. *Nonsampling Errors in Surveys.* New York: Wiley.

If you are interested in more information on questionnaire design and or on procedures for taking social surveys, the following references are a good place to start. All are clearly written and list other references. In addition, many issues of the journal *Public Opinion Quarterly* have articles dealing with questionnaire design.

Asher, H. 1992. *Polling and the public: What every citizen should know.* Washington, D.C.: Congressional Quarterly Press.

Belson, W. A. 1981. *The design and understanding of survey questions.* Aldershot, Hants., England: Gower.

Blair, J., and S. Presser. 1993. Survey procedures for conducting cognitive interviews to pretest questionnaires: A review of theory and practice. *Proceedings of the Section on Survey Research Methods, American Statistical Association,* 370–375.

Converse, J. M., and S. Presser. 1986. *Survey questions: Handcrafting the standardized questionnaire.* Beverly Hills, Calif.: Sage University Press.

Dillman, D. 1978. *Mail and telephone surveys.* New York: Wiley.

Fienberg, S. E., and J. M. Tanur. 1992. Cognitive aspects of surveys: Yesterday, today, and tomorrow. *Journal of Official Statistics* 8: 5–17.

Fowler, F. J. 1993. *Survey research methods.* 2d ed. Newbury Park, Calif.: Sage.

Khurshid, A., and H. Sahai. 1995. A bibliography on telephone sampling methodology. *Journal of Official Statistics* 11: 325–367.

Lavrakas, P. J. 1993. *Telephone survey methods: Sampling, selection, and supervision.* 2d ed. Newbury Park, Calif.: Sage.

Lepkowski, J. M. 1988. Telephone sampling methods in the United States. In *Telephone survey methodology.* Edited by R. M. Groves, P. P. Biemer, L. E. Lyberg, J. T. Massey, W. L. Nicholls II, and J. Waksberg, 73–99. New York: Wiley.

Parten, M. 1950. *Surveys, polls, and samples: Practical procedures.* New York: Harper and Brothers. Some of the material in this book is outdated, but it has useful insights.

Potthoff, R. F. 1994. Telephone sampling in epidemiologic research: To reap the benefits, avoid the pitfalls. *American Journal of Epidemiology* 139: 967–978.

Scheaffer, N. C. 1995. A decade of questions. *Journal of Official Statistics* 11: 79–92. Contains a summary and bibliography of recent work in designing survey instruments.

Schuman, H., and S. Presser. 1981. *Questions and answers in attitude surveys: Experiments on question form, wording, and context.* New York: Academic Press.

Schwarz, N. 1995. What respondents learn from questionnaires: The survey interview and the logic of conversation. *International Statistical Review* 63: 153–177.

Schwarz, N., and S. Sudman, eds. 1990. *Context effects in social and psychological research.* New York: Springer-Verlag.

Sudman, S., and N. M. Bradburn. 1982. *Asking questions: A practical guide to questionnaire design.* San Francisco: Jossey-Bass.

Sudman, S., N. M. Bradburn, and N. Schwarz. 1995. *Thinking about answers: The application of cognitive processes to survey methodology.* San Francisco: Jossey-Bass.

Tanur, J., ed. 1993. *Questions about questions: Inquiries into the cognitive bases of surveys.* New York: Sage.

Chapter 2

If you would like more insight into the structure of probability sampling, you should read

Stuart, A. 1984. *The ideas of sampling.* New York: Oxford University Press.

For a more rigorous mathematical treatment, see Raj (1968) and Thompson (1997). One of the most important papers in the development of sampling theory is

Neyman, J. 1934. On the two different aspects of the representative method: The method of stratified sampling and the method of purposive selection. *Journal of the Royal Statistical Society* 97: 558–606.

Neyman's paper pretty much finished off the idea that results from purposive samples could be generalized to the population. He presented an example of the purposive sample taken by Gini and Galvani in the late 1920s. Gini and Galvani chose 29 districts that gave the averages of all 214 districts in the 1921 Italian census, on a dozen variables. But Neyman showed that "all statistics other than the average values of the controls showed a violent contrast between the sample and the whole population."

Another paper of historical interest is

Bowley, A. L. 1906. Address to the economic science and statistic section of the British Association for the Advancement of Science. *Journal of the Royal Statistical Society* 69: 540–558.

Bowley argues that samples could completely replace censuses and that the central limit theorem could be used to evaluate the precision of the sample.

Chapter 3

Raj (1968) and Cochran (1977) have good treatments of ratio and regression estimation in SRSs. For regression models in a general framework, discussed in this book in Chapter 11, see Särndal et al. (1992). Some papers for further reading include the following:

Bellhouse, D. R. 1987. Model-based estimation in finite population sampling. *American Statistician* 41: 260–262.

Brewer, K. R. W. 1963. Ratio estimation and finite populations: Some results deducible from the assumption of an underlying stochastic process. *Australian Journal of Statistics* 5: 93–105. Brewer first proposed a model-dependent approach to survey inference.

Hansen, M. H., W. G. Madow, and B. J. Tepping. 1983. An evaluation of model-dependent and probability-sampling inferences in sample surveys (with discussion). *Journal of the American Statistical Association* 78: 776–807. Discusses the relative merits of design- and model-based approaches to inference. An excellent place to start further exploration.

Laplace, P. S. 1814. *Essai philosophique sur les probabilités*. Paris: MME VE Courcier, Imprimeur-Libraire pour les Mathématiques, quai des Augustins, no. 57. (An English translation was published by Dover in 1951.) This is the first instance of ratio estimation that I am aware of.

Rao, J. N. K. 1997. Developments in sample survey theory: An appraisal. *Canadian Journal of Statistics* 25: 1–21.

Royall, R. M. 1970. On finite population sampling theory under certain linear regression models. *Biometrika* 57: 377–387. Royall further developed the theory of model-based inference in this and a series of subsequent papers, which are listed in the technical references.

———. 1976. Current advances in sampling theory: Implications for human observational studies. *American Journal of Epidemiology* 104: 463–474. Gives a clear exposition of a model-based approach to survey inference.

The papers by Ericson (1969a; 1969b; 1988) present model-based inference in a Bayesian framework.

Chapter 4

Raj (1968, chap. 4) gives a rigorous and concise treatment of stratification. Some papers to start you on your journey include the following:

Bethel, J. 1989. Sample allocation in multivariate surveys. *Survey Methodology* 15: 47–57. Considers optimal allocation for stratification when there is more than one variable of interest.

Cochran, W. G. 1939. The use of analysis of variance in enumeration by sampling. *Journal of the American Statistical Association* 34: 492–510. Cochran describes using ANOVA tables to give relative precision of suitable sampling designs.

Neyman, J. 1934. On the two different aspects of the representative method: The method of stratified sampling and the method of purposive selection. *Journal of the Royal Statistical Society* 97: 558–606. If you did not read this paper for Chapter 2, read it now!

Chapters 5 and 6

Any of the books in the general references offer further discussion on cluster sampling. Stuart (1984) gives a great deal of intuition into cluster sampling.

Brewer, K. R. W., and M. Hanif. 1983. *Sampling with unequal probabilities.* New York: Springer-Verlag. Presents over 50 different methods for drawing with- and without-replacement samples with unequal probabilities.

Overton, W. S., and S. V. Stehman. 1995. The Horvitz–Thompson theorem as a unifying perspective for probability sampling: With examples from natural resource sampling. *American Statistician* 49: 261–268. Gives a clearly written overview of unequal-probability sampling and includes examples.

Some interesting papers for the historical development of cluster sampling are the following:

Godambe, V. P. 1955. A unified theory of sampling from finite populations. *Journal of the Royal Statistical Society,* ser. B, 17: 269–278. Presents a mathematical framework for inference in finite population sampling. In this paper, Godambe also shows that optimal estimators do not exist. Must reading for the student interested in the theoretical foundations of survey sampling.

Hansen, M. H., and W. N. Hurwitz. 1943. On the theory of sampling from a finite population. *Annals of Mathematical Statistics* 14: 333–362. In this paper, Hansen and Hurwitz developed the theory of pps sampling with replacement.

Horvitz, D. G., and D. J. Thompson. 1952. A generalization of sampling without replacement from a finite universe. *Journal of the American Statistical Association* 47: 663–685. Horvitz and Thompson extended the work of Hansen and Hurwitz to unequal-probability sampling without replacement.

Mahalanobis, P. C. 1946. Recent experiments in statistical sampling in the Indian Statistical Institute. *Journal of the Royal Statistical Society* 109: 325–70. One of the classics in the development of survey sampling, this paper gives insight into many different issues. Among other concepts, Mahalanobis developed the technique of interpenetrating subsampling, in which the sample is drawn as two smaller, independent subsamples. In Chapter 5, we mentioned this technique briefly for estimating the variance of systematic samples. Ultimately, Mahalanobis's idea led to the replication methods (discussed in Sections 9.2 and 9.3) now commonly used for variance estimation in complex surveys.

Royall, R. M. 1976. The linear least-squares prediction approach to two-stage sampling. *Journal of the American Statistical Association* 71: 657–664. Applies best linear unbiased estimation to finite population sampling problems with naturally occurring clusters. Essential reading for those wishing to pursue a model-based approach.

Scott, A. J., and T. M. F. Smith. 1969. Estimation in multi-stage surveys. *Journal of the American Statistical Association* 71: 657–664.

Two papers on optimal design for sampling are the following:

Bellhouse, D. R. 1984. A review of optimal designs in survey sampling. *Canadian Journal of Statistics* 12: 53–65.
Rao, J. N. K. 1979. Optimization in the design of sample surveys. In *Optimizing methods in statistics: Proceedings of an international conference.* Edited by J. S. Rustagi, 419–434. New York: Academic Press.

Chapter 7

The book edited by Skinner et al. (1989) is a good place to start your reading about complex surveys. Thompson (1997) presents general theory for estimation in complex surveys. Two papers by Kish (1992; 1995) further explain the ideas behind weighting and design effects: The idea of using design effects for sample-size estimation was introduced by Cornfield (1951); the paper gives an interesting example of sampling in practice.

Chapter 8

The most complete work to date on incomplete data is the following three-volume set:

Madow, W. G., I. Olkin, and D. B. Rubin, eds. 1983. *Incomplete data in sample surveys.* New York: Academic Press.

Groves (1989) is another useful reference for methods of dealing with nonresponse. Another general reference for missing data (not necessarily in surveys) is Little and Rubin (1987). Dalenius (1981) emphasizes the importance of dealing with nonsampling as well as sampling errors.

The following references on experiment design and quality improvement for surveys may be useful when designing a survey:

Alwin, D. F. 1991. Research on survey quality. *Sociological Methods and Research* 20: 3–29.
Biemer, P., and R. Caspar. 1994. Continuous quality improvement for survey operations: Some general principles and applications. *Journal of Official Statistics* 10: 307–326.
Colledge, M., and M. March. 1993. Quality management: Development of a framework for a statistical agency. *Journal of Business and Economic Statistics* 11: 157–165.
Deming, W. E. 1986. *Out of the crisis.* Cambridge: MIT Press.
Fienberg, S. E., and J. M. Tanur. 1988. From the inside out and the outside in: Combining experimental and sampling structures. *Canadian Journal of Statistics* 16: 135–151.
Fisher, R. A. 1925. *Statistical methods for research workers.* London: Oliver and Boyd.
Frankel, L. R. 1983. The report of the CASRO task force on response rates. In *Improving data quality in sample surveys.* Edited by F. Wiseman, 1–11. Cambridge, Mass.: Marketing Science Institute.
González, M. E. 1994. Improving data quality awareness in the United States federal statistical agencies. *American Statistician* 48: 12–17.
Gower, A. R. 1979. Nonresponse in the Canadian Labour Force Survey. *Survey Methodology* 5: 29–58.

Hidiroglou, M. A., J. D. Drew, and G. B. Gray. 1993. A framework for measuring and reducing nonresponse in surveys. *Survey Methodology* 19: 81–94. Gives definitions of various response rates. The place to start if you are concerned about how to define nonresponse for your survey.

Joiner, B. L. 1994. *Fourth generation management: The new business consciousness*. New York: McGraw-Hill.

Kempthorne, O. 1952. *The design and analysis of experiments*. New York: Wiley.

Morganstein, D., and M. Hansen. 1990. Survey operations processes: The key to quality improvement. In *Data quality control: Theory and pragmatics*. Edited by G. E. Lepins and V. R. R. Uppuluri, 91–104. New York: Marcel Dekker.

Platek, R. 1977. Some factors affecting non-response. *Survey Methodology* 3: 191–214.

Ryan, T. P. 1989. *Statistical methods for quality improvement*. New York: Wiley.

Salant, P., and D. A. Dillman. 1994. *How to conduct your own survey*. New York: Wiley.

Spisak, A. W. 1995. Statistical process control of sampling frames. *Survey Methodology* 21: 185–190.

The journals *Survey Methodology, Journal of Official Statistics,* and *Public Opinion Quarterly* publish many articles on experiments that have been done to reduce nonresponse in surveys of persons.

Chapter 9

Binder, D. A. 1983. On the variances of asymptotically normal estimators from complex surveys. *International Statistical Review* 51: 279–292. Presents a general theory for using the linearization method of estimating the variance, even when the quantities of interest are defined implicitly.

McCarthy, P. J. 1969. Pseudo-replication: Half-samples. *Review of the International Statistical Institute* 37: 239–264. Describes the BRR method.

Rao, J. N. K., and C. F. J. Wu. 1985. Inference from stratified samples: Second-order analysis of three methods for nonlinear statistics. *Journal of the American Statistical Association* 80: 620–630. Gives theory (and references to earlier work) showing the asymptotic equivalence of different variance estimators.

Shao, J., and D. Tu. 1995. *The jackknife and bootstrap*. New York: Springer-Verlag. Presents theory for the jackknife and bootstrap methods used in complex surveys.

Wolter, K. M. 1985. *Introduction to variance estimation*. New York: Springer-Verlag. Describes many of the methods used to estimate variances in sample surveys and summarizes research on these methods up to 1985. The place to start your exploration of variance estimation methods.

Chapter 10

The first two books are general references on categorical data analysis:

Agresti, A. 1990. *Categorical data analysis*. New York: Wiley.

Christensen, R. 1990. *Log-linear models*. New York: Springer-Verlag.

Skinner, C. J., D. Holt, and T. M. F. Smith, eds. 1989. *Analysis of complex surveys*. New York: Wiley. Contains chapters on categorical data analysis on complex survey data.

Chapter 11

Regression analysis (not in complex surveys):

Cook, R. D., and S. Weisberg. 1994. *An introduction to regression graphics*. New York: Wiley.

Draper, N. R., and H. Smith. 1998. *Applied regression analysis*. 3d ed. New York: Wiley.

Graybill, F. A. 1976. *Theory and application of the linear model.* North Scituate, Mass.: Duxbury Press.

Graybill, F. A., and H. K. Iyer. 1994. *Regression analysis: Concepts and applications.* Belmont, Calif.: Duxbury Press.

Neter, J., M. H. Kutner, C. J. Nachtsheim, and W. Wasserman. 1996. *Applied linear statistical models.* 4th ed. Chicago: Irwin.

Schafer, J. L. 1997. *Analysis of incomplete multivariate data.* London: Chapman & Hall. Presents methods and models for statistical inference in multivariate data in which some observations have missing data in all or some items. One example given is from the National Health and Nutrition Examination Survey.

Searle, S. 1971. *Linear models.* New York: Wiley.

Weisberg, S. 1985. *Applied linear regression.* 2d ed. New York: Wiley.

Regression in complex surveys:

Brewer, K. R. W., and R. W. Mellor. 1973. The effect of sample structure on analytical surveys. *Australian Journal of Statistics* 15: 145–152. You can't tell from the title, but this paper is an insightful and entertaining debate between "Harry," a design-based survey statistician, and "Fred," who is fresh from graduate school and promotes a model-based approach. This is the paper to start with if you want to learn more about different approaches to inference in sample surveys.

Cassel, C.-M., C.-E. Särndal, and J. H. Wretman. 1977. *Foundations of inference in survey sampling.* New York: Wiley. This book summarizes and presents results on design-based and model-based inference in finite population surveys. It requires familiarity with theory of statistical inference. Put on your theory glasses before you pick it up.

Kish, L., and M. R. Frankel. 1974. Inference from complex samples (with discussion). *Journal of the Royal Statistical Society,* ser. B, 36: 1–37. One of the first papers to show that the sample design affects estimates of regression parameters.

Robinson, J. 1987. Conditioning ratio estimates under simple random sampling. *Journal of the American Statistical Association* 82: 826–831. Robinson studies another approach to inference in survey sampling, conditional design-based inference; references to earlier work are given in the paper.

Smith, T. M. F. 1994. Sample surveys 1975–1990: An age of reconciliation? (with discussion). *International Statistical Review* 62: 5–34. An interesting review of philosophies of inference, by a statistician whose previous work adhered to a model-based approach.

The theory for regression estimation in complex surveys has been developed by many people. Other references for further reading include Konijn (1962), Royall (1970; 1976a, b), Fuller (1975; 1984), Holt et al. (1980), Särndal (1980), Hansen et al. (1983), Kalton (1983), Rubin (1985), Smith (1988), Skinner (1989), Rao and Bellhouse (1990), Kott (1991), Little (1991), Särndal et al. (1992), Pfeffermann (1993), Brewer (1995), and Hidoroglou et al. (1995).

Chapter 12

See Cochran (1977) for more discussion on two-phase sampling with SRSs; Särndal et al. (1992, chap. 9) give a theoretical development for general probability sampling designs. Other references are as follows:

Armstrong, J., C. Block, and K. P. Srinath. 1993. Two-phase sampling of tax records for business surveys. *Journal of Business and Economic Statistics* 11: 407–416. An example of two-phase sampling in the 1990s.

Ghosh, M., and J. N. K. Rao. 1994. Small area estimation: An appraisal (with discussion). *Statistical Science* 9: 55–93.

International Working Group for Disease Monitoring and Forecasting. 1995. Capture-recapture and multiple-record systems estimation. *American Journal of Epidemiology* 142: 1047–1068. Gives a good overview and bibliography for capture-recapture estimation.

Kalton, G., and D. W. Anderson. 1986. Sampling rare populations. *Journal of the Royal Statistical Society,* ser. A, 149: 65–82.

Neyman, J. 1938. Contribution to the theory of sampling human populations. *Journal of the American Statistical Association* 33: 101–116. Develops the theory for two-phase sampling.

Watson, D. J. 1937. The estimation of leaf area in field crops. *Journal of Agricultural Science* 27: 474–483. An early example of two-phase sampling for regression.

All References Cited in Book

Agresti, A. 1990. *Categorical data analysis*. New York: Wiley.

Alexander, C. H. 1991. Discussion of papers on survey weights. In *Proceedings of the Section for Survey Methods, American Statistical Association,* 643–645.

Altham, P. M. E. 1976. Discrete variable analysis for individuals grouped into families. *Biometrika* 63: 263–269.

Alwin, D. F. 1991. Research on survey quality. *Sociological Methods and Research* 20: 3–29.

Arizona Office of Tourism. 1991. *In-state travel patterns of Arizona residents*. Phoenix: Arizona Office of Tourism.

Armstrong, J., C. Block, and K. P. Srinath. 1993. Two-phase sampling of tax records for business surveys. *Journal of Business and Economic Statistics* 11: 407–416.

Arnold, T. W. 1991. Intraclutch variation in egg size of American coots. *The Condor* 93: 19–27.

"The ASCAP Advantage." 1992. *ASCAP in action* (spring): 10–11.

Asher, H. 1992. *Polling and the public: What every citizen should know.* Washington, D.C.: Congressional Quarterly Press.

Aye Maung, N. 1995. Survey design and interpretation of the British Crime Survey. In *Interpreting crime statistics*. Edited by M. Walker, 207–227. Oxford, Eng.: Oxford University Press.

Azuma, D. L., J. A. Baldwin, and B. R. Noon. 1990. Estimating the occupancy of spotted owl habitat areas by sampling and adjusting for bias. U.S. Forest Service General Technical Report PSW-124. Berkeley, Calif.: Pacific Southwest Research Station, Forest Service, Department of Agriculture.

Babbie, E. R. 1973. *Survey research methods.* Belmont, Calif.: Wadsworth.

Bain, L., and M. Engelhardt. 1992. *Introduction to probability and mathematical statistics,* 2d ed. Boston: PWS-Kent.

Basow, S. A., and N. T. Silberg. 1987. Student evaluations of college professors: Are female and male professors rated differently? *Journal of Educational Psychology* 79: 308–314.

Basu, D. 1971. An essay on the logical foundations of survey sampling, part 1. In *Foundations of Statistical Inference.* Edited by V. P. Godambe and D. A. Sprott, 203–242. Toronto: Holt, Rinehart & Winston.

Beck, A. J., S. A. Kline, and L. A. Greenfeld. 1988. *Survey of youth in custody.* Bureau of Justice Statistics Special Report NCJ-113365. Washington, D.C.: Bureau of Justice Statistics.

Bedrick, E. J. 1983. Adjusted chi-squared tests for cross-classified tables of survey data. *Biometrika* 70: 591–596.

Bellhouse, D. R. 1984. A review of optimal designs in survey sampling. *Canadian Journal of Statistics* 12: 53–65.

———. 1987. Model-based estimation in finite population sampling. *American Statistician* 41: 260–262.

Belson, W. A. 1981. *The design and understanding of survey questions.* Aldershot, Hants., England: Gower.

Bethel, J. 1989. Sample allocation in multivariate surveys. *Survey Methodology* 15: 47–57.

Biderman, A. D., and D. Cantor. 1984. A longitudinal analysis of bounding, respondent conditioning and mobility as sources of panel bias in the National Crime Survey. *Proceedings of the Section for Survey Research Methods, American Statistical Association,* 708–713.

Biemer, P., and R. Caspar. 1994. Continuous quality improvement for survey operations: Some general principles and applications. *Journal of Official Statistics* 10: 307–326.

Biemer, P. P., R. M. Groves, L. E. Lyberg, N. A. Mathiowetz, and S. Sudman. 1991. *Measurement errors in surveys.* New York: Wiley.

Binder, D. A. 1983. On the variances of asymptotically normal estimators from complex surveys. *International Statistical Review* 51: 279–292.

———. 1996. Linearization methods for single phase and two phase samples: A cookbook approach. *Survey Methodology* 22: 17–22.

Bisgard, K. M., A. R. Folsom, C. P. Hong, and T. A. Sellers. 1994. Mortality and cancer rates in nonrespondents to a prospective study of older women: 5-year follow-up. *American Journal of Epidemiology* 139: 990–1000.

Blair, J., and S. Presser. 1993. Survey procedures for conducting cognitive interviews to pretest questionnaires: A review of theory and practice. *Proceedings of the Section on Survey Research Methods, American Statistical Association,* 370–375.

Bowden, D. C., A. E. Anderson, and D. E. Medin. 1984. Sampling plans for mule deer sex and age ratios. *Journal of Wildlife Management* 48: 500–509.

Bowley, A. L. 1906. Address to the economic science and statistic section of the British Association for the Advancement of Science. *Journal of the Royal Statistical Society* 69: 540–558.

Box, G. E. P., W. G. Hunter, and J. S. Hunter. 1978. *Statistics for experimenters: An introduction to design, data analysis, and model building.* New York : Wiley.

Brackstone, G. J., and J. N. K. Rao. 1976. Raking ratio estimators. *Survey Methodology* 2: 63–69.

Bradburn, N., and S. Sudman. 1979. *Improving interview method and questionnaire design.* San Francisco: Jossey-Bass.

Breiman, L. 1994. The 1991 census adjustment: Undercount or bad data? *Statistical Science* 9: 458–475.

Brewer, K. R. W. 1963. Ratio estimation and finite populations: Some results deducible from the assumption of an underlying stochastic process. *Australian Journal of Statistics* 5: 93–105.

———. 1975. A simple procedure for sampling πpswor. *Australian Journal of Statistics* 17: 166–172.

———. 1995. Combining design-based and model-based inference. In *Business survey methods.* Edited by B. G. Cox, D. A. Binder, B. N. Chinnappa, A. Christianson, M. J. Colledge, and P. S. Kott, 589–606. New York: Wiley.

Brewer, K. R. W., E. K. Foreman, R. W. Mellor, and D. J. Trewin. 1977. Use of experimental design and population modelling in survey sampling. *Bulletin of the International Statistical Institute* 3: 173–190.

Brewer, K. R. W., and M. Hanif. 1983. *Sampling with unequal probabilities.* New York: Springer-Verlag.

Brewer, K. R. W., and R. W. Mellor. 1973. The effect of sample structure on analytical surveys. *Australian Journal of Statistics* 15: 145–152.

Brewer, K. R. W., and C. Särndal. 1983. Six approaches to enumerative survey sampling (with discussion). In *Incomplete data in sample surveys.* Vol. 3. Edited by W. G. Madow, I. Olkin, and D. B. Rubin, 363–405. New York: Academic Press.

Brick, J. M., P. Broene, P. James, and J. Severynse. 1996. *A user's guide to WesVarPC.* Rockville, Md.: Westat. (The Web site www.westat.com has the software available for downloading free of charge.)

Brier S. E. 1980. Analysis of contingency tables under cluster sampling. *Biometrika* 67: 591–596.

Brown, G. H., and F. D. Harding. 1973. A comparison of methods of studying illicit drug usage. *Human Resources Research Organization, Technical Report 73-9.* Alexandria, VA: Human Resources Research Organization.

Bryk, A. S., and S. W. Raudenbush. 1992. *Hierarchical linear models: Applications and data analysis methods.* Newbury Park, Calif.: Sage.

Bryk, A. S., S. W. Raudenbush, M. Seltzer, and R. T. Congdon. 1988. *An introduction to HLM: Computer program and user's guide.* Chicago: University of Chicago Press.

Buckland, S. T. 1984. Monte Carlo confidence intervals. *Biometrics* 40: 811–817.

Buckland, S. T., and P. H. Garthwaite. 1991. Quantifying precision of mark-recapture estimates using the bootstrap and related methods. *Biometrics* 47: 255–268.

Burnard, P. 1992. Learning from experience: Nurse tutors' and student nurses' perceptions of experiential learning in nurse education: Some initial findings. *International Journal of Nursing Studies* 29: 151–161.

Bye, B. V., and S. J. Gallicchio. 1993. Sampling variance estimates for SSA program recipients from the 1990 Survey of Income and Program Participation. *Social Security Bulletin* 56: 75–86.

Calahan, D. 1989. The *Digest* poll rides again. *Public Opinion Quarterly* 53: 129–133.

Carlson, B. L., A. E. Johnson, and S. B. Cohen. 1993. An evaluation of the use of personal computers for variance estimation with complex survey data. *Journal of Official Statistics* 9: 795–814.

Casady, R. J., and R. Valliant. 1993. Conditional properties of post-stratified estimators under normal theory. *Survey Methodology* 19: 183–192.

Cassel, C.-M., C.-E. Särndal, and J. H. Wretman. 1977. *Foundations of inference in survey sampling.* New York: Wiley.

Catlin, G., and S. Ingram. 1988. The effects of CATI on cost and quality. In *Telephone survey methodology.* Edited by R. M. Groves, P. P. Biemer, L. E. Lyberg, J. T. Massey, W. L. Nicholls II, and J. Waksberg, 437–450. New York: Wiley.

Chambers, J. M., W. S. Cleveland, B. Kleiner, and P. A. Tukey. 1983. *Graphical techniques for data analysis.* Belmont, Calif.: Duxbury Press.

Chambless, L. E., and K. E. Boyle. 1985. Maximum likelihood methods for complex sample data: Logistic regression and discrete proportional hazards models. *Communications in Statistics—Theory and Methods* 14: 1377–1392.

Chang, T., S. Lohr, and C. G. McLaren. 1992. Teaching survey sampling using simulation. *American Statistician* 46: 232–237.

Chapman, D. G. 1951. Some properties of the hypergeometric distribution with applications to zoological sample censuses. *University of California Publications in Statistics* 1: 131–160.

Christensen, R. 1990. *Log-linear models.* New York: Springer-Verlag.

Cleveland, W. 1994. *Visualizing data.* Summit, N.J.: Hobart Press.

Cochi, S. L., L. E. Edmonds, K. Dyer, W. L. Greaves, J. S. Marks, E. Z. Rovira, S. R. Preblud, and W. A. Orenstein. 1989. Congenital rubella syndrome in the United States: 1970–1985. *American Journal of Epidemiology* 129: 349–361.

Cochran, W. G. 1939. The use of the analysis of variance in enumeration by sampling. *Journal of the American Statistical Association* 34: 492–510.

———. 1977. *Sampling techniques.* 3d ed. New York: Wiley.

———. 1978. Laplace's ratio estimator. In *Contributions to survey sampling and applied statistics.* Edited by H. A. David, 3–10. New York: Academic Press.

Cohen, J. E. 1976. The distribution of the chi-squared statistic under clustered sampling from contingency tables. *Journal of the American Statistical Association* 71: 665–670.

Cohen, S. B. 1997. An evaluation of alternative PC-based software packages developed for the analysis of complex survey data. *American Statistician* 51: 285–292.

Colledge, M., and M. March. 1993. Quality management: Development of a framework for a statistical agency. *Journal of Business and Economic Statistics* 11: 157–165.

Converse, J. M. 1987. *Survey research in the United States: Roots and emergence 1890–1960.* Berkeley: University of California Press.

Converse, J. M., and S. Presser. 1986. *Survey questions: Handcrafting the standardized questionnaire.* Beverly Hills, Calif.: Sage University Press.

Converse, P. E., and M. W. Traugott. 1986. Assessing the accuracy of polls and surveys. *Science* 234: 1094–1098.

Cook, R. D., and S. Weisberg. 1994. *An introduction to regression graphics.* New York: Wiley.

Cormack, R. M. 1992. Interval estimation for mark-recapture studies of closed populations. *Biometrics* 48: 567–576.

Cornfield, J. 1944. On samples from finite populations. *Journal of the American Statistical Association.* 39: 236–239.

———. 1951. Modern methods in the sampling of human populations. *American Journal of Public Health* 41: 654–661.

Cox, D. R. 1952. Estimation by double sampling. *Biometrika* 39: 217–227.

Crewe, I. 1992. A nation of liars? Opinion polls and the 1992 election. *Parliamentary Affairs* 45: 475–495.

Cullen, R. 1994. Sample survey methods as a quality assurance tool in a general practice immunisation audit. *New Zealand Medical Journal* 107: 152–153.

Czaja, R., and J. Blair. 1990. Using network sampling in crime victimization surveys. *Journal of Quantitative Criminology* 6: 185–206.

Dalenius, T. E. 1981. The survey statistician's responsibility for both sampling and measurement errors. In *Current topics in survey sampling.* Edited by D. Krewski, R. Platek, and J. N. K. Rao, 17–29. New York: Academic Press.

D'Alessandro, U., M. K. Aikins, P. Langerock, S. Bennett, and B. M. Greenwood. 1994. Nationwide survey of bednet use in rural Gambia. *Bulletin of the World Health Organization* 72: 391–394.

deLeeuw, J., and I. Kreft. 1986. Random coefficient models for multilevel analysis. *Journal of Educational Statistics* 11: 57–85.

Deming, W. E. 1950. *Some theory of sampling.* New York: Dover.

———. 1956. On simplification of sampling design through replication with equal probabilities and without stages. *Journal of the American Statistical Association* 51: 24–53.

———. 1977. An essay on screening, or on two-phase sampling, applied to surveys of a community. *International Statistical Review* 45: 29–37.

———. 1986. *Out of the crisis.* Cambridge: MIT Press.

Deming, W. E., and F. F. Stephan. 1940. On a least squares adjustment of a sample frequency table when the expected marginal totals are known. *Annals of Mathematical Statistics* 11: 427–444.

Deville, J.-C. 1991. A theory of quota surveys. *Survey Methodology* 17: 163–181.

DeVries, W., W. Keller, and A. Willeboordse. 1996. Reducing the response burden: Some developments in the Netherlands. *International Statistical Review* 64: 199–213.

Dillman, D. 1978. *Mail and telephone surveys.* New York: Wiley.

Dillman, D. A., J. R. Clark, and M. D. Sinclair. 1995a. How prenotice letters, stamped return envelopes and reminder postcards affect mailback response rates for census questionnaires. *Survey Methodology* 21: 159–165.

Dillman, D. A., D. E. Dolsen, and G. E. Machlis. 1995b. Increasing response to personally-delivered mail-back questionnaires. *Journal of Official Statistics* 11: 129–139.

Ding, Y., and S. E. Fienberg. 1994. Dual system estimation of census undercount in the presence of matching error. *Survey Methodology* 20: 149–158.

———. 1996. Multiple sample estimation of population and census undercount in the presence of matching errors. *Survey Methodology* 22: 55–64.

Dippo, C. S., R. E. Fay, and D. H. Morganstein. 1984. Computing variances from complex samples with replicate weights. *Proceedings of the Section on Survey Research Methods, American Statistical Association,* 489–494.

Dobishinski, W. M. 1991. ASCAP/BMI primer. In *The musician's business and legal guide.* 4th ed. Edited by M. Halloran, 176–221. Englewood Cliffs, N.J.: Prentice-Hall.

Domingo-Salvany, A., R. L. Hartnoll, A. Maquire, J. M. Suelves, and J. M. Anto. 1995. Use of capture-recapture to estimate the prevalence of opiate addiction in Barcelona, Spain, 1989. *American Journal of Epidemiology* 141: 567–574.

Droege, S. 1990. The North American Breeding Bird Survey. *Biological Report* 90: 1–4.

Duce, R. A., J. G. Quinn, C. E. Olney, S. R. Piotrowicz, B. J. Ray, and T. L. Wade. 1972. Enrichment of heavy metals and organic compounds in the surface microlayer of Narragansett Bay, Rhode Island. *Science* 176: 161–163.

Duffy, J. C., and J. J. Waterton. 1988. Randomized response vs. direct questioning: Estimating the prevalence of alcohol related problems in a field survey. *Australian Journal of Statistics* 30: 1–14.

DuMouchel, W. H., and G. J. Duncan. 1983. Using sample survey weights in multiple regression analysis of stratified samples. *Journal of the American Statistical Association* 78: 535–543.

Duncan, J. W., and W. C. Shelton. 1978. *Revolution in United States government statistics, 1926–1976.* Washington, D.C.: Department of Commerce.

Durbin, J. 1953. Some results in sampling theory when the units are sampled with unequal probabilities. *Journal of the Royal Statistical Society,* ser. B, 15: 262–269.

Durrett, R. 1994. *The essentials of probability.* Belmont, Calif.: Duxbury Press.

Efron, B. 1979. Bootstrap methods: Another look at the jackknife. *Annals of Statistics* 7: 1–26.

———. 1982. *The jackknife, the bootstrap, and other resampling plans.* Philadelphia: SIAM.

Efron, B., and R. J. Tibshirani. 1993. *An introduction to the bootstrap.* London: Chapman & Hall.

Egeland, G. M., K. A. Perham-Hester, and E. B. Hook. 1995. Use of capture-recapture analyses in fetal alcohol syndrome surveillance in Alaska. *American Journal of Epidemiology* 141: 335–341.

Ericson, W. A. 1969a. Subjective Bayesian models in sampling finite populations. *Journal of the Royal Statistical Society,* ser. B, 31: 195–233.

———. 1969b. Subjective Bayesian models in sampling finite populations: Stratification. In *New developments in survey sampling.* Edited by N. L. Johnson and H. Smith, 326–357. New York: Wiley.

———. 1988. Bayesian inference in finite populations. In *Handbook of statistics.* Vol. 6. Edited by P. R. Krishnaiah and C. R. Rao, 213–246. New York: North Holland.

Ezzati-Rice, T. M., and R. S. Murphy. 1995. Issues associated with the design of a national probability sample for human exposure assessment. *Environmental Health Perspectives,* 103 (suppl. 3): 55–59.

Fay, R. E. 1985. A jackknifed chi-squared test for complex samples *Journal of the American Statistical Association* 80: 148–157.

———. 1990. VPLX: Variance estimates for complex samples. *Proceedings of the Section for Survey Research Methods, American Statistical Association,* 266–271. (Updated documentation is available on the Bureau of the Census Web site at ftp.census.gov/ftp/pub/sdms/www/vwelcome.html.)

———. 1996. Alternative paradigms for the analysis of imputed survey data. *Journal of the American Statistical Association* 91: 490–498.

Fay, R. E., J. S. Passel, and J. G. Robinson. 1988. *The coverage of population in the 1980 census.* Evaluation and Research Report PHC80-E4. Washington, D.C.: Department of Commerce.

Fellegi, I. P., and D. Holt. 1976. A systematic approach to automatic edit and imputation. *Journal of the American Statistical Association* 71: 17–35.

Fienberg, S. E. 1972. The multiple recapture census for closed populations and incomplete 2^k contingency tables. *Biometrika* 59: 591–603.

———. 1979. The use of chi-squared statistics for categorical data problems. *Journal of the Royal Statistical Society,* ser. B, 41: 54–64.

———. 1980. The measurement of crime victimization: Prospects for panel analysis of a panel survey. *Statistician* 29: 313–350.

———. 1992. Bibliography on capture-recapture modelling with application to census undercount adjustment. *Survey Methodology* 18: 143–154.

Fienberg, S. E., and J. Tanur. 1987. Experimental and sampling structures: Parallels diverging and meeting. *International Statistical Review* 55: 75–96.

———. 1988. From the inside out and the outside in: Combining experimental and sampling structures. *Canadian Journal of Statistics* 16: 135–151.

————. 1992. Cognitive aspects of surveys: Yesterday, today, and tomorrow. *Journal of Official Statistics* 8: 5–17.

————. 1996. Reconsidering the fundamental contributions of Fisher and Neyman on experimentation and sampling. *International Statistical Review* 64: 237–253.

Fisher, R. A. 1925. *Statistical methods for research workers.* London: Oliver and Boyd.

————. 1938. Presidential address. In *Proceedings of the Indian Statistical Conference.* Calcutta: Statistical Publishing Society.

Ford, D., D. F. Easton, D. T. Bishop, S. A. Narod, D. E. Goldgar, and the Breast Cancer Linkage. 1994. Risks of cancer in *BRCA1*-mutation carriers. *Lancet* 343: 692–695.

Fowler, F. J. 1993. *Survey research methods.* 2d ed. Newbury Park, Calif.: Sage.

Fox, J. A., and P. E. Tracy. 1986. *Randomized response: A method for sensitive surveys.* Beverly Hills, Calif.: Sage.

Francisco, C. A., and W. A. Fuller. 1991. Quantile estimation with a complex survey design. *Annals of Statistics* 19: 454–469.

Frank, A. 1978. The contingency table approach to mark-recapture population estimate. University of Minnesota Department of Applied Statistics Plan B Paper.

Frankel, L. R. 1983. The report of the CASRO task force on response rates. In *Improving data quality in sample surveys.* Edited by F. Wiseman, 1–11. Cambridge, Mass.: Marketing Science Institute.

Freedman, D. A., and W. C. Navidi. 1992. Should we have adjusted the U.S. Census of 1980? *Survey Methodology* 18: 3–74.

Fuller, W. A. 1975. Regression analysis for sample survey. *Sankyhā,* ser. C, 37: 117–132.

————. 1984. Least squares and related analyses for complex survey designs. *Survey Methodology* 10: 97–118.

Fuller, W. A., W. Kennedy, D. Schnell, G. Sullivan, and H. J. Park. 1989. *PC CARP.* Ames: Iowa State University, Statistical Laboratory.

Ghosh, M., and J. N. K. Rao. 1994. Small area estimation: An appraisal (with discussion). *Statistical Science* 9: 55–93.

Gill, R. D., Y. Vardi, and J. A. Wellner. 1988. Large sample theory of empirical distributions in biased sampling models. *Annals of Statistics* 106: 1069–1112.

Gnap, R. 1995. Teacher load in Arizona elementary school districts in Maricopa County. Ph.D. diss., Arizona State University.

Godambe, V. P. 1955. A unified theory of sampling from finite populations. *Journal of the Royal Statistical Society,* ser. B, 17: 269–278.

Goldstein, H. 1987. *Multilevel models in educational and social research.* London: Griffin.

Goldstein, H., and R. Silver. 1989. Multilevel and multivariate models in survey analysis. In *Analysis of complex surveys.* Edited by C. J. Skinner, D. Holt, and T. M. F. Smith, 221–235. New York: Wiley.

González, M. E. 1994. Improving data quality awareness in the United States federal statistical agencies. *American Statistician* 48: 12–17.

González, M. E., D. Kasprzyk, and F. Scheuren. 1994. Nonresponse in federal surveys: An exploratory study. *Amstat News* 208: 1ff.

Goren, S., L. Silverstein, and N. Gonzales. 1993. A survey of food service managers of Washington state boarding homes for the elderly. *Journal of Nutrition for the Elderly* 12: 27–36.

Gower, A. R. 1979. Nonresponse in the Canadian Labour Force Survey. *Survey Methodology* 5: 29–58.

Graybill, F. A. 1976. *Theory and application of the linear model.* North Scituate, Mass.: Duxbury Press.

Graybill, F. A., and H. K. Iyer. 1994. *Regression analysis: Concepts and applications.* Belmont, Calif.: Duxbury Press.

Gross, S. 1980. Median estimation in sample surveys. *Proceedings of the Section on Survey Research Methods, American Statistical Association,* 181–184.

Groves, R. M. 1989. *Survey errors and survey costs.* New York: Wiley.

Hájek, J. 1960. Limiting distributions in simple random sampling from a finite population. *Publication of the Mathematical Institute of the Hungarian Academy of Sciences* 5:

361–374.

Hansen, M. H., T. Dalenius, and B. J. Tepping. 1985. The development of sample surveys of finite populations. In *A celebration of statistics*. Edited by A. C. Atkinson and S. E. Fienberg, 327–354. New York: Springer-Verlag.

Hansen, M. H., and W. N. Hurwitz. 1943. On the theory of sampling from a finite population. *Annals of Mathematical Statistics* 14: 333–362.

———. 1946. The problem of non-response in sample surveys. *Journal of the American Statistical Association* 41: 517–529.

———. 1949. On the determination of the optimum probabilities in sampling. *Annals of Mathematical Statistics* 20: 426–432.

Hansen, M. H., W. N. Hurwitz, and W. G. Madow. 1953. *Sample survey methods and theory.* Vols. I and II. New York: Wiley.

Hansen, M. H., and W. G. Madow. 1976. Some important events in the historical development of sample surveys. In *On the history of statistics and probability.* Edited by D. B. Owen, 75–102. New York: Marcel Dekker.

Hansen, M. H., W. G. Madow, and B. J. Tepping. 1983. An evaluation of model-dependent and probability-sampling inferences in sample surveys (with discussion). *Journal of the American Statistical Association,* 776–807.

Hanson, R. H. 1978. *The Current Population Survey: Design and methodology.* Technical paper 40. Washington, D.C.: Department of Commerce, Bureau of the Census.

Harris, R. 1992. We are a nation of liars. (London) *Sunday Times,* April 12.

Hartley, H. O. 1946. Discussion of paper by F. Yates. *Journal of the Royal Statistical Society* 109: 37–38.

———. 1962. Multiple frame surveys. *Proceedings of the Social Statistics Section, American Statistical Association,* 203–206.

Hartley, H. O., and A. Ross. 1954. Unbiased ratio estimators. *Nature* 174: 270–271.

Hedayat, A. S., and B. K. Sinha. 1991. *Design and inference in finite population sampling.* New York: Wiley.

Hidiroglou, M. A., J. D. Drew, and G. B. Gray. 1993. A framework for measuring and reducing nonresponse in surveys. *Survey Methodology* 19: 81–94.

Hidoroglou, M. A., C.-E. Särndal, and D. A. Binder. 1995. Weighting and estimation in business surveys. In *Business survey methods.* Edited by B. G. Cox, D. A. Binder, B. N. Chinnappa, A. Christianson, M. J. Colledge, and P. S. Kott, 477–502. New York: Wiley.

Hite, S. 1987. *Women and love: A cultural revolution in progress.* New York: Knopf.

Hogan, H. 1993. The 1990 post-enumeration survey: Operations and results. *Journal of the American Statistical Association* 88: 1047–1060.

Holt, D., and D. Elliot. 1991. Methods of weighting for unit non-response. *Statistician* 40: 333–342.

Holt, D., A. J. Scott, and P. O. Ewings. 1980. Chi-squared tests with survey data. *Journal of the Royal Statistical Society,* ser. A, 143: 303–320.

Horvitz, D. G., B. V. Shah, and W. R. Simmons. 1967. The unrelated question randomized response model. *Proceedings of the Social Statistics Section, American Statistical Association,* 65–72.

Horvitz, D. G., and D. J. Thompson. 1952. A generalization of sampling without replacement from a finite universe. *Journal of the American Statistical Association* 47: 663–685.

Hosmer, D. W., and S. Lemeshow. 1989. *Applied logistic regression.* New York: Wiley.

Iachan, R., and M. L. Dennis. 1993. A multiple frame approach to sampling the homeless and transient population. *Journal of Official Statistics* 9: 747–764.

International Working Group for Disease Monitoring and Forecasting. 1995. Capture-recapture and multiple-record systems estimation. *American Journal of Epidemiology* 142: 1047–1068.

Jackson, K. W., I. W. Eastwood, and M. S. Wild. 1987. Stratified sampling protocol for monitoring trace metal concentrations in soil. *Soil Science* 143: 436–443.

Jacoby, J., and A. H. Handlin. 1991. Non-probability sampling designs for litigation surveys. *Trademark Reporter* 81: 169–179.

Jessen, R. J. 1978. *Statistical survey techniques.* New York: Wiley.

Jinn, J. H., and J. Sedransk. 1989a. Effect on secondary data analysis of common imputation methods. *Sociological Methodology* 19: 213–241.

———. 1989b. Effect on secondary data analysis of the use of imputed values: The case where missing data are not missing at random. *Proceedings of the Section for Survey Research Methods, American Statistical Association,* 51–59.

Joiner, B. L. 1994. *Fourth generation management: The new business consciousness.* New York: McGraw-Hill.

Kalton, G. 1983. Models in the practice of survey sampling. *International Statistical Review* 51: 175–188.

Kalton, G., and D. W. Anderson. 1986. Sampling rare populations. *Journal of the Royal Statistical Society,* ser. A, 149: 65–82.

Kalton, G., and D. Kasprzyk. 1982. Imputing for missing survey responses. *Proceedings of the Section for Survey Research Methods, American Statistical Association,* 22–33.

———. 1986. The treatment of missing survey data. *Survey Methodology* 12: 1–16.

Kempthorne, O. 1952. *The design and analysis of experiments.* New York: Wiley.

Khurshid, A., and H. Sahai. 1995. A bibliography on telephone sampling methodology. *Journal of Official Statistics* 11: 325–367.

Kiaer, A. 1897. The representative method of statistical surveys [1976 translation of the original Norwegian]. Oslo: Central Bureau of Statistics of Norway.

Kinsley, M. 1981. The art of polling. *New Republic* 184: 16–19.

Kish, L. 1965. *Survey sampling.* New York: Wiley.

———. 1987. *Statistical design for research.* New York: Wiley.

———. 1992. Weighting for unequal P_i. *Journal of Official Statistics* 8: 183–200.

———. 1995. Methods for design effects. *Journal of Official Statistics* 11: 55–77.

Kish, L., and M. R. Frankel. 1974. Inference from complex samples (with discussion). *Journal of the Royal Statistical Society,* ser. B, 36: 1–37.

Koch, G. G., D. H. Freeman, and D. L. Freeman. 1975. Strategies in the multivariate analysis of data from complex surveys. *International Statistical Review* 93: 59–78.

Konijn, H. S. 1962. Regression analysis in sample surveys. *Journal of the American Statistical Association* 57: 509–606.

Korn, E. L., and B. I. Graubard. 1990. Simultaneous testing of regression coefficients with complex survey data: Use of Bonferroni t-statistics. *American Statistician* 44: 270–276.

———. 1995a. Analysis of large health surveys: Accounting for the sampling design. *Journal of the Royal Statistical Society,* ser. A, 158: 263–295.

———. 1995b. Examples of differing weighted and unweighted estimates from a sample survey. *American Statistician* 49: 291–295.

———. 1998. Scatterplots with survey data. *American Statistician* 52: 58–69.

Kosmin, B. A., and S. P. Lachman. 1993. *One nation under God: Religion in contemporary American society.* New York: Harmony Books.

Kott, P. S. 1991. A model-based look at linear regression with survey data. *American Statistician* 45: 107–112.

Kovar, J. G., J. N. K. Rao, and C. F. J. Wu. 1988. Bootstrap and other methods to measure errors in survey estimates. *Canadian Journal of Statistics* 16 (suppl.): 25–45.

Krewski, D., and J. N. K. Rao. 1981. Inference from stratified samples: Properties of the linearization, jackknife, and balanced repeated replication methods. *Annals of Statistics* 9: 1010–1019.

Kruskal, W., and F. Mosteller. 1980. Representative sampling, IV: The history of the concept in statistics, 1895–1939. *International Statistical Review* 48: 169–195.

Kruuk, H., A. Moorhouse, J. W. H. Conroy, L. Durbin, and S. Frears. 1989. An estimate of numbers and habitat preferences of otters *Lutra lutra* in Shetland, UK. *Biological Conservation* 49: 241–254.

Kuk, A. Y. C. 1990. Asking sensitive questions indirectly. *Biometrika* 77: 436–438.

Lahiri, D. B. 1951. A method of sample selection providing unbiased ratio estimates. *Bulletin of the International Statistical Institute* 33: 133–140.

Landers, A. 1976. If you had it to do over again, would you have children? *Good Housekeeping* 182 (June): 100–101, 215–216, 223–224.

Laplace, P. S. 1814. *Essai philosophique sur les probabilités.* Paris: MME VE Courcier, Imprimeur-Libraire pour les Mathématiques, quai des Augustins, no. 57. [An English translation was published by Dover in 1951.]

Lavrakas, P. J. 1993. *Telephone survey methods: Sampling, selection, and supervision.* 2d ed. Newbury Park, Calif.: Sage.

Lehnen, R. G., and W. G. Skogan. 1981. *The National Crime Survey: Working papers.* Washington, D.C.: Department of Justice.

Lehtonen, R., and E. J. Pahkinen. 1995. *Practical methods for design and analysis of complex surveys.* New York: Wiley.

Lenski, G., and J. Leggett. 1960. Caste, class, and deference in the research interview. *American Journal of Sociology* 65: 463–467.

Lepkowski, J. M. 1982. The use of OSIRIS IV to analyse complex sample survey data. *Proceedings of the Section on Survey Research Methods, American Statistical Association,* 38–43.

————. 1988. Telephone sampling methods in the United States, In *Telephone survey methodology.* Edited by R. M. Groves, P. P. Biemer, L. E. Lyberg, J. T. Massey, W. L. Nicholls II, and J. Waksberg, 73–99. New York: Wiley.

Lepkowski, J., and J. Bowles. 1996. Sampling error software for personal computers. *Survey Statistician* 35 (December): 10–17.

Lessler, J. T., and W. D. Kalsbeek. 1992. *Nonsampling errors in surveys.* New York: Wiley.

Lincoln, F. C. 1930. Calculating waterfowl abundance on the basis of banding returns. *Circular of the U.S. Department of Agriculture* 118: 1–4.

Lindgren, B. 1993. *Statistical theory.* 4th ed. New York: Chapman & Hall.

Link, H. C., and H. A. Hopf. 1946. *People and books: A study of reading and book-buying habits.* New York: Book Manufacturer's Institute.

Literary Digest. 1932. Roosevelt bags 41 states out of 48. *Literary Digest* 114 (November 5): 8–9.

————. 1936a. "The Digest" presidential poll is on: Famous forecasting machine is thrown into gear for 1936. *Literary Digest* 122 (August 22): 3–4.

————. 1936b. Landon, 1,293,669: Roosevelt, 972,897. *Literary Digest* 122 (October 31): 5–6.

————. 1936c. What went wrong with the polls? *Literary Digest* 122 (November 14): 7–8.

Little, R. J. A. 1986. Survey nonresponse adjustments for estimates of means. *International Statistical Review* 54: 139–157.

————. 1991. Inference with survey weights. *Journal of Official Statistics* 7: 405–424.

Little, R. J. A., and D. B. Rubin. 1987. *Statistical analysis with missing data.* New York: Wiley.

Lohr, S. 1990. Accurate multivariate estimation using triple sampling. *Annals of Statistics* 18: 1615–1633.

Lohr, S., and J. Liu. 1994. A comparison of weighted and unweighted analyses in the NCVS. *Journal of Quantitative Criminology* 10: 343–360.

Lush, J. L. 1945. *Animal breeding plans.* Ames: Iowa State College Press.

Lydersen, C., and M. Ryg. 1991. Evaluating breeding habitat and populations of ringed seals *Phoca hispida* in Svalbard fjords. *Polar Record* 27: 223–228.

Macdonell, W. R. 1901. On criminal anthropometry and the identification of criminals. *Biometrika* 1: 177–227.

Madow, W. G., I. Olkin., and D. B. Rubin, eds. 1983. *Incomplete data in sample surveys.* New York: Academic Press.

Mahalanobis, P. C. 1946. Recent experiments in statistical sampling in the Indian Statistical Institute. *Journal of the Royal Statistical Society* 109: 325–370.

Mayr, J., M. Gaisl, K. Purtscher, H. Noeres, G. Schimpl, and G. Fasching. 1994. Baby walkers–an underestimated hazard for our children? *European Journal of Pediatrics* 153: 531–534.

McAuley, R. G., W. M. Paul, G. H. Morrison, R. F. Beckett, and C. H. Goldsmith. 1990. Five-year results of the peer assessment program of the College of Physicians and Surgeons of Ontario. *Canadian Medical Association Journal* 143: 1193–1199.

McCarthy, P. J. 1966. Replication: An approach to the analysis of data from complex surveys. In *Vital and Health Statistics,* ser. 2, no. 14. Washington, D.C.: National Center for Health Statistics.

———. 1969. Pseudo-replication: Half-samples. *Review of the International Statistical Institute* 37: 239–264.

———. 1993. Standard error and confidence interval estimation for the median. *Journal of Official Statistics* 9: 691–703.

McFarland, S. G. 1981. Effects of question order on survey responses. *Public Opinion Quarterly* 45: 208–215.

McGuiness, R. A. 1994. Redesign of the sample for the Current Population Survey. *Employment and Earnings* 41 (May): 7–10.

McIlwee, J. S., and J. G. Robinson. 1992. *Women in engineering: Gender, power, and workplace culture.* Albany: State University of New York Press.

Mitofsky, W. J. 1970. Sampling of telephone households. Unpublished CBS News memorandum.

Morganstein, D., and M. Hansen. 1990. Survey operations processes: The key to quality improvement. In *Data quality control: Theory and pragmatics.* Edited by G. E. Lepins and V. R. R. Uppuluri, 91–104. New York: Marcel Dekker.

Morton, H. C., and A. J. Price. 1989. *The ACLS survey of scholars: Final report of views on publications, computers, and libraries.* Washington, D.C.: University Press of America.

Mosteller, F., H. Hyman, P. J. McCarthy, E. S. Martis, and D. B. Truman. 1949. *The pre-election polls of 1948.* New York: Social Sciences Research Council.

Nathan, G., and T. M. F. Smith. 1989. The effect of selection on regression analysis. In *Analysis of complex surveys.* Edited by C. J. Skinner, D. Holt, and T. M. F. Smith, 149–163. New York: Wiley.

National Center of Educational Statistics. 1977. *National longitudinal study of the high school class of 1972: First followup survey design, instrument preparation, data collection, and file development.* NCES 77-262. Washington, D.C.: Government Printing Office.

National Center for Health Statistics. 1987. *Vital statistics of the United States.* Washington, D.C.: Government Printing Office.

Neter, J. 1978. How accountants save money by sampling. In *Statistics: A guide to the unknown.* Edited by J. M. Tanur, 249–258. San Francisco: Holden-Day.

Neter, J., M. H. Kutner, C. J. Nachtsheim, and W. Wasserman. 1996. *Applied linear statistical models.* 4th ed. Chicago: Irwin.

Neter, J., R. A. Leitch, and S. E. Fienberg. 1978. Dollar unit sampling: Multinomial bounds for total overstatement and understatement errors. *Accounting Review* 58: 77–93.

Neyman, J. 1934. On the two different aspects of the representative method: The method of stratified sampling and the method of purposive selection. *Journal of the Royal Statistical Society* 97: 558–606.

———. 1938. Contribution to the theory of sampling human populations. *Journal of the American Statistical Association* 33: 101–116.

Nusser, S. M., A. L. Carriquiry, K. W. Dodd, and W. A. Fuller. 1996. A semiparametric transformation approach to estimating usual daily intake distributions. *Journal of the American Statistical Association* 91: 1440–1449.

O'Brien, L. A., J. A. Grisso, G. Maislin, K. LaPann, K. P. Krotki, P. J. Greco, E. A. Siegert, and L. K. Evans. 1995. Nursing home residents' preferences for life-sustaining treatments. *Journal of the American Medical Association* 274: 1775–1779.

Oh, H. L., and F. J. Scheuren. 1983. Weighting adjustment for unit nonresponse. In *Incomplete data in sample surveys.* Vol. 2. Edited by W. G. Madow, I. Olkin, and D. B. Rubin, 143–184. New York: Academic Press.

Overton, W. S., and S. V. Stehman. 1995. The Horvitz–Thompson theorem as a unifying perspective for probability sampling: With examples from natural resource sampling. *American Statistician* 49: 261–268.

Parten, M. 1950. *Surveys, polls, and samples: Practical procedures.* New York: Harper and Brothers.

Paulin, G. D., and D. L. Ferraro. 1994. Imputing income in the Consumer Expenditure Survey. *Monthly Labor Review* (December): 23–31.

Peart, D. 1994. Impacts of feral pig activity on vegetation patterns associated with *Quercus agrifolia* on Santa Cruz Island, California. Ph.D. diss., Arizona State University.

Petersen, C. G. J. 1896. The yearly immigration of young plaice into the Limfjord from the German Sea. *Reports of the Danish Biological Station* 6: 5–84.

Pfeffermann, D. 1993. The role of sampling weights when modeling survey data. *International Statistical Review* 61: 317–337.

Pfeffermann, D., and D. J. Holmes. 1985. Robustness considerations in the choice of method of inference for the regression analysis of survey data. *Journal of the Royal Statistical Society,* ser. A, 148: 268–278.

Pincus, T. 1993. Arthritis and rheumatic diseases: What doctors can learn from their patients. In *Mind/body medicine: How to use your mind for better health.* Edited by D. Goleman and J. Gurin, 177–192. Yonkers, N.Y.: Consumer Reports Books.

Plackett, R. L., and J. P. Burman. 1946. The design of optimum multifactorial experiments. *Biometrika* 33: 305–325.

Platek, R. 1977. Some factors affecting non-response. *Survey Methodology* 3: 191–214.

Politz, A., and W. Simmons. 1949. An attempt to get the "not at homes" into the sample without callbacks. *Journal of the American Statistical Association* 44: 9–31.

Pollock, K. H. 1991. Modeling capture, recapture, and removal statistics for estimation of demographic parameters for fish and wildlife populations: Past, present, and future. *Journal of the American Statistical Association* 86: 225–238.

Potthoff, R. F. 1994. Telephone sampling in epidemiologic research: To reap the benefits, avoid the pitfalls. *American Journal of Epidemiology* 139: 967–978.

Potthoff, R. F., K. G. Manton, and M. A. Woodbury. 1993. Correcting for nonavailability bias in surveys by weighting based on number of callbacks. *Journal of the American Statistical Association* 88: 1197–1207.

Prentice, R. L., and R. Pyke. 1979. Logistic disease incidence models and case-control studies. *Biometrika* 66: 403–411.

Prosser, R., J. Rabash, and H. Goldstein. 1992. *ML3 software for three-level analysis: User's guide for V2.* London: University of London, Institute of Education.

Quenouille, M. H. 1949. Problems in plane sampling. *Annals of Mathematical Statistics* 20: 355–375.

———. 1956. Notes on bias in estimation. *Biometrika* 43: 353–360.

Raj, D. 1968. *Sampling theory.* New York: McGraw-Hill.

———. 1971. *The design of sample surveys.* New York: McGraw-Hill.

Rao, J. N. K. 1963. On three procedures of unequal probability sampling without replacement. *Journal of the American Statistical Association* 58: 202–215.

———. 1973. On double sampling for stratification and analytical surveys. *Biometrika* 60: 125–133, 669.

———. 1979a. On deriving mean square errors and their non-negative unbiased estimators in finite population sampling. *Journal of the Indian Statistical Association* 17: 125–136.

———. 1979b. Optimization in the design of sample surveys. In *Optimizing methods in statistics: Proceedings of an international conference.* Edited by J. S. Rustagi, 419–434. New York: Academic Press.

———. 1988. Variance estimation in sample surveys. In *Handbook of statistics.* Vol. 6. Edited by P. R. Krishnaiah and C. R. Rao, 427–447. Amsterdam: Elsevier.

———. 1996. On variance estimation with imputed survey data. *Journal of the American Statistical Association* 91: 499–506.

———. 1997. Developments in sample survey theory: An appraisal. *Canadian Journal of Statistics* 25: 1–21.

Rao, J. N. K., and D. R. Bellhouse. 1990. History and development of the theoretical foundations of survey based estimation and analysis. *Survey Methodology* 16: 3–29.

Rao, J. N. K., H. O. Hartley, and W. G. Cochran. 1962. A simple procedure for unequal probability sampling without replacement. *Journal of the Royal Statistical Society,* ser. B, 24: 482–491.

Rao, J. N. K., and A. J. Scott. 1981. The analysis of categorical data from complex sample surveys: Chi-squared tests for goodness of fit and independence in two-way tables. *Journal of the American Statistical Association* 76: 221–230.

———. 1984. On chi-squared tests for multi-way tables with cell proportions estimated from survey data. *Annals of Statistics* 12: 46–60.

———. 1987. On simple adjustments to chi-squared tests with sample survey data. *Annals of Statistics* 15: 385–397.

Rao, J. N. K., A. J. Scott, and C. J. Skinner. 1998. Quasi-score tests with survey data. *Statistica Sinica* 8: 1059–1070.

Rao, J. N. K., and R. R. Sitter. 1995. Variance estimation under two-phase sampling with application to imputation for missing data. *Biometrika* 82: 453–460.

Rao, J. N. K., and D. R. Thomas. 1988. The analysis of cross-classified categorical data from complex sample surveys. *Sociological Methodology* 18: 213–269.

———. 1989. Chi-squared tests for contingency tables. In *Analysis of complex surveys*. Edited by C. J. Skinner, D. Holt, and T. M. F. Smith, 89–114. New York: Wiley.

Rao, J. N. K., and C. F. J. Wu. 1985. Inference from stratified samples: Second-order analysis of three methods for nonlinear statistics. *Journal of the American Statistical Association* 80: 620–630.

———. 1987. Methods for standard errors and confidence intervals from sample survey data: Some recent work. *Bulletin of the International Statistical Institute* 52: 5–21.

———. 1988. Resampling inference with complex survey data. *Journal of the American Statistical Association* 83: 231–241.

Rao, J. N. K., C. F. J. Wu, and K. Yue. 1992. Some recent work on resampling methods for complex surveys. *Survey Methodology* 18: 209–217.

Rao, P. S. R. S., and J. N. K. Rao. 1971. Small sample results for ratio estimators. *Biometrika* 58: 625–630.

Remafedi, G., M. Resnick, R. Blum, and L. Harris. 1992. Demography of sexual orientation in adolescents. *Pediatrics* 89: 714–721.

Roberts, G., J. N. K. Rao, and S. Kumar. 1987. Logistic regression analysis of sample survey data. *Biometrika* 74: 1–12.

Roberts, R. J., Q. D. Sandifer, M. R. Evans, M. Z. Nolan-Farrell, and P. M. Davis. 1995. Reasons for non-uptake of measles, mumps, and rubella catch up immunisation in a measles epidemic and side effects of the vaccine. *British Medical Journal* 310: 1629–1632.

Robinson, J. 1987. Conditioning ratio estimates under simple random sampling. *Journal of the American Statistical Association* 82: 826–831.

Rosenbaum, P. R., and D. B. Rubin. 1983. The central role of the propensity score in observational studies for causal effects. *Biometrika* 70: 41–55.

Ross, S. 1998. *A first course in probability*. 5th ed. Upper Saddle River, N.J.: Prentice-Hall.

Rothenberg, R. B., A. Lobanov, K. B. Singh, and G. Stroh. 1985. Observations on the application of EPI cluster survey methods for estimating disease incidence. *Bulletin of the World Health Organization* 63: 93–99.

Royall, R. M. 1970. On finite population sampling theory under certain linear regression models. *Biometrika* 57: 377–387.

———. 1976a. The linear least-squares prediction approach to two-stage sampling. *Journal of the American Statistical Association* 71: 657–664.

———. 1976b. Current advances in sampling theory: Implications for human observational studies. *American Journal of Epidemiology* 463–474.

———. 1988. The prediction approach to sampling theory. In *Handbook of statistics*. Vol. 6. Edited by P. R. Krishnaiah and C. R. Rao, 399–413. New York: North Holland.

———. 1992a. The model based (prediction) approach to finite population sampling theory. In *Current issues in statistical inference: Essays in honor of D. Basu*. Edited by M. Ghosh and P. K. Pathak, 225–240. Hayward, Calif.: Institute of Mathematical Statistics.

————. 1992b. Robustness and optimal design under prediction models in finite population sampling. *Survey Methodology* 18: 179–195.

Royall, R. M., and W. G. Cumberland. 1981a. An empirical study of the ratio estimator and estimators of its variance. *Journal of the American Statistical Association* 76: 66–77.

————. 1981b. The finite-population linear regression estimator and estimators of its variance—an empirical study. *Journal of the American Statistical Association* 76: 924–930.

Royall, R. M., and K. R. Eberhardt. 1975. Variance estimates for the ratio estimator. *Sankhyā C* 37: 43–52.

Royall, R. M., and J. H. Herson. 1973. Robust estimation in finite populations, I. *Journal of the American Statistical Association* 68: 880–889.

Rubin, D. B. 1985. The use of propensity scores in applied Bayesian inference. In *Bayesian statistics 2*. Edited by J. M. Bernardo, M. H. DeGroot, D. V. Lindley, and A. F. M. Smith. Amsterdam: North Holland.

————. 1987. *Multiple imputation for nonresponse in surveys*. New York: Wiley.

————. 1996. Multiple imputation after 18+ years. *Journal of the American Statistical Association* 91: 473–489.

Ruggles, S. 1995. Sample designs and sampling errors. *Historical Methods* 28: 40–46.

Russell, H. J. 1972. Use of a commercial dredge to estimate a hardshell clam population by stratified random sampling. *Journal of the Fisheries Research Board of Canada* 29: 1731–1735.

Rust, K. F., and J. N. K. Rao. 1996. Variance estimation for complex surveys using replication techniques. *Statistical Methods in Medical Research* 5: 283–310.

Ryan, A. S., D. Rush, F. W. Krieger, and G. E. Lewandowski. 1991. Recent declines in breast-feeding in the United States, 1984–1989. *Pediatrics* 88: 719–727.

Ryan, T. P. 1989. *Statistical methods for quality improvement*. New York: Wiley.

Salant, P., and D. A. Dillman. 1994. *How to conduct your own survey*. New York: Wiley.

Samuels, C. 1996. Full-time vs. part-time instructors. *Arizona AAUP Advocate* 45: 1–3.

Sande, I. G. 1983. Hot-deck imputation procedures. In *Incomplete data in sample surveys*. Vol. 3. Edited by W. G. Madow, I. Olkin, and D. B. Rubin, 339–349. New York: Academic Press.

Sanderson, M., P. J. Placek, and K. G. Keppel. 1991. The 1988 National Maternal and Infant Health Survey: Design, content, and data availability. *Birth* 18: 26–32.

Sanzo, J. M., M. A. Garcia-Calabuig, A. Audicana, and V. Dehesa. 1993. Q fever: Prevalence of antibodies to *Coxiella burnetii* in the Basque country. *International Journal of Epidemiology* 22: 1183–1188.

Särndal, C.-E. 1978. Design-based and model-based inference in survey sampling. *Scandinavian Journal of Statistics* 5: 27–52.

————. 1980. On π inverse weighting versus best linear unbiased weighting in probability sampling. *Biometrika* 67: 639–650.

Särndal, C.-E., and B. Swensson. 1987. A general view of estimation for two phases of selection with applications to two-phase sampling and nonresponse. *International Statistical Review* 55: 279–294.

Särndal, C.-E., B. Swensson, and J. Wretman. 1992. *Model assisted survey sampling*. New York: Springer-Verlag.

Satterthwaite, F. E. 1946. An approximate distribution of estimates of variance components. *Biometrics* 2: 110–114.

Saville, A. 1977. *Survey methods of appraising fishery resources*. FAO Fisheries technical paper no. 171. Rome: Food and Agriculture Organization of the United Nations.

Schafer, J. L. 1997. *Analysis of incomplete multivariate data*. London: Chapman & Hall.

Scheaffer, N. C. 1995. A decade of questions. *Journal of Official Statistics* 11: 79–92.

Schei, B., and L. S. Bakketeig. 1989. Gynaecological impact of sexual and physical abuse by spouse: A study of a random sample of Norwegian women. *British Journal of Obstetrics and Gynaecology* 96: 1379–1383.

Scheuren, F. J. 1973. Ransacking CPS tabulations: Applications of the log linear model to poverty statistics. *Annals of Economic and Social Measurement* 2: 159–182.

Schnabel, Z. E. 1938. The estimation of the total fish population of a lake. *American Mathematical Monthly* 45: 348–352.

Schreuder, H. T., J. Sedransk, and K. D. Ware. 1968. 3-P sampling and some alternatives, I. *Forest Science* 14: 429–453.

Schuman, H., and J. M. Converse. 1971. The effects of black and white interviewers on black responses in 1968. *Public Opinion Quarterly* 35: 44–68.

Schuman, H., and S. Presser. 1981. *Questions and answers in attitude surveys: Experiments on question form, wording, and context.* New York: Academic Press.

Schwarz, N. 1995. What respondents learn from questionnaires: The survey interview and the logic of conversation. *International Statistical Review* 63: 153–177.

Schwarz, N., and S. Sudman, eds. 1990. *Context effects in social and psychological research.* New York: Springer-Verlag.

Scott, A. J., and T. M. F. Smith. 1969. Estimation in multi-stage surveys. *Journal of the American Statistical Association* 71: 657–664.

———. 1974. Analysis of repeated surveys using time series methods. *Journal of the American Statistical Association* 64: 674–678.

———. 1975. Minimax designs for sample surveys. *Biometrika* 62: 353–357.

Scott, A. J., and C. J. Wild. 1989. Selection based on the response variable in logistic regression. In *Analysis of complex surveys.* Edited by C. J. Skinner, D. Holt, and T. M. F. Smith, 191–208. New York: Wiley.

Scott, D. W. 1992. *Multivariate density estimation: Theory, practice, and visualization.* New York: Wiley.

Searle, S. 1971. *Linear models.* New York: Wiley.

Seber, G. A. F. 1970. The effects of trap response on tag-recapture estimates. *Biometrics* 26: 13–22.

———. 1982. *The estimation of animal abundance and related parameters.* 2d ed. London: Griffin.

Sen, A. R. 1953. On the estimate of the variance in sampling with varying probabilities. *Journal of the Indian Society of Agricultural Statistics* 5: 119–127.

Seng, Y. P. 1951. Historical survey of the development of sampling theories and practice. *Journal of the Royal Statistical Society,* ser. A, 114: 214–231.

Senturia, Y. D., K. K. Christoffel, and M. Donovan. 1994. Children's household exposure to guns: A pediatric practice-based survey. *Pediatrics* 93: 469–475.

Serdula, M., A. Mokdad, E. Pamuk, D. Williamson, and T. Byers. 1995. Effects of question order on estimates of the prevalence of attempted weight loss. *American Journal of Epidemiology* 142: 64–67.

Shah, B. V., B. G. Barnwell, and G. S. Bieler. 1995. *SUDAAN user's manual: Software for the statistical analysis of correlated data.* Research Triangle Park, N.C.: Research Triangle Institute. [Also see www.rti.org/patents/sudaan/sudaan.html, the SUDAAN Web page.]

Shah, B. V., M. M. Holt, and R. E. Folsom. 1977. Inference about regression models from sample survey data. *Bulletin of the International Statistical Institute* 47: 43–57.

Shao, J., and D. Tu. 1995. *The jackknife and bootstrap.* New York: Springer-Verlag.

Shao, J., and C. F. J. Wu. 1992. Asymptotic properties of the balanced repeated replication method for sample quantiles. *Annals of Statistics* 20: 1371–1393.

Silverman, B. W. 1986. *Density estimation for statistics and data analysis.* London: Chapman & Hall.

Siniff, D. B., and R. O. Skoog. 1964. Aerial censusing of caribou using stratified random sampling. *Journal of Wildlife Management* 28: 391–401.

Sitter, R. R. 1992. Comparing three bootstrap methods for survey data. *Canadian Journal of Statistics* 20: 135–154.

———. 1997. Variance estimation for the regression estimator in two-phase sampling. *Journal of the American Statistical Association* 92: 780–787.

Skelly, F., R. Goldberg, and D. Yankelovich. 1968. Women's attitudes toward cotton and other fibers used in wearing apparel. U.S. Department of Agriculture Statistical Reporting Service, Marketing Research Report no. 820. Washington, D.C.: Government Printing Office.

Skinner, C. J. 1989. Domain means, regression, and multivariate analysis. In *Analysis of complex surveys*. Edited by C. J. Skinner, D. Holt, and T. M. F. Smith, 59–88. New York: Wiley.

Skinner, C. J., D. Holt, and T. M. F. Smith, eds. 1989. *Analysis of complex surveys*. New York: Wiley.

Skinner, C. J., and J. N. K. Rao. 1996. Estimation in dual frame surveys with complex designs. *Journal of the American Statistical Association* 91: 349–356.

Smith, T. M. F. 1988. To weight or not to weight: That is the question. In *Bayesian statistics 3*. Edited by J. M. Bernardo, M. H. DeGroot, D. V. Lindley, and A. F. M. Smith, 437–451. Oxford, Eng.: Oxford University Press.

———. 1994. Sample surveys 1975–1990: An age of reconciliation? (with discussion). *International Statistical Review* 62: 5–34.

Snedecor, G. W. 1939. Design of sampling experiments in the social sciences. *Journal of Farm Economics* 21: 846–855.

Spisak, A. W. 1995. Statistical process control of sampling frames. *Survey Methodology* 21: 185–190.

Squire, P. 1988. Why the 1936 *Literary Digest* poll failed. *Public Opinion Quarterly* 52: 125–133.

Stasny, E. A. 1991. Hierarchical models for the probabilities of a survey classification and nonresponse. *Journal of the American Statistical Association* 86: 296–303.

StataCorp. 1996. *Stata statistical software: Release 5.0*. College Station, Tex.: Stata Corporation.

Statistics Canada. 1993. *Standards and guidelines for reporting of nonresponse rates*. Ottawa, Canada.

Stein, C. 1945. A two-sample test for a linear hypothesis whose power is independent of the variance. *Annals of Mathematical Statistics* 37: 36–50.

Stockford, D. D., and W. F. Page. 1984. Double sampling and the misclassification of Vietnam service. *ASA Proceedings of the Social Statistics Section,* 261–264.

Strunk, W., and E. B. White. 1959. *The elements of style*. New York: Macmillan.

Stuart, A. 1984. *The ideas of sampling*. New York: Oxford University Press.

Student. 1908. The probable error of the mean. *Biometrika* 6: 1–25.

Sudman, S. 1976. *Applied sampling*. San Diego: Academic Press.

Sudman, S., and N. M. Bradburn. 1982. *Asking questions: A practical guide to questionnaire design*. San Francisco: Jossey-Bass.

Sudman, S., N. M. Bradburn, and N. Schwarz. 1995. *Thinking about answers: The application of cognitive processes to survey methodology*. San Francisco: Jossey-Bass.

Sudman, S., M. G. Sirken, and C. D. Cowan. 1988. Sampling rare and elusive populations. *Science* 240: 991–996.

Sukhatme, P. V., B. V. Sukhatme, S. Sukhatme, and C. Asok. 1984. *Sampling theory of surveys with applications*. 3rd ed. Ames: Iowa State University Press.

Tanur, J., ed. 1993. *Questions about questions: Inquiries into the cognitive bases of surveys*. New York: Sage.

Taylor, B. M. 1989. *Redesign of the National Crime Survey*. Washington, D.C.: Department of Justice.

Taylor, H. 1995. Horses for courses: How different countries measure public opinion in very different ways. *Public Perspective* 6: 3–7.

Teichman, J., D. Coltrin, K. Prouty, and W. Bir. 1993. A survey of lead contamination in soil along Interstate 880, Alameda County, California. *American Industrial Hygiene Association* 54: 557–559.

Thomas, D. R. 1989. Simultaneous confidence intervals for proportions under cluster sampling. *Survey Methodology* 15: 187–201.

Thomas, D. R., and J. N. K. Rao. 1987. Small-sample comparisons of level and power for simple goodness-of-fit statistics under cluster sampling. *Journal of the American Statistical Association* 82: 630–636.

Thomas, D. R., A. C. Singh, and G. R. Roberts. 1996. Tests of independence on two-way tables under cluster sampling: An evaluation. *International Statistical Review* 64: 295–311.

Thompson, M. E. 1997. *Theory of sample surveys.* London: Chapman & Hall.

Thompson, S. K. 1992. *Sampling.* New York: Wiley.

Thompson, S. K., and G. A. F. Seber. 1996. *Adaptive sampling.* New York: Wiley.

Thomsen, I., and E. Siring. 1983. On the causes and effects of nonresponse: Norwegian experience. In *Incomplete data in sample surveys.* Vol. 3. Edited by W. G. Madow, I. Olkin, and D. B. Rubin, 25–59. New York: Academic Press.

Traugott, M. W. 1987. The importance of persistence in respondent selection for preelection surveys. *Public Opinion Quarterly* 51: 48–57.

Treder, R. P., and J. Sedransk. 1993. Double sampling for stratification. *Survey Methodology* 19: 95–101.

Tukey, J. W. 1958. Bias and confidence in not-quite large samples. *Annals of Mathematical Statistics* 29: 614.

———. 1968. Discussion of "Balanced repeated replications for analytical statistics" by L. Kish and M. Frankel. *Proceedings of the Social Statistics Section, American Statistical Association,* 32.

U.S. Bureau of the Census. 1994. *County and city data book.* Washington, D.C.: Department of Commerce.

U.S. Department of Justice. Bureau of Justice Statistics. 1989. Survey of Youths in Custody, 1987. United States computer file. Conducted by Department of Commerce, Bureau of the Census. 2d ICPSR ed. Ann Arbor, Mich.: Inter-University Consortium for Political and Social Research.

———. 1991. National Crime Surveys: National sample, 1986–1990. Near-term data computer file. Conducted by Department of Commerce, Bureau of the Census. 3d ICPSR ed. Ann Arbor, Mich.: Inter-University Consortium for Political and Social Research.

———. 1992. *Criminal victimization in the United States, 1990* (NCJ-134126). Washington, D.C.: Government Printing Office.

U.S. Environmental Protection Agency. 1990a. *National Pesticide Survey: Survey design.* Washington, D.C.: Government Printing Office.

———. 1990b. *National Pesticide Survey: Summary results of EPA's National Survey of Pesticides in Drinking Water Wells.* Washington, D.C.: Government Printing Office.

Valliant, R. 1987. Generalized variance functions in stratified two-stage sampling. *Journal of the American Statistical Association* 82: 499–508.

Venables, W. N., and B. D. Ripley. 1994. *Modern applied statistics with S-PLUS.* New York: Springer-Verlag.

Waksberg, J. 1978. Sampling methods for random digit dialing. *Journal of the American Statistical Association* 73: 40–46.

Wald, A. 1943. Tests of statistical hypotheses concerning several parameters when the number of observations is large. *Transactions of the American Mathematical Society* 54: 426–482.

Warner, S. L. 1965. Randomized response: A survey technique for eliminating evasive answer bias. *Journal of the American Statistical Association* 60: 63–69.

Watson, D. J. 1937. The estimation of leaf area in field crops. *Journal of Agricultural Science* 27: 474–483.

Webb, W. B. 1955. The illusive phenomena in accident proneness. *Public Health Reports* 70: 951.

Weisberg, S. 1985. *Applied linear regression.* 2d ed. New York: Wiley.

Wetherill, G. B., and K. D. Glazebrook. 1986. *Sequential methods in statistics.* 3d ed. London: Chapman & Hall.

Wilk, S. J., W. W. Morse, D. E. Ralph, and T. R. Azarovitz. 1977. *Fishes and associated environmental data collected in New York bight, June 1974–June 1975.* NOAA Technical Report NMFS SSRF-716. Washington, D.C.: Government Printing Office.

Wilson, J. R., and K. J. Koehler. 1991. Hierarchical models for cross-classified overdispersed multinomial data. *Journal of Business and Economic Statistics* 9: 103–110.

Wisconsin Department of Natural Resources. 1993. *The fisher in Wisconsin.* Technical Bulletin no. 183. Madison: Department of Natural Resources.

Wolter, K. M. 1985. *Introduction to variance estimation.* New York: Springer-Verlag.

Woodruff, R. S. 1952. Confidence intervals for medians and other position measures. *Journal of the American Statistical Association* 47: 635–646.

———. 1971. A simple method for approximating the variance of a complicated estimate. *Journal of the American Statistical Association* 66: 411–414.

Wright, J. 1988. The mentally ill homeless: What is myth and what is fact? *Social Problems* 35: 182–191.

Wynia, W., A. Sudar, and G. Jones. 1993. Recycling human waste: Composting toilets as a remedial action plan option for Hamilton Harbour. *Water Pollution Research Journal of Canada* 28: 355–368.

Yates, F. 1946. A review of recent statistical developments in sampling and sampling surveys. *Journal of the Royal Statistical Society* 109: 12–30.

———. 1981. *Sampling methods for censuses and surveys*. 4th ed. New York: Macmillan.

Yates, F., and P. M. Grundy. 1953. Selection without replacement from within strata with probability proportional to size. *Journal of the Royal Statistical Society,* ser. B, 15: 253–261.

Zehnder, G. W., D. M. Kolodny-Hirsch, and J. J. Linduska. 1990. Evaluation of various potato plant sample units for cost-effective sampling of Colorado potato beetle *(Coleoptera chrysomelidae). Journal of Economic Entomology* 83: 428–433.

Author Index

Agresti, A., 321, 466
Alexander, C., 367
Altham, P., 335
Alwin, D.F., 465
Anderson, D.W., 400, 468
Arizona Office of Tourism, 18
Armstrong, J., 467
Arnold, T., 148, 440
ASCAP, 107
Asher, H., 10, 462
Asok, C., 460
Aye Maung, N., 252
Azuma, D.L., 278

Babbie, E.R., 281
Bailey, P., 95
Bain, L., 460
Bakketeig, L.S., 341
Basow, S.A., 134, 328
Basu, D., 212
Beck, A.J., 235
Bedrick, E.J., 333
Bellhouse, D.R., 155, 168, 461, 463, 465, 467
Belson, W.A., 11, 462
Bethel, J., 464
Biderman, A.D., 257
Biemer, P., 461, 465
Binder, D., 290, 310, 317, 371, 466
Bisgard, K.M., 6
Blair, J., 10, 402, 462
Block, C., 467
Boswell, J., 423
Bowden, D.C., 155
Bowles, J., 314

Bowley, A., 463
Box, G.E.P., 300
Boyle, K.E., 371, 470
Brackstone, G., 270
Bradburn, N., 12, 462
Breiman, L., 392
Brewer, K., 198, 202, 217, 363, 463, 464, 467
Brick, J.M., xvi, 315
Brier, S.E., 335
Brown, G.H., 406
Bryk, A.S., 369
Buckland, S.T., 390, 396
Burman, J.P., 300, 303
Burnard, P., 124
Bye, B.V., 303

Calahan, D., 7
Cantor, D., 257
Carlson, B.L., 314
Casady, R.J., 313
Caspar, R., 465
Cassel, C.-M., 225, 467
Catlin, G., 261
Chambers, J.M., 235, 236
Chambless, L.E., 371
Chang, T., xvi, 413
Chapman, D.G., 388
Christensen, R., 321, 466
Cleveland, W., 235
Cochi, S.L., 410
Cochran, W.G., 49, 59, 156, 218, 363, 459, 463, 464, 467
Coffman, J., 20
Cohen, J.E., 335, 336

Cohen, S.B., 314
Colledge, M., 465
Converse, J.M., 9, 143, 456, 462
Cook, R.D., 355, 466
Cormack, R.M., 390, 395
Cornfield, J., 44, 239, 241, 465
Cox, D.R., 403
Crewe, I., 117
Cullen, R., 53
Czaja, R., 402

Dalenius, T., 465
D'Alessandro, U., 223
de Tocqueville, A., 1
deLeeuw, J., 369
Deming, W., 17, 270, 401, 459, 465
Dennis, M.L., 402
Deville, J.-C., 117
DeVries, W., 261
Dillman, D., 259, 260, 462, 466
Ding, Y., 392
Dippo, C., 314
Dobishinski, W.M., 107
Domingo-Salvany, A., 393, 410
Draper, N.R., 466
Drew, J.D., 466
Droege, S., 9
Duce, R.A., 8
Duffy, J.C., 404, 406
DuMouchel, W., 365
Duncan, G.J., 365
Duncan, J.W., 461
Durbin, J., 197
Durrett, R., 409, 423, 434, 461

Eberhardt, K.R., 211
Efron, B., 55, 306, 307, 450
Egeland, G.M., 411
Elliot, D., 257, 266
Engelhardt, M., 460
Ericson, W.A., 463
Ezzati-Rice, T., 228

Fay, R., 5, 278, 315, 328, 329, 338
Fellegi, I., 275
Ferraro, D.L., 276
Fienberg, S., 247, 322, 378, 388, 391, 392, 461, 462, 465
Fisher, R.A., 248, 262, 465
Ford, D., 19
Fowler, F.J., 462
Fox, J.A., 404
Francisco, C., 312
Frank, A., 411
Frankel, L.R., 465
Frankel, M.R., 351, 397, 467
Freedman, D.A., 392
Fuller, W., 312, 314, 467

Gallicchio, S.J., 303
Garthwaite, P.H., 396
Ghosh, M., 398, 467
Gill, R.D., 234
Glazebrook, K.D., 403
Gnap, R., 174, 286, 446
Godambe, V.P., 464
Goldstein, H., 369
González, M., 282, 465
Goren, S., 18
Gower, A.R., 260, 465
Graubard, B.I., 355, 360, 365-367
Gray, G.B., 466
Graybill, F.A., 467
Gross, S., 307
Groves, R., 10, 260, 264, 461, 465
Grundy, P.M., 196, 197

Hájek, J., 37
Handlin, A., 170, 442
Hanif, M., 198, 202, 464
Hansen, M.H., 88, 248, 263, 459, 461, 463, 464, 466, 467
Hanson, R.H., 242
Harding, F.D., 406
Harris, R., 10
Hartley, H.O., 66, 218, 271, 401
Hidiroglou, M.A., 259, 281, 466, 467

Hite, S., 1-3, 257
Hogan, H., 391
Holmes, D.J., 365
Holt, D., 257, 266, 275, 328, 329, 460, 466, 467
Hopf, H.A., 116
Horvitz, D.G., 197, 205, 404, 464
Hosmer, D.W., 370
Hurwitz, W.N., 263, 460, 464

Iachan, R., 402
Ingram, S., 261
International Working Group for Disease Monitoring and Forecasting, 396, 468
Iyer, H.K., 467

Jackson, K.W., 125
Jacoby, J., 170, 442
Jefferson, T., 379
Jessen, R., 460
Jinn, J.H., 274
Joiner, B.L., 466
Juster, N., 131

Kalsbeek, W.D., 461
Kalton, G., 272, 400, 467, 468
Kasprzyk, D., 272
Kempthorne, O., 248, 466
Khurshid, A., 462
Kiaer, A., 16
Kinsley, M., 19
Kipling, R., 179
Kish, L., 228, 239, 241, 351, 365, 397, 460, 465, 467
Koch, G., 329
Koehler, K., 335
Konijn, H., 467
Korn, E., 355, 360, 365-367
Kosmin, B., 284
Kott, P., 365, 467
Kovar, J.G., 310, 311
Kreft, I., 369
Krewski, D., 100, 302, 306, 310
Kruskal, W., 461
Kruuk, H., 124, 443
Kuk, A.Y.C., 412
Kutner, M., 467

Lachman, S., 284
Lahiri, D.B., 187
Landers, A., 19
Laplace, P.-S., 59, 463

Lavrakas, P., 200, 462
Leggett, J., 9
Lehnen, R.G., 11
Lehtonen, R., 460
Lemeshow, S., 370
Lenski, G., 9
Lepkowski, J., 200, 314, 462
Lessler, J.T., 461
Lewis, S., 347
Lincoln, F.C., 387
Lindgren, B., 321, 461
Link, H.C., 116
Literary Digest, 7
Little, R.J.A., 229, 264, 268, 280, 365, 465, 467
Liu, J., 365
Lohr, S., 365, 403
Lush, J.L., 377
Lyberg, L.E., 461
Lydersen, C., 123, 444

Macdonell, W.R., 350, 438
Madison, J., 22
Madow, W.G., 264, 460, 461, 463, 465
Mahalanobis, P.C., 161, 293, 464
March, M., 465
Mathiowetz, N.A., 461
Mayr, J., 53
McAuley, R., 261
McCarthy, P., 299, 311, 466
McFarland, S.G., 14
McGinley, P., 289
McGuiness, R.A., 242
McIlwee, J.S., 96
McKibben, B., 413
Mellor, R.W., 363, 467
Mitofsky, W.J., 200
Morganstein, D., 466
Morton, H.C., 102
Mosteller, F., 117, 461
Murphy, R.S., 228

Nachtsheim, C.J., 467
Nathan, G., 353
National Center for Health Statistics, 440
National Center of Educational Statistics, 277
Navidi, W.C., 392
Neter, J., 106, 203, 467
Neyman, J., 379, 463, 464, 468
Nightingale, F., 319

Nusser, S., 234

O'Brien, L.A., 179
Oh, H.L., 266, 268, 270
Olkin, I., 460, 465
Overton, W., 202, 464

Page, W.F., 380
Pahkinen, E., 460
Parten, M., 462
Paulin, G.D., 276
Peart, D., 72
Petersen, C.G.J., 387
Pfeffermann, D., 229, 365, 467
Pincus, T., 12
Plackett, R.L., 300, 303
Platek, R., 259, 260, 466
Politz, A., 271
Pollock, K., 396
Potthoff, R., 200, 272, 278, 462
Prentice, R., 372
Presser, S., 10, 14, 462
Price, A.J., 102
Prosser, R., 369
Pyke, R., 372

Quenouille, M., 304

Raj, D., 460, 462, 463, 464
Rao, J.N.K., 100, 145, 168, 210,
    218, 270, 278, 290, 302, 306,
    307,
    310-312, 317, 329, 333-335, 338,
    344, 345,
    371, 384, 385, 398, 402, 407, 461,
    463, 465-467
Rao, P.S.R.S., 145
Raudenbush, S.W., 369
Remafedi, G., 6
Ripley, B.D., 234
Roberts, G., 371
Roberts, R.J., 175, 442
Robinson, J., 467
Robinson, J.G., 100
Rosenbaum, P., 264
Ross, A., 66
Ross, S., 423, 434, 461
Rothenberg, R.B., 251
Roush, W., 21
Royall, R., 168, 211, 248, 463, 464,
    467
Rubin, D.B., 264, 277, 280, 365,
    460, 465, 467

Ruggles, S., 213, 214
Russell, H.J., 119
Rust, K., 290, 310
Ryan, A.S., 282
Ryan, T.P., 466
Ryg, M., 123, 444

Sahai, H., 462
Salant, P., 466
Samuels, C., 342
Sande, I.G., 272
Sanderson, M., 347
Sanzo, J., 118
Sarndal xvii, 198, 212, 313, 363,
    372, 381, 384, 460, 463, 467
Satterthwaite, F.E., 334
Saville, A., 110
Sayers, D., 255
Schafer, J.L., 467
Scheaffer, N.C., 462
Schei, B., 341
Scheuren, F., 170, 266, 268, 270,
    338
Schnabel, Z., 393
Schreuder, H., 201
Schuman, H., 9, 14, 462
Schwarz, N., 462
Scott, A.J., 163, 333-335, 344, 372,
    465
Scott, D., 234
Searle, S., 467
Seber, G.A.F., 388, 396, 404
Sedransk, J., 274, 408
Sen, A.R., 197
Seng, Y.P., 461
Senturia, Y.D., 169
Serdula, M., 15
Shah, B.V., xvi, 314, 360
Shao, J., 303, 304, 306, 466
Shelton, W.C., 461
Silberg, N.T., 134, 328
Silver, R., 369
Silverman, B.W., 234
Simmons, W., 271
Siniff, D.B., 101
Siring, E., 256
Sitter, R., 307, 310, 311, 384
Skelly, F., 12
Skinner, C., 229, 362, 402, 460,
    465-467
Skogan, W.G., 11
Skoog, R.O., 101
Smith, H., 466

Smith, T.M.F., 163, 248, 353, 365,
    460, 465-467
Snedecor, G.W., 461
Sorensen, T., 23
Spisak, A.W., 466
Squire, P., 7
Srinath, K.P., 467
Stasny, E.A., 280
StataCorp, 314
Statistics Canada, 281
Stehman, S., 202, 464
Stein, C., 403
Stephan, F., 270
Stockford, D., 380
Strunk, W., 12
Stuart, A., 460, 462, 464
Student, 350
Sudman, S., 12, 401, 460-462
Sukhatme, B.V., 460
Sukhatme, P.V., 460
Sukhatme, S., 460
Swensson, B., 381, 460

Tanur, J.M., 10, 247, 261, 461, 462,
    465
Tarbell, I., 221
Taylor, B., 11
Taylor, H., 117
Teichman, J., 8
Tepping, B.J., 463
Thomas, D.R., 328, 329, 331, 332,
    338, 344, 345
Thompson, D.J., 197, 205, 464
Thompson, M.E., 87, 212, 248, 313,
    460, 462, 465
Thompson, S., 404, 460
Thomsen, I., 256
Tibshirani, R., 55, 306, 450
Tracy, P.E., 404
Traugott, M., 262
Treder, R.P., 408
Tu, D., 304, 306, 466
Tukey, J., 241, 304
Twain, M., 179

U.S. Bureau of the Census, 90, 440
U.S. Department of Justice, 235,
    308, 443, 445
U.S. Environmental Protection
    Agency, 111-113

Valliant, R., 309, 313
Venables, W.N., 234

Waksberg, J., 200
Wald, A., 329
Warner, S.L., 404
Wasserman, W., 467
Waterton, J.J., 404, 406
Watson, D.J., 468
Webb, W.B., 323
Weisberg, S., 355, 466, 467
Wetherill, G.B., 403

White, E.B., 12
Wild, C.J., 372
Wilk, S.J., 126, 443
Wilson, J., 335
Wisconsin Department of Natural
    Resources, 409
Wolter, K.M., 290, 300, 310, 466
Woodruff, R.S., 290, 311
Wretman, J., 460, 467

Wright, J., 18
Wu, C.F.J., 302, 303, 307, 310, 312,
    317, 466
Wynia, W., 284

Yates, F., 196, 197, 247, 460, 461

Zehnder, G.W., 155

Subject Index

Accounting, sampling in, 77, 108, 170, 202–204
Accurate estimator, 28–29
Adaptive cluster sampling, 403–404
Adjusted R-squared, 140–141, 156–158
Advantages of sampling, 15–17
Agriculture, U.S. Census of, 31–32, 34, 38, 41, 63–65, 68–69, 83–85, 96–99, 104, 175
Allocation, in stratified sampling, 104–109
American Statistical Association, 284, 314, 465
Analysis of Variance (ANOVA)
    in cluster sampling, 138–142, 156–158, 163
    in stratified sampling, 105–106, 113, 248, 468
Asymptotic results, 37
Autocorrelation, 160
Auxiliary variable, 60

Balanced repeated replication (BRR), 298–303, 305, 308, 310–311, 314–317, 356
Balanced sampling, 211
Best linear unbiased estimate, 349, 351, 468
Best linear unbiased predictor, 47
Bias
    estimation, 27–28
    measurement, 8–10, 28
    nonresponse and, 257–258
    ratio estimator, 66–71

Bias (continued)
    regression estimator, 74
    selection, 4–8, 15, 25, 28, 181
Binomial distribution, 72–73
Bonferroni test
    chi-square test, 331–333
    loglinear models, 338–339
    regression parameters, 360
Bootstrap, 55, 306–308, 310, 313–314, 450

Callbacks, 262–263
Call-in surveys, 6–7, 248
Capture-recapture estimation, 387–395
    assumptions, 387, 391
    confidence intervals, 388–390, 395–396
    contingency tables in, 388–390
    loglinear models, 394–396
    multiple samples, 393–396
Case-control study, 372
Categorical data analysis, 319–341, 470. See also Chi-square test
Census, compared with sample, 16–17, 33, 467
Census, U.S. decennial, 5, 16, 21–22
    undercount in, 5, 257, 391–392
Central limit theorem, 37–38, 100, 269, 300
Chi-square test
    complex surveys, 329–336
    design effects and, 326–327, 333–335
    likelihood ratio, 320, 322

Chi-square test (continued)
    model-based inference, 335–336
    moment matching, 333–335
    multinomial sampling, 319–324
    Pearson, 320, 322
    software, 341
    survey design and, 324–329
Choice-based sampling, 372
Cluster, 24, 131
Cluster sampling, 23–25, 43, 50, 131–168, 229, 468. See also Unequal-probability sampling
    adaptive, 403–404
    chi-square test and, 325–329
    confidence intervals, 159
    design, 155–159, 168
    design-based inference, 209–210
    design effect, 240
    estimating means, 136, 144–145, 147–148, 168
    estimating totals, 136, 143–144, 147–148, 168
    model-based inference, 163–168
    notation for, 134–135
    one-stage, 134, 136–146
    precision and, 131–133, 240
    reasons for use, 132–133
    sample size, 158–159
    simple random sampling compared with, 24, 50, 131–134, 138–141
    stratified sampling compared with, 132–133, 138–139
    three-stage, 210, 219, 226
    two-stage, 134, 145–152

Cluster sampling *(continued)*
  variance, 136–137, 139–140, 147
  variance estimation, 137,
    144–145, 147–148, 168–169
  weights in, 138, 144–145,
    153–154, 226
Coefficient of determination (R2),
    350, 377
Coefficient of variation
  estimator, 33
  population values, 29–30, 41
  random variable, 427
Cognitive psychology, 10, 261
Cold-deck imputation, 276
Combined ratio estimator, 225,
    252–253
Complex surveys, 221–247, 469
  building blocks, 221–223
  confidence intervals, 241,
    310–313
  design, 221–224
  estimating means, 226
  estimating totals, 226
  notation for, 227
  ratio estimation in, 222, 224–225
  reasons for use, 225
  sample size, 241–242
  variance, 238–240, 290–293
  variance estimation in, 221,
    289–315
  weights in, 221, 225–239
Composite estimator, 399
Computer assisted telephone inter-
    viewing (CATI), 261
Computer software. *See* Software
Conditional expectation, 204,
    432–435
Conditional inference, 268, 313
Conditional probability, 430–432
Confidence interval, 26, 35–38,
    48–49, 100–101, 159, 310–313
  cluster sampling, 159
  complex surveys, 241, 310–313
  conditional, 313
  design effect and, 241
  interpretation of, 35–36, 48–49
  log odds ratio, 321, 327
  model-based, 48–49
  for population size, 388–390,
    395–396
  quantiles, 311–313
  regression coefficients, 356, 360
  sample size and, 40

Confidence interval *(continued)*
  simple random sampling, 35–38,
    48, 49
  stratified sampling, 100–101
Contingency tables, 319–326
  in capture-recapture, 388–390
Convenience sample, 5, 26, 117
Correlation, 427
Correlation coefficient
  intraclass, 139–143, 159, 171,
    240, 325
  Pearson, 60, 139
  population, 60, 316
  sample, 74
Covariance, 45, 427, 429–430
Cumulative-size method, 185–187
Current Population Survey, U.S.
    (CPS), 110, 242, 255, 269,
    273, 308, 328

Degrees of freedom (df), 101, 294,
    297, 310, 332, 334, 356
Design
  cluster sampling, 155–159, 168
  complex surveys, 221–224
  importance of, 8, 248–249, 258,
    262
  model-based approach and, 87,
    168
  nonsampling error reduction,
    258–262
  optimal, 469
  simple random sampling, 39
  stratified sampling, 104–110
  unequal-probability sampling,
    189–192, 211
Design-based inference, 43–46, 423
  cluster sampling, 209–210
  regression coefficients, 354–368
  simple random sampling, 43–46
  unequal-probability sampling,
    204–210
Design effect, 239–242
  in chi-square tests, 326–327,
    333–335
  cluster sampling, 240
  confidence intervals and, 241
  domains, 397
  regression coefficients, 353,
    361–362
  sample size estimation and,
    241–242, 469
  stratified sampling, 240

Difference estimation, 77
Distribution function, 229
Dollar stratification, 108
Dollar unit sampling, 202–204
Domains, 77–81, 96, 309, 363, 377,
    396–400
Double sampling. *See* Two-phase
    sampling
Dual-system estimation, 391–392

Element, 3
Empirical distribution function,
    230–234
Empirical probability mass func-
    tion, 230–234
Error
  nonsampling, 15–17, 23, 42,
    258–262, 465, 469
  sampling, 15–17
Estimation bias, 27–28
Event, 423
Expected value, 27, 47–48,
    427–428
Experiment, designed
  contrasted with survey, 49–50,
    247–249
  nonresponse investigation,
    259–262, 469–470

Finite population correction (fpc),
    33, 45
First-order correction to chi-square
    test, 333–335
Fixed-effects ANOVA model, 113,
    248
Frame, sampling, 3, 5, 23, 31

Generalized regression (GREG), 88,
    372–374
Generalized variance function
    (GVF), 308–310, 313–314
Goodness-of-fit tests, 280, 323–324.
    See also Categorical data
    analysis
Graphs
  cluster samples, 148–150,
    190–191, 235
  complex samples, 235–239
  design of surveys with, 42,
    156–157
  regression, 351–352, 355, 364
  simple random samples, 32
  stratified samples, 97

Hierarchical linear model, 369
Homogeneity
   measure of, 139–140, 155
   test of, 322–323
Horvitz-Thompson estimator,
   196–199, 205–210, 212, 222,
   453
Horvitz-Thompson theorem,
   205–207
Hot-deck imputation, 275

Ignorable nonresponse, 265
Imputation, 272–278
Inclusion probability, 196
Incomplete data. *See* Capture-
   recapture estimation;
   Nonresponse
Independence
   chi-square test for, 321–322
   cluster sampling and, 143, 163
   events, 425
   random variables, 44
Indicator variable, 44, 264–266, 428
Internet addresses for survey re-
   sources, 22, 31, 250, 314, 413,
   453, 465, 476
Interpenetrating subsampling, 294,
   468
Interpenetrating systematic sam-
   pling, 161
Interviewers, effect on survey accu-
   racy, 9–10, 260–261, 406
Intraclass correlation coefficient
   (ICC), 139–143, 159, 171, 240,
   325
Item nonresponse, 255. *See also*
   Nonresponse

Jackknife, 169, 304–306, 310,
   313–315
   regression coefficients, 356, 359
Judgment sample, 5, 8

Lahiri's method, 187, 216
Leading question, 2, 14
Likelihood ratio test, 320, 322. *See*
   also Chi-square test
Linearization, 290–293, 303, 310,
   313–315
   regression coefficients, 356–361
Linear regression. *See* Regression analysis; Regression
   estimation

Literary Digest Survey, 7–8, 15–16,
   23, 257
Logistic regression, 276, 370–372
Loglinear models
   capture-recapture, 394–396
   complex surveys, 336–341

Margin of error, 15, 39, 49
Mark-recapture estimation. *See*
   Capture-recapture estimation
Maternal and Infant Health Survey
   (MIHS), 347–348, 353,
   365–367, 400
Maximum likelihood estimation,
   279, 336
Mean
   population, 29
   sample, 32
Mean-of-ratios estimator, 92, 224
Mean squared error (MSE)
   design-based, 28
   model-based, 47
Measurement bias, 8–10
Measure of homogeneity, 139–140,
   155
Median, estimating, 230, 302–303,
   306–307, 311–313
Missing at random (MAR), 265
Missing completely at random
   (MCAR), 264
Missing data, 255. *See also*
   Nonresponse
Mitofsky-Waksberg method,
   200–201
Mixed models, 368–370
Model-assisted inference, 363,
   372
Model-based inference, 46–49,
   81–88, 113–114, 163–168,
   335–336, 348–352, 362–370,
   467
   chi-square tests, 335–336
   cluster sampling, 163–168
   confidence intervals, 48–49
   design and, 87, 168
   quota sampling, 116–117
   ratio estimation, 81–85, 87–88
   regression analysis, 348–352,
   356–360, 362–370
   regression estimation, 86–88
   simple random sampling, 38,
   46–49
   stratified sampling, 113–114

Model-based inference *(continued)*
   unequal-probability sampling,
   211–212
   weights and, 228–229
Model-unbiased estimator, 47
Multilevel linear model, 369
Multinomial distribution, 56, 321
Multinomial sampling
   chi-square tests with, 319–325
   definition, 321
   loglinear models and, 336–338
Multiple frame surveys, 401–402
Multiple imputation, 277
Multiple regression, 359–361. *See*
   also Regression analysis

National Assessment of Educational
   Progress, 368–370, 398
National Crime Victimization Survey, 3–4, 9, 11–12, 23, 221,
   242–247, 252
   chi-square tests with, 327
   design of, 242–244
   domains in, 396–398
   nonresponse in, 255, 257, 267,
   269
   questionnaire design, 11–12
   regression, 348
   variance estimation in, 246–247,
   308–309
   weights in, 244–247
National Health and Nutrition Examination Survey, 228–229,
   471
National Pesticide Survey, 4,
   110–113
Network sampling, 402–403
Neyman allocation, 108
Nonignorable nonresponse, 265
Nonparametric inference, 44
Nonresponse, 6, 63, 249, 255–282
   bias, 257–258
   effects of ignoring, 256–258
   factors affecting, 259–262
   guidelines for reporting, 281–282
   ignorable, 265
   imputation for, 272–278
   item, 255
   mechanisms, 264–265
   missing at random, 265
   missing completely at random,
   264
   models for, 264, 278–280

Nonresponse *(continued)*
  nonignorable, 265
  rare events and, 404
  rate, 281–282
  survey design and, 258–262
  unit, 255
  weight adjustments for, 244–245, 265–272
Nonsampling error, 15–17, 23, 42, 465, 469. *See also* - Nonresponse
Normal distribution, 37–38, 41, 310
Normal equations, 349, 354
Not-at-homes, 4. *See also* Nonresponse
Notation
  cluster sampling, 134–135
  complex surveys, 227
  ratio estimation, 60
  simple random sampling, 27–30
  stratified sampling, 99

Odds ratio, 320, 322, 325–327, 343
One-stage cluster sampling. *See* Cluster sampling
Optimal allocation, 106–108
Ordinary least squares, 74, 349

Pearson's chi-square test. *See* Chi-square test
Percentiles, 229, 313–313
Pilot sample, 41
Plots. *See* Graphs
Poisson sampling, 202
Politz-Simmons method, 271–272
Polls, public opinion, 6–8, 13–15, 40, 56
Population
  estimating the size of, 387–395
  finite, 25, 29
  sampled, 3–4
  target, 3–5
Poststratification, 63, 114–115, 313, 316
  as generalized regression, 372, 374
  for nonresponse, 268–269
Precise estimator, 28–29
Presley, Elvis, 10
Primary sampling unit (psu), 131
Probability distribution, 26, 426
Probability mass function, 229, 304

Probability proportional to size (pps) sampling, 190, 211–212. *See also* Unequal probability sampling
Probability sampling, 23–30, 423, 466
Probability theory, 17, 23, 423–435
Product-multinomial sampling, 322, 327
Propensity score, 264
Proportion, 30
Proportional allocation, 104–106
Public Use Microdata Samples, 213–214
Purposive sample, 8, 467

Quality improvement, 259–262, 469–470
Quantiles, estimation of, 311–313
Question order, effect of, 10, 14–15
Questionnaire design, 9–15, 261, 466
Quota sampling, 115–118

Raking, 269–271
Random-coefficient regression model, 369
Random digit dialing, 199–201
Random effects, 369
Random-effects ANOVA model, 163–168, 176–177, 211, 248
Random group methods, 293–297, 313–314, 356
Randomization inference, 43–46. *See also* Design-based inference
Randomized response, 404–406
Random numbers
  table, 457–458
  use in selecting sample, 23, 26, 31, 52
Random variable, 44, 46, 423, 426
Rao-Hartley-Cochran estimator, 218
Rare events, sampling for, 400–404
Ratio estimation, 61–71, 81–85, 373, 467
  bias, 66–71
  capture-recapture and, 387–388
  combined, 225, 252–253
  complex surveys, 222, 224–225
  design-based inference, 82
  estimating means, 61

Ratio estimation *(continued)*
  estimating proportions, 72–73, 225, 252–253
  estimating ratios, 61
  estimating totals, 61, 66
  mean squared error, 67–71
  model-based inference, 81–85, 87–88
  reasons for use, 71
  separate, 225, 253
  two-phase sampling and, 383–384
  variance, 66–68
  variance estimation, 68, 292–293
Realization, of random variable, 47, 426
Refusals, 4. *See also* Nonresponse
Regression analysis, 347–374, 470–471
  causal relationships and, 362–363
  complex surveys, 352–362
  confidence intervals, 356, 360
  design-based inference, 354–368
  design effects, 353, 361–362
  effects of unequal probabilities, 352–353
  estimating coefficients, 349, 351, 354–355, 360
  graphs, 351–352, 355, 364
  model-based inference, 348–352, 362–370
  purposes of, 362
  software, 361–362, 364
  straight-line model, 348–359
  variance, 357
  variance estimation, 356–360
Regression estimation, 74–77, 348, 467
  bias, 74
  estimating means, 74
  estimating totals, 88
  generalized, 88, 372–374
  mean squared error, 74–75
  model-based inference, 86–88
  reasons for use, 74
  variance, 74–75
  variance estimation, 75, 373
Regression imputation, 275–276
Replication for variance estimation, 298–308, 313–315, 468
Resampling for variance estimation, 298–308, 313–315

Residuals
 plotting, 85, 351–352
 use in variance estimation, 68
Respondent burden, 261
Response rate, 258, 281–282. *See
 also* Nonresponse

Sample. *See also* specific sample
 design
 cluster, 23–25, 131–168
 convenience, 5, 26, 117
 definition of, 3
 judgment, 5, 8
 probability, 23–30
 purposive, 8, 467
 quota, 115–118
 representative, 2–3
 self-selected, 2
 self-weighting, 105
 simple random, 24, 26, 30–50
 stratified, 23–24, 95–118
 systematic, 42–43, 159–162
Sampled population, 3–4
Sample size, 39–42, 241–242
 accuracy and, 8
 cluster sampling, 158–159
 complex surveys, 241–242
 decision-theoretic approach, 55
 design effect and, 241–242, 469
 importance of, 41–42
 simple random sampling, 39–42,
  55
 stratified sampling, 109
Sampling, advantages of, 15–17
Sampling distribution, 26–27
Sampling error, 15–17
Sampling fraction, 33
Sampling frame, 3, 5, 23, 31
Sampling unit, 3
Sampling weight. *See* Weights
Secondary sampling unit, 131
Second-order correction to chi-
 square test, 334
Selection bias, 4–8, 15, 25, 28, 181
Self-representing psu, 242
Self-selected sample, 2
Self-weighting sample, 105, 138,
 153, 180, 228
 advantages of, 228
 complex surveys, 228, 230, 232,
  233, 236, 244
Sen-Yates-Grundy variance, 197
Separate ratio estimator, 225, 253

Sequential sampling, 403–404
Simple random sampling, 24, 26,
 30–50, 466–467
 cluster sampling contrasted with,
  24, 50, 131–134, 138–141
 confidence intervals, 35–38, 48,
  49
 design, 39
 design-based inference, 43–46
 design effect and, 240
 estimating means, 32, 49
 estimating proportions, 34–35
 estimating totals, 33–34, 49
 model-based inference, 38, 46–49
 notation for, 27–30, 33
 probability of, 31
 reasons for use, 50
 sample size, 39–42, 55
 selection of, 30, 52, 56, 449
 stratified sampling contrasted
  with, 24, 50, 106
 systematic sampling contrasted
  with, 43
 variance, 32, 34, 44–45
 variance estimation, 33–34, 45–46
 with replacement (SRSWR), 30,
  40, 424–425
 without replacement (SRS),
  30–50, 56, 425–426
Small area estimation, 397–400
Snowball sampling, 403
Software
 chi-square tests, 341
 regression, 361–362, 364
 variance estimation, xvi, 314–315
Standard deviation, 29
Standard error, 33. *See also*
 Variance
Stochastic regression imputation,
 275
Strata, 24, 95
Stratification variable, choice of, 110
Stratified random sampling,
 95–118, 322
 allocating observations to strata,
  104–109
 cluster sampling contrasted
  with, 132–133, 138–139
 confidence intervals, 100–101
 defining strata, 109–113
 design, 104–110
 design effects, 240
 estimating means, 100

Stratified random sampling
 (continued)
 estimating proportions, 102
 estimating totals, 100, 102
 model-based inference, 113–114
 notation for, 99
 precision and, 96, 240
 reasons for use, 95–96, 109–110,
  112–113
 sample size, 109
 simple random sampling contrasted with, 106
 variance, 100
 variance estimation, 100
 weights in, 103, 226
Stratified sampling, 23–24, 32, 50,
 95–118, 227, 322, 468
 chi-square tests and, 327
 complex survey component, 222
 rare events, 128, 400–401
 two-phase sampling and,
  384–386
Subdomains. *See* Domains
Subsidiary variable, 60
Substitution, for nonrespondents,
 5–6, 276–277
Successive conditioning, 204, 434
Superpopulation, 37
Survey of Youth in Custody,
 235–239, 251–252, 296–297,
 330–334, 339–341
SURVEY program, 57–58, 92–93,
 129–130, 177–178, 219, 253,
 286–287, 317–318, 345, 378,
 413–421
Synthetic estimator, 398–399
Systematic sampling, 42–43,
 159–162, 185–186, 197–198

Tag-recapture estimation. *See*
 Capture-recapture estimation
Target population, 3–5
Taylor series, 290–293
Telephone surveys, 3, 199–201,
 261
Telescoping of responses, 9
3-P sampling, 201–202
Three-stage cluster sampling, 210,
 219, 226
Total, population, 26, 29
Two-phase sampling, 263, 379–386,
 401, 471–472
 for nonresponse, 263, 381

Two-phase sampling *(continued)*
   for rare events, 401
   for ratio estimation, 383–384
   for stratification, 384–386
Two-stage cluster sampling. *See*
      Cluster sampling

Unbiased estimator
   design-based, 28–29, 43–44
   model-based, 47, 87
Undercount, in U.S. census, 5, 257,
      391–392
Undercoverage, 5
Unequal-probability sampling,
      179–212, 221–222
   complex surveys and, 221–222
   design, 189–192, 211
   design-based inference, 204–210
   estimating means, 198
   estimating totals, 183, 185, 188,
      192
   examples of, 179–180, 199–204
   model-based inference, 211–212
   with one psu, 181–184
   one-stage, 184–192
   reasons for use, 179–181
   with replacement, 184–194,
      221–222
   without replacement, 194–199,
      222
   selecting psu's, 185–187
   simple random sampling con-
      trasted with, 183–184
   stratified sampling contrasted
      with, 180–181, 199
   two-stage, 192–194
   variance, 183, 188, 197
   variance estimation, 188, 192,
      197

Unequal-probability sampling
      (continued)
   weights, 180, 183, 198
Unit
   observation, 3
   primary sampling (psu), 131
   sampling, 3
   secondary sampling (ssu), 131
Unit nonresponse, 255. *See also*
      Nonresponse
Universe, 25

Variance
   cluster sampling, 136–137,
      139–140, 147
   complex surveys, 238–240,
      290–293
   model-based, 87
   population, 29, 33
   random variable, 427
   ratio estimation, 66–68
   regression coefficients, 357
   regression estimation, 74–75
   sample, 33
   sampling distribution, 28
   simple random sampling, 32, 34,
      44–45
   stratified sampling, 100
   unequal-probability sampling,
      183, 188, 197
Variance estimation
   cluster sampling, 137, 144–145,
      147–148, 168–169
   complex surveys, 221, 289–315,
      470
   insufficiency of weights for, 104,
      226, 234, 367
   ratio estimation, 68, 292–293
   regression coefficients, 356–360

Variance estimation *(continued)*
   regression estimation, 75, 373
   replication methods, 298–308,
      313–315, 468
   simple random sampling, 33–34,
      45–46
   software, xvi, 314–315
   stratified sampling, 100
   unequal-probability sampling,
      188, 192, 197

Wald test, 329–331, 338
Weighted least squares, 81, 355,
      360–361
Weighting-class adjustments for
      nonresponse, 266–268
Weights, 103–104, 144–145,
      153–154, 225–239, 265–272,
      360–368
   cluster sampling, 138, 144–145,
      153–154, 226
   complex surveys, 221,
      225–229
   contingency tables, 326
   epmf and, 230–234
   graphs and, 235–239
   insufficiency for variance estima-
      tion, 104, 226, 234, 367
   model-based analysis and,
      228–229
   nonresponse adjustments,
      265–272
   regression and, 354–355,
      360–368
   stratified sampling, 103–104,
      226
   truncation of, 227
   unequal–probability sampling,
      180, 183, 198